Compromise Planning : A Theoretical Approach from a Distant Corner of Europe

Louis C. Wassenhoven

Compromise Planning : A Theoretical Approach from a Distant Corner of Europe

Louis C. Wassenhoven
National Technical University of Athens
Athens, Greece

ISBN 978-3-030-94333-2 ISBN 978-3-030-94331-8 (eBook)
https://doi.org/10.1007/978-3-030-94331-8

This Springer imprint is published by the registered company Springer Nature Switzerland AG
The registered company address is: Gewerbestrasse 11, 6330 Cham, Switzerland

Dedicated to the memory of my parents, Dino, who taught me the art of thinking and imagining, and Coula, who showed me the practical art of living, and to the memory of my close friend Dimitris, an opera lover. If they were to ask me now, what I am doing, I could answer, like Rodolfo in Act I of Puccini's La bohème: Che cosa faccio? Scrivo. E come vivo? Vivo.

Foreword

Planning theory has traditionally originated mainly in the countries of North and West and its relevance for those of South and East has been repeatedly doubted and disputed. It is this plain truth, admitted by the author, that is the starting point of this book by Louis Wassenhoven. The book is divided into three parts of which the first is a thorough review of existing theories classified in four dominant currents. The first and second, i.e., the rational/comprehensive and the communicative/collaborative currents, constitute the main and well-established stream in the theoretical literature. The third, which the author labels radical current, emanates mainly from the Global South and from a transformative/insurgent approach upholding the views of disadvantaged groups in society. A fourth current, according to Louis Wassenhoven, is rapidly emerging in response to the threat of climate change and environmental deterioration. To these currents, Louis proposes the addition of a fifth one, namely compromise theory, already touched upon by theoreticians of the pragmatist tradition. In the second part of his book, he presents the example of Greece, a country sitting on the interface between North-West and South-East. The "empirical reality" of Greece is what compromise theory is meant to explain; a reality created out of specific historical, social, geopolitical and geographic conditions, in which the role of the state has been of paramount importance. The final part of the book is where the author develops compromise theory and links it to a Southern perspective and a pragmatist approach. Compromise is found to exist in the elaborate nexus of clientelist and patronage bonds between state and civil society, particularly evident in land use and spatial planning. In neoliberal times, intimately associated with the recent economic crisis, this nexus is extended to new forms of high-profile planning. The radical perspective, to which the author incorporates past efforts for development planning and the realities of informal settlement, is now faced with the additional challenge of climate change, which he links with the growing interest in resilience planning. However, political compromise between the state and the electoral clientele of small landowners over the use of land in areas outside the official planning system perpetuates unstable conditions which are inimical to environmental protection and climate change adaptation, let alone reversal.

The conjunction of the existential threat of global climate change, the continuing Covid-19 pandemic and a new awareness of social injustice, largely stimulated by the international success of the Black Lives Matter (BLM) movement, particularly in the Global North and West, has occasioned a significant disruption and reappraisal in prevailing planning theory. As is extensively recounted and analysed in Louis Wassenhoven's scholarly book, planning theory is itself an ongoing dynamic process of debate and change, currently broadly embraced by the collaborative/communicative and rational theories of planning, more especially falling under Louis' classification category of climate current, particularly notable in coastal locations, threatened by rising sea levels and inland demographic changes, occasioned by farmers (with few, if any, urban skills) escaping the loss of livelihoods caused by extremes of heatwave causing aridity or excessive precipitation and flooding (Roberts 2010). The significance of the Covid-19 pandemic has shown the extent to which it rests on the transfer of viral pathogens, particularly within and between urban residential neighbourhoods and places of public gathering, including transport systems and recreation facilities – issues central to spatial planning and environmental management. In large measure, social justice stimulated by the alienation of ethnic or racial groups, particularly of socially marginalised youths and other 'minority' groups, has been shown to be ameliorated by engaging their participation in planning and physical development processes.

Operational partnerships of stakeholders, notably local government and national line ministries and service agencies, planning agencies, and user groups, present an effective antidote to the frequently, and often erroneously, perceived inefficiencies of participatory spatial planning, for instance, 'City Development Strategies' (CDS) (Freire and Stren 2001), promulgated by the Word Bank and Cities Alliance, with some emphasis on urban economic development, and by United Nations agencies[1] and some bilateral aid donors that are principally concerned with social development and poverty alleviation and reduction (Wakely 2020). Also, examples of successful national development partnerships exist, such as the 'Arms-length Management Organisations' (ALMO) programme for the planning and management of local government housing, established by the UK (Labour) government in 2000 (Broughton 2018) and the Government of Kenya Local Authority Service Delivery Action Plan (LASDAP) partnership programme, launched in 2012 (Riley and Wakely 2005, 66).

Such partnerships between local government and community organisations provide opportunities for the engagement of young people and social minority group members[2] with their residential or workplace communities and in wider governance and planning activities with formal agencies of government, thereby enhancing assimilation in the majority community and appraising the latter of their cultural assets, social norms and aspirations. Environmental planning and management

[1] E.g., UN-Habitat, UNICEF.

[2] Including political and economic refugees and other migrant groups.

partnerships dispel the customary divide between the producer[3] and users of spatial plans that frequently are antagonistic, on occasions conflictive. They are central to addressing the new and urgent demands of planning in the twenty-first century, such as minimising dependencies on non-renewable energy (e.g., for motorised transport) and planning open spaces to minimise the transfer of viral pathogens and vegetation to enhance the absorption of carbon dioxide.

Operational risk- and benefit-sharing partnerships, embracing (local) government agencies and civil society organisations, fit comfortably in Louis Wassenhoven's 'Theory of Compromise Planning'.

Professor Emeritus of Urban Development
University of London
London, UK
November 2021

Patrick Wakely

References

Boughton J (2018) Municipal dreams: the rise and fall of council housing. Verso, London

Freire M, Stren R (eds) (2001) The challenge of urban government: policies and practices. World Bank Institute of Development Studies, Washington DC

Riley E, Wakely P (2005) Communities and communication: building urban partnerships. ITDG Publishing, Rugby

Roberts D (2010) Prioritizing climate change adaptation and local level resilience in Durban, South Africa. Environ Urban 22(2). Sage, London

Wakely P (2020) Partnership: a strategic paradigm for the production and management of affordable housing and sustainable urban development. Int J Urban Sustain Dev 12(1). Taylor & Francis, Abingdon

[3] Public sector planning authorities.

Preface

This book is the product of my long-standing interest in the theory of planning. As a young university assistant with planning ambitions, I was perplexed by the familiar questions facing all aspiring planners. The first question was about the steps between the diagnosis of a problem and the end result of planning. How does one go about it? What does this intermediate process consist of? The second question concerned the object under study that planning is required to address, i.e., the village, the neighbourhood, the city or region. How much does one need to know about this object and how is this knowledge arrived at? Trying to answer these questions in theoretical terms involved me, although I was then unaware of the planning theory idiom, in procedural and substantive planning theory. A few years later, the proper terminology entered my language. Several decades after this initial familiarization, I felt ready to make a personal contribution to the spatial planning theory debate.

I am writing this book as a planner and planning academic with only amateurish and rudimentary knowledge of other loftier disciplines. I must make it clear that I do not intend to resort to the original (philosophical, sociological, scientific) sources that inspired planning theories in the past. Their influence reached me through intermediaries, i.e., distinguished planning theorists whom I studied. Some of them had a philosophy, social science, and/or law background which I am totally unable to match. Planning theorists often dig out the writings of famous thinkers of remote centuries or turn to relatively more obscure intellectuals of the recent past, in order to base on their work a path-breaking planning theory. They could have easily imagined even a non-existent philosopher, writer or artist. In fact, I am surprised that no planning theories have been based on works of art, e.g., on paintings by Giorgione, Bosch, Bruegel, Vermeer, or de Chirico, although, on occasion, they have been based on literary works. Planners who read the planning theory literature, in their vast majority, would never dare read the originals that inspire planning theorists. Still, they acquire precious second-hand knowledge, for which they are grateful, as they are for the theoretical work on planning. I recall that when I was still at high school in Athens, a classmate of mine wrote a paper on the life of a foreign Philhellene, who came to join the Greek Revolution against the Ottoman Empire.

He fought heroically and died in the battle of Peta. His name was Van der Mohl. Our teacher praised the paper and the diligent work of my friend, but everyone else in the classroom knew that this Philhellene had never existed. It did not really matter, because the presentation sparked a lively discussion about an event which resembled real sacrifices of real persons.

What convinced me that I could and should enter the planning theory debate was my realization that planning theory is not independent of the conditions of its country of origin and is, therefore, not appropriate for another country into which it is imported. This is a theme that crops up regularly in the so-called Southern theory and that I develop repeatedly in the book, especially in Chap. 9, in a section on my "personal journey". The dichotomy between North-West and South-East is always present. Working in the South, as a planning professional or theorist, with tools and assumptions of foreign extraction is a difficult task. It also has practical, albeit less scientific, difficulties. It has disadvantages which are both debilitating and costly. Access to libraries with a decent collection of books on planning theory is non-existent; my personal library, which I donated to my old university department, was a haven. Working conditions, especially in the years of the Covid-19 pandemic, tended to be a problem. I was naturally working at home, but disruptions were frequent. Power failures, because of storms, snowfall, or heat waves were unnerving but perhaps gave me one more reason to include among the theoretical planning currents one that is the direct result of climate change.

Once engaged in the effort to look for an alternative "southern" perspective, I found myself immersed in the study of the conditions of a country I know well, namely Greece, which serves me as a case study. I studied intensely the country's history, political system, social structure, the functioning of the state and the planning system. I present my findings extensively in Chaps. 6, 7, and 8, because they are essential for putting together a theory of "compromise planning" which in my view is best suited to explain the empirical reality of Greek planning. The Greek word *symvivasmos* is translated into English as reconciliation, compromise, agreement, settlement or accommodation.[1] It describes, I believe, the essence of Greek planning, in spite of the rational character of the official urban and regional planning system. I find it amusing that the term has impeccable classical credentials, since it can be found in Plato' Protagoras. Intervening in the argument between Socrates and Protagoras, the wise man Hippias advises them: "I do pray and advise you, Protagoras, and you, Socrates, to agree upon a compromise. Let us be your peacemakers".[2]

It is up to the reader to judge whether my "compromise planning" is a useful addition to the theory of spatial planning. Writing about it has certainly been a source of immense pleasure for me. I recognize that it has taxed the patience of Nafsica – or Nausicaa for English-speaking readers of the Odyssey. While working

[1] *Korais Greek-English Dictionary* (2008) Patras University Press, Patras.

[2] *Plato's Protagoras* (337a–338b). Translation by Benjamin Jowett www.gutenberg.org/files/1591/1591-h/1591-h.htm

on my manuscript I had the privilege of an excellent collaboration with Margaret Deignan, Senior Publishing Editor at Springer Publishers, for which I am deeply grateful. Sincere thanks are due to my former colleague Prof. Patrick Wakely for his foreword. I would also like to thank the reviewers of my initial book proposal for their constructive comments and those friends and colleagues who offered me advice. Especially valuable were the opinions and comments of fellow academics Georgia Giannakourou, Maria Giaoutzi, Lila Leontidou, Thanos Pagonis and Kalliopi Sapountzaki, who read parts of the book. Needless to add, they are by no means responsible for my errors and omissions.

Athens, Greece
December 2021

Louis C. Wassenhoven[3]

[3] Loudovikos K. Vassenhoven in Greek.

About the Book

This is a book about the theory of spatial planning which departs from the tradition of theory inspired by the conditions of developed countries of the North and West. It is not a manual of urban and regional planning and does not claim universal validity which is independent of a country's history, social structures, state organization, geography, and geopolitical position. In a sense, it is inscribed in the stream of ideas originating in the Global South, although it draws its empirical material from a country, Greece, located "at the corner of Europe", at the interface of East and West, something which historically has left an indelible mark and still affects the empirical reality of planning. In fact, it is this empirical reality that the proposed "theory of compromise planning" is intended to explain. The book is the product of the author's interest in the theory of planning, ever since, as a young professional and university assistant, he engaged in a search of the theoretical base of key issues such as the process of plan making and the understanding of the object of spatial planning. Later experience of teaching students from a variety of nations convinced him that the transfer of knowledge from North-West to South-East was filled with flaws and questionable assumptions. His familiarity with the Greek planning system taught him that there existed a gap between the official rational planning system and reality in the trenches of planning practice. The outcome on the ground was rather the result of constant bargaining and compromise between the state and political powers, on one hand, and the public at large, on the other, i.e., between two worlds linked together by a nexus of clientelism and political patronage. Compromise was not just a version of consensus or rational deliberation but rather an accommodation of interests. It was not the outcome of the implementation of statutory planning, although such planning exists, but the deal arrived at by ad hoc legislative initiatives and discreet interpretations of the law at the level of land use and building activity. This covert planning may be receding under the influence of modernizing forces but still survives and acquires a modernistic high profile through the sanctioning of a parallel system of planning favouring large scale development projects, a system, however, already incorporated in mainstream planning legislation.

The author does not negate the value and continued influence of existing spatial planning theoretical work. On the contrary, he seeks to understand the ties between his compromise theoretical model and other theoretical currents, particularly of a pragmatist perspective, to which he dedicates Chaps. 2, 3, 4, and 5 of the book, currents which he labels as rational, communicative, radical and "climate", in recognition of the influence of climate change on the future directions of planning. In Chaps. 6, 7, and 8, the author concentrates on the case study of Greece, first to explain the effect of historical forces, second to analyse the role of the state, and, finally, to dissect the planning system. This is an important task, because it is the reality of social and political forces, dependence conditions, imported practices and location peculiarities, as a bridge between East and West, that ultimately help to construct a theoretical model. This model of compromise planning is elaborated in the final Chaps. 9, 10, and 11, which contain the constitutive elements of a novel planning theory.

Contents

List of Boxes

List of Figures

About the Author

Louis C. Wassenhoven

Professor Emeritus, Urban and Regional Planning, National Technical University of Athens (NTUA).

Studies:

- Diploma in Architectural Engineering, NTUA (1961)
- Certificate in Planning Techniques, Ministry of Housing, France (1967)
- Postgraduate Diploma in Planning, Architectural Association School of Architecture, London (1971)
- PhD, London School of Economics and Political Science, University of London (1980)

Career:

- Senior Assistant, NTUA (1960s)
- Lecturer – Senior Lecturer, Development Planning Unit, Bartlett School of Architecture and Planning, University College London (1973–1982).
- Visiting teacher, London School of Economics and Political Science, MSc in Social Administration
- Professor of Urban and Regional Planning (NTUA) and Director, Spatial Planning and Urban Development Research Laboratory (1982–2005)
- Visiting Professor, Bandung Institute of Technology/Indonesia
- Visiting Scholar, University of Melbourne, on an Australian Government fellowship
- Visiting Professor, National School of Public Administration, Greece

Past activities:

- Chairman, National Council of Spatial Planning and Sustainable Development, Greece
- Vice Chairman, Greek State Scholarships Foundation
- Former member, Royal Town Planning Institute (UK)
- Chairman, Joint Activity on Urban Management, OECD

- Member, Association of Greek Urban and Regional Planners
- Member, Board of the Company for the Redevelopment of the Hellinikon International Airport in Athens
- Member of the Council of the Crete Technical University

Current activity:
- Research and writing
- Member of the special law-drafting commission of the Greek Ministry of the Environment and Energy for the production of a code of urban and regional planning legislation

Author of several articles and of the following books [GR: In Greek]
- *Maritime spatial planning* (Crete University Press, 2017, GR)
- *The Ancestry of Regional Spatial Planning: A planner's look at history* (Springer Publishers, 2019)
- *EU Compendium of Spatial Planning Systems and Policies: Greece* (European Commission, 2000)
- *Territorial governance* (with collaborators, Kritiki, 2010, GR)
- *Hellenic … passion: Development of the former Hellinikon airport* (Sakkoulas, 2018, GR)
- *Putting order in our country: A history of spatial planning in Greece after World War II* (Kritiki, 2022, GR)

Acronyms[1]

APE	Renewable energy sources
CEC	Commission of European Communities
CEMAT	Conference of Council of Europe Spatial Planning Ministers
CPER	Centre of Planning and Economic Research
CRU	Centre de Recherche d'Urbanisme
DATAR	Délégation à l'Aménagement du Territoire et à l'Action Régionale
DEPOS	Public Company of Town Planning, Settlement and Shelter
diaNEOsis	Non-profit Organization of Research and Analysis
DPU	Development Planning Unit UCL
EChM	Special Spatial Planning Study
EChOA	Laboratory of Spatial Planning and Urban Development, National Technical University of Athens
EEC	European Economic Community
EKDDAA	National Centre for Public Administration and Self-Government
EKKE	National Centre of Social Research
EKPAA	National Centre for the Environment and Sustainable Development
ELSTAT	(see ESYE)
EPA	Operation of Town Planning Re-organization (1980s)
EPChS	Special (nationwide) Spatial Planning Framework
EPS	Special Town Plan
ESCHADA	Special Plan of Spatial Development of Public Properties
ESCHASE	Special Plan of Spatial Development of Strategic Investments
ESPA EU	Development Framework Partnership Agreement
ESPON	European Spatial Planning Observatory Network
ESYE	National Statistical Service (renamed ELSTAT)
ETVA	National Bank of Industrial Development (see HBID)
EVEA	Athens Union of Industrial Enterprises
FAO	Food and Agriculture Organization

[1] Note: Most abbreviations refer to Greek ministries, agencies and types of plans.

GGET	Secretariat General of Research and Technology
GOK	General Building Regulation
GPChS	General (national) Spatial Planning Framework
GPS	General Town Plan (later replaced by TPS)
HBID	Hellenic Bank of Industrial Development (see ETVA)
ICLEI	International Council for Local Environmental Initiatives
ICOMOS	International Council on Monuments and Sites
ISTAME	Institute of Strategic and Development Studies
ITA	Local Government Institute
IUCN	International Union for the Conservation of Nature
KEDKE	Central Union of Municipalities and Communes (now simply, of municipalities)
KEPE	See CPER
MinCoord	Ministry of Coordination[2]
MinDevComp	Ministry of Development and Competitiveness
MinEcDev	Ministry of Economic Development
MinEcDevT	Ministry of Economic Development and Tourism
MinEcFin	Ministry of Economy and Finance
MOD	Organization and Management Unit of Development Programmes
MPW	Ministry of Public Works
MSD	Transfer of Building Coefficient
NTUA	National Technical University of Athens
OECD	Organization of Economic Cooperation and Development
OKE	Economic and Social Commission
ORSA	Organization of the Athens Master Plan
PASPE	Panteios High School of Political Science (later Panteion University)
PChS	(see SPF)
PERPO	Area of Specially Regulated Urbanization
POTA	Integrated Tourism Development Area
PUCSA/Env	Greek Association of Personnel of the Ministry of Environment, Spatial Planning and Public Works
RPChS	Regional Spatial Planning Framework
RTPI	Royal Town Planning Institute
SADAS	Association of Greek Architects of University Education
SCHOOAP	Plan of Spatial and Urban Organization of Open Towns
SD	Building Coefficient (*syntelestis domisis*)
SEP	Association of Greek Regional Scientists
SEPOCH	Association of Greek Urban and Regional Planners
SPE	Town planning research laboratory, National Technical University of Athens

[2]At various stages, Greek economic development planning was the responsibility of different ministries.

SPF	Spatial Planning Framework
StE	Council of State (supreme administrative court)
TEE	Technical Chamber of Greece
TPS	Local Town Plan
UN	United Nations
UNEP	United Nations Environment Programme
UNESCO	United Nations Educational, Scientific and Cultural Organization
VIPETVA	Industrial Estates Company of ETVA (see above)
WWF	World Wildlife Fund (now World-Wide Fund for Nature)
YCHOP	Ministry of Spatial Planning, Settlement and Environment[3]
YPECHODE	Ministry of Environment, Spatial Planning and Public Works
ZEP	Active Town Planning Zone
ZOE	Zone of Settlement Control

[3] The Greek ministry responsible for urban and regional planning changed name frequently.

Chapter 1
Introduction – Defining the Problem

Abstract This chapter is a short introduction to the book. The author explains the rationale and structure of the book and looks back at the emergence of a theoretical literature purporting to explain the activity of planning. He outlines the dilemmas faced by planning researchers and practitioners in places where theories are imported from foreign countries. He introduces the task of researchers keen to build the foundations of indigenous theory. The concepts of space and spatial planning are approached in a cursory fashion, before an analytical presentation of the book chapters.

Keywords Space · Spatial planning · Planning theory · Imported theories · Structure of the book

Prologue

In the last half-century, the growth of planning theory literature has been spectacular. The pioneers of pre-Second World War town planning could not have imagined the volume and breadth of the body of theory which would follow and in accordance to which their work would be classified and judged. They would be probably astounded, at least most of them, if told that their profession would evolve into a discipline which needs, like all other respectable scientific disciplines, a theoretical base borrowed from a vast field of knowledge. As mentioned in the preface, even literary works have served as a basis of planning theories (Harrison 2001). What happened of course, is that the very success of the work of the pioneers caused a veritable explosion of academic education in town and country planning, city planning, urban and regional planning, *urbanisme*, *urbanistica*, or *Städtebau*, depending on the tradition. The intervening experiences of the interwar and early post-Second World War are no doubt part of the explanations; so is the growing number of university planning courses, the demand for academic status and the "entryism" of other disciplines in the body of planning professionals. Academic status demanded the injection of a variety of other scientific and professional perspectives. It also

L. C. Wassenhoven, *Compromise Planning : A Theoretical Approach from a Distant Corner of Europe*, https://doi.org/10.1007/978-3-030-94331-8_1

attracted students who had started their university education in the fields of law, geography, economics, sociology and natural science. Soon the interdisciplinary character of planning activity and education required the reliance on a body of theory which vastly exceeded the boundaries of traditional town planning. Aspiring planners were now undertaking postgraduate studies and were producing doctoral theses. Their teachers were experts in a wide gamut of learning and research.

By the end of the twentieth century, any researcher with a theoretical interest in planning had to become familiar with an enormous volume of theoretical treatises and publications. What this ambitious researcher was taking for granted, or, if he/she was unaware of, would soon discover, is that the sort of planning practice and its associated theory, he/she was being taught had clear historical, social, economic and geographical roots. It reflected the conditions of specific national contexts and would be different if the conditions and contexts were different. If our imaginary researcher went beyond simple learning or beyond the production of a paper and a dissertation and had the audacity to formulate a new theory, he/she would soon be faced with the critical question of what conditions the new theory would be founded on. Even a mere sketch of the proposed theory would have to rest on a thorough study and comprehension of these conditions. This implies that apart from an in-depth study of pre-existing theoretical work, this ambitious theorist has to understand and assimilate all the intricacies of the empirical reality that his/her theory will hopefully explain. This requires additional hard work. The researcher whose national context is identical with that of the established and recognized theoreticians has an advantage. These theoreticians, in their great majority, live and work in economically advanced countries of the West, usually in North America or West Europe. If our researcher has the same origins, he/she has at his/her disposal a vast volume of native literature on cities and regions, on space and spatial planning. There are some exceptions naturally of countries outside this elite geographical space. Here, however, one more parameter plays a role, i.e., language. The dominance of the English language in planning studies is a fact which favours countries the spoken language of which is English. There is more to it, however, because there are still countries, of which the language is widely spoken in a huge part of the world. The real language disadvantage is more evident when the spoken language of a country is practically limited to that country alone. The problem of a researcher originating in such a country is not limited to the relative difficulty of access to international literature, often written in a very sophisticated language, but also to the limited literature of his/her own country in a specialized subject, such as planning and, even more so, planning theory, not just practice. To this, one can add the practical problems of access to supporting material, e.g., available in libraries, or the cost of purchasing foreign publications.

When the researcher concerned still insists to make a novel theoretical contribution, e.g., on spatial planning which is our interest here, the foundation work normally available to his/her colleagues of the "developed world" is simply insufficient. When he/she seeks to delve deeply into the "contexts and conditions" which have created the practice of planning he or she is keen to explain theoretically, even the literature available in his/her native language does not suffice. If, for instance, he/

she studies the history, geography, politics, or social structure of his/own country, he/she soon discovers that very little has been written from his/her own specific planning perspective. In reading a text on history, he/she has to understand how history exercised influence on the evolution of planning. In addition, if he/she has the even more daring ambition to write a theoretical study on planning addressed to an international (e.g., English-speaking) audience his/her analysis of the planning system which is to be theoretically explained has to be adapted to the knowledge of that audience. This explains why the present author found himself forced to dedicate a large part of the book on the history, the state and the planning system of Greece, which he uses as a case study to support his proposal for a new theory of "compromise planning".

1.1 Spatial Planning

This is not a book about planning in the generic sense; nor is it a manual of urban and regional planning. It is rather a book about the theory of spatial planning, and, to put it more explicitly, public planning. Therefore, it is considered imperative to list all possible definitions of planning which one can find in the relevant literature. Nevertheless, it is essential to quote some of them, on the basis of two criteria. The first is the assistance one can get from sources about spatial planning in practice, of which the authors do not pretend to develop an explanatory or prescriptive theory. The second is the author's, admittedly selective, choice of works of theorists who, in his opinion, provide useful guidance in the effort to put forward a personal theoretical view. Besides, it is true that for a long time he has adopted a definition of planning which has served him well in previous writings and in teaching: Spatial planning is the premeditated transformation of socio-economic geographical space which results from the intention of an organized society and responds to its objectives (Wassenhoven 2009). The author has however admitted that the difficulty of defining spatial planning is due to the difficulty of defining its object, i.e., geographical space, which is far more complex than a mere unit of land. Space, as conceived by a planner, is an entity created by a large number of factors, from environmental and natural processes to the historic inheritance of the human past, and from the interactions of economic activities to the hierarchical structures of political power. Society, economy and the natural environment have shaped a complex space, which bears the marks, improvements, scars and disfigurements caused by the efforts of humankind, in all stages of its history, to tame nature and adapt it to its needs and to its, frequently short-sighted, desires (Wassenhoven 2007, 518–519).

Having laid his cards on the table, the author can dispense of general definitions of planning as an individual human action, i.e., "as a basic activity pervading human behavior at the individual and every social level" (Alexander 1992, 69). Ernest Alexander in fact provides an account of these generic definitions of planning as a general human activity, as anticipation and foresight, or as individual sequencing of actions. The emphasis on action, but also on its foundation on prior knowledge, had been of course paramount in a much earlier classical definition: "A useful way to look at planning is to consider it as an activity centrally concerned *with the linkage between knowledge and organized action*" (Friedmann and Hudson 1974, 2). Seven years before his article with Hudson, Friedmann (1973, 346–347) had written that "planning will be considered as the *guidance of change within a social system*. Specifically, this means a process of self-guidance that may involve *promoting differential growth* of subsystem components (sectors), *activating the transformation of system structures* (political, economic, social), and *maintaining system boundaries* during the course of change. Accordingly, the idea of planning involves a confrontation of expected with intended performance, the application of controls to accomplish the intention when expectations are not met, the observation of possible variances from the prescribed path of change, and the repetition of this cycle each time significant variations are perceived".

Widening the perspective from individual activity to collective action, Alexander stresses planning's orientation to the future, its non-routine character, the illusion of relying on trial-and-error methods, the utopian character of future imaging, the slippery path of mistaking planning for simple plan-making, and the difference between problem-solving and strategic frame-setting. He concludes with his own definition of planning, as "the deliberate social or organizational activity of developing an optimal strategy of future action to achieve a desired set of goals, for solving novel problems in complex contexts, and attended by the power and intention to commit resources and to act as necessary to implement the chosen strategy" (Alexander 1992, 73). He is quick to acknowledge that for some this definition is too broad, and that planning is usually about a narrower range of activities, i.e., urban design, development control, and regulation of the built environment. Such conceptions, Alexander responds, are dependent on the practices and perceptions which are dominant in particular countries. In the USA planning is perceived as encompassing a range from economic development to environmental planning. "In Western Europe, on the other hand, planning is more narrowly understood, reflecting what people do there who think of themselves as planners: national, regional, or areawide spatial-environmental planning and local development control. And in many countries of Southern Europe, Latin America and Asia planning is still viewed as an activity of architects specializing in land-use planning and urban design" (op.cit., 73–74). The country-specific character of planning is no doubt correctly underlined and the observation regarding Southern Europe, which applied to Greece too, has been made by other commentators as well (European Commission 1997, 37; Newman και Thornley 1996, 57–60). However, since 1992, when Alexander was writing, a great deal has changed, as it will be pointed out later.

1.2 Space

In simple terms, space is the object of planning work, except that the meaning of space is open to different interpretations, particularly when what is under examination is its interaction with society. "Perhaps the simplest way to conceptualize the interaction of society with space is to regard spatial arrangements as a straightforward reflection of social divisions" (Smith 1999, 12). Understanding what space is all about has a direct effect on what planning is and on how theory approaches planning. The analysis of Parker and Doak (2012, 158) helps us appreciate this complex relationship: "Space has traditionally been thought to be a universal, abstract phenomenon and subject to scientific laws, often viewed as a definable area. In this traditional, or Euclidean account, space is measurable, observable and limited or bounded. Efforts to define space in this way, however, have been criticized as a simplification. They are often supported by an instrumental rationality that can fail to understand the social construction of space, or value the richness of meanings and cultures of 'places' or the shrinking, folding or pleating of space that competing conceptualisations provide, as with the notion of time-space compression …". Parker and Doak then turn to Edward Relph to dig deeper in the complexities of space: "Relph … identified four sorts of 'layers' of space, starting with 'pragmatic spaces' relating to human location and orientation within a territory and how we navigate through space. Secondly, he labels 'perceptual space' reflecting that what we experience or observe is largely centred on the self and will be an individualised experience. Thirdly 'existential space' that is informed by the culture and meanings that we perceive from place and, lastly the idea of 'cognitive space' as a combination of spatial relations or uses and interrelations between elements". We have to note the distinction between space and place. Bolan discusses space and place in detail (Bolan 2017, 225–239). He refers to the various specializations of urban planners in the USA. "From a philosophical standpoint, these specialties do in fact have one thing in common. They are all involved with *space*. Noticeable in literature from European and other international urban planners is their tendency to talk about their activity as 'spatial planning'. There is, however, a tendency in North American urban planning to take space for granted. There is also a tendency to view space primarily from one's field of specialization. Looking back at the thinking of Kant and Descartes, the *a priori* dimensions of human existence – time and space and their interactions – are critically important aspects of urban planning" (Bolan 2017, 226). He quotes Edward Soja who wrote that "there are three rather than two fundamental or ontological qualities of human existence, from which all knowledge follows: the social/societal, the temporal/historical, and the spatial/geographical" (op. cit., 228).

1.3 Structure of the Book

After this introductory chapter, the book is organized in three parts and ten chapters. Part I (Chaps. 2, 3, 4 and 5) is a thorough review of existing planning theories. Part II (Chaps. 6, 7 and 8) is a presentation of the case study of Greece. It is in Part III (Chaps. 9, 10 and 11) that the proposed theory of compromise planning is finally introduced.

In **Chap. 2**, it was deemed necessary to outline the typologies of planning theory as proposed by theoreticians of planning. But in the attempt to simplify a rather confusing landscape the author sketched what he calls currents of ideas that feed into respective theories. Four currents are thus proposed, labelled as rational, communicative, radical and "climate", the latter implying a new strand of theory influenced by climate change. A fifth current could correspond to the author's compromise planning theory, which has a "southern" flavour.

In **Chap. 3**, the focus is on what can be considered as the "mainstream" theories of, on one hand, the rational – comprehensive – systemic current, and, on the other, the communicative – collaborative – deliberative current. These theories have dominated the theoretical field for a long time. In the same chapter, a separate section is reserved for the theoretical approach of pragmatism.

Chapter 4 is about the radical – transformative – insurgent current which encompasses a variety of approaches, many of which originate in "Southern" theory, which stresses the divide between North-West and South-East. Work related to social movements, activism, militancy, justice and ethics has a prominent place in this current, which represents a challenge to mainstream theories.

Developments in theory associated with the environment and climate change are the subject of **Chap. 5**. The climate current incorporates elements of natural science which signify a return to rationality, while at the same time the relevant theoretical contributions underline ethical issues. This theoretical current emphasizes the conditions created in the Anthropocene era and calls for an adaptation of human behaviour and for a revision of planning in line with the necessary response to the threat of climate change. Concepts like risk, vulnerability and resilience occupy a central position.

Having completed the panoramic overview of theory, the author turns to the Greek case study, first to the historic past of modern Greece (**Chap. 6**), with the aim of highlighting the developments which in due course had to be tackled by spatial planning. The gradual territorial growth of the country, geographical divisions, political schisms and civil conflicts, the land question, fragmentation of land property, illegal land development, bankruptcies, foreign control, the adoption of alien administrative practices, refugee inflows, war damages and reconstruction are some of the parameters which formed the background against which urban and regional planning had to solve the resulting problems.

The role of the state is of crucial importance in any attempt to put forward a theory which explains the empirical reality of a country like Greece. In **Chap. 7**, the author discusses the arguments regarding the construction of the new nation-state in the nineteenth century, the impact of the state, its controversial relations with the citizenry, the political nexus of patronage and clientelism and the compromise reached between political powers and voters. The size of the state, the structure of the administration and its efficiency are some of the issues that affect planning. Populism has been a force which modified the nature of the clientelist system and therefore influenced the state-citizen relations, especially in matters of land use and illegal building. Legislation, the plethora of laws and erratic legislative activity also affect seriously the practice of planning.

Chapter 8 is dedicated exclusively to the Greek planning system, which compromise theory will try later to explain. It is a long chapter because it is this system which is the "empirical reality" which compromise theory will deal with. The gradual development of urban and regional planning activity is presented and the successive reforms explained. The distance separating the official system and practice on the ground is repeatedly underlined. Planning legislation, although having a promising start in the 1920s, follows essentially the constitutional reform of 1975. The contrast between areas with a statutory plan and out-of-plan land is explained, to demonstrate how it is in the latter that compromise is primarily manifested. The goal of a balanced urban network, the problem of regional inequalities and the role of Athens are a constant concern, but spatial planning is often divorced from other policy sectors. Economic crises have affected the structure of the planning system, especially after 2010, when foreign pressures were unprecedented. Nevertheless, an impressive official system is in place.

Chapter 9 is the introduction to Part III, in which compromise theory is developed. After a personal statement written to explain how the author came to realize the impossible task of transferring foreign knowledge to a different environment, the main aim in this chapter is to outline the Greek planning culture and the transfer of foreign expertise, e.g., in the form of imported types of plans and instruments. It was decided to include in this chapter a section on the profile of planners, their education and the knowledge they bring back when they are trained in other countries. The chapter ends with an attempt to sketch the "empirical reality", to which reference was made earlier, and to prepare the ground for the following chapters.

The main feature of **Chap. 10** is arguably twofold, on one hand to develop a new theoretical model, and, on the other, to sketch the person, the individual, who enters a negotiation process with the state and the political power, in the context of a clientelist relationship ending in compromise, which is a far cry from a simple consensus. To do this, the author developed a dialogue with human geography to trace the transmutations of the classical *homo economicus* into what he describes as *homo individualis*, after a succession of different types of *homo*. Pragmatism, rationalism, the public interest, an analysis of the theorist himself as a "product", and the factors influencing the compromise model, are all brought into the discussion.

The concluding, and short, **Chap. 11** brings the book to a close. It contains an analysis of the faces of compromise, sometimes low-profile and populist, on other occasions high-profile of the neoliberal variety. The role of the state is reiterated and the unpleasant aspects of compromise discussed, while admitting that compromise guarantees a form of stability. The chapter and the book end with a summary of the compromise model.

References[1]

Alexander ER (1992) Approaches to planning: introducing current planning theories, concepts and issues. Gordon and Breach, Luxembourg

Bolan RS (2017) Urban planning's philosophical entanglements: the rugged, dialectical path from knowledge to action. Routledge, London

European Commission (1997) The European Union compendium of spatial planning systems and policies. Office for Official Publications of the European Communities, Luxembourg

Friedmann J (1973) A conceptual model for the analysis of planning behavior. In: Faludi A (ed) A reader in planning theory. Pergamon Press, Oxford, pp 345–370 (first published 1967)

Friedmann J, Hudson B (1974) Knowledge and action: a guide to planning theory. J Am Inst Plann 40(1):2–16

Harrison P (2001) Romance and tragedy in (post) modern planning: a pragmatist's perspective. Int Plan Stud 6(1):69–88

Newman P, Thornley A (1996) Urban planning in Europe. Routledge, London

Parker G, Doak J (2012) Key concepts in planning. Sage, London

Smith SJ (1999) Society – space. In: Cloke P, Crang P, Goodwin M (eds) Introducing human geographies. Arnold, London, pp 12–23

Wassenhoven L (2007) Spatial planning (*chorotaxia*). Lemma in the Encyclopaedia Papyrus – Larousse – Britannica (new edition 2007), vol 55. Annex on Greece. Papurus, Athens, pp 518–532 (in Greek)

Wassenhoven L (2009) Regional spatial and urban planning. Teaching manual. EKDDA, Department of Regional Government, Athens (in Greek)

[1] **Note**: The titles of publications in Greek have been translated by the author. The indication "in Greek" appears at the end of the title. For acronyms used in the references, apart from journal titles, see at the end of the table of diagrams, figures and text-boxes at the beginning of the book.

Part I
Review of Theories

Chapter 2
Planning Theories: Typologies and Overcrowding

Abstract This chapter is a first effort to make sense of the over-extensive and often perplexing field of planning theory. It is reasonable to start with a taxonomy of theories, i.e., a classification of the various theories in meaningful categories. Planning theory typologies have been produced by most major theorists and they are condensed in this chapter in the hope that they will render easier the reading of the following chapters of Part I of the book. Apart from a presentation of the typologies proposed in the past, this chapter also contains a new re-organization of the body of theories into dominant currents of ideas, based on the rational, communicative-collaborative and radical traditions and on the emerging field of theory instigated by climate change. A new current is hesitantly suggested containing the author's ideas of compromise planning. The theoretical approach which is founded on pragmatism is also included, because, in the author's opinion, it offers additional support to the compromise current.

Keywords Typologies · Planning theory currents · Rational · Communicative · Radical · Southern theory · Climate · Pragmatism · Compromise

Prologue
The aim of the first part of this book is to review the body of planning theory, as an essential prerequisite of the attempt to embark on a venture of enriching it. The following chapters deal with the main theories that have appeared in the decades after the Second World War and tried to provide a basis for the discipline of planning. The breadth and volume of theory that has been accumulated over the years is awesome and even its study, let alone its condensation, is a real challenge. However, in an effort to take control of this task the author resorts to a familiar tool, i.e., classification in order to make sense of diversity. Producing typologies of planning theory has been already the work of eminent theorists, who, as expected, disagree among themselves. This can cause considerable disarray, but in fact is but a reflection of the confusion which many of them confess that it exists. This chapter is a bird's eye view of the attempted typologies, followed by an attempt to consolidate the field

L. C. Wassenhoven, *Compromise Planning : A Theoretical Approach from a Distant Corner of Europe*, https://doi.org/10.1007/978-3-030-94331-8_2

into what are called "currents of ideas", that generate theories. The author thus tries to narrow the field into cohesive domains, the rational, communicative, radical and "climate" currents, before making a suggestion of an additional "compromise" current. A section on pragmatism at the end of the chapter lends support to this suggestion.

2.1 Origins of Planning Theory

In the introductory chapter, the phenomenon of the growing number of planning theories since, roughly, the 1950s has been underlined. An interesting question is whether the competing theories that have seen the light of day since the period of post-war reconstruction were in fact in competition. Did they really exclude one-another? Were they incompatible? Was it justifiable to classify them in diverging groups, or to describe them as theories in, of, for, on, or about planning, let alone in opposing camps? One is entitled to ask first another key question. How far in the past do we have to return to trace the beginnings of what we now understand as planning theory, in fact a theory of urban and regional planning? One well discussed answer is that the appropriate turning point is the years of birth of a modern movement reacting to the ravages of the industrial revolution, i.e., the early twentieth century or even late nineteenth. These are the years of a vision of a modern city liberated from smog and slums, a dream embedded in the designs of visionary architects, although ultimately the most influential dreamers and utopian thinkers were not members of the architectural profession.[1] Such thinkers left a legacy the loss of which is mourned, e.g., in the view of Ellis (2015), in England, where "visionary, socially progressive town planning" has been abandoned. The question of course, from the present author's point of view, is whether the work of dreamers and utopians and their proposed urban models amount to a theory. In his perspective they do not, although in some cases they may have limited their ideas to the formulation of a future image, but they also touched on the planning process itself. Patrick Geddes (1968) and his process of "survey – analysis – plan", which he proposed in 1915, is obviously a case in point. The reason why it is important not to neglect their contribution is not simply because it can be contrasted with much more recent planning theory conceptions, but also to stress the absence of such precedents in countries on which we can focus later, e.g., Greece. This is not to claim that, for instance, Ebenezer Howard or Tony Garnier, had not influenced the thinking of architect-planners who played an important role in Greece in the interwar years, but rather to

[1] The figures referred to include Ebenezer Howard, Arturo Soria y Mata, Tony Garnier, Clarence Stein, Frank Lloyd Wright, Le Corbusier, Robert Owen, Charles Fourier, Pierre Joseph Proudhon, or Pyotr Kropotkin. For bibliographical sources on these formative years see: Abercrombie 1959; Alexander 1964, 1992; Ashworth 1954; Bairoch 1988; Benevolo 1967, 1980; Bolan 2017; Cherry 1996, 1980, 1981; Choay 1965, 1969; Friedmann 2003; Galantay 1975; Geddes 1968; Gohier 1965; Hall 1998; Hall and Ward 1998; Howard 1965; Joly 1991; Lavedan 1959; Meller 1981; Mumford 1961; Pinol 2000; Ragon 1986; Roncayolo and Paquot 1992; Sutcliffe 1981; Taylor 1999.

stress the absence of local contributions such as that of Geddes. It is not to be forgotten that the architectural civic design conception of town planning, represented in several post-Second World War textbooks,[2] was seen later as a theoretical model responsible for the failures of planning, and was contrasted with theories of either the rational or the collaborative variety. The visionary, design-oriented planner was scorned for his arrogance and his failure to build his visions on systematic research and on rational proof. Rational, comprehensive planning was called upon to replace this approach. Strangely, both these approaches were later condemned and lumped together in a single, allegedly reason-obsessed bourgeois tradition, which elevated the values of the Enlightenment and modernism to unassailable heights.

The design tradition has never ceased to be a major component of urban planning, particularly where the "grand projects" are a tool for various forms of creative or sustainable planning, smart growth, compact city planning, and the new urbanism movement (Grant 2018). In the case of new urbanism, design was elevated to the level of entire regions, ranging from street layout, to districts and corridors, and ultimately to regions and ecological zones. There is here a novel approach, even if "its ideas may seem obvious and old hat" (Kelbaugh 2002). The importance of urban form and the contribution of urban design have been downgraded in the sociological and philosophical rhetoric of both rational and postmodern theory, but deserve to be seen as a rich source of visioning in the twenty-first century. As Bolan put it, "we not only design things – we design relationships and experience!" (Bolan 2017, 241). Palermo and Ponzini (2010, 102–130) are staunch supporters of this direction. Nigel Taylor too advocates such a visionary outlook when he reinstates key concepts like amenity and aesthetics among the habitual normative aims of planning, which tend to dominate most planning theory. These concepts include protection of amenity, quality of place, development, regeneration, environmental goods, equity, social inclusion, public interest, democratic collaboration, and sustainability (Taylor 2003, 93). This is a carefully balanced statement of the goals of planning, but also of the theory which is aiming at explaining, justifying and reconciling it with the lasting values of design-oriented planning, e.g., amenity (Abercrombie 1959). "The term amenity", Parker and Doak (2012, 218 and 226) explain, "centres on some subjective quality of being pleasing or agreeable in terms of the lived environment". It is, they admit "a rather nebulous idea".

2.2 Urban Planning Under Attack

Modernist urban planners in the postwar period attracted fierce criticism for having succumbed to a logic of indiscriminate urban redevelopment, the notorious "bulldozer" mentality, for ignoring the interests of the "man in the street", and for failing to appreciate the creative potential of cities (Woods 1975; Gans 1972; Goodman

[2] E.g., Bardet 1963, Gibberd 1967, Keeble 1969, or Catanese and Snyder 1979.

1972; Cockburn 1977; Jacobs 1965, 1969). They were blamed for behaving like profit-minded businessmen and acting as "natural heirs of nineteenth-century entrepreneurs". The plans they produced were merely accommodating big business (Davies 1982). Was it urban planning that should shoulder the blame? Not according to a view expressed in 1977, because the failures of planning were due to the operation of market forces, not to the work of idealist planners (Pickvance 1982).

This attack has been renewed in the days of early twenty-first century neoliberalism, but this time design is not only about the *grands projets* of capitalism and the regeneration of cities, so popular since the 1990s (Robers and Sykes 2000). It is also about environmental and aesthetic quality, the "spirit of place" (see Chap. 10), and the perception of the urban and natural environment by ordinary people. The quality of the environment reinforces the sense of belonging and, therefore, encourages the desire for sharing experiences, for participation and consensus. All forms of planning have the focus on "place" as a central concern. The problem is that although a multiplicity of actors and agencies emphasize place, their conceptions, purposes and values are not consistent. "We may be sharing the same ball, but we are playing quite different games", wrote Upton (2012). The lingering doubt which Upton implies is unfortunately the point that makes an enormous difference.

2.3 Planning Theory Typologies: Theories in, of, for, on and about Planning and Faludi's Legacy

Allmendinger (2002a, 35), used the very entertaining expression "fixation with prepositions" when he presented the frequent taxonomic ordering of theories *in*, *of*, *for*, *on*, and *about* planning. It is worth focusing on this typology not simply because several scholars refer to it, but because it is very instructive. It highlights the obsession with *alternative* theories when a closer look reveals the impasse that results from the insistence to separate the object, the process and the social environment of planning. This impasse will be discussed again at the end of this section. In his seminal book on planning theory, Andreas Faludi shows his hand from the start: "Theories provide explanations. Explanations are responses to states of tension resulting from observing unexpected events … They represent efforts to reduce surprise caused by such events by giving plausible accounts of how they have come about, accounts which must not contradict anything that the subject knows" (Faludi 1973a, 1). In simple terms, theories are meant to explain situations and events. As to planning theory in particular, he is quick to identify "two types of theory which currently come under planning theory: *procedural* and *substantive* theory. The latter helps planners to understand whatever their area of concern may be. The former can be seen as planners understanding themselves and the ways in which they operate which, at present, are less clearly seen as problematic. I shall argue that planning theory should be concerned with this rather than with substantive theory" (op.cit., 3; see also Parker and Doak 2012, 20). The substantive v. procedural argument has fueled a lively debate over the years. Faludi considers that procedural theory is theory *of* planning, while substantive theory is theory *in* planning (op.cit., 4), which encompasses theories and models (e.g., location models) used as inputs and ancillary analytical tools (Faludi 1973b, 1–3). Faludi added that in making this distinction he was paraphrasing a statement by Britton Harris, that "we have a great need of a science *of* planning in order to determine what is science *in* planning" (Faludi 1973a, 4). Many years later, Bolan's reading of Faludi's dichotomy was that "theory in planning dealt with the theoretical knowledge planners borrowed and relied on, such as economics, sociology, geography, etc. Theory of planning was concerned about a theory of action – or the theory of knowing what to do and how to do it. This dichotomy has persisted; it also contributes in some respects to the current schism between planning theory and planning practice" (Bolan 2017, 5).

Archibugi (2008, 5) makes the point that Faludi made an "excessive division" and that by restraining planning theory to the procedural concept introduced a "misleading factor". Allmendinger (2002a, 2) gives a somewhat different, and rather perplexing, interpretation, viz. that theories of planning are about "why it exists and what it does" and theories in planning are about "how to go about it". McConnell (1981, 14) added a third type of theories, i.e., "the *social theories <u>for</u> planning* which explain why society and planning is as it is and how it should be in future". Paris, who made it clear that he adopts a Marxist perspective, postulated that "the theory of planning rests on a distinction between (a) theories used to comprehend the milieu within which planning operates and (b) theories of how planning itself works. The theory of planning is concerned with the latter as opposed to, for example, 'economic theory' and 'social theory' which planners and others use to structure their understanding of economy and society" (Paris 1982, 5).

Although the earlier presentation of theories *in*, *of*, and *for* planning risks becoming a "prepositional game", it is worth quoting Archibugi (2008, 5) again who refers to theories *on* planning. Faludi's division, he argues, "has probably been at the root of the fact that planning theory, instead of becoming a theory of planning including the difficult problem of defining *interrelationships* between procedural and substantive planning, has become a sort of theory on planning...; and, moreover, a theory on planning which is limited mainly to the experience of town planners, missing the involvement of planners in other substantive plans". Archibugi had however strongly defended Faludi, "the scholar who has contributed more than anyone else to the animation of the wide reflection on planning", but deplored the fact that "an explosion has been produced of a sort of theory on planning (or about planning)", which has "prevented any real progress in the theory of/in planning" (Archibugi 2004, 426).

Brooks (2017, 24–25) suggested yet another prepositional typology with his own version of the use of prepositions, thus adding theories *about* planning, which "focus on its role in a particular milieu – the community, nation, society, or political economy". This differentiates these theories from "theories *of* planning, which seek to explicate characteristics of planning practice" and "theories *for* planning, which propose models or strategies for consideration by practitioners". It is not clear how Brooks' theory about planning differs from Archibugi's theory on planning. At the end of his book, Brooks makes the point that the approach he proposes is a combination of the second and third types, i.e., theories of and for planning. If he was forced to choose, the present author would rather leave out the third approach.

2.4 Planning Theory Yes, but What About Practice?

Starting in the 1960s the epithets used to describe types of planning theory poured out of the literature in increasing numbers. Faludi had referred to substantive and procedural theory, but this was only one such categorization. McConnell (1981, 14–15) protested against what he called "academic jargon mongering" and argued that each of the "prepositional" categories (in, of, for planning) could be simultaneously explanatory or normative. Theories, in his view, can be prescriptive, explanatory, or predictive, yet all theory is impotent unless used in practice (ibid., 21). The theory-practice relationship is a constant source of concern for all theoreticians, because the so-called "translation model", i.e., the process whereby basic scientific theory is translated into technological development, cannot be automatically replicated in the field of spatial planning. "As we know", Alexander points out, "in planning it doesn't happen that way … For … fields, which include those based (to the extent they are science-based at all) on the social and policy sciences, another process, which they call the 'enlightenment model', offers a better account of the diffusion of relevant knowledge from theorists and investigators to its relevant publics. These may include decision-makers, professionals and practitioners, and the media for translation to the consuming or affected public-at-large" (Alexander 2010, 99–100).

The underlying disagreement is related to the perennial issue of the "scientific" claims of planning. "The relationship between the scientific justification of spatial planning (spatial planning as a discipline) and its practical application fields (spatial planning as a profession) can be described as eternal tension of the field of spatial planning itself" (Behrend and Levin-Keitel 2020). Hillier (2005, 272) speaks of "an apparent 'lack of fit' with developments in both planning theory and practice" and ascribes it to a "poststructuralist abyss … between transcendence and immanence as planners struggle with regulatory strategies seeking to tame processes of simultaneous fragmentation and coalescence".

Unavoidably, this analysis leads to the question of what, after all, is planning. E.R. Alexander (2016) recalls an old definition that planning is what planners do! This may sound like a joke. It "looks like a tautology, but it offers a pragmatic answer to the question. One of its merits is that it closes an infinite regress of debate. As an old party in this debate, I have come to the conclusion that the effort to define 'planning' is futile … 'Planners' are the people who a particular community acknowledges are involved in a process it recognizes as 'planning'". What does this "planning" do, beyond applying the state-imposed rules and procedures? An idealistic answer is that "planning fulfils the wishes and demands of different groups by negotiating among conflicting interests and trying to reach agreement between stakeholders. It is a tool for social transformation, social justice and reform, and is a means of political struggle that engages one both inside and outside the workplace" (Tasan-Kok et al. 2016).

The routinization of the work of planning practitioners and the claustrophobic attitude of theoreticians may be blamed for the erection of barriers between them. A distance developed between practitioners and theorists, who seemed to become isolated, as Beauregard (1996, 225–226) observed. The object of planning, the city, he says, was somehow lost and attention was turned to processes, which practitioners were only tangentially interested in. The danger was that for them theorizing was becoming an irrelevant and tedious activity far removed from real practice (March 2010). For a large number of planners this situation is a source of bewilderment. Robin Thompson (2000) was equally puzzled, and for good reason, for the segregation of theory and practice in a supposedly "applied" discipline. Theory seemed obsessed with the accumulation of knowledge from other disciplines. H. Thomas (2004) spoke of an erosion of the discipline of planning within both planning practice and universities. The uneasiness felt by Thompson still persists. Vogelij (2015) drew attention to the undesirable divorce between "the reflective perspective of theorists and the more pragmatic perspective of practitioners", which weakens the profession. Low's suggestion that the work of planners must be placed within the "care structure of humanity" is in the present author's view pertinent here. He means caring "in the sense of wanting to understand and master the world, to understand and look after other humans' bodies with a view to the future, to understand and look after cities with a view to the future" (Low 2020, 68). It is worth keeping in mind this conundrum as we steer in the complex webs of typologies, which take us from a simple procedural or substantive classification to other complicated matrices. The procedural – substantive dichotomy is also discussed by Beauregard (1989, 384–385). Procedural assumptions, he pointed out, assume that the world is malleable provided we understand its internal logic. If we formulate a theoretical paradigm, we can then view the planning process as its "plot". All we need is "comprehensive solutions that have a unitary logic". This implies a holistic perspective which is a-historic.

2.5 The Traditions Underpinning Planning Theory and the Early John Friedmann

Faludi's book on planning theory practically coincided with John Friedmann's book on a theory of transactive planning (Friedmann 1973) and his joint article with Barclay Hudson on knowledge and action, which offered "a guide to planning theory" (Friedmann and Hudson 1974). Faludi's book had just appeared and is not mentioned in it. This article is a turning point in the development of planning theory. In fact, what Friedmann and Hudson did was to document the four traditions, which, as they saw it, provided the foundation on which planning theory was built: Philosophical synthesis, rationalism and systems theory, organization development, and empiricism. It is worth summarizing, as briefly as possible, the key concepts which these traditions bequeathed to planning theory. It is the present author's position that the second and third of these traditions (rationalism systems theory and organization development) belong to the same strand of thought and scientific analysis and cannot be treated separately. As to the tradition of empiricism, which, according to Friedmann and Hudson, is represented by studies of national and urban planning focusing "on the functioning of large-scale political and economic systems", it is nearer the core of planning theories. In other words, it is not so much a tradition which enriches the intellectual content of planning theory, as it is planning theory *per se*, even though it relates to its infancy and deals rather with planning-in-practice. Contributors to this tradition, according to the article, include figures like Banfield, Dror, Wildavsky, Altshuler, or Waterston, who are credited with work which squarely belongs to the body of planning theory.

Part of the philosophical tradition, according to Friedmann and Hudson, revolved around the question whether planning enhances freedom or, instead, generates coercion. Planning is supposed to benefit from the enlightening force of reason, especially in conditions of crisis, but this has to be seen against a background of continuous learning with society operating as a self-organizing or learning system. Critical issues arise regarding the replacement of a central technocracy by a network of learning groups and the existence of comprehensive social knowledge as culmination of history. Was planning, as it evolved after the Second World War, contributing to democratic reconstruction or was it also producing enslavement? Are rationality and freedom compatible? Could we aspire to aggregate welfare through incremental planning, societal guidance, social mobilization, consensus-building, and encouragement of an active society, all combined with central control? Pure rationality now seemed an absurdity, with planning being based on networking, learning, innovation, and human development. This analysis foreshadows the emergence of communicative and collaborative planning theory.

The essence of the rationalist and systems tradition, which, for the author's purposes, are combined with the organization development tradition, was obviously in direct opposition to the main thrust of argument outlined in the blueprint design tradition. Planning of complex spatial systems is envisaged here as advance decision-making using methods presumed to enable rational decisions. The trouble is that inadequate and unreliable knowledge leads to probabilistic and uncertain judgments and to limitations in the use of quantitative models employed to estimate the impact of decisions. Difficulties increase as rapid and disjointed change causes a "turbulent" environment (Godschalk 1974). Efforts to maximize a community's welfare stumble on the impossibility of devising the necessary formulae. The coordination required to effect change is lacking because of administrative inefficiencies. The application of a whole range of methods and evaluation tools is thus handicapped by such weaknesses and by data unreliability. This explains the emergence of less ambitious, watered-down alternatives, of a sub-optimal variety, like bounded rationality, satisficing, incrementalism, mixed scanning, or Lindblom's version with the most picturesque title of all: muddling-through (Allmendinger 2002a, 196; Alexander 1992, 48–49; Beauregard 2020, 25–30; Palermo and Ponzini

2010, 31; Friedmann 1996, 15–22; Brooks 2017, 65, 84, 97–100, 104; Low 2020, 80; Lindblom 1973; Dryzek and Dunleavy 2009, 50–51). In spite of these handicaps, the whole edifice of the so-called policy science rests on perfect rationality premises, which risks causing over-management and social alienation. A similar rationale prevailed in the tradition of organization development, equally influenced by systems theory, where planning is called upon to induce organizational transformation and act as a change-agent. The "system" here is no longer the city or region, but rather an organizational structure. Organizations were seen as sociotechnical systems, composed of communication networks (Pugh 1971; Pugh et al. 1971).

2.6 Yiftachel's Typology

The subsequent proliferation of theories led to a major attempt of producing a typology of *urban* planning theories, which we owe to Yiftachel (1989). In the words of Allmendinger (2002b, 82), "the first real typology that sought to provide a deeper understanding was developed by Oren Yiftachel (1989) … Yiftachel's typology sought to frame planning theory around the three questions – the analytical debate (what is urban planning?); the urban form debate (what is a good urban plan?); and the procedural debate (what is a good planning process?)". As Yiftachel (1989, 36–37) himself put it: "The typology … shows that theories of urban planning can be organised usefully along three clearly defined debates, which should form a framework for a more coherent and consistent body of knowledge aimed at explaining the phenomenon of urban planning, advocating methods of decision making, and at examining the merits of different solutions of urban form". The three debates lead to a paradigm which can be broken down into three strands. The first has a socio-political character influenced by Weberian, Marxist, neo-Marxist, pluralistic, or managerial analyses. The second has a distinct spatial and environmental flavour and speaks of containment, expansion, corridors, decentralization, renewal, and sustainability. The third refers to the procedural dimension and to types of planning: systems, incremental, mixed scanning, advocacy, rational comprehensive, pragmatist, or positively discriminatory (Yiftachel 1989, 27; Allmendinger 2002b, 83). Yiftachel maintained that his typology delimits the parameters of planning inquiry, identifies avenues for research, helps to bridge the gap between theory and practice, clarifies existing confusions regarding the distinction between, on one hand, explanatory and prescriptive theories, and, on the other, substantive and procedural theories, and illuminates major trends in the evolution of planning thought (op. cit., 36–37).

Yiftachel's classification can be described as being ahead of its time and foreshadows later developments. In the second of his "debates" he remains in the realm of the physical, to a large extent design-orientation of planning. In the third debate he gives precedence to a methodological viewpoint. In the first, he touches on the socio-political dimension, which was due to receive increased emphasis in the following years. Although it is not easy to find an exact equivalence between Yiftachel's categories and those of Alexander of, more or less, the same period, there are similarities. Alexander refers to planning approaches (or rather "models") which, in his first category, are defined in terms of *substantive* or sectoral characteristics, being, for instance, economic, physical, transportation, or health planning. In the second category, considered as *conceptual*, the point of view is *instrumental*, i.e., planning's aims and choice of tools, depending on the task of control or regulation. Typologies in his third approach, which he seems to favour, require both a "historical retrospect" and contemporary observation: "We could call this approach *contextual*. It relates different types of planning to different socio-political contexts and ideologies, asking: Who plans for whom with what ideas in mind; These 'ideas' encompass both overt planning rationales and the hidden social and political agenda that are a model's strengths and weaknesses" (Alexander 1992, 94). We may observe, that it is this "contextual" approach to theory that finds an expression in communicative theories, which challenge the supremacy of "instrumental" approaches. It should not escape our attention that the use of terms such as "substantive" is not identical to that adopted by Faludi and others. This caution is required throughout the study of the full range of planning theories. Friedmann and Hudson's philosophical tradition and Alexander's contextual approach find an echo in Britton Harris's reference to moral, ethical, political and social theories, which formed an underlying theoretical stratum of planning theory: "Planning is the premeditation of action. Because it is oriented to action in a defined area of expertise, city planning is a profession, and any theory that aspires to support it must illuminate the dynamics of social action. There are thus many distinct types of theory involved in planning: moral philosophy or a theory of ethics; political and social theory; a theory of the operation of the system being planned; a theory of the technology used in or available to the profession; a theory of management and decision making; and a theory of planning itself" (Harris 1996, 484).

2.7 The "Communicative" Challenge

The typologies discussed so far soon gave their place to a new categorization of planning and theory, better suited to what might be described as the postmodern "political correctness" period of theorizing. Leontidou (1993) explains the effect of post-modernism on the study of cities. Henceforth, traditional goal-oriented planning is differentiated from democratic or egalitarian planning. The following classification spells out this direction: "The four kinds of planning we discuss are (1) traditional, (2) democratic, (3) equity, and (4) incremental (although we shall attempt to demonstrate that incrementalism, while often presented as a *de facto* planning model, is not truly planning" (Fainstein and Fainstein 1996, 266). It is then explained that "equity and democratic planning are overlapping types, but while democratic planning emphasizes the participatory process, the thrust of equity planning is on the substance of programs. The issue thus shifts from who governs to who gets what" (op.cit., 270). The insinuation that the other kinds of planning (traditional and incremental) are not, or cannot be, democratic causes concern. The earlier "design" approach and that of rational comprehensive planning, which rejected it, are practically included in the same "traditional" elitist school of goal-and-means prescriptions. "Planners, by virtue of their expertise and experience, know the correct path, can exercise unbiased judgment, and can be trusted to use their technical knowledge to discover the public interest ... [P]lanners have generally advocated policies that fit the predispositions of the upper classes but not those of the rest of the population" (op.cit., 266–267). In a different formulation, "the rational process model describes only the 'form' (the 'procedure') of the reasoning involved in making rational decisions; it says nothing about the actual 'substantive' ends or goals planning should aim at" (Taylor 1999, 71). It is also deficient in the sense that it ignores the need to communicate with the "planned" and deploy skills of communication and negotiation. Thus, Taylor points out, a new communicative planning theory emerged (op.cit., 122).

The communicative turn, according to Innes (1995, 183), amounts to a new theoretical paradigm, i.e., to a new set of "assumptions and practices of thought and research that a group of people share among themselves ... The communicative action theorists refer to a different set of intellectual mentors and literatures than do the systematic thinkers. They cross-reference each other and build on each other's work ... while referring only infrequently to the work of the systematic thinkers". A key element of a communicative approach is the value placed on information and knowledge and the emphasis on the means used for acquiring them. Although the value of sound information is not disputed, a distinction from knowledge is regularly made and requires attention. Information is frequently considered as a bundle of data which does not amount to knowledge. Information, as opposed to knowledge, does not shed light on causal relations. Alexander takes issue with this assertion: "Almost everything we think of as factual information, from processed data such as demographics, social and economic statistics, and physical-spatial geographic information, down to elementary sensory and cognitive perceptions, rests on implied causal relationships – if only the assumed relationship between the reality of the phenomenon and the validity of its presentation as data or information" (Alexander 2008, 207–208).

Reviewing information as a source of power for the planner, in a text which first appeared in 1989, Forester distinguished between "the perspectives of the technician, the incrementalist or pragmatist, the liberal-advocate, the structuralist, and ... the progressive" (Forester 1995, 438–440; see also Forester 1989). It is on the last two types that he draws attention: "The structuralist paradoxically supposes that the planner's information is a source of power because it serves necessarily ... to legitimize the maintenance of existing structures of power ... The progressive approaches information as a source of power because it can enable the participation of citizens and avoid the legitimizing functions of which the structuralist warns". Forester's analysis was fundamental for the formulation of the communicative theory, which attempts to "counter the effects of interests that threaten to make a mockery of a democratic planning process by misrepresenting cases, improperly invoking authority, making false promises, or distracting attention from key issues" (op.cit., 451).

Writing shortly afterwards, in 1992, Patsy Healey noted that "the conception of planning as a communicative enterprise holds most promise for a democratic form of planning in the contemporary context" (Healey 1996, 234–236). Her observations are triggered primarily from the experience of the neoliberal Thatcherite period in Britain (op.cit., 234–235), when land use planning was judged as a misguided attempt to correct market failures, planners were accused of being far from altruistic, and alternative approaches were advocated (Pennington 2002). Still, Healey's remarks on the then prevailing debate and the "routes for invention" of a new planning this debate has taken are of great interest because she clarifies the route leading to the communicative ideal.

The integration of this new approach becomes clearer in the typologies that Healey developed later. She identified first of all three earlier planning traditions, viz. economic planning, physical planning, and administration management/policy analysis (Healey 1997, 10). For Healey these traditions are succeeded by an argumentative, interpretive, communicative turn in planning theory, born out of a tension between a neoliberal, instrumental rationality, and an interpretation of the relationship of knowledge and action as a social interactive process (op.cit., 28–29). The communicative turn is the foundation of collaborative planning that she advocated (op.cit., 7). In fact, as Sager explains, in due course the terms communicative and collaborative planning ended by being used interchangeably (Sager 2018, 96). He gave his own most illuminating definition, which I will quote in Chap. 3. Communicative planning theory was in fact a movement that succeeded that of the rational, comprehensive and systemic school of the 1950s and 1960s, although Forsyth argues that it took over from the public interest and political economy approaches. The questions that planning theory examines, in the USA at any rate, spanned, in her view, "a range of procedural, political, economic, epistemological, ethical, and cultural dimensions" (Forsyth 2002, 203–205). The influence of social sciences on planning education, she admits, had played a great role. Forsyth seems to suggest, that only the political economy and communicative planning theories reached the status of a "paradigm or at least dominant perspective", which flowed from critical theory, phenomenology, and linguistic analysis. This contrasts with regulation theory, aimed at dealing with the crises and contradictions of capitalism (Painter 2002, 93; see also Low 2020, 127). It equally differs from regime theory which refers to governing coalitions, i.e., informal governance networks made up of public agencies and private interests (Taylor 1999, 139–144).

The discussion about the various strands of planning theory was picked up by John Friedmann in an article aimed at countering the assertion that planning theory is of no value for practitioners, an issue to which we will come back. Friedmann (2003, 7–8) returns to the classification of theories in, of and about planning. For him, theories in planning were specific to various planning specializations, while theories of planning sought to discover their common elements. Theories about planning take a critical look at planning "as it is actually practiced". Such critiques may come from Marxist, political economy, or sustainability perspectives, among others. Commenting on the theory of planning, Friedmann makes the point that "when we argue that planning *ought to be* in or *reflect* a general or public interest, we have in mind a theory of planning. And the same is true when we speak of *comprehensive* or, alternatively, of *advocacy* planning". He also reminds the reader that years earlier, in 1973 (incidentally the year of Faludi's opus), he had challenged the rational decision-making approach as obsolete and that he had proposed "an epistemology of social learning", which was followed by Donald Schön (reflective practice), John Forester (communicative planning), and Patsy Healey (collaborative planning). At the beginning of the new century, in a period marked by the transition "from globalization and neoliberalist ideologies to multiculturalism and postmodernity", planning theory turns to different directions. Here, Friedmann points to the work of Leonie Sandercock (2002) and the emerging radical and insurgent planning theories. In essence, he refers to the next paradigm. To be noted that the distinction that Friedmann makes between theory of planning (his theory 2) and theory about planning (his theory 3) is considered unjustified by Ernest Alexander (Alexander 2003, 179).

2.8 Planning Theory as Textbook Material

The 1970s and early 80s had been already marked by textbooks expressly devoted to planning theory, of which some have been mentioned already and will reappear in this book (Friedmann 1973; Faludi 1973a; McConnell 1981; Cooke 1983). The time had come, after the communicative incursion (e.g., Healey 1997), when efforts could be undertaken to categorize theories of planning in major textbooks dedicated to the whole theoretical spectrum of the planning activity, something which Taylor (1999) and Healey (1997) bravely attempted, as we saw earlier. Allmendinger (2002a) followed suit. He took the view that Yiftachel's typology (see above) had outlived its usefulness and that the procedural-substantive distinction was a false dichotomy: "Taking any one of the current schools of theory it is impossible to separate substance and procedure. Postmodern planning theory, for example, starts from the premise of incommensurability between private languages as well as the notion of *consensus as 'terror'*. Both are normative positions but both could clearly influence any procedures or approaches to planning (though it could be argued that *postmodernism precludes planning at all* …" (Allmendinger 2002a, 35) (my italics). Allmendinger identifies four bodies of theory on which planning theory *per se* (labelled "indigenous theory") is built. They include, first, exogenous theories, e.g., on democracy, psychology, regulation, nationalism etc.; second, framing theories aimed at placing planning in a scientific paradigm or conceptual schema; third, social theories contributed by sociologists or ethnographers; fourth, philosophical understandings. Indigenous theories are planning specific and are founded on various perspectives, such as Marxism, New Right, advocacy, systems thinking, rational comprehensiveness, communicative-collaborative turn, or neo-pragmatism (op.cit., 38–39). Indigenous theories, enriched with temporal, historical and spatial elements, are the field which Allmendinger develops within a post-positivist perspective (op.cit., 213). A most interesting comment by Barrie Needham must be quoted here: "It continues to surprise me how little attention is paid to law in most planning thought ... Outside the US, law is often regarded as difficult, or boring, best left to planning lawyers" (Needham 2017, 177).

Allmendinger returned to the question of typologies in the expanded 2017 edition of his book, to stress once again the importance of typologies, which "provide a 'frame' or a common understanding of subject area, methodologies, language and history of the development of ideas and practice beyond the random" (Allmendinger 2017, 39). He referred to relativism, "particularly in the collaborative, postmodern and pragmatic approaches and, to a lesser extent, some post-structuralist understandings of space", and warned against the dangers of nihilism and of the pitfalls of reliance on exogenous, framing and social theories (op.cit., 291–293). He gives as examples the emergence of regime theory, actor-network theory, and assemblage theory (see Chap. 3) which, he suggests, "do not represent an approach to planning in the same way that the systems or collaborative approaches do" (op.cit., 294). Possible future scenarios for planning theory include, first, an insistent postmodern approach, with its prevalent "celebration of difference" and fragmented interpretations; second, a perpetuation of neo-modern communicative and collaborative theories, with its emphasis on the dubious role of the planners and on the "dark side" of planning; and, third, not surprisingly, a "middle-way" (op.cit., 302–307). Allmendinger and Gunder (2005, 89) explain this often quoted and ominous attribute of planning, its "dark side": "We will suggest that much of planning's control and regulation is posited on a desire for security and certainty that can be perceived, dependent on whose interests are being served, as either progressive or regressive. This may be particularly so for those in a hegemonic position of dominance who wish to maintain their positions of financial and/or social capital".

What is surprising is the final paragraph in the conclusions of Allmendinger's book, which was already present in the 2002 edition: "Options other than those outlined above exist, of course. One area that seems to be gaining increased attention from academics and practitioners is systems theory. Work on retail impact assessment, environmental impact assessment and strategic environmental assessment emphasize an instrumentally rational and systematic approach" (op.cit., 308). This admission, after a long journey across the landscape of postmodern and critical theory, is indeed unexpected, but no doubt sincere, although Allmendinger does his best to soften the impression in the following phrase: "There are, however, differences between such new approaches and the 'classic' systems and rational understandings ... Significantly, these approaches are embedded within more open and accountable understandings of the role and purpose of planning and are not ends in themselves" (ibid.). The admission remains that embedding the time-honoured rational planning theory, as initially expounded by Faludi, in a broader accountability and consultative context is after all possible.

2.9 Critical Stances

The opportunity could not be missed by subsequent writers, such as Archibugi, who returned to the issue of typology. As indicated earlier, he had criticized Faludi's "misleading" and rigid division between substantive and procedural planning. But he also considered Faludi's work as a valid foundation and deplored the fact that it was received as "an essay of political philosophy", and not as major opportunity to renovate the discipline of planning. Planning theorists, whom Archibugi (2008, 3) accuses of "philosophical talkativeness", missed this opportunity. We have already seen Archibugi's remark that the substantive-procedural division (*in* and *of* planning) "has probably been at the root of the fact that planning theory, instead of becoming a theory of planning including the difficult problem of defining *interrelationships* between procedural and substantive planning, has become a sort of theory *on* planning …; and, moreover, a theory on planning which is limited mainly to the experience of town planners, missing the involvement of planners of other substantive plans" (op.cit., 5). Archibugi refers here to macro-economic, social, development, and operational planning. He proceeds to propose his own theory of planology (op.cit., 9, 38–39, 52–55, 72, 101), without missing the opportunity to make a scornful comment on the "Pollyannaesque interpretation of theory". In the forms of planning which urban and regional planning theory hardly touches he equally includes the range of engineering studies that often constitute the spine of urban development planning, without resorting to planning theories. They belong, to use Talvitie's phrase, to a "theoryless" domain: "Despite the well-articulated shortcomings of *modernity*, 'rational' men and women are needed to do modernist planning at its best: to make land use plans; design and build buildings, roads, heating, ventilation, water and sewage systems, drainage systems, road structures and pavements, and environmental mitigation measures, and so on. Modernist planning is also needed to schedule investments and maintenance to make up a meaningful and economic program" (Talvitie 2009, 175).

Palermo and Ponzini too have as their starting point the conviction that much of planning theory is plainly irrelevant: "[M]uch planning theory literature over the last 20 years has, in our opinion, been too abstract and self-referential, raising many doubts about the relevance of the different studies … If today many trends in *planning theory* seem excessively academic and exhortatory, there is no doubt that the issues under discussion are of great importance" (Palermo and Ponzini 2010, 5–6). In terms of theory classification, they refer to the political economy tradition, to the collaborative paradigm, which, in their view, "runs the risk of becoming an ideological simplification or even a fallacy" and of taking "irrationalistic forms", to critical pragmatism, to the structuralist framework, and to Marxist radicalism (op. cit., 9–10 and 24). With a clear intention to defend the "Mediterranean paradigm", to which we will come back in Chap. 4, they insist that "we have reached a paradox in a planning culture that might forget, or relegate to a marginal role, the regulatory and design issues" (op.cit., 20). The structure of their book follows a series of conceptions of planning: Public decision-making process, structural determinism, multiple interactions, communicative turn, critical possibilism, and design culture (op.cit., 24–25).

Allmendinger's conclusion that environmental concerns and methods of analysis are reinstating an interest in systems planning was mentioned earlier. This is not to underplay the continued concern for other parameters. The title of a 1996 article by Scott Campbell (Green cities, growing cities, just cities?) says it all and, in itself, implies a typology seen from a different angle. In a "triangle model", S. Campbell (2016) underlines the conflicts facing the planner between social justice and equality, economic growth and environmental protection. Ideally, planners should stand at the centre of the triangle, but this could turn out to be a naïve demand like the belief in comprehensive planning for a single public interest. Sustainability critics will probably develop arguments against this triangle logic remindful of attacks on comprehensiveness. According to Campbell, "the incrementalists will argue that one cannot achieve a sustainable society in a single grand leap, for it requires too much social and ecological information and is too risky. The advocacy planners will argue that no common social interest in sustainable development exists, and that bureaucratic planners will invariably create a sustainable development scheme that neglects the interests both of the poor and of nature". Campbell's warning must be carefully heeded, now that adaptation to climate change gradually leads to a new major current in planning thought. His triangle model is in fact multiple triangles of interactions and choice-dilemmas facing the planner, which the present author tried to picture in a tentative figure (see Fig. 2.1).

2.10 An Inheritance of Confusion

Where does all this leave us in terms of a typology of theories? Planning, depending on the approach, was found to be – and will be found again in subsequent chapters – rational, systemic, comprehensive, complexity-based, positivist and postpositivist, empiricist, pragmatic, utilitarianist, state-dominated, class-unequal, even servile, choice-based, technofix-obsessive, muddling-through, incremental, mixed scanning, substantive, normative, procedural, optimizing or satisficing, participatory, advocacy, deliberative, discursive, relativistic, interpretive, storytelling and narrative-based, communicative, collaborative, interactive, Marxist, structure- or agency-oriented, equity-targeted, critical, radical, modern and postmodern, sustainable, socially just and fair, assemblage-based, feedback-strategy, actor-network-structured, neoliberal, populist, clientelist, multiplanar, reflexive, innovatory, feminist, postcolonial, transformative, "dark side"-focused, antifragile, ethical, deontic, post-political, post-truth, resilient, "panarchic", insurgent, subversive, agonistic and other variants, which one cannot possibly summarize.

The philosophical foundations of much of planning theory explain the growing infiltration of extraneous fields of thinking in the theoretical explanations of planning. This was becoming obvious already in the 1970s. It is explained vividly by Taylor, when he refers to the endless streams of philosophical thinking invoked in

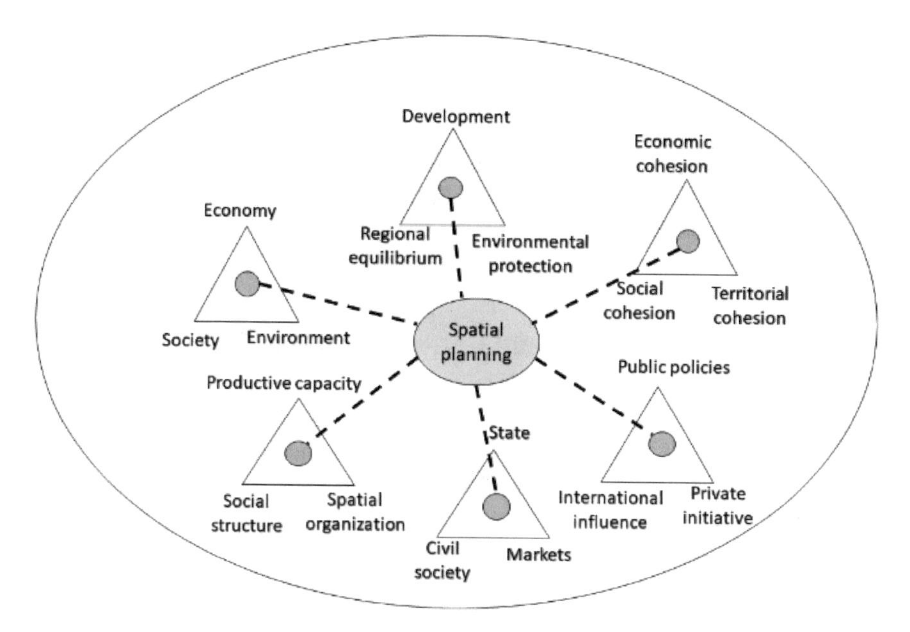

Fig. 2.1 Spatial planning in the middle of interactions and choice-dilemmas. (Figure drawn by the author)

planning theory: "But, as far as planning is concerned, there are two areas where philosophical inquiry is particularly important, and these areas concern respectively, the ethical judgments, and the knowledge, upon which planning is based" (Taylor 1980). From then onwards planning theory expanded in all directions absorbing greedily inputs from philosophy and social science. "Large areas of the social sciences and humanities – read as reflections on the links between planning processes and their outcomes – are incorporated into the domain of the community even if it sometimes seems that a minnow has swallowed a whale" (Mandelbaum 1996, xi). Mandelbaum refers here to the community of academic planning theoreticians.

Several years later, Lord (2014) commented on the explosion in planning theory and observed that this trend continues in the context of post-positivism, the communicative/deliberative approach, American pragmatism, while doubts are also expressed about the purpose, adequacy and clarity of this hectic theoretical activity: "Such doubts about the value of planning theory have been accompanied by the ever-present spectre of the 'theory–practice gap' … with the result that planning theory has been 'widely criticised by academics and practitioners as being confused and impractical'". Law-Yone (2007) is right in describing the flood of planning theories as a "literary genre", which gives birth to a string of essays and reports. He speaks of a theory bandwagon, of fuzzy discourses, of cacophony, of fragmentation, and of a murky and formless field, while practitioners remain indifferent: "Planning theory texts have in fact become a recognizable genre somewhat akin to 'fast-capitalist texts', which are described as 'a mix of history and description, prophesy, warning, proscriptions and recommendations, parables (stories of success and failure) and large doses of utopianism'". In spite of this proliferation of theories and their taxonomies, limited effort has been made to classify planning theories in terms of theoretical approaches to the state, like pluralism, elite theory, Marxism or Foucault's postmodern discourse analysis (Hay and Lister 2006, 3). This does not mean that planning theories ignore these schools of thought, but rather that the latter are not used as a basis of a planning theory typology.

2.11 Towards a More Lucid Categorization of Theories?

In their introduction to a recent volume they edited, Gunder et al. (2018a, b) provide a summary overview of this torrent of theories. The theories represented in their reader complete the picture. To some of these theoretical perspectives we will return later in this or the subsequent chapter, but to close this panorama of typologies the question must be asked whether we can satisfy ourselves with a simpler and all-embracing categorization. This is what Brooks (2017) attempts when he asserts that, ultimately, we can speak of either positive (empirical or descriptive) or normative (ethical or functional) theories. He adds his own version of planning, the Feedback Strategy of public planning (op.cit., 160–180). Bolan too adopts the descriptive – normative distinction. It is an important dilemma "whether a theory is primarily developed to describe reality or to prescribe those behaviors that are appropriate within the (presumed) framework of reality … As the term implies, a descriptive theory seeks primarily to describe or explain and, in principle, makes no statement of how things ought to be" (Bolan 2017, 51). As it will become evident in Part III of this book, it is this form of "descriptive" theory which is followed, although "explanatory" would be a more appropriate word.

The basic dichotomy between positive and normative theories seems to be endorsed by Beauregard (2020, 2) as well, who attributes the domination of the normative element to the mode of thinking of planning academics: "For the most part, their goal is not to explain how and why planning is actually done, but to indicate to practicing planners what they *should* be doing so that cities and regions can prosper and people can live well in a just and democratic world. Normative theory, not positive theory, is what dominates. Moreover, the interests of planning theorists are both scholarly and practical: they want their ideas to be recognized by their academic colleagues, absorbed by the students who sit attentively in their classrooms, and adopted by practitioners". Planning theory, in his view, is the intellectual project of planning education. Brooks (1995, 560–561), referring to the USA, had added his own touch in 1988: "The rapid growth of the planning profession during the 1960s and 1970s had a dramatic impact on planning education as well". The result was that scholars "stopped doing significant [practice] work after acquiring tenure", while "practitioners who joined university facilities … ceased growing professionally". Beauregard, who presents a most useful and concise overview of theoretical currents, has more to add to his assessment of the proliferation of theories:

"Regardless of whether theory is intended to motivate, guide, or enlighten, most of the debate about closing the theory-practice gap focuses on which approach best fits planning practice. This became quite explicit when, in the 1990s, 'armchair' theorists – the reference being to the Marxists and postmodernists – were criticized by Judith Innes for developing perspectives so distant from planning practice as to be of little relevance" (op.cit., 111). Innes was of course one of the protagonists of the communicative turn, which, supposedly, was meant to remedy this situation. One can doubt whether theorists who were quoted earlier, like Archibugi, Brooks or Palermo and Ponzini, would share her optimism.

2.12 An Effort to Simplify the Prepositional Game: Currents of Ideas and Types of Planning

The nightmare of the theoretical landscape, for which several writers expressed their despair, as it was indicated earlier and will be stressed again later, forces the author to return for a while to the confusing overlaps and to the puzzling typology of theories in, of, for, on and about planning. For the sake of the reader, the main views are summarized here:

- Theories *of* planning are procedural and "can be seen as planners understanding themselves and the ways in which they operate" (Faludi); they are about "why [planning] exists and what it does" (Allmendinger); they are "concerned about a theory of action – or the theory of knowing what to do and how to do it" (Bolan)"; they are theories "in which the processes and operations of planning are analysed and explained" (McConnell); or, finally, they "seek to explicate characteristics of planning practice (for example its communicative effects ...)" (Brooks).
- Theories *in* planning are substantive and help planners "to understand whatever their area of concern may be" (Faludi); they consist of "theoretical knowledge planners borrowed and relied on, such as economics, sociology, geography, etc." (Bolan); they are about "how to go about it" (Allmendinger); or, they are theories "derived from many disciplines" (McConnell).
- Theories *for* planning are social theories "which explain why society and planning is as it is and how it should be in future" (McConnell); such theories "propose models or strategies for consideration by practitioners" (Brooks).
- In a sense, a theory *on* planning is what Faludi's division amounts to, "and, moreover, a theory on planning which is limited mainly to the experience of town planners ..." (Archibugi).
- There are also "theories *about* planning, which focus on its role in a particular milieu – the community, nation, society, or political economy" (Brooks).

It is difficult, to begin with, to distinguish between theories for, on and about planning and to explain how and why they differ. They all focus ultimately on the wider context in which planning takes place and within which planners operate. Brooks's view of theory for planning (models and strategies for consideration by practitioners) may be an exception, but it could easily fit into a theory in planning, or even of planning. Equally, Allmendinger' conception of theory about "how to go about it", could be better accommodated in the "of planning" category, instead of the "in planning" one. But beyond the possible consolidation of theories for, on and about planning in a single class, they could easily be merged into the "in planning" group. We would then get back where we started, with just two types of planning theory, one on the context of planning and one on its process and mode of action.

It is essential at this point to distinguish between *types* of planning theory, as outlined above, and the *currents* of ideas, preoccupations, interpretations, concerns, and socio-political conditions (currents of ideas, for short) that feed these types. What is implied here is not only philosophical and sociological ideas, but also issues that create a *problematique* which forms the background of planning, especially in the years after the decline of the notion of planning as mere civic design. Four currents of ideas seem, in the author's view, to stand out in the constant stream of theorizing to which a fifth is added. Each current feeds into a corresponding group of particular types of planning and planning theory, which are tentatively listed here.

- The first group of types of planning would include planning approaches described as rational, positive, comprehensive, organizational, managerial, policy-oriented, economic development, and systemic, with their offshoots like incremental, satisficing, mixed-scanning forming a closely related branch, which overlaps with the proposed fifth current. Structuralist interpretations could also find their place here, in spite of different ideological origins.
- The second group is manifested in forms of planning described as participatory, relational, communicative, collaborative, equity, discursive, deliberative, argumentative, storytelling, and interpretive approaches.
- The third group encompasses radical, insurgent, subversive, agonistic, minoritarian, post-structuralist, postcolonial, and transformative types of planning, as well as forms of the "Global South" variety.
- Current preoccupations lead to a fourth group, resulting from the need to promote environmental sustainability, restore ecological equilibrium, reconcile humans and nature, resist climate change, mitigate its consequences,[3] and reverse the effects of the Anthropocene age. Commenting on the shifts of planning concerns, Allmendinger identified certain broad changes of which one is of relevance for the fourth group: [It] "concerns the broadening and fuzzing of the objectives of planning and the implications of this upon expectations of what planning could actually achieve. A major factor in this is the central role that planning has been given in environmental protection and addressing climate change" (Allmendinger 2017, 206).

[3] These lines were written at the peak of a heat wave episode, which was later followed by devastating forest fires around Athens!

If we were forced to find one key word for each group of planning approaches, and the respective current of ideas, the best terms would be rational, communicative, radical, and climate. The "climate" current could be called "environmental", but it is the mobilization, that climate change is giving rise to, which provides the content of the new tendency. It is here that the author is tempted to propose a fifth current in which would be amalgamated pragmatism and his personal view of compromise planning, hence the name "compromise current", in the hope that his presumption will be forgiven. The first and second currents will be further explored in Chap. 3, the third in Chap. 4, the fourth in Chap. 5, and the fifth current in Chaps. 9, 10 and 11.

An effort is being made, in Text-Box 2.1, to describe the five currents in terms of key-words which, in the author's view, express their exclusive emphasis. There is no doubt that the various currents share certain characteristics, as they differ in others. However, there has been a deliberate intention to single out those features which are typical of each case. The text-box is a kind of portrait of each current's particular physiognomy. The key-words are repeated later at the beginning of the relevant chapter or sub-chapter dedicated to each theoretical current (Chaps. 3, 4, 5 and 10).

The forms of planning that these currents of ideas give rise to belong no doubt to certain periods, which appeared in succession; but this is not just a matter of chronological ordering and it would be misleading to remain trapped in a scheme of periodization. There has been a constant interpenetration of ideas and forms of planning in all currents. Environmental concerns and warnings, as well as resistance to unbridled urbanization were present even in the "glorious fifties and sixties",[4] at the apex of rationalism in planning, and so were countervailing forces aimed at securing increased participation or to protect minority groups from the excesses of "bulldozer" urbanism. Rationalism, of the neoliberal version, resurfaced while participatory and communicative planning theory was flourishing. Alternative modes of urban development, especially in less developed countries of the South and East, were advocated side-by-side with modernist projects even in the days when comprehensive planning ideologies reigned supreme in the North and West. Self-help, site-and-services, informal, and/or spontaneous urban development practices became a routine in the so-called Third World at a time when rational statutory planning was the official version of planning in industrial countries, but also in non-industrialized ones. Action- or speed-planning was advocated in lieu of time-consuming planning processes. Such models of planning will be discussed further in Chap. 4. Dubai-style urban developments and flagship projects are currently in fashion, while transformative and insurgent planning is advocated across the globe (see Chap. 4). Ecologically-conscious planning and settlements were advocated long before the current wave of climate change and the appearance of the Ecosystem Approach, albeit in an experimental form.[5] Cross-fertilization and

[4] See the pathbreaking work of Rachel Carson and Barbara Ward (Carson 1989 and 1999, first published in 1950 and 1962; Ward and Dubos 1972; Ward 1966, 1976).

[5] See: Allison 1975; Barton 2002; Blowers 1993; Devuyst 2001; Elliott 1999; Expert Group on the Urban Environment 1996, Girardet 1999, 2004; Hardin 1968; IUCN 1980, 1991; Kenny and Meadowcroft 1999; Lafferty and Meadowcroft 2000; Landry 2000; Layard et al. 2001; Meadows et al. 1972; Pearce and Barbier 2000; Raban 1974; Riddell 2004; Satterthwaite 1999; Silberstein and Maser 2000; Skea 1999; Solomon 1980; Urban Task Force 1999.

mutual infiltration were a reality which negates the over-simplification of water-tight theoretical constructs, taking over from one another in quick succession, inspired, in most cases, by the experience of a very small number of developed countries.

Text-Box 2.1: Keywords of the Main Theoretical Currents
[Ideas, concepts, preoccupations, interpretations, concerns, and socio-political conditions embedded in each current of ideas, which inspire the corresponding theory and practice]

The rational current

Objectivity, rationality, absolute values, choice, goal attainment, process, space, observation, structure, technofix solutions, authority, calculation, utility, state power, expertise, idealism, scientific truth, facts, action, end-state planning, order, development, alternatives, systems thinking, management, regulation, explanation, stability, futurology, law rule, methodology, comprehensiveness, innovation, evaluation, formal knowledge, future trajectories, policy making, reality simulation, professionalism, foresight, feedback loops, open systems.

The communicative current

Subjectivity, relative values, diversity, collaboration, deliberation, reflexivity, place, recognition, hermeneutic approach, agency, opinion variations, group values, devolution, mediation, understanding, storytelling, cognitive truth, dialogical learning, arbitration, difference, structuration, search, stakeholders, actor roles, truth relativity, visioning, audiences, multiple publics, advocacy, grassroots movements, consensus, ethics, participation, contradictory views, equity, discourse, argumentation, situatedness, multiple knowledges, debate arenas, interpretation.

The radical current

Transformation, otherness, subversion, insurgency, agonism, antagonism, revolt, minoritarian values, becoming, resistance, collective values, feminism, activism, challenge, systemic defiance, class interests, race, self-defense, power overthrow, ethnicity, radical politics, Marxism, freedom, oppression, unemployment, policing, poverty, exclusion, Global South, underdevelopment, colonialism, dependence, informality, refugees, periphery, dispossession, prison syndrome, unsettling, enclaves, terror, fear.

(continued)

Text-Box 2.1 (continued)

The "climate" current

Ecologism, belonging, natural values, perception, protection, trend reversal, adaptation and adaptability, restoration, coexistence, circular economy, socio-eco-systems, balanced action, being in nature, technical adjustment, equilibrium, life values, self-organization, earth care, corrective action, practice re-evaluation, ecological cooperation, risk, resilience, climate change, natural disasters, vulnerability, mitigation, redundancy, Anthropocene heritage, spaceship earth, sustainability, maritime space, biosphere, pollution, species extinction, landscape disfigurement, biodiversity loss, economic repercussions, resource exhaustion, solidarity, preservation, energy exchange, unity, contamination, earth science progress, responsibility, resourcefulness.

The tentative "compromise" current

Pragmatism, incrementalism, agreement, instrumentalism, experience, synthesis, settlement, accommodation, blending values, individuality, mutuality, moderation, pluralism, practical knowledge, realism, inquiry, togetherness, testing, learning-by-doing, disruption avoidance, self-realization, pioneer spirit, opportunism, contingency, correlations, experimentation, truth-illusion, reciprocity, human as doer, coping, transparency, robustness, negotiation, clientelism, patronage, corruption, favouritism, improvising, complicity, knowledge accumulation, land property, private interests, storage of experience, precedent.

2.13 Charles Hoch's Pragmatism and Ernest Alexander's Contingency Model

But then, who can claim that this mind-boggling range of theories can be squeezed into a handful of types? Who can still remain faithful to a set of restrictive typologies? What is perhaps needed is a totally novel formulation of alternative perspectives. Charles Hoch tried to do that in his 2019 book, in a chapter adapted from a 2011 article. To be more precise, he distinguishes not between theories, but between what he calls "discourse arenas linking each of three planning domains" (Hoch 2019, 39), i.e., field, movement, and discipline: "The spatial planning field includes purposeful efforts to anticipate, influence and cope with urbanization and its effects … The planning movement refers to collective efforts to develop and promote the practice of spatial planning … The planning discipline describes efforts to study and teach spatial planning on the job, in the profession and at the university" (op.cit., 40). It is the interactions between movement and field, field and discipline, or movement and discipline that give birth to different theoretical explanations. Notwithstanding brave efforts to synthesize these currents, by eminent theoreticians like Friedmann, Healey, Alexander, Faludi, or Archibugi, Hoch claims, "the contest for a theory of planning across these domains will likely never end with one grand theory achieving intellectual hegemony" (op.cit., 42). Arguing for his own *pragmatist* approach, Hoch points out that the task of theory could be rejuvenated if practice and theory could be brought closer. "Practitioners do this kind of appropriation all the time by selectively combining policies, regulations, designs, innovations and other forms of planning activity that they find useful for their own work. Such practical synthesis fits the demands of professional work and of course my own pragmatist conception of the discipline. But I think it offers a more fruitful role for theory" (op.cit., 46). Hoch is a disciple of the American pragmatist philosophy and it is in this spirit that he approaches planning. We have to keep constantly in mind, at least from the perspective of this book, what Low calls the "fundamentals of planning": "First of all collective 'public' planning requires institutions of the state. It is essentially a 'public sector' task … Second, planning must be based on facts and truth … Third, planning must be for a specified 'place',,," (Low 2020, 261). We shall come back to these terms (public planning, facts, truths, state, place) at various points in the book, but it must be made clear at this stage that a pragmatist interpretation – let alone a compromise one – is dependent on these elements of the planning endeavour.

Hoch had already embarked on his pragmatist adventure much earlier: "A number of theoreticians in the 1980s broke the grip of the severe dichotomy of 'technical' and 'political' planning to adopt what I have called a 'pragmatic orientation'. Pragmatism allows us to imagine individuals acting in communities (and communities in individuals). It fosters an appreciation of the search for a communal welfare that is not a bargain among strangers but that also cannot be tarred with the brush of a specious utilitarian rationality. Instead of treating professional expertise (the technical) and social values (the political) as exclusive and even antagonistic domains of action, the critical pragmatists recast the two as distinct aspects of practical advice giving" (Hoch 1996, 229). From the present author's perspective, Hoch's pragmatism is a key component of *compromise* theory, as proposed in this book. Elsewhere (Chap. 9), his own views on compromise will be quoted again. His view also lends support to this author's claim that practitioners generate theory through their own professional work. They endow theory with flesh and bones. The need for synthesis had been the central call in a much earlier article by Ernest Alexander (1996, 45–46). who disputed the real difference between the alternative rational and communicative models and introduced his conception of a contingency model: "Planning theoreticians are in a state of turmoil. Nothing is accepted; everything is questioned. The requirements that theory must meet have been well stated … but no coherent body of thought has been accepted as meeting these criteria. Indeed, Mandelbaum … and others have questioned whether a 'general theory of planning' is even possible. Given these cogent reservations about the feasibility of a general theory of planning, it is clear that a more modest approach is indicated. Consequently, a contingency framework is suggested here as a way of addressing the problems planning theoreticians face today". For Alexander, the theory of rational decision making and the theory of communicative action can be juxtaposed within the contingency model, as they are a mirror image of each other. The complementarity of the rational and communicative models is accepted by Baum (1996, 372) too: "Communicative theorists share with their predecessors a rational view of planners (and others) as actors. This assumption limits their ability to interpret actions and to advise planners on how to act … It is important to interpret these theorists' writings as the texts of authors whose practice requires them to publish theoretical texts. Their books, articles, and chapters are words published in specific contexts for specific purposes".

Epilogue: An Opening for a Theory of Compromise Planning?

This chapter was a panorama of planning theory typologies. It was a journey along the rugged path of the typologies of theories of, in, for, on or about planning; procedural and substantive, normative or explanatory, rational and comprehensive, collaborative and deliberative, radical and subversive and so on. The classificatory contributions of several theorists were presented and then an effort was made to simplify this range of typologies by narrowing it down to the most important currents, first the rational, communicative and radical currents and then one inspired from the conditions of climate change. This effort was finished with the author's own view of a fifth current described as compromise planning, to be developed in detail in Part III of the book. This chapter ended with a section on pragmatism.

Charles Hoch's pragmatist approach and Ernest Alexander's contingency model, as well as views on the process of compromise which is inherent in planning practice, are good theoretical precedents on which to base this author's theoretical compromise planning approach. Hoch, Alexander and other theorists, whose opinions were quoted, have illustrated convincingly the futility of the vast range of alternative theories. We shall return to their views, as well as to those of Sanyal and Vidyarthi, in Chaps. 3 and 9. Bur apart from theoretical models it is the contextual reality and the history of a territorial entity which must inform a novel form of interpretation. This will occupy us extensively in Part II of the book, in which the example of Greece will be examined. However, we need first to delve deeper into the currents of ideas – and of the corresponding claims of planning theory types – which were labelled as rational, communicative, radical, and "climate". This will be the object of Chap. 3, 4, and 5.

References[6]

Abercrombie P (1959) Town and country planning. Oxford University Press, Oxford (first edition: 1933)

Alexander C (1964) Notes on the synthesis of form. Harvard University Press, Cambridge, MA

[6] **Note**: The titles of publications in Greek have been translated by the author. The indication "in Greek" appears at the end of the title. For acronyms used in the references, apart from journal titles, see at the end of the table of diagrams, figures and text-boxes at the beginning of the book.

Abbreviations of journal titles: AT Architektonika Themata (Gr); DEP Diethnis kai Evropaiki Politiki (Gr); EpKE Epitheorisi Koinonikon Erevnon (Gr); EPS European Planning Studies; PeD Perivallon kai Dikaio (Gr); PoPer Poli kai Perifereia (Gr); PPR Planning Practice and Research; PT Planning Theory; PTP Planning Theory and Practice; SyT Synchrona Themata (Gr); TeC Technika Chronika (Gr); UG Urban Geography; US Urban Studies; USc Urban Science. The letters Gr indicate a Greek journal. The titles of other journals are mentioned in full.

Alexander ER (1992) Approaches to planning: introducing current planning theories, concepts and issues. Gordon and Breach, Luxembourg
Alexander ER (1996) After rationality: towards a contingency theory for planning. In: Mandelbaum SJ, Mazza L, Burchell RW (eds) Explorations in planning theory. Center for Urban Policy Research/Rutgers, New Brunswick NJ, pp 45–64 (earlier version published in 1988)
Alexander ER (2003) Response to 'why do planning theory?'. PT 2(3):179–182
Alexander ER (2008) The role of knowledge in planning. PT 7(2):207–210
Alexander ER (2010) Introduction: does planning theory affect practice, and if so, how? PT 9(2):99–107
Alexander ER (2016) There is no planning – only planning practices: notes for spatial planning theories. PT 15(1):91–103
Allison L (1975) Environmental planning: a political and philosophical analysis. George Allen and Unwin, London
Allmendinger P (2002a) Planning theory. Palgrave, Basingstoke
Allmendinger P (2002b) Towards a post-positivist typology of planning theory. PT 1(1):77–99
Allmendinger P (2017) Planning theory, 3rd edn. Palgrave, Basingstoke
Allmendinger P, Gunder M (2005) Applying Lacanian insight and a dash of Derridean deconstruction to planning's "dark side". PT 4(1):87–112
Archibugi F (2004) Planning theory: reconstruction or requiem for planning? EPS 12(3):425–445
Archibugi F (2008) Planning theory: from the political debate to the methodological reconstruction. Springer Verlag Italia, Milan
Ashworth W (1954) The genesis of modern British town planning. Routledge and Kegan Paul, London
Bairoch P (1988) Cities and economic development: from the dawn of history to the present. The University of Chicago Press, Chicago (original: De Jéricho à Mexico, 1985)
Bardet G (1963) L'urbanisme. Presses Universitaires de France – Collection Que Sais-je?, Paris (first published 1945)
Barton H (ed) (2002) Sustainable communities: the potential for eco-neigbourhoods. Earthscan, London
Baum H (1996) Practicing planning theory in a political world. In: Mandelbaum SJ, Mazza L, Burchell RW (eds) Explorations in planning theory. Center for Urban Policy Research/Rutgers, New Brunswick, p 365–382 (revised version of 1988 publication)
Beauregard RA (1989) Between modernity and postmodernity: the ambiguous position of U.S. planning. Environ Find Plan D Soc Space 7:381–395
Beauregard RA (1996) Between modernity and postmodernity: the ambiguous position of U.S. planning. In: Campbell S, Fainstein S (eds) Readings in planning theory. Blackwell, Oxford, pp 213–233 (first published 1989)
Beauregard RA (2020) Advanced introduction to planning theory. Elgar, Cheltenham
Behrend L, Levin-Keitel M (2020) Planning as scientific discipline? Digging deep toward the bottom line of the debate. PT 19(3):306–323
Benevolo L (1967) The origins of modern town planning. Routledge and Kegan Paul, London (Italian edition: 1963)
Benevolo L (1980) The history of the city. Scolar Press, London (Italian original: Storia della Città, 1975)
Blowers A (ed) (1993) Planning for a sustainable environment. A report by the Town and Country Planning Association, Earthscan, London
Bolan RS (2017) Urban planning's philosophical entanglements: the rugged, dialectical path from knowledge to action. Routledge, London
Brooks MP (1995) Four critical junctures in the history of the urban planning profession: an exercise in hindsight. In: Stein JM (ed) Classic readings in urban planning. McGraw-Hill, New York, pp 555–572 (first published 1988)
Brooks MP (2017) Planning theory for practitioners. Routledge, London (first published 2002)

Campbell S (2016) Green cities, growing cities, just cities? In: Fainstein SS, DeFilippis J (eds) Readings in planning theory. Wiley – Blackwell, Chichester, pp 214–240 (first published 1996)
Carson R (1989) The sea around us. Oxford University Press, New York (first published 1950)
Carson R (1999) Silent spring. Penguin, London (first published 1962)
Catanese AJ, Snyder JC (eds) (1979) Introduction to urban planning. McGraw-Hill, New York
Cherry G (1996) Town planning in Britain since 1900: the rise and fall of the planning ideal. Blackwell, Oxford
Cherry G (ed) (1980) Shaping an urban world. Mansell, London
Cherry G (ed) (1981) Pioneers in British planning. The Architectural Press, London
Choay F (1965) L'urbanisme: Utopies et réalités. Seuil, Paris
Choay F (1969) The modern city: planning in the 19th century. Studio Vista, London
Cockburn C (1977) The local state: management of cities and people. Pluto Press, London
Cooke P (1983) Theories of planning and spatial development. Hutchinson, London
Davies JG (1982) The oppression of progress . In: Paris C (ed) (1982) Critical readings in planning theory. Pergamon Press, Oxford, pp 123–127 (originally published 1972)
Devuyst D (ed) (2001) How green is the city?: sustainability assessment and the management of urban environments. Columbia University Press, New York
Dryzek JS, Dunleavey P (2009) Theories of the democratic state. Macmillan International/Red Globe Press, London
Elliott JA (1999) An introduction to sustainable development. Routledge, London
Ellis H (2015) The re-creation of social town planning? PTP 16(3):436–440
Expert Group on the Urban Environment (1996) European sustainable cities. European Commission, Brussels
Fainstein SS, Fainstein N (1996) City planning and political values: an updated view. In: Campbell S, Fainstein S (eds) Readings in planning theory. Blackwell, Oxford, pp 265–287
Faludi A (1973a) Planning theory, 1984th edn. Pergamon Press, Oxford
Faludi A (1973b) Introduction to part I (what is planning theory?). In: Faludi A (ed) A reader in planning theory. Pergamon Press, Oxford, pp 1–10
Forester J (1989) Planning in the face of power. University of California Press, Berkeley
Forester J (1995) Planning in the face of power. In: Stein JM (ed) Classic readings in urban planning. McGraw-Hill, New York, p 437–456. (first published 1989)
Forsyth A (2002) Trajectories of planning theory: the 2001 ACSP anniversary round table. PT 1(3):203–208
Friedmann J (1973) Retracking America: a theory of transactive planning. Anchor Press/Doubleday, Garden City
Friedmann J (1996) Two centuries of planning theory: an overview. In: Mandelbaum SJ, Mazza L, Burchell RW (eds) (1996) Explorations in planning theory. Center for Urban Policy Research/Rutgers, New Brunswick NJ, pp 10–29 (first published in Friedmann J (1987) Planning in the public domain: from knowledge to action. Princeton University Press, Princeton)
Friedmann J (2003) Why do planning theory? PT 2(1):7–10
Friedmann J, Hudson B (1974) Knowledge and action: a guide to planning theory. J Am Inst Plan 40(1):2–16
Galantay EY (1975) New towns: antiquity to the present. Braziller, New York
Gans HJ (1972) People and plans: essays on urban problems and solutions. Penguin, Harmondsworth
Geddes P (1968) Cities in evolution. Ernest Benn, London (first edition 1915)
Gibberd F (1967) Town design. The Architectural Press, London
Girardet H (1999) Creating sustainable cities. Green Books, Totnes
Girardet H (2004) Cities, people, planet: liveable cities for a sustainable world. Wiley-Academy, Chichester
Godschalk DR (ed) (1974) Planning in America: learning from turbulence. American Institute of Planners, Washington, DC
Gohier J (1965) L'evolution de l'urbanisme en France. CRU65: Études et Essais. Centre de Recherche d'Urbanisme, Paris

Goodman R (1972) After the planners. Penguin, Harmondswoth
Grant JL (2018) Growth management theory: from the garden city to smart growth. In: Gunder M, Madanipour A, Watson V (eds) The Routledge handbook of planning theory. Routledge, London, pp 41–52
Gunder M, Madanipour A, Watson V (2018a) Planning theory: an introduction. In: Gunder M, Madanipour A, Watson V (eds) The Routledge handbook of planning theory. Routledge, London, pp 1–12
Gunder M, Madanipour A, Watson V (eds) (2018b) The Routledge handbook of planning theory. Routledge, London
Hall P (1998) Cities in civilization. Weidenfeld and Nicolson, London
Hall P, Ward C (1998) Sociable cities: the legacy of Ebenezer Howard. Wiley, Chichester
Hardin G (1968) The tragedy of the commons. Science 162(3859):1243–1248
Harris B (1996) Planning technologies and planning theories. In: Mandelbaum SJ, Mazza L, Burchell RW (eds) Explorations in planning theory. Center for Urban Policy Research/Rutgers, New Brunswick, pp 483–496
Hay C, Lister M (2006) Introduction: theories of the state. In: Hay C, Lister M, Marsh D (eds) The state: theories and issues. Macmillan International/Red Globe Press, London, pp 1–20
Healey P (1996) Planning through debate: the communicative turn in planning theory. In: Campbell S, Fainstein S (eds) Readings in planning theory. Blackwell, Oxford, p 234–257 (first published 1992) (see Healey P 1992)
Healey P (1997) Collaborative planning. Macmillan, London
Hillier J (2005) Straddling the post-structuralist abyss: between transcendence and immanence? PT 4(3):271–299
Hoch C (1996) What do planners do in the United States? In: Mandelbaum SJ, Mazza L, Burchell RW (eds) Explorations in planning theory. Center for Urban Policy Research/Rutgers, New Brunswick, pp 225–240 (earlier version published in 1994)
Hoch C (2019) Pragmatic spatial planning: practical theory for professionals. Routledge, New York
Howard E (1965) Garden cities of to-morrow. Faber and Faber, London (first edition 1898)
Innes JE (1995) Planning theory's emerging paradigm: communicative action and interactive practice. J Plan Educ Res 14(3):183–189
IUCN (1980) World conservation strategy (available on the web)
IUCN (1991) Caring for the earth: a strategy for sustainable living. IUCN, UNEP and WWF, Gland
Jacobs J (1965) The death and life of great American cities. Penguin, Harmondsworth
Jacobs J (1969) The economy of cities. Penguin, Harmondsworth
Joly R (1991) City and urban civilization. Synchroni Skepsi, Athens (translated in Greek; original: La ville et la civilisation urbaine, 1985)
Keeble L (1969) Principles and practice of town and country planning. The Estates Gazette Ltd, London
Kelbaugh D (2002) The New Urbanism. In: Fainstein SS, Campbell S (eds) Readings in urban theory. Blackwell, Oxford, pp 354–361 (first published 1997)
Kenny M, Meadowcroft J (eds) (1999) Planning sustainability. Routledge, London
Lafferty WM, Meadowcroft J (eds) (2000) Implementing sustainable development: strategies and initiatives in high consumption societies. Oxford University Press, Oxford
Landry C (2000) The creative city: a toolkit for urban innovations. Earthscan, London
Lavedan P (1959) Histoire de l'urbanisme: Renaissance et temps modernes, 2ème edn. Henri Laurens, Paris
Law-Yone H (2007) Another planning theory? Rewriting the meta-narrative PT 6(3):315–326
Layard A, Davoudi S, Batty S (eds) (2001) Planning for a sustainable future. Spon Press, London
Leontidou L (1993) Postmodernism and the city: Mediterranean versions. US 30(6):949–965
Lindblom CE (1973) The science of "muddling through" In Faludi A (ed) A reader in planning theory. Pergamon Press, Oxford, pp 151–169 (first published 1959)
Lord A (2014) Towards a non-theoretical understanding of planning. PT 13(1):26–43
Low N (2020) Being a planner in society: for people, planet, place. Edward Elgar, Cheltenham

Mandelbaum SJ (1996) Introduction: the talk of the community. In: Mandelbaum SJ, Mazza L, Burchell RW (eds) Explorations in planning theory. Center for Urban Policy Research/Rutgers, New Brunswick NJ, pp xi–xix
March A (2010) Practising theory: when theory affects urban planning. PT 9(2):108–125
McConnell S (1981) Theories for planning. Heinemann, London
Meadows DH et al (1972) The limits to growth. Potomac Associates
Meller H (1981) Patrick Geddes 1854–1932. In: Cherry G (ed) Pioneers in British planning. The Architectural Press, London, pp 46–71
Mumford L (1961) The city in history. Penguin, Harmondsworth
Needham B (2017) A renegade economist preaches good land-use planning. In: Haselsberger B (ed) Encounters in planning thought: 16 autobiographical essays from key thinkers in spatial planning. Routledge, New York and London, pp 165–183
Painter J (2002) Regulation theory, post-Fordism and urban politics. In: Fainstein SS, Campbell S (eds) Readings in urban theory. Blackwell, Oxford, pp 92–109 (first published 1995)
Palermo PC, Ponzini D (2010) Spatial planning and urban development: critical perspectives. Springer, Dordrecht
Paris C (1982) A critique of pure planning: introduction. In: Paris C (ed) Critical readings in planning theory. Pergamon Press, Oxford, pp 3–11
Parker G, Doak J (2012) Key concepts in planning. Sage, London
Pearce D, Barbier E (2000) Blueprint for a sustainable economy. Earthscan, London
Pennington M (2002) Liberating the land. Institute of Economic Affairs, London
Pickvance C (1982) Physical planning and market forces in urban development. In: Paris C (ed) Critical readings in planning theory. Pergamon Press, Oxford, pp 69–82(originally published 1977)
Pinol J-L (2000) The world of cities in the 19th century. Plethron, Athens (translated in Greek; original: Le monde des villes au XiXe siècle, 1991)
Pugh DS (ed) (1971) Organization theory. Penguin, Harmondsworth
Pugh DS, Hickson DJ, Hinings CR (1971) Writers on organizations. Penguin, Harmondsworth
Raban J (1974) Soft city. Fontana/Collins, Glasgow
Ragon M (1986) Histoire de l'architecture et de l'urbanisme modernes, vol 3 volumes. Casterman, Paris
Riddell R (2004) Sustainable urban planning. Blackwell, Oxford
Roberts P, Sykes H (eds) (2000) Urban regeneration: a handbook. Sage, London
Roncayolo M, Paquot T (eds) (1992) Villes et civilisation urbaine : XVIIIe – XXe siècle. Larousse, Paris
Sager T (2018) Communicative planning. In: Gunder M, Madanipour A, Watson V (eds) (2018) The Routledge handbook of planning theory. Routledge, London, pp 93–104
Sandercock L (2002) Towards cosmopolis. Wiley, Chichester
Satterthwaite D (ed) (1999) Sustainable cities. Earthscan, London
Silberstein J, Maser C (2000) Land-use planning for sustainable development. Lewis Publishers, Boca Raton
Skea J (1999) The environment and the European economy. In: Dyker DA (ed) (1999) The European economy. Addison Wesley Longman, Harlow, pp 262–281
Solomon AP (ed) (1980) The prospective city: economic, population, energy, and environmental developments. The MIT Press, Cambridge, MA
Sutcliffe A (1981) The history of urban and regional planning: an annotated bibliography. Mansell, London
Talvitie A (2009) Theoryless planning. PT 8(2):166–190
Tasan-Kok T et al (2016) 'Float like a butterfly, sting like a bee': giving voice to planning practitioners. PTP 17(4):621–651
Taylor N (1980) Planning theory and the philosophy of planning. US 17(1):159–172
Taylor N (1999) Urban planning theory since 1945. Sage, London

Taylor N (2003) More or less meaningful concepts in planning theory (and how to make them more meaningful): a plea for conceptual analysis and precision: an essay in memory of Eric Reade: 1931-2002. PT 2(2):91–100
Thomas H (2004) What future for British planning theory. PT 3(3):189–198
Thompson R (2000) Re-defining planning: the roles of theory and practice. PTP 1(1):126–133
Upton R (2012) On the genealogy of planning. PTP 13(2):189–192
Urban Task Force (1999) Towards an urban renaissance. Final report. Department of the Environment, Transport and the Regions, London
Vogelij J (2015) Is planning theory really open for planning practice? PTP 16(1):128–132
Ward B (1966) Spaceship earth. Columbia University Press, New York
Ward B (1976) The home of man. Penguin, Harmondsworth
Ward B, Dubos R (1972) Only one earth: the care and maintenance of a small planet. An unofficial report commissioned by the Secretary-General of the United Nations Conference on the Human Environment. Penguin, Harmondsworth
Woods S (1975) The man in the street: a polemic on urbanism. Penguin, Harmondsworth
Yiftachel O (1989) Towards a new typology of urban planning theories. Environ Plan B Plan Design 16:23–39

Chapter 3
Mainstream Theories: The Rational and Communicative Currents

Abstract This chapter is dedicated to the presentation of the main corpus of planning theory, as it appears in most theoretical writing. This corpus in made up of what are here considered as the mainstream theories of rational/comprehensive and communicative/collaborative planning. These currents are presented mainly through the work of their representative advocates and/or critics. Arguments for or against them are outlined to help the reader appreciate their explanatory power. The presentation of these theories provides the opportunity to introduce a large number of important concepts which are extensively used in the respective theoretical writing. The chapter sheds light on the opposition between these currents, their mutual criticisms, their inadequacies, their overlaps, but also the possible conciliation of their conflicting paradigms and arguments.

Keywords Rational · Comprehensive · Systems · Communicative · Collaborative · Pragmatism · Deliberative · Actor networks · Structure · Agency · Participation

Prologue

The chapter on planning theory typologies has provided a panoramic view of the field of theory. In it, the author had the opportunity to introduce what he considers as the dominant currents of ideas. It is time to expand this investigation to individual currents, which is attempted in this chapter on the "rational" and "communicative" currents. The "radical" current will be left for Chap. 4 and the "climate" current for Chap. 5. In the present chapter the author has endeavoured to condense the main tenets of rational-comprehensive and communicative-collaborative planning theory, and the arguments for and against them. He is naturally aware that the intention to achieve that in a single chapter may be considered presumptuous. It is true that the main bulk of past theoretical writing is concentrated in these two currents which constitute the mainstream of planning theory. In spite of the difficulty involved, the rest of the book would be meaningless without the review attempted in this chapter. The theoretical work introduced in subsequent chapters is, in a sense, developed as a reaction, or in opposition, to the rational/communicative corpus, as indeed was the

L. C. Wassenhoven, *Compromise Planning : A Theoretical Approach from a Distant Corner of Europe*, https://doi.org/10.1007/978-3-030-94331-8_3

communicative approach in relation to the rational tradition. Reviewing the mainstream theories in this chapter provides the opportunity to quote from the work of eminent theorists and to introduce essential concepts. Rationality, systems thinking, goal-oriented planning, incrementalism, methods, public interest, group diversity, communication, deliberation, actor networks, pragmatism, structure, agency, structuration are some of them, but they suffice to demonstrate both the difficulty of the task and the importance of the chapter.

3.1 The Rational Current

Keywords of the Rational Current

Ideas, concepts, preoccupations, interpretations, concerns, and socio-political conditions embedded in the rational current, which inspire rational planning theory and practice:

Objectivity, rationality, absolute values, choice, goal attainment, process, space, observation, structure, technofix solutions, authority, calculation, utility, state power, expertise, idealism, scientific truth, facts, action, end-state planning, order, development, alternatives, systems thinking, management, regulation, explanation, stability, futurology, law rule, methodology, comprehensiveness, innovation, evaluation, formal knowledge, future trajectories, policy making, reality simulation, professionalism, foresight, feedback loops, open systems.

3.1.1 The Heritage of Rational Planning

The rational current deserves attention not only because it dominated for long the dialogue for or against it, but also because it seems destined for revival, as indicated in comments quoted earlier, under the influence of environmental concerns, methods and technologies. This influence brings back to the fore the conception of systems, which had played such a key role in the initial thrust of rational planning. Such developments are no longer viewed as an inheritance of modernism but rather as an outcome of new priorities closely related to what was earlier described as the "climate current". The awakening to the need for an appropriate planning model leads planning theory and practice to hitherto unexplored paths, only remotely related to the modernist traditions still influential in the 1960s. It has been made clear that Faludi's theoretical construct was considered from the beginning as inherited from modernism and as a typical example of rational planning theory, a term which in due course caused a storm of opposition.

We cannot separate the rise of the rational planning movement from the mounting interest, in the same period, for the urban phenomenon in general and the structure of cities as they were evolving rapidly in a period of economic growth and reconstruction. The literature on cities, urbanization, and urban organization was altering the perspectives of planners. Classical readings appeared, often from the field of human geography, which became standard stock of planning education (Webber et al. 1964; Hauser and Schnore 1965; Hall 1967; Haggett 1975; Berry and Horton 1970; Hall et al. 1973a, b; and many others). Their contribution is acknowledged by later planning theorists (Cooke 1983, 88; Palermo and Ponzini 2010, 46–47). As in the case of urban studies and human geography, rational planning in general, and Faludi's approach in particular, cannot be easily divorced from the systems approach to planning, i.e., the whole edifice of systems logic, operations research, their branches of choice theory, and goal-oriented planning (Burns 2004; Castles et al. 1971; Friend and Jessop 1969; Friend 1996; Faludi 2004; Hickling 1974; Low 2020, 93; Dryzek and Dunleavey 2009, 103–108). Notwithstanding Bolan's verdict that "the traditional rational choice theory for urban planning is outmoded today and has been for some time" (Bolan 2017, 16), these approaches are exemplified in papers written before Faludi's book, e.g., by Banfield (1973) on "ends and means" or Davidoff and Reiner (1973) on choice-based planning, first published in 1959 and 1962 respectively, by Young (1966) on "goals and goal-setting", and by Altshuler (in a paper which first appeared in 1965) on the goals of comprehensive planning (Altshuler 1995). In 1959, when Banfield was developing what in effect was an early version of the rational model, Lindblom was putting forward an incrementalist approach of step-by-step planning, not for attaining goals but for solving problems, as and when they appear (Lindblom 1973). Lindblom, an associate of Robert Dahl, followed a pluralist tradition of political philosophy, well known for its emphasis on diversity and often accused for failing to account for inequality. His approach might be better placed in the tradition of neo-pluralism and pragmatism (Dryzek and Dunleavey 2009, 132–134). A little later, in 1967, Etzioni proposed his middle-way form of planning, known as "mixed scanning" (Etzioni 1973). However, none of these counter-proposals adopted a progressive, radical ideology, like those that were to follow.

In 1969, Bolan described what he considered as the "classical model", which included the repeatedly mentioned phases: scanning the environment, defining goals, formulating alternative actions, predicting probable consequences, evaluating available means, assessing possible achievement of goals. This "rational process becomes a part of the individual's experiential makeup" (Bolan 1973, 373). With remarkable foresight, Bolan, a major force in planning theory, proposed his own version, in which he insisted on an open decision-decision making process, involving relevant audiences, negotiation, situated contingent action, and legitimation (op. cit., 377). Parenthetically, it must be mentioned that, many years later, he elaborated his own Model of the Design Process, which still had as its major phases (a) problem framing and goal setting, (b) alternative possibilities/formulation and exploration, (c) heuristic testing and focusing, and (d) agreed-upon design and action plan (Bolan 2017, 255). To close the parenthesis, it is apparent that the theoretical contributions listed earlier naturally precede that of Faludi, but deserve attention because their lasting impact is still visible in the planning systems and actions in the real world. In an equally older paper, of 1963, on the facets of planning, Dror defined planning as “the process of preparing a set of decisions for action in the future, directed at achieving goals by preferable means”. In his view (Dror 1973, 332), “the four primary facets of planning appear to be the following … (a) the general environment of the planning process, (b) the subject matter of the planning process, (c) the planning unit, (d) the form of the plan to be arrived at”. In viewing planning in this manner, he blended together practically all the aspects of the theories of, in, for, on and about planning. He thus raised the vexed issue of the typology of theories which has been widely discussed in the literature and to which the previous chapter was dedicated.

3.1.2 Rationalism and Its Critics

The point has been made already that Faludi’s theory cannot be divorced from the rationalist tradition and systems thinking. Commenting on the former, as inherited from the late nineteenth century, Lancaster (1968, 266) wrote that “in politics the rationalist tradition required that programmes of action should ‘make sense’ in terms of consistency between the relevant facts, clear objectives, and appropriate legislative and administrative devices” (see also Wilson 1977). Lancaster’s definition, which could not be more appropriate for the rational model of planning, places rationalism in the philosophical context of positivism, which demands the application of agreed methodological rules in order to reach conclusions through a logical sequence of reasoning. “[P]ositivists base knowledge upon empirical or mathematical observations trying to uncover ‘truths’ or relationships between objects. In planning the high point of positivism was to be found in the systems and rational approaches of the 1960s” (Allmendinger 2002, 28). It was the period marked, to quote Safier (1983), by the “passage to positive planning”. “A rationalist”, Allmendinger adds, “is normally taken to be someone who emphasizes logical capacities and can give reasons for a particular view. This is normally contrasted with more emotive or intuitive reasoning” (op.cit., 53). The assumption of rationality had a deep effect on planning and the related literature (Breheny and Hooper 1985). Rationalism is also considered “a sub-category of idealism and as such, shares the general criticism that in giving precedence to ideas over real processes it is unable to avoid an ultimately hopeless relativism. However, rationalistic explanation is of some interest because it postulates a logical conceptual structure to social processes” (Cooke 1983, 22).

The influence of the tradition of rationalism was evident in the very operation of the town planning system of advanced countries like the UK, with its reputed statutory framework. This has bred, in the view of critics, a managerial mentality and a transformation of urban planning into urban management (Wassenhoven 1978). The practice of local management, not limited to the work of urban planners as such, rested on rationalistic assumptions, which, from a sociological perspective, could be found wanting. This was the verdict of Pahl in 1975: "[T]he planners, social workers, housing managers and so forth are very often trying to turn the taps of their resources to favour the most disadvantaged; but either through a mistaken belief in the validity of their data, a lack of awareness of the unintended consequences of their actions or simply through human error, the results of their activity fail to improve, and possibly add to, the plight of the poor …" (Pahl 1982, 48).

The most pointed attack on rationalism emanates from the analysis which links rationalism with power structures. In his well-known book on rationality and power, Flyvbjerg, using material from a Danish case study, makes this approach quite clear: "In the Enlightenment tradition, rationality is typically seen as a concept that is well-defined and context-independent. We know what rationality is, and rationality is supposed to be constant over time and place. This study, however, demonstrates that rationality is context-dependent and the context of rationality is power. Power blurs the dividing line between rationality and rationalization. Rationalization presented as rationality is shown to be a principal strategy in the exercise of power" (Flyvbjerg 1998, 2). The reader, especially a planning professional, may perhaps wonder why an extended theoretical analysis is needed to prove this obvious truth, omnipresent in planning practice in the real world. It is not a secret, that it is political power that not only determines the outcome of planning initiatives, but also sets the norms and rules that constrain them. The "norms and rules" approach is the standard process associated with countries with rational and hierarchical planning systems. The Netherlands is typically presented as a case in point, considered as the "planning paradise", to quote Boelens. "Planning practice in the Netherlands seems particularly to value network skills in engaging with multiple interest groups to arrive, collaboratively, at a plan. But the collaborative approach seems ingrained in that society's political culture" (Low 2020, 33). Needham, comparing his ample experience in this country with that which he had acquired in England, concurs: "There is … widespread public agreement that land-use planning is good and useful … Dutch people think much more positively about land-use planning: It is about creating places and spaces which could not be realised without public planning (Needham 2017, 167). Yet, Boelens casts doubts on the Dutch system: "That kind of planning turned mostly into highly regulatory and prescriptive operations, resulting in syrupy planning processes, which were very costly, inflexible and inefficient, and even in the suppression of all new and creative initiatives that did not fit within the framework. Nevertheless, due to its supposedly visionary regulating promises, it still remains attractive to many of the spatial disciplines (architects, urbanists, planners and the like), as well as to social-democratic politicians" (Boelens 2010, 55). Boelens proposed a new "actor-relational approach", to which we will return in this chapter.

In Faludi's approach, according to Thomas, "the essence of planning is rationality or the application of reason in human affairs. Rationality is presented as the method by which an intelligent human approaches the problem of taking action to secure his goals. Rationality as a style of acting generally refers to decisions being made in a particular way according to technical rules for choice between a number of different strategies" (Thomas 1982, 14; see also Camhis 1979). To which Paris added his own critique: "The use of rationality as a guiding principle of planning is … tautological: *of course* planning should be rational, what else could we have it be? That, however, tells us nothing specific about 'planning' as it exists, as particular institutional or 'professional' activities. Indeed, it obscures what is specific about planning as an occupation, i.e., the very institutional contexts in which it operates, which vary considerably over time and between societies …" (Paris 1982a, 6).

Rationalism and what we could call the "scientification" of planning are placed by Dryzek and Dunleavey (2009, 50) in the context of US postwar pluralism: "The growth of the post-war US state produced many attempts to reform government through centralized planning and goal-oriented policy analysis. Popular rationalist techniques included cost-benefit analysis from economics, systems-analysis from engineering, decision analysis from game theory, management by objectives, and programme budgeting … Pluralist theorists such as Charles Lindblom … ridiculed such a 'rational comprehensive' approach to public policy as essentially impossible, imposing an unfeasible burden of calculation on policy makers". Methodological rigidity, alleged scientific orthodoxy and, in some cases, obsession with mathematical techniques exposed the rational model to accusations that it was "too infused with narrow, technical rationality and missed out on the more spontaneous aspects of humanity" (Allmendinger 2002, 63). If Faludi's work "is searched for an underlying view of society, none is found", claimed M.J. Thomas (1982, 20). Faludi, as expected, rejected the view that rational planning should be labelled "technical" (Faludi 1982, 32) and he added later (1996, 72–73): "[T]he rational planning model is a means of sharing responsibility for the consequences of actions, and this is the most important reason for advocating it. Planning involves risks, so there must be a way of legitimizing decisions". John Friedmann (1995, 78) acknowledged the rational model's contribution but, already in 1989, announced its eclipse: "Rational planning was one of the triumphs of the modern age. It prided itself in being the scientific way to guide society's future course of progress. But in the end, like so many myths of modernity, planning, too, had to bite the dust. Science, it turned out, was not the One True Way; and neither was planning". Campbell and Fainstein (1996, 6) seem to allude to vain expectations. "For some", they point out, "the hope of rational planning was simply to equate the market with uncertainty and to believe that the logic of the plan would therefore replace the chaos of the market". This discussion inevitably involves the Marxist critique which connects the role of the state and the use made of urban planning to facilitate the reproduction of labour (Taylor 1999, 101–107; Allmendinger 2002, 69–80; Foglesong 1996).

This is a vast subject, but David Harvey provides a useful summary of the Marxist position: "The proper conception of the role of the state in capitalist society is controversial. I shall simply take the view that state institutions and the processes whereby state powers are exercised must be so fashioned that they too contribute, insofar as they can, to the reproduction and growth of the social system. Under this conception we can derive certain basic functions of the capitalist state. It should (a) help to stabilize an otherwise erratic economic and social system by acting as a 'crisis manager'; (b) strive to create the conditions for 'balanced growth' and a smooth process of accumulation; (c) contain civil strife and factional struggles by repression (police power), cooptation (buying off politically and economically), or integration (trying to harmonize the demands of warring classes or factions)" (Harvey 1996, 185). The state is then viewed as an instrument of the dominant capitalist class, a thesis that determines the role of planning as well. "From the Marxist perspective, the role of planning in contemporary society can only be understood by recognising the structure of modern capitalism as it relates to the physical environment. That is, it is argued, the fundamental social and economic institutions of capitalist society systematically promote the interests of those who control society's productive capital over those of the remainder of society" (Klosterman 1985, 160). However, the state has two faces. "State intervention is sometimes seen as complete *laissez-faire* and sometimes as authoritarian, the latter realising a monopolistic strategy. It is the state intervention which has come to be called urban planning" (Preteceille 1982, 129). The impact of urban planning on land is what consolidates the necessary nexus with state functions (Scott 1980). It is in the process of land use planning that space is demarcated and private properties designated as subject to restrictions of use or protected against trespassing. Legislation is enacted to buttress these limitations and rules, which are embedded in public perceptions (Bolan 2017, 233).

The co-existence of planning and market and the opposition between them was a stimulus for a good deal of analysis when the influence of rational thinking was already being challenged (Hodgson 1984), although the "plan-market dichotomy" is not universally endorsed, even regarded as "fallacious" (Lai 2016). For the planning profession the key interest lies in the role of the planner as a "servant" of the state. To quote from Harvey again, "the urban planner occupies just one niche within the total complex of the instrumentalities of state power. The internalization of conflicting interests and needs within the state typically puts one branch of the bureaucracy at loggerheads with another, one level or branch of government against another, and even different departments at odds with one another within the same bureaucracy … I shall simply suggest that the planner's task is to contribute to the processes of social reproduction and that in so doing the planner is equipped with powers vis-à-vis the production, maintenance, and management of the built environment that permit him or her to intervene in order to stabilize, to create the conditions for 'balanced growth', to contain civil strife and factional struggles by repression, cooptation, or integration" (op.cit., 186).

3.1.3 Rational Planning, Its Historical Roots and Its Resilient Persistence

The debate about planning v. market forces was already raging in the days of Friedrich von Hayek and Karl Mannheim (Allmendinger 2002, 45, 53–55, 95–97; Beauregard 2020, 79; Taylor 1999, 132–134; McConnell 1981, 173, 180; Faludi 1973, 171–173; Klosterman 1985; Low 2020, 85–90; Fainstein and Fainstein 1996, 273–274; Dryzek and Dunleavey 2009, 103–104). Fainstein and Fainstein blame technocratic thinking inherited from the industrial era: "The technocrats desire to unleash the power of reason and science … Through rational planning, the power of the state will be employed to regulate the economy and advance the lower classes, as well as to ensure the position of the productive ones … According to the technocrats, social change must be engineered from the top". This, Tait (2016, 336) asserts, was an illusion: "The figure of the rational planner, who was able through theoretical knowledge and patient reflection, to identify a transcendent interest that rose above sectional attachments has been identified as born from Enlightenment thought, reaching its apogee in the post–Second World War period. Yet, it is often argued that this was (and remains) deeply problematic". The image of the rational planner, as sketched here, is a reflection of his/her bureaucratic role and his/her supposed subordination to his/her boss, the politician, which makes him/her vulnerable. As put in a 1964 paper, "primarily this is because it is in the nature of his job to become deeply and inevitably involved in politics and the political process. In the field of government, every administrative action, from surly behavior of a licensing clerk to a health department decision to prohibit individual wells and septic tanks in an urban area, is weighed on a political scale. Those actions which benefit or hurt individuals produce a reading on that scale" (Beckman 1973, 256).

This situation can only be appreciated in a historical context, involving the growth of the modern nation state and, as repeated above, the evolution of Enlightenment thinking. The Enlightenment roots cannot be doubted: "The Enlightenment and modernist perspective was crystallized in the rational comprehensive model, an explicitly normative approach to how planning should be done" (Beauregard 2020, 9). The question is whether one is really justified to declare the demise of rational planning. Not really, if we follow Ernest Alexander: "[A]s a normative model that prescribed how decisions can be made that conform to certain norms of logic and consistency, the rational model has yet to be superseded" (Alexander 1992, 58). He repeated his reaction against the onslaught on rational planning, at one more point: "Though planning practice is changing, and recent theories have shown sensitivity to many issues which the 'classical' rational planning model fails to address, this model of what should be done has yet to be superseded" (op.cit., 86). Beauregard (1996a, 107) has emphasized his admiration for Alexander and Faludi's work. He considered them "skilful practitioners of rationalist planning theory" and "sensitive to its limitations". Over the years, he stressed, "their work has evolved, becoming more and more sophisticated and pushing against the constraints of high theory". The resilience of the rational model is not due only to the adaptability of the views of its theoreticians. It is also due to its domination in planning practice. Brooks has a very convincing interpretation: "If, as noted earlier, planning theorists have indeed thoroughly discredited the notion of planning as an exercise in rationality, why spend time on that notion here? The answer rests in the fact that, the theorists' views notwithstanding, the rational model is still widely invoked in the world of planning practice … Ironically, planning schools often display a split personality on this matter – thrashing rationality in the planning theory class, while continuing to teach it in all its glory in the methods and studio classes. Much like the creatures in horror movies, rationality is dead – but keeps showing up in public places. Despite its purported flaws, rationality is still the dominant paradigm in planning practice, and therefore continues to deserve careful scrutiny" (Brooks 2017, 81).

3.1.4 Statutory Planning and Its Rational Foundations

In the days when the rational model was flourishing, either in its traditional design form, or in its procedural "goals and means" variety, a general optimism seemed to predominate and was reflected in a large number of publications on urban and regional planning, although the problems were not neglected,[1] no doubt due, on one hand, to successive statutory planning reforms, at least (and especially) in the UK, and, on the other, to the desire to bolster the theoretical foundations of public planning in the USA. The British system, we must not forget, had been very influential and amply documented, e.g., in repeated editions of a book by Cullingworth, rewritten later with Nadin (Cullingworth and Nadin 2006; see also Rydin 1998). There was a similar interest and optimistic attitude in France, where the concepts of an "active" or "voluntary" geography and of "inventing the future", or of planning as negation of "randomness" were popular in geography and planning literature.[2] The recognition of the rational model's flaws was also evident. The lack of sensitivity of the rational model may well be due to being, at least initially, "politically naïve" (Brooks 1996, 117), but this does not imply that it is beyond rescue. Archibugi, e.g., fully adopts Alexander's position on the content and meaning of planning itself and on the planning process as the core of planning theory (Archibugi 2008, 10). At the same time, he supports the broadening of the model's scope into what he calls a "unitary methodological scheme" (op.cit., 56). Given that the planning process does not end with the formulation of an end-image but continues into a full cycle of implementation and interaction, the prospect of the rational model's continued

[1] See, e.g., Alden and Morgan 1974; Bruton 1974; Chapin and Kaiser 1979 (and later Kaiser et al. 1995); Cowan 1973; Donnison and Soto 1980; Friedmann and Alonso 1964, 1975; Gillingwater 1975; Glasson 1978; Godschalk 1974; Greed 1993 and Greed 1999; Morley et al. 1980; Pile et al. 1999; Ravetz 1986; Reissman 1970; Self 1981, 1982; Solesbury 1974; Stewart 1972; Wannop 1995.

[2] See Derycke 1979a, b; George et al. 1964; Hayward and Watson 1975; Labasse 1966; Massé 1965; Williams 1984.

usefulness seems possible. "Decision-making rationality is not defined beforehand through synoptic evaluation and analysis; it emerges a posteriori through social interaction. The multiplicity of points of view and the interactive, incremental method may involve risks of fragmentation and short-sightedness, but they can also guarantee pluralistic, compatible, cautious and effective choices" (Palermo and Ponzini 2010, 71). The incorporation of an interactive implementation cycle into the planning process was in fact stressed even during the heyday of rational planning, as McConnell (1981, 197) pictures it in diagrammatic form (see also Lichfield et al. 1975), although it is true that initially it was underestimated, an omission that was quickly criticized (Pressman and Wildavsky 1979; Barrett and Fudge 1981). "Protagonists of the rational process model … gave little attention to how plans and policies were, or might be, implemented. However, the rational process model did not necessarily imply that implementation was glossed over or ignored. Implementation was explicitly identified as part of any rational planning process …, and so the rational process model should always have been seen as a model of rational action" (Taylor 1999, 111). It is worth mentioning that Healey, credited with the major theoretical departure from the rational mould in the form of her communicative-collaborative approach, acknowledges the contribution of the rational model, of which of course she underlines the shortcomings: "[I]t is worth stressing the innovations which the rational planning model brought to the discussion of policy processes" (Healey 1997, 251–252). These innovations concerned the recognition of interconnections, the process of arriving at strategic proposals, the policy making activity, the definition of problems, the choice of strategies, the systematic use of available knowledge, and the stages of testing and evaluation.

The strategic dimension is the element on which Albrechts builds his vision of strategic spatial planning. For Faludi (2000, 299), "strategic spatial planning concerns major spatial development issues. Such issues may arise on any planning scale, but it is more common for them to be addressed at the regional and even more so on national level". For Albrechts, however, strategic spatial planning has a much broader meaning and incorporates the priorities of the communicative-collaborative approach, that had dominated the theoretical field in the meantime, hence his reference to bridging the gap (Albrechts 2006, 1491). Albrechts (2008) elaborated further his argument on strategic planning by focusing on four different types of reaction to developments and challenges: reactive (the rear-view mirror), inactive (going with the flow), preactive (preparing for the future) and proactive (designing the future and making it happen). His thesis is that only the 'proactive' reaction is appropriate. When he returned to the subject some years later, he clarified that the dimensions of strategic planning are context, visioning, relational, legitimacy; the key concepts are envisioning, becoming, and legitimacy (Albrechts 2018, 36). In simple terms, Albrechts speaks about "inventing the future" and advocates a synthetic approach which incorporates a whole range of concepts, from visioning an image for the future to incorporating alternative perspectives. Ultimately, it can be said, he does not depart from the basic tenets of a rational approach as it can be practiced in a democratic society. However, because of his reference to "becoming" we will return to this notion later, when a reference is made to Giddens and Habermas.

3.1.5 Planning, Power and the State

Critics of rational planning emphasize the existence of a "hidden world of power in planning", its so-called "dark side", and of "nefarious forces" using planning as a tool of domination: "Planning has the potential to oppress subordinate groups and is structurally devised to exert control and oppression" (Allmendinger 2002, 172–173). This malevolent aspect of planning has imbued a good deal of theoretical work. "Much of the writing we find in planning theory literature tends to argue that power is a negative, oppressive force – the 'dark' side of planning. It is important for urban planners to understand not only the negative influences of power but also the positive forces that can help in planning episodes" (Bolan 2017, 215). The position of Flyvbjerg on the rationality-power connection was quoted earlier. This connection, in other words the planner's bureaucratic position, is a source of authority, in fact more important than his or her professional expertise, the ability to serve diverse interests or the values he/she is supposed to hold. As put by Ernest Alexander, "in the bureaucratic role, planners get their authority from the elected officials of the government they serve. This legitimacy, however, is limited by several factors. Political officials may not be as broadly representative as they should be, so that they respond to the interest groups that elected them rather than to the 'public interest'. Furthermore, the theoretical distinction between the policy-making master and the administrative servant is blurred in practice, where many policy decisions are in fact made at the administrative level rather than by the legislative body" (Alexander 1992, 120). In the framework of this book's analysis of planning as a perpetual compromise, the situation presented by Alexander is the background of compromise planning.

Can the planner be accused of complicity or is he/she an innocent, unwilling victim of the political system? Brooks (2017, 43) summarizes this ethical dilemma as follows: "Like the progressives, the 'dark side' theorists tend to view planning behavior as a logical and necessary outcome of capitalist property relations. The only significant difference is the question of intent: to the progressives, the planner is basically an unwitting dupe; to the current group of critical theorists, the planner is a conscious and willing participant in oppressive government practices. The only solution, apparently, would be to shift from capitalism to another form of political economy … It is a major (and, in my view, untenable) leap, however, to the notion that planners are purposeful and malevolent oppressors". Regardless of the planner's culpability, the fact remains that the planning-power nexus is due to social control tools, such as administrative rules and regulations, which planners employ to impose a form of normality, as perceived by those in power. This turns the planners into a class of "judges of normality" (Allmendinger 2017, 22). In such a fuzzy situation rationality is bound to suffer, especially its theoretical democratic variety.

The restrictions that administrative and political rules impose on planning do not concern only the civil servant, but also the free-lance consultant whose work depends on political decisions.

Criticisms were not limited to planning's association with the power nexus. Part of the critique, as Healey (1997, 250–251) argues, "derived from those who shared a commitment to instrumental rationality, while recognizing the limits to knowledge", but also from those who challenged the whole edifice of instrumental rationality and positivist social science. It is true that confidence in the planner's ability to monopolize knowledge has done a lot to undermine the credibility of rational planning. Yet, planners "must also be concerned with the space of knowledge, with the entanglement of knowledge and power" (Roy 2010, 41). Bolan speaks of "an ineradicable gap between formal knowledge and a decision (and capacity) to act" (Bolan 2017, 32). He also adds, that "when we examine expert knowledge more closely, in many instances we find that much of what guides the specialist has not really been put through the crunch of rigorous proof of validity. In many cases, it is nothing more than speculative hypothesis or tacit intuition" (op.cit., 40). Critics "noted that the rational-comprehensive model posed an endless task when it asked planners to thoroughly define the problem, account for all possible responses, and assess every response against all possible objectives and conditions ... Research became a fetish" (Beauregard 2020, 26). Still, in Baum's words, "as description, the rational theory is unrealistic, though ..., on an unconscious level it makes for compelling prescription" (Baum 1996, 368).

The meaning of rational action takes us to the tradition of Max Weber who classified all action into four types: "purposively rational, action where means are correctly chosen to obtain ends, value rational, where action is in accord with conscious value standards, affectual, and traditional, the last two types being regarded as deviations from rational action" (Mitchell 1968). One reason why all this is relevant to the present argument, as it will become evident in Chap. 10, is because rationality is supposed to be the main quality of the "economic man" (*homo economicus*), i.e., of an abstract construct of neoclassical economics, representing the average man, who is "concerned with the immediate aim of obtaining the largest possible command over resources, with the minimum of sacrifice" (Seldon and Pennance 1965). Carr (1964), in his fascinating study on the nature of history, wrote, perhaps optimistically, that nobody believes any longer in the existence of economic man: "Today economics has become either a series of mathematical equations, or a practical study of how some people push others around". Leaving aside, at least for the time being, the question whether the economic, or the rational, man are illusions, we may concentrate again on the rational planning model, although quite a lot has been said in Chap. 2 on typologies.

3.1.6 Planning's "Scientific" Methods and Techniques

Within the framework of "purposively rational action", it was the ambiguity of notions like "relevant facts" and "clear objectives", mentioned above, that led to the search for safe scientific methods. As expressed by an experienced planner of the 1960s, who himself rejected the argument, "these premises have to be validated by scientific methods in order to secure rational objective consensus. These methods should be either the same as applied in physical sciences or through scientifically devised or organised opinion polls" (Kriesis 1970, 18). Incidentally, one of the most typical outcomes of this approach was the effort of planning theoreticians and practitioners, in the 1950s and 1960s, to devise objective, "scientific" techniques of plan formulation, design, analysis and plan evaluation.[3] These techniques are usually perceived as independent of theoretical assumptions, at least in practice. Based on his Australian experience, March wrote that "to the Australian practitioner, scientifically and quasi-scientifically posited and tested hypotheses, however contentious the issues they inform, are not typically understood as contributing to 'theory'. Rather, they are seen as 'facts', from which debates can stem, or from which practical planning conclusions might be drawn according to specific circumstances" (March 2010, 110).

"Facts" are supposed to be discovered through an objective analysis of information and data. They are the "evidence" on the basis of which judgments are later made. The term "evidence" returned to fashion in the UK in the late 1990s, when the then government pursued a policy of "evidence-based planning" and "expert advice", which attracted intense criticism in the ensuing debate (Solesbury 2002). Although confined to a particular national context and a specific period, the arguments are of general and never-ending significance. Solesbury links the necessity of evidence to policy development in general and argues that while there is always room for evaluative "academic" research, the need to answer questions about the nature of problems demands "an evidence base to policy in all stages of the policy cycle, in shaping agendas, in defining issues, in identifying options, in making choices of action, in delivering them and in monitoring their impact and outcomes" (op.cit., 94). A descriptive and analytical information base retains its value. Heather Campbell responded by pointing out that policy making had changed character since the days of rational planning supremacy and the unchallenged acceptance of expert knowledge. "The planning community has long grappled with the conundrum of finding a process which will increase the possibility of policies producing desirable outcomes. Policy making is now generally viewed as a process of argumentation …, however, beyond that, debate persists as to how and in what form knowledge should be enrolled into such a process. The traditional rational notion

---

[3] See: Alexander 1992, 78–84; Bolan 2017, 76–80; Chapin 1965; Cowling and Steeley 1973; Dasgupta and Pearce 1972; Hill 1968; Layard 1972; Lichfield 1956, 1964; Lichfield et al. 1975; Lean 1969; Levin 1967; Mishan 1972; Morgan 1971; Perroux 1968; Prest and Turvey 1966; Wannop 1970; Wassenhoven 1974.

that evidence (usually quantitative in nature) *should* in theory or *does* in practice determine the most appropriate policy options has been widely derided. It is acknowledged that knowledge comes in a variety of forms and is certainly no longer the exclusive domain of the `expert'" (Campbell 2002, 89). This debate may be concealing a more profound change in the nature of research which masquerades behind a mask of professional expert analysis. It would seem that Adrian Healy detected a shift of this nature, which leads to "researchers" assuming a role, which is prone to external manipulation, although he does not use this designation: "[R]esearchers are now more amenable to undertaking policy related research, professional cadres are more interested in commissioning studies and making use of evidence and … our politicians too are more interested in basing policy upon evidence" (Healy 2002, 97). Such direction denies the expectations voiced by Campbell.

Decision-making quantitative models and evaluation techniques are not of course the exclusive feature of the rational planning period of the 1960s and 1970s. Guyadeen and Seasons (2016) claim that "in recent years, plan evaluation methods are not commonly used in practice", although "evaluation could be used to increase the legitimacy of planning, improve decision-making, and foster continuous learning … Despite these apparent benefits, the planning profession has been unable to embrace and apply evaluation methods, particularly the evaluation of plan outcomes". This seems to be an erroneous claim, given that quantitative methods and techniques continue to inform decision making in a variety of planning tasks and remain a valuable instrument for understanding the forces which shape a given territorial situation. Quantitative models reign supreme in transportation or infrastructure planning. They benefit from a new generation of web technologies (Web 3.0), which gave birth to what Potts (2020) considers as a new planning paradigm: "Planning 3.0 is defined as an emerging planning paradigm in which the systems and structures of planning are innately 'smart', drawing on artificial and systemic intelligence to support more responsive and interconnected planning processes". Quantitative models are applied in a number of urban and regional contexts, particularly in the days of digital technologies, foresight studies, and the extensive use of geographical information systems (Giaoutzi and Nijkamp 1993; Giaoutzi and Sapio 2013). Plan formulation and evaluation methods are still being borrowed from economics and business studies. Multi-criteria techniques, which are not limited to traditional cost-benefit methods, are one example (Department for Communities and Local Government 2009). Another is SWOT analysis, which has been adopted in European Union development studies: "SWOT analysis (Strengths – Weaknesses – Opportunities – Threats) is a strategy analysis tool. For example, it combines the study of the strengths and weaknesses of an organisation, a geographical area, or a sector, with the study of the opportunities and threats to their environment. As such, it is instrumental in development strategy formulation. For strategy, this approach takes into account internal and external factors, with a view to maximising the potential of strengths and opportunities, while minimising the impact of weaknesses and threats".[4] The roots of the method, as in the past those of cost-benefit analysis, are obviously in economics. Vogelij (2015, 129) has an entertaining comment: "In the 1990s I proposed a SWOT analysis when working in a multi-national team on strategic planning. The pertinent reaction of academic team members was: 'a SWOT belongs to economic science, not to planning, so drop it.' Of course, I did not. Apparently, the scientific claims of theory require a long time for new concepts to be accepted". Foresight studies also have an old ancestry in what in the 1940 and 1950s was called long-range forecasting or future studies or, in France, *prospective*. They are back with a vengeance: "Foresight is a professional practice that supports significant decisions, and as such it needs to be more assured of its claims to knowledge (methodology). Foresight is practiced across many domains and is not the preserve of specialized 'futurists' or indeed of foresight specialists … The present trends towards methodological concerns indicate a move from 'given' expert-predicted futures towards a situation in which futures are nurtured through the dialogue between 'stakeholders' …" (Giaoutzi and Sapio 2013, 3–4). The exploration, or "invention", of the future is a central task of planning. The typical rational model approach referred to the search of alternatives for attaining goals. Scenarios is now a more acceptable term. "The scenario concept was adapted by planners seeking to remedy the limits that accompanied the application of rational scientific methods to decision making within modern global corporations" (Hoch 2019, 69). The concept was transferred to spatial planning and the author of this book used it with his colleagues in a plan for the redevelopment of the former Hellinikon international airport of Athens (Wassenhoven 2018; see Text-box 8.11).[5] It has found widespread use in urban planning in a variety of countries. We read in a paper from Serbia: "Scenario planning techniques are increasingly gaining attention in the process of spatial and urban planning because of their usefulness in times of uncertainty and complexity. Scenario planning encourages strategic thinking and helps to overcome thinking limitations by creating multiple futures. In this way, it can help to shape the future according to the values and desires of society" (Stojanović et al. 2014). And in a paper from Poland: "The upcoming future should be identified and recognized on an objective basis. To accomplish that, certain prognostic methods can be applied. One of the most useful methods, especially in the case when the social and institutional behaviors play a certain, more or less crucial role, is a scenario method. Scenarios could help to reduce the uncertainty, make the future more clear, or even build a structure based on an uncertain image of the future. That's their main task" (Soltys and Lendzion n.d.). Let us not forget that the formulation of alternatives is a standard procedure in present-day Strategic Environmental Impact Assessment studies (Greiving et al. 2008).

[4] See https://europa.eu/capacity4dev/evaluation_guidelines. Capacity4dev is the European Commission's knowledge sharing platform for International Cooperation and Development.

[5] This plan was never implemented, when the decision was taken to develop the site as a private project.

3.1.7 Systems Thinking

A great deal of the rationale of planning techniques derives from the rational model's affinity with systems thinking. The bonds between rational planning and systems analysis were underlined earlier; bonds which environmentalism brings back into planning practice. Thinking in terms of systems involves thinking in terms of wholes and not just of relations. The structure of wholes cannot be described simply in terms of relations. In a paper first published in 1941 it is stressed that this structure "may be described in terms of some more adequate logical unit, representing an entirely logical unit", i.e., a system (Angyal 1969, 17). The holistic quality is fundamental for the understanding of a system. "A system is an organized or complex whole: an assemblage or combination of things or parts forming a complex or unitary whole" (Kast and Rosenzweig 1972, 14). A hierarchical structure of subsystems nesting into each other is inevitable in complex systems. Their boundaries are not simple lines of separation, but lines demarcating system activity. This is particularly true of complex organizations, which are hierarchical systems of a very high order. Such organizations consist of a technical level, an organizational level, where tasks are coordinated, and an institutional level, where the organization's activities are related to the external environment, of which the organization is itself a sub-system (ibid., 24; see also Etzioni 1975). Systemic analysis is a close relative of complexity theory: "Complexity science and planning are not strangers. There is a considerable interest in representing urban systems through simulation and related procedures" (Byrne 2003, 173). The recognition of systemic properties in institutions and organizations uncovers a link with the work of planners, in which "institutional design" is a central task. "Institutional design means designing institutions: the devising and realization of rules, procedures, and organizational structures that will enable and constrain behavior and action so as to accord with held values, achieve desired objectives, or execute given tasks. By this definition institutional design is pervasive at all levels of social deliberation and action, including legislation, policymaking, planning and program design and implementation" (Alexander 2005, 213). This is a useful reminder that planning work is not limited to producing plans; it is extended to the devising of regulations and guidelines for plan production.

We are but one step away from interpreting geographical units (cities or regions) and decision-making organizations, such as a planning department, as complex systems. A system is any entity, conceptual or physical, which consists of interdependent parts (Ackoff 1969, 332). Holistic systems of this nature can only be approached through a study of their systemic properties, because they are more than mere sums of their parts. "In an aggregation the parts are added, in wholes the parts are arranged in a system. The system cannot be derived from the parts; the system is an independent framework in which the parts are placed" (Angyal 1969, 26–27). In this sense, biological systems, but also human and social ones, are totally different from simple mechanical systems, because they are open to their environment and highly interactive. In human and social systems certain factors acquire special significance: communication, language, symbols, values, ideas, and emotions.

Including these paragraphs on general systems theory and open systems is not only in order to shed light on the insights borrowed from physics and biology which had a great impact on rational planning theories; nor is it to recall the influence of systems thinking on futures research and long-range forecasting (Jantsch 1969a, b; Gabor 1969). The reason is the far more recent return to an environmental – ecological conception of planning theory after a long interlude of sailing in other waters. It is true however that in the formative years of rational systems planning, loans of natural science concepts like "entropy" were regular, because cities and regions were interpreted as open systems (Wilson 1970). According to McLoughlin's bold statement of 1969, "within a decade … we moved from a view of the city as a machine-like system – a system that *works* – to a view of the city as a system that evolves. To analyse the city as a complex system that evolves has profound consequences for many aspects of planning thought and practice" (McLoughlin 1969, 81; see also Lagopoulos 1973). "In physics", Chadwick (1971, 54) wrote in his equally influential book on systems planning, "entropy is a measure of the disbalance of energy in a system, its disorder, or randomness of organisation, as systems tend to move from a less to a more probable state". Inputs from the environment into an open system (energy, materials, even information) allows the system to arrest the process of entropy and survive. An increase in entropy means loss of information and eventual collapse. Identifying an urban system with a living organism may seem, at first sight, far-fetched, but the concepts of systems analysis undoubtedly had a strong impact on planning (Batty and Hutchinson 1983). Is such a parallel out of fashion? In the days of awakening to the effects of the Anthropocene, of climate change, and of the Eco-System Approach, it is not. "A striking contrast between inanimate and animate nature, seems to exist", wrote Von Bertalanffy (1969, 79) a long time ago: "In organic development and evolution, a transition toward states of higher order and differentiation seems to occur". Is this a more stable state? "Living systems, maintaining themselves in a steady state, can avoid the increase of entropy, and may even develop towards states of increased order and organization" (Kast and Rosenzweig 1972, 21). "In open systems which are exchanging materials with the environment", as cities do, we may remind ourselves, "in so far as they attain a steady state, the latter is independent of the initial conditions; it is equifinal … Equifinality is found also in certain inorganic systems, which, necessarily, are open ones" (Von Bertalanffy 1969, 76–77; see also Von Bertalanffy 1973). The system in question, through its boundary with the environment and its interface with adjacent systems receives positive feedbacks, which maximize deviations and eventually ruin the system, or negative feedbacks, which indicate deviations and cause corrective actions. No matter how much we remain doubtful of the wisdom of transferring such knowledge into the theory and practice of urban and regional planning, how is it possible to ignore such analogies in the days of the diagnosis of the Ecological Footprint of human activity and of the observation of Heat Islands over urban agglomerations? We are forced to return to the roots of general systems theory and its effect on rational planning models. It is to the former that we owe the conception of a city or region as a transformation system which receives inputs and produces outputs which consist of a flow of materials, energy, and information (Kast and Rosenzweig 1972, 18). In the heyday of the general systems theory, such a transformation system was seen as typical of complex organizations (Churchman et al. 1957, 76; Perrow 1970; Pugh 1971). The close ties of the tradition of systems analysis and operations research with rational planning theory have been discussed time and time again, e.g., with regard to policy analysis (Friedmann 1996, 21–24; Mintzberg 1994). The entire public policy process in modern capitalism has been subjected to analysis which owes a great deal to the theory of complex organizations (Ham and Hill 1993). Systems thinking, in the era of climate change and global relations, seems to be refusing to die. Theorists return to it, as does Moroni in a definition reminiscent of those by Kast and Rosenzweig: "A complex system is a *structure* of phenomena, not a *mass* of phenomena: Its very characteristics are those of a 'general order' whose specific elements are perpetually changing. Self-organisation is certainly one of the most characteristic features of a complex system: this kind of system spontaneously seeks out some form of order, with articulated structures being created randomly; no one deliberately imposes order on its numerous components – the system spontaneously exhibits synchrony" (Moroni 2015, 250). It is significant that Moroni suggests the adoption of a systems logic to deal with cities and land-use planning. His approach "proposes that the problem of complexity and intrinsic unpredictability requires a radical overhaul of the tools used to regulate land use – especially by exploring the idea of shifting the emphasis to framework-instruments (such as urban codes) … To avoid misunderstandings: neither generic deregulation nor laissez-faire 'liberalism' is suggested in this case; simply, a different way to take seriously the need for regulation … Indeed, I believe that *law* must once again assume the central role that it has regrettably lost over time" (op. cit., 263).

3.1.8 Rational Planning Abandoned, but for How Long?

The world of systems analysis and rational planning seemed "too good to be true" and the very fact that it was so closely associated with the theories of organizations and economic efficiency inevitably led Doubting Thomases to question its supposed independence from the realities of power politics. As the ideology of a universal consensus, when and where it really existed, faded, so did the trust in impeccable technocratic, scientific answers to problems. The direction of promised transformations was no longer obvious and easy to grasp. Beauregard (2005, 205–206) put it very elegantly: "As the sheen of technocratic solutions dulled and resistances multiplied, power itself, as embodied in states and markets, came to be seen as less benevolent … Eventually, planning theorists turned to civil society as a source of ideas and legitimacy … Then there is 'transformation'. What does it mean to change something in a world where change is relentless? Theorists interested in institutional transformation have a normative intent embodied in their concern with purposive change. Most want to know where change might lead, although a few … seem committed to change simply for its own sake, a position that, despite its agnosticism, is no less value-laden". Perhaps the mistake of theoreticians was that they did not ask the right questions from the start. This is the diagnosis reached by Dobrucká (2016) in a recent article: "This article explores the five fundamental question categories (*what, who, when/where, how* and *why*) in the context of planning theories and planning practice. It is shown that current theories of planning are based on an understanding of the interactions between these questions that is not consistent with their relationship in practice, and that the theories overstate the importance of the ends–means discussion".

As the modernist view of society faded, planning inevitably followed suit. Modernist planning was "besieged", in Beauregard's expression in a 1989 article reviewing the situation in the United States: "Modernist planning began to come apart in the 1970s and 1980s … Novel political forms, economic relations, and restructured cities posed new difficulties for the premises that underlie practice" (Beauregard 1996b, 221). Even so, he pointed out, "theoretically, planning remains in a modernist mode. The literature on planning theory is devoid of attempts to view planning theory through the lens of the postmodern cultural critique. Rather, this theoretical investigation has been initiated by urban geographers …" (op.cit., 224).

Thus, the complexity of the real world came to the fore. Paris had rightly warned that "the central problem for planning theory as an abstract 'theory of planning' is that it cannot reconcile technical dimensions of decision making with the forces which shape and reshape the social world" (Paris 1982b, 308).

Over-optimistic or over-simplistic versions of rationalism had certainly underestimated the complexity not just of achieving goals but even of solving problems. Problems tended often to be, to use a popular designation found in planning theory, "wicked". Following an older definition, Brooks describes them as follows: "Wicked problems … are ill- and variously defined; often feature a lack of consensus regarding their causes; lack obvious solutions – or even agreement on criteria for determining when a solution has been achieved; and have numerous and often unfathomable links to other problems" (Brooks 2017, 12). The assumption of perfect knowledge of social reality and of automatic consensus is undermined by this realization. This stubborn reality made the job of planners all the more complex, as Bolan and Nuttal (1975, 7) emphasized while rationalism in planning was still dominant: "[U]rban planning and community decision making is a complex process involving an intricate combination of social, political, and psychological factors". The complexity and diversity of the real world, which, according to its critics, rational planning could not possibly grasp, was bound to encourage the birth of another theoretical current for which multiplicity of views was a key point of departure.

3.2 The Communicative Current

Keywords of the Communicative Current
Ideas, concepts, preoccupations, interpretations, concerns, and socio-political conditions embedded in the communicative current, which inspire communicative and collaborative planning theory and practice:

Subjectivity, relative values, diversity, collaboration, deliberation, reflexivity, place, recognition, hermeneutic approach, agency, opinion variations, group values, devolution, mediation, understanding, storytelling, cognitive truth, dialogical learning, arbitration, difference, structuration, search, stakeholders, actor roles, truth relativity, visioning, audiences, multiple publics, advocacy, grassroots movements, consensus, ethics, participation, contradictory views, equity, discourse, argumentation, situatedness, multiple knowledges, debate arenas, interpretation.

3.2.1 Communicative and Collaborative Planning

The communicative current stressed the awareness of multiple voices in society and the existence of citizen groups with diverging opinions. It was evident that social realities warranted a different approach to planning, to make room at least for meaningful participation, already celebrated in Arnstein's famous 1969 article on the "ladder of citizen participation" (Arnstein 1995), but also in concrete statutory initiatives (Fagence 1977). This is how Albrechts formulated this demand: "Citizens must claim their role in the political system … The purpose of this claim is to promote structural change in order to improve individual and collective potential, to respond to problems, needs, challenges and to take an active part in the processes of decision making, plan making and implementation processes aimed at solving problems, and realizing visions and potentials" (Albrechts 2002, 331). He added later: "Planning is not an abstract analytical concept but a concrete socio-historical practice, which is indivisibly part of social reality. As such, planning is in politics, and cannot escape politics, but is not politics. Since planning actions are clear proof that they are not only instrumental, the implicit responsibility of planners can no longer simply be to 'be efficient' …" (Albrechts 2003, 251). Changes in the understanding of what planning is about were brought to the fore by new interpretations of the city. "There are two ways of analyzing cities, neither incorrect. The first, or global, approach scrutinizes the international system of cities (and its national and regional subsystems …). In contrast, the second approach, which works from the inside out, examines the forces creating the particularities of a specific place – its economic base, its social divisions, its constellation of political interests, and the actions of participants" (Fainstein 2002, 110). The second approach, as expressed by Fainstein, was of decisive importance in the search for a new theoretical interpretation.

A new route was searched and arguably found in a communicative perspective. Patsy Healey's chosen route was "a communicative conception of rationality, to replace that of the self-conscious autonomous subject using principles of logic and scientifically formulated empirical knowledge to guide actions. This new conception of reasoning is arrived at by an inter-subjective effort at mutual understanding. This refocuses the practices of planning, to enable purposes to be communicatively discovered" (Healey 1992, 238–239). Healey again: "In this way, knowledge for action, principles of action and ways of acting are actively constituted by the members of an inter-communicating community, situated in the particularities of time and place" (op.cit., 243). The existence of various forms of knowledge is a reality that constitutes the basis of understanding, as advocated by Healey. "Given the multitude of choices, dilemmas, and pitfalls associated with developing theory, it is not surprising that it is possible to speak of a wide variety of different 'types' of knowledge" (Bolan 2017, 61). Healey's earlier remarks were the foundation of her collaborative or communicative planning theory, which she based on Jürgen Habermas' concepts of communicative rationality and deliberative democracy (Dryzek and Dunleavey (2009, 216 and 220). Her intention is clarified in her later book: "The starting point of this book has been the challenge of how to deal with matters of collective concern which arise from the problems and opportunities of the co-existence in shared spaces of relational groups, or cultural communities, often with very different priorities and ways of looking at things. This is the focus of the practices of what is variously called urban and regional planning, spatial planning and local environmental management" (Healey 1997, 310). Clearly, this leads to a new portrait of the urban and regional planner. Low paints the portrait with a broad brush: "What role for the planner does Healey's pluralist vision prescribe? It would seem to be that of facilitator of 'place-shaping' through an inclusive process … Healey's prescriptions for inclusive and collaborative planning are somewhat utopian. I see nothing wrong in that. They provide a benchmark for the facilitator of collaboration, and a counterfactual against which to measure actual practices. As Healey herself says, an inclusive collaborative approach does not do away with unequal power relations or conflict over local environments … Some conflicting interests cannot be reconciled" (Low 2020, 35–36). Place-making occupies a central position in this approach and is, in fact, a major innovation in planning thinking (Vigar et al. 2020).

Brooks, who curiously makes no reference whatsoever in his book to the work of Patsy Healey, claims that a collaborative planning approach had been proposed as early as 1966 (Brooks 2017, 150), and quotes from a 1997 article by Judith Innes who offers her own definition of "collaborative, communicative planning": "This 'postmodern' planning involves making connections among ideas and among people; setting in motion joint learning; coordinating among interests and players; building social, intellectual and political capital; and finding new ways to work on the most challenging tasks. This kind of planning, when it is done well, builds its own support and changes the world" (op.cit., 120). This "when it is done well" is a formidable proviso! If it is done well and deliberation is sincere, then the change in the participants' minds would be spectacular. "In deliberative settings", Forester observes, "we could see, parties sometimes entered as 'I' and later emerged with more developed, more informed, more acute senses of 'we'. *Relationships* between people changed and developed (for better or worse, to be sure), not just individuals' organizing frames or theories. *Senses of identity* could change, not just utilities. But more than that, parties could also learn about value and significance, what *mattered* as well as how things worked" (Forester 2013, 9).

It is arguable that the best introduction for this conception of planning is a quotation from Tore Sager (2018, 93 and 95): "Communicative planning (CP) is seen here as a participatory and dialogical endeavour involving a broad range of stakeholders and affected groups in socially oriented and fairness seeking development of land, infrastructure, or public services. It is guided by a process exploring the potential for cooperative ways of settling planning disputes and designed to approach the principles of discourse ethics, demanding processes to be open, undistorted, truth-seeking, and directed at mutual understanding … An important difference between typical citizen participation and critical CP is illustrated by the concepts 'invited space' and 'invented space'. … Citizen participation is about inviting groups and individuals from civil society in the official planning process and creating arenas – *invited* space – for expression of opinion and information exchange between professional planners and affected lay people. CP entails citizen participation, but acknowledges the need that some protest groups and social movements have for standing outside the official planning process. Communicative planners meet these actors in *invented* spaces external to the planning framework of the authorities. Communication in these arenas can come closer to discussion on an equal footing and bring a wider set of interests into the planning dialogue" (my italics). "Inventing" the proper spaces for ideal communication is of course no easy task. "The big question for the pragmatic analysts is how practitioners construct the free space in which democratic planning can be institutionalized" (Hoch 1996, 42; see also Vidyarthi 2019). Practical experience of communication in 'invited spaces', let alone 'invented' ones, does not provide much comfort. Campbell and Marshall comment on the "capacity of a rights-based approach to public participation to virtually paralyse the decision-making process. Whether by accident or by design, the focus on the right of individuals or communities to articulate their self-interests appears to reduce local democracy to confusion and noise" (Campbell and Marshall 2000, 339–340; Campbell and Marshall 1999). In his definition, Sager speaks of discourse ethics. Ethics is an enormous subject in planning practice and theory, which is not limited to the planning discourse but rather includes the entire field of the planning and public policy-making activity. Doubts regarding its ethical foundation can undermine the confidence of planners, a danger that led Heather Campbell to write "the planning community needs to rediscover its ethical voice and its confidence in *the idea of planning*" (Campbell 2012, 379). She placed her plea in the context of the communicative enterprise: "[A]s a community of academics and practitioners, planners need to discover the art of conversation" (op.cit., 394). We must not overlook the fact that planners, especially in administration, may experience fear when they feel that their insistence on ethical standards threatens their position and promotion. Sturzaker and Lord (2018) warn that "many public sector planners in England are unconsciously motivated by fear—fear of losing control over development, fear of being blamed for unsatisfactory development, and ultimately fear of losing their jobs".

As in the case of the rational period, the overall climate of the days when the communicative approach was emerging was marked by a literature dealing with the new conditions of urban and regional development, planning and governance, and rising theoretical interests.[6] This literature had a serious effect on planning theory. Communicative and/or collaborative planning is extensively discussed in several basic planning theory textbooks, above all in that by Healey (Allmendinger 2002, 21, 29–42, 63–66, 182–191; Beauregard 2020, 36–37, 53–58; Palermo and Ponzini 2010, 65–78; Taylor 1999, 117–125; Healey 1997, 5–7, 28–53; Brooks 2017, 120–128). A point frequently made is that this form of planning is not merely "advocacy" in the sense of Davidoff's classical advocacy planning paper of 1965 (Davidoff 1973). Collaborative planning is not simply about a group of planners representing a group, e.g., a poor neighbourhood, as its advocate. In such advocacy ventures, in the USA, "the planners found themselves acting as community organisers, drawing residents into their organisation's activities. But they could never claim to represent the 'neighbourhood' or 'community' and they lacked legitimacy" (Low 2020, 34).

3.2.2 Conceptual Foundations

A large number of theorists have debated on communicative planning's value, originality and contribution to planning practice, as well as on its shortcomings. It marks undoubtedly a most important stage in theoretical thinking and owes a lot to the writings of Jürgen Habermas and Anthony Giddens (Bolan 2017, 193–195). "Habermas argues that, far from giving up on reason as an informing principle for contemporary societies, we should shift perspective from an individualised, subject-object conception of reason, to reasoning formed within inter-subjective communication. Such reasoning is required where 'living together but differently' in shared space and time drives us to search for ways of finding agreement on how to 'act in

[6] This is a purely indicative list: Allmendinger 2001; Allmendinger and Tewdwr-Jones 2002; Benko and Strohmacher 1997; Bridge and Watson 2002, 2003; Brindley et al. 1996); Chambers 1997; Davies and Hall 1978; Dear and Flusty 2002; DiGaetano and Klemanski 1999; Forester 2000; Friedmann 1987; Hall and Pfeiffer 2000; Harvey 1973; Haywood 1979; Healey et al. 1982; Krumholz and Forester 1990; Lauria 1997; Lefebvre 1968; LeGates and Stout 1996; Marcou et al. 1994; McAuslan 1980; Pahl 1964, 1968, 1970, 1975; Parfitt 2002; Sassen 1991, 2002; Schön 1973; Short 1996; Simmie 1974, 1981, 1993; Smith 1974, 1980; Stephens and Wikstrom 2000.

the world' to address our collective concerns. Habermas' communicative rationality has parallels within conceptions of practical reasoning, implying an expansion from the notion of reason as pure logic and scientific empiricism to encompass all the ways we come to understand and know things and to use that knowledge in acting. Habermas argues that without some conception of reasoning, we have no way out of fundamentalism and nihilism" (Healey 1992, 242–243). A psychoanalytical root has been detected in the analysis of Habermas (Gunder 2011, 299; see also Hillier and Gunder 2003; Hillier 2003). The "subject-object" dichotomy is of interest not only to interpret the planner's perception of reality, but also to understand that of the stakeholders. It is a "bipolar construct", in Low's words, "which has dominated scientific and philosophical thought since antiquity: subject-object, or subjective-objective. This dualism *constructs* an absolute boundary between what is inside our heads and what is outside" (Low 2020, 65–66).

Reasoning through communication is in itself a call for escaping from the dominant role of political and social structures in dictating individual human behaviour and thought. On his part, "Giddens presents what he calls 'three intersecting moments' of 'difference'. Every social practice occurs in space or time. Indeed, even routinized habitual action is not independent of space or time. Its very repetition, or its rhythm, occurs in a specific, designated place and moment. The third moment of 'difference' in Giddens theory is what he calls the 'paradigmatic difference'. He elaborates this as the transformations of a virtual order of differences, that emerge having the 'meaning' of rules" (Bolan 2017, 194). Bolan adds to his analysis an extensive reference to the work of Piotr Sztompka and his concept of "social becoming": "[S]ocial structures are not, *a priori*, externally constituted entities binding and regulating all human action. Social reality is a dynamic process – composed of *events*, not objects. The social world *occurs* rather than *exists*" (op.cit., 196). Thus, "the modes of potentiality and actuality are linked by mutual feedback loops. There is an incessant, dialectical back-and-forth oscillation between what is possible and what actually occurs, extending in time. This Sztompka defines as *becoming*. For him, social reality is a '*living socio-individual field in the process of becoming*'" (op.cit., 198). The meaning of becoming is further explored by Albrechts et al. (2019) in the context of transformative practices (see Chap. 4): "Transformative practices focus on the structural problems in society. They construct images of preferred outcomes and how to implement them. Transformative practices become the activity whereby that which might become is imposed on that which is, and it is imposed for the purpose of changing what is into what might become. This means a shift from an ontology of being, which privileges outcome and end-state, towards an ontology of becoming. Becoming takes into account the unconscious, emotional, and social relationships. In this way, it shapes perception, attention, assessment, intention, and commitment … in which actions, movement, relationships, process, and emergence are emphasized". Hillier (2005) borrows Deleuze and Guattari's concept of becoming and observes that "as planning theorists and practitioners we seem to have had a pervasive commitment to an ontology of *being* which privileges end-states and outcomes, rather than an ontology of *becoming* which emphasizes movement, process and emergence" (see also Chap. 4).

The importance of structures and their open, or sometimes indirect, impact on the action of individual agents is stressed in Marxist analysis, or that of Michel Foucault's post-structuralism (Dryzek and Dunleavey 2009, 289–293). Structures "police the social body" and prevent the "self-definition of individuals" and the "assertion of difference" (Allen 1996, 331). Structures are a reflection of power: "In order to overcome the bias in favor of powerful social groups, an emphasis on democratic deliberation has become central to discussions within planning theory" (Fainstein 2016, 258). Yet, human agency is itself a factor in the creation of structures. This structure-agency opposition, a field of analysis so familiar in geographical studies (Goodwin 1999; Leontidou 2011), was given a new twist in Giddens' theory of structuration which Healey has built upon. "For me", she has written, "it was the approach to the interaction between structural dynamics and active agency of Giddens's structuration theory which helped me make the connection between the worlds of practice and their broader context. I was struck not only by his emphasis on how agency created structures which shaped the way people as active agents thought and acted; he also presented three ways in which this powerful interaction was accomplished – through the flows of material resources, through regulatory practices, both formal and informal, and through framing ideas" (Healey 2017, 115). She had written earlier that "the planning field, with its concern for practices as well as ideas, is also interested in planners as actors, that is, in the work of agency" (Healey 2010, 16). McDougall has explained this interaction by pointing out that there has been a "sociological orthodoxy … which insisted that structures are both constraining and enabling and are the conduit and product of human action … Giddens defined *structures* as organized sets of rules and resources that are produced and reproduced through human action. Structural continuity is dependent and on the knowledge of human actors who concretize and reproduce through their thought and action the routinized patterns of social life. This does not mean that single actors can create or control their social systems, for these stretch in both time and space beyond the boundaries of an individual's action space. But through their thought and action, human actors are the agents of structural reproduction and this role, in association with others gives them the capability for transformation" (McDougall 1996, 189–190). Bolan quotes Giddens' own words: "The concept of structuration involves that of the *duality of structure*, which relates to the *fundamentally recursive character of social life, and expresses the mutual dependence of structure and agency*" (Bolan 2017, 194). For Sztompka, the link between structure and agency is *praxis*. It is there that operation and action meet: "Praxis is conceived as the nexus of 'operating' structures and 'acting' agents … The concept of *agency* is complementary to *praxis*" (op.cit., 199).

Structural reproduction by human actors is no easy task. It means that planning in the broad sense of the word must embrace the design of institutions that set the framework within which plans are produced. This widens the role of the planning community. Healey (2005, 301–302) draws attention to this new direction: "In recent years, a new emphasis on the role of institutional parameters in structuring social action is appearing like an intellectual wave across economics, political science, organizational sociology and management, public administration, policy analysis and planning". Planning for the setting-up of an institution and for the mode of its operation is totally different from conventional plan-making but can have much wider repercussions. The importance of institutional planning lies in the relationship of institutions with individuals, a matter which is central in institutional theory. Two definitions of institutional theory make this clear: "A theory that examines how an individual uses his social process to accustom himself in an organization that has its own norms, practices, rules, and conventions", or, a theory defined as "the guidelines for social behavior in the form of accepted structures, schemas, rules, norms, and routines influenced by other members of the collective network of actors".[7] Bolan (1996, 510) makes his own position very clear: "Designing institutions calls for a different kind of skill and imagination than planning discrete projects, programs and policies … [This] is a plea for a new sort of specialist practitioner and the cultivation of a new body of theory … [We] planners cannot afford to leave institutional design to a new group of specialists. It is inevitable, part of the fabric of our daily work lives. Theorists, therefore, must attend to institutionalization".

3.2.3 The Role of Plans and the Public Interest

Communicative action, in the opinion of its supporters turns planning into a process of argumentation and plans into "texts" which help redefine reality and unveil alternative truths (Baum 1996). This implies a process "involving negotiation and mediation in working through a problem with those directly affected. It requires life experience, communicative skills and, in multicultural or multi-ethnic contexts, cross-cultural understanding" (Sandercock 2000, 23, 2002). An essential prerequisite is the recognition of "others" and of "situated" processes, hence the emphasis on otherness, cross-culturalism, and places. The plan itself is assigned a role in communication action. "Plans are seen to perform multiple roles, sending quite different messages to different 'audiences'. The plan, in turn, may mean different things to different groups". The planners must project its "textual meanings" and be themselves "characters in a drama" (Healey 1996, 263 and 265). This is a phase preceding the activity of choice. "[B]efore the rationality of choice comes the prior practical rationality of careful attention, critical listening, setting out issues, and exploring working relationships as pragmatic aspects of problem construction. Planners build up such structures of value in their stories in institutionally and ideologically staged ways … By telling practical stories, for better or worse, planners bring to bear moral imagination and shape the moral imaginations of others". (Forester 1996a, 210 and 220). Their plans can act as texts of "storytelling", which functions "as a method for revealing how formal planning practices may be destabilized by more vernacular narratives seeking to subvert dominant discourses and processes" (Bulkens et al. 2015). Storytelling seems to have become a most popular

[7] What is Institutional Theory/IGI Global (igi-global.com).

description, aiming at setting on fire the imagination of both planners and the planned: "Planners should tell future-oriented stories that help people imagine and create sustainable places" (Throgmorton 2003, 125; see also Van Hulst 2012 and Zitcer 2017). With imagination and a strong measure of irrationality planners can be creative, Ferraro declares: "[A]cting 'irrationally' does not mean acting incoherently, randomly, or chaotically. It means acting to be free from strict rationality prescriptions, not to be 'rational fools' … that is, slaves of one's own untouchable, unreflected individual preferences. Planning is irrational because it is creative, insofar as it influences people's preferences, shapes their perceived needs and expectations, and finally produces new values" (Ferraro 1996, 316).

Planners must be capable of empathy and inner reflection, which will help them understand the situations they are facing. Empathy is not far from Heidegger's 'care structure' of humanity, which I mentioned in the previous chapter, in a quotation from Nicholas Low. Empathy includes kindness, respect, recognition, tolerance, civility and care (Low 2020, 262). The planners must "develop sophisticated empathies with the multiple and contradictory codings", as Harvey (2002, 387) put it. Forester finds here an equivalence with the "sociological imagination" once extolled by C. Wright Mills, who wrote that "the sociological imagination enables its possessor to understand the larger historical scene in terms of its meaning for the inner life and the external career of a variety of individuals" (Wright Mills 1959, 5). Forester notes that "the planners' practical judgments are particularly challenging and fascinating because they are peculiarly anticipatory in nature. These are not the judgments that a judge must make: 'Guilty' or 'Not guilty'" (Forester 1996b, 248). The planner must be capable to act as "reflective practitioner" as Donald Schön (1983) would like him to be, although for De Leo and Forester (2017) this is only a half-truth: "[It] is certainly right, but not right enough: it over-psychologizes planning practices and under-theorizes their social and political interactivity. Although reflective practice can occur alone, deliberative practice requires us to examine the precariousness and vulnerabilities of trust and relationship building … Also, because reflective practice can face ambiguity, deliberative practice requires practitioners to anticipate and respond to other actors' interpretations, biases, ideologies, and presumptions". To reach this level of mutual understanding, Howe and Langdon (2002, 212) recommend the employment of "reflexive analysis" to help uncover the participants' understanding of the world, i.e., "their predispositions and orientations", which are "pivotal in understanding land use planning, development and policy-making". Such demands undoubtedly strike terror in the heart of the average practitioner. "This is the fundamental problem of planning today; knowledge requirements have expanded so significantly that one can easily visualize the planner as needing to be a true twenty-first century 'Renaissance person'" (Bolan 2017, 36). What is no doubt essential for the planner is to delve modestly into his inner personality and bring out his own self-image and vision, as Low stresses in an effort to comprehend "personal construct theory" and to hypothesize what a planner's "personal construct" might be. "[I]n order to get along in the world", Low (2020, 11) writes, "a person generates an internal picture of the outside world based on experience which she uses to predict what will happen next – and next after that, and so on into the future, so that the world is seen to have continuity and is not composed of one-off events". But being akin to a soothsayer is not enough for the planner, who has also to mobilize endless resources of reform, like a miracle-worker who wishes to create a brave new world. "Central to the being of 'planning' and thus to the being of the 'planner', in whatever local context they find themselves, is what we might call the transformative impulse: the desire to change the conditions of life of people for the better. In planning, the transformative impulse is purposive" (op.cit., 77).

For a planner, this task of building his/her "personal construct", to which he/she is supposedly juxtaposing claims to professional knowledge, and then to mobilize his/her transformative drive is already a very demanding enterprise. But there is more to come. Storytelling and persuasion skills require in addition a rhetorical competence which the technocratically trained professional probably lacks. The combination of these qualities, as presented by Throgmorton, adds to the already colossal qualities that the planner must be endowed with: "Planners, as I have heard them speak, tend to describe rhetoric as 'mere words' that simply add gloss to the important stuff, to the objective methods that we use to discover the 'facts'. On other occasions, they describe rhetoric with suspicion as the use of seductive language to manipulate others into embracing a speaker's preferred values, beliefs, and behaviors. We are not sure, in effect, whether rhetoric is trivial or insidious. There is a good way to bring together these two definitions. Rather than thinking of rhetoric as gloss or seduction, let us regard it as the study and practice of persuasion, and let us recognize that persuasion is constitutive. Rather than divorcing planning tools from their contexts of application or treating them simply as political ammunition, let us think of surveys, models, and forecasts as rhetorical tropes, as figures of speech and argument that give meaning and power to the larger narratives of which they are part" (Throgmorton 1996, 345). The planner must combine technical expertise, oratorical skills, and mediating capacities, all in one and the same person. "[S]peech is very much the principal form of action that he or she employs" (Bolan 2017, 172). Planners must act as mediators and thus discover the multiplicity of interests and not just a "monolithic" public interest: "The recent interest in communicative action – planners as communicators rather than as autonomous, systematic thinkers – also reflects this effort to renew the focus of planning theory on the public interest" (Campbell and Fainstein 1996, 10–11). Planners, Rivero (2017) adds, must be cast "as skilled mediators in a democratized process of knowledge production".

The existence of a public interest is regularly asserted in politics and policy making. "Planning policies are often justified in terms of their overall benefit for the wider population. In this view, the outcomes of planning actions *should* be beneficial in terms of an aggregate social good or in some utilitarian sense. This role is seen widely as a key justification for planning systems and as part of this, it is assumed that planners will seek to minimise detrimental outcomes. However, the key question here is *how* is the public interest evinced and maintained through principled positions, institutional design and policy detailing?" (Parker and Doak 2012,

111–112). The belief in the existence of a public interest is further discussed in Chap. 10 with reference to the case of Greece. The problem is, as Moroni points out, that not everyone agrees that the public interest exists as an *a priori* criterion. “This is precisely the argument advanced by planning theorists who stress the centrality of communicative practices” (Moroni 2018, 71). It is true that the existence of a public interest has been repeatedly disputed (see Chap. 10). It is also true that political powers easily resort to it as a justification for their actions. Planning, by its sheer existence, is supposed to serve the public interest. “However”, Salet (2019) explains, “the public good is not a possession; it is an abstract but real kind of justification and, already for this reason, always contested. It does not provide sources of power but requires a particular sort of arguing, certainly from those in a powerful position. The justification of public interest is not a monopoly of the powers in the state; it is a challenge of the full public”. Communicative planning, in line with Habermas’ theoretical premises (Habermas 1971), is supposed to uncover a diversity of interests which cannot be summed up in a “generalizable interest”. It is worth remembering here the pioneering work of Jane Jacobs (1965, 1969) on cities: “In hindsight, Jacobs was to have a long-lasting and profound effect on city planning with her attack on the male visionaries and her call for planning for diversity” (Low 2020, 9). We must note that students of Habermas’ work claim that in his later work his position had shifted. As Mattila (2016) wrote, “in the light of Habermas’ recent works, it seems that the idea of generalizable interests is not necessarily in conflict with the respect of concrete others and situated reasoning”.

The public interest as a goal and as evaluation criterion deserves a further elaboration. Ernest Alexander provides a succinct comment: “When discussing the ontology of the public interest, we need to distinguish between its procedural and substantive aspects. In its procedural sense the public interest is identified with the political decision-making or planning process … In its procedural sense, the public interest is undisputed. The same is not true of the public interest as a principle that has substantive content. This, however, is the aspect that must command our attention … Three kinds of arguments have been made against the existence of a substantive public interest: it does not exist as a fact, it cannot exist as a holistic interpersonal value, and it should not exist as a privileged substantive value” (Alexander 2002, 234). The “holistic” interpretation of the value of public interest is rejected by Campbell and Marshall (2002, 176–177) as a sad relic of utilitarianism. They argue that their own conceptualization “differs from the utilitarian concept in that, while individual interests are seen to be an important part of determining the public interest, it also gives emphasis to an ‘outsider’ perspective or an ‘objective’ evaluation. The necessity for this is based upon three important considerations. First, there is the existence of inequalities of various kinds … which requires someone … to compensate for the differences. The second consideration is the notion that individuals may be mistaken in their interests … Third, it acknowledges the existence of collective values and principles which transcend private interests and their summation”.

3.2.4 Objections to, and Weaknesses of, Collaborative Planning

The onus placed on the shoulders of the communicative planner is enormous but it is clear that his task is hampered by obstacles which the communicative approach cannot easily overcome, hence a variety of criticisms have been levelled against it. To her credit, Healey (2003, 108–109) was quick to identify, acknowledge and address them: “The first two criticisms echo those made of the rational planning process itself. These claim that both my treatment of ‘collaborative planning’ and the diverse enterprise of communicative planning theory more generally neglect ‘context’ and focus too much on process, divorced from ‘substantive content’. The third takes a broader aim and claims that both collaborative planning and communicative planning theory lack an adequate base in social theory. The fourth also echoes a criticism of the rational model and centres on the neglect of ‘power’ in the approach. I then make a criticism of my own with respect to the social theory I use, namely ‘institutionalism’, as my treatment of it in the book is partial. Finally, I turn to two strong and deliberate biases in my work, the search for more inclusionary governance processes in a ‘multicultural’ context and the emphasis on place quality as a policy focus”. Along with joint authors, she had earlier observed that “at the heart of the planning tradition is the promotion of the quality of places as a key ingredient of quality of life. As planners well know, it is not enough just to focus on jobs, houses, health, education and welfare services. How these are linked in the relations of daily life and the impacts of such activities in place and time are also important. For much of the second part of the last century, it was the functional service agendas that flourished and the `place quality’ agendas languished, except where there was strong local autonomy and commitment. But now, policy-makers and pressure groups are increasingly arguing for more attention to the qualities of places, to `area integration’, to the vitality of cities and the future of sustainable rural areas” (Healey et al. 2000, 7). Healey wrote later: “Planning ideas and planning activity express, and contribute to, the way people understand and feel about places. They may come to affect and express people’s sense of identity as well as their material conditions” (Healey 2016, 143). Places are not mere geographical entities; they are not just spaces. “[T]he term ‘place’ generally refers to specific socio-spatial arrangements where physical forms entwine with social patterns at different time-space scales …, institutional rules with daily practices and uses … and technical representations with narratives and social imaginaries …” (Lieto 2013, 144). It is apparent that the breadth of this definition leaves plenty of room to the imagination and to subjectivity. “In my view”, wrote Bolan (2017, 227), “the definition of a *place* can be pretty vague, while *space* can indeed be precisely measured – for place, boundaries are often arbitrary, subject to the vagaries of theme and horizon for any form of human experience. For planners, the two terms are pretty equivalent, although for most planning issues or episodes, one is invariably caught up in measuring space in one way or another. For philosophers and human geographers, the argument is that place comes ahead of space”. The usual emphasis on space in terms of functionality and efficiency is inadequate, for "space (and place) is far broader than that – space (and place) presents complexities regarding impacts on communal consciousness and interactions with time. It may serve to create social inequalities in spatial use and character; it also may create feelings of boredom and excitement, welcoming and threat" (op.cit., 237). If we recall Doreen Massey's dictum that "space matters", it is not surprising to find Klaus Kunzmann exclaiming that "places matter" (Kunzmann 2017, 219). The slogan becomes a sort of radical rallying cry: "Theorists taking a southern perspective have argued that 'place matters' (whether in global North or South), and that the degree of abstraction assumed in concepts claiming to be applicable everywhere sweeps away the possibility of a thorough understanding of cities and regions, and directly constrains potentially meaningful and effective planning intervention" (de Satgé and Watson 2018, 11). Fabrizio Barca, in his well-known report on a "place-based approach", compiled for the European Commission, would certainly agree with this position (Barca 2009). Place, *topos* and topophilia are concepts about which more will be said in Chap. 10.

It is true that even theorists who sought a radical departure from the rational model have been critical of some aspects of the communicative literature: "I have two critical differences with this literature", wrote Sandercock (2000, 23–24), with notable perspicacity. "One is that the so-called communicative action or collaborative planning approach is a model that has assumed (following Habermas) that *rational discourse* among stakeholders is both appropriate and achievable. The other is that outcomes from this approach, even when consensus is reached on a way forward, are not necessarily (or even intended to be) *transformative*". This is certainly not due to ignorance, but could well be fostered by the introversion of the communicative community, as Innes (1995, 183) has remarked (see Chap. 2). Rydin was more specific: "Some concerns are general to the use of deliberative and collaborative processes within planning. These have been well rehearsed … and can be summarized as: a lack of specificity as to how the theory of these processes should be put into practice; doubts as to the abilities of planners to undertake such processes successfully; the potential for powerful interests to subvert the processes; and the inability of such processes to handle conflicts of interests and generate a consensus or agreement in the face of such conflicts" (Rydin 2007, 55). The essential component of participation can be manipulated in advance. This criticism is summarized by Thorpe (2017): "Planners exercise power in what they choose to include as matters for debate, which stakeholders they engage with, which participatory tools and techniques they use, which background materials they research and which they provide to participants, in how they use knowledge, and in how they advise the government of the outcomes and range of options produced through participatory processes. Despite the collaborative rhetoric, such critics argue, planning remains a closed process". Equally revealing is a comment by Bolan on the process involving a planner and a group of stakeholders: "The members of this group are very likely to be concerned about what they conceive to be a problem in living or working conditions (or in social relations among them). But the group may be quite varied in their perception of the problem being addressed and of the trustworthiness of individual roles of other participants in the pursuit of a solution to their problem. The members of the group may be of different social class, may involve both male and female outlooks on problem definition, and may be of quite different experience, educational background, and talent. Many of the authors writing about planning theory present case studies of just such groups" (Bolan 2017, 207–208). In a group internally differentiated with divisions of great magnitude "the potential for conflict may be very high" (op.cit., 211).

The next generation of inquirers, that Innes was looking for (see Chap. 2), is represented by adherents to the radical school and to the post-politics approach who denounced the consensus approach as a sham. As Ozdemir and Tasan-Kok (2019) explain, "the consensus approach has been disparaged as a tool for taming people, accused of strengthening established agendas and giving them a pseudo-democratic look, whereby disagreements are neither heard nor accommodated but rather circumvented or ignored. Consensus-building approaches in urban policy and planning are criticised for excluding and marginalising contestation and conflict, but also for lending itself to neoliberal instrumentalisation, which leads to exclusionary practices". The authors propose "a constructive approach to consensus-building, based on the precept of planning as an instrument for actual democracy. The planner is perceived as a human being, is situated in a consensus building context and understood in terms of how s/he responds to disagreement". They draw their example from the Dutch experience of the "'polder model', which is defined as "harmonious patterns of interaction between social partners".

We shall have the opportunity in Chap. 4 to explore further the claims of the radical current. Much more serious than the comments on consensus is the attack on the communicative theorists that they fail to grapple with the omnipresent effects of power structures, which ostensibly will be the concern of the "next generation of inquirers", to use again Innes' phrase. Holden and Scerri (2015), from a pragmatic perspective, put their fingers precisely on this omission: "Critical realist voices hit pragmatists and communicative action theorists with the warning that to focus on the creative and innovative potential of social groups engaged in experimentation and experience is to ignore the heavy hand of power that represses all such efforts in the 'real world'. In this view, compromises, when they occur in real instances, are typically plagued by systematic exclusions and imbalances of representation and voice". From the author's personal view in this book, what is most interesting is the suggestion for a theory of compromise, which Holden and Scerri borrow from Iris Young and Bish Sanyal. Young offered a new perspective on the conceptualization of groups, their identity, and their differentiation by historic or cultural connections: "A structural social group is a collection of persons who are similarly positioned in interactive and institutional relations that condition their opportunities and prospects" (Young 2016, 397).

Sanyal (2002) and, much later, Hoch (2019) offer a most welcome opportunity to link the notion of compromise theory with a pragmatist strand of thought. *Symvivasmos* (the Greek term for compromise) is invoked even in a Platonic dialogue, which makes the ambition to erect on it a new theoretical approach even more challenging. Without lofty moral pretentions, the present author will have the chance to show how adopting a "compromise planning" attitude best explains the shoddy reality of planning in a given historical, social, political and economic context (see Chaps. 9, 10 and 11). It is naturally true that "context" can be such an all-embracing term that may drive one into a cul-de-sac, in which the notion of planning would be hard to explain. Forester has jokingly written that he used to think "that 'context' was a weasel word, invoked – either out of dubiously thin air or allegedly vast historical experience – to end a conversation or a line of argument" (Forester 2016, 169). He referred to situations where the success or failure of planning was seen as dependent "on context".

In a discussion on compromise and on deliberative communication there is a danger of walking along the slippery path of populism, which, in practice, is not far from the truth. Sager discusses communicative planning and communicative planning theory (CPT) in this light: "Communicative planning is a participatory and dialogical endeavour involving a broad range of stakeholders and affected groups in socially oriented and fairness-seeking developments of land, infrastructure or public services … According to CPT, the process should explore the potential for cooperative ways of settling planning disputes and be open, undistorted, truth-seeking and directed at mutual understanding. Little in this definition of communicative planning is in line with [some of] the characteristics of authoritarian populism. What nevertheless suggests an association of CPT with populism is the opportunities given for local lay people to take part in processes and negotiations concerning plans for their area. They can play out nativist cultivation of local values in the process, and besides air anti-elitist sentiment against investors, developers and technocratic expert planners" (Sager 2020, 93). Populism can be a major ingredient of state-society relations, as in the case of Greece (Chap. 7). The fear has been expressed by the present author that, in a populist environment, we imperceptibly internalize the message, in the name of a non-autocratic and bottom-up approach, that a more democratic planning governance is better served not by elected representatives but by a host of non-universally elected representatives of various organizations, pressure groups, or even private associations. In this manner we might open the back door to allow in a perception of planning which dilutes its essence and makes it colourless, tasteless and devoid of substance (Wassenhoven 2002).

If communicative planning does not provide the basis for compromise, does it pave the way towards a really transformative form of planning, as Sandercock advocates? The answer for Baeten (2018, 112) is negative, because of the tendency for communicative and collaborative planning to become a vehicle for, and a lubricant of, neoliberalism, with planners acting as its mandarins. Mäntysalo and Bäcklund (2018, 243), too, ask the embarrassing question: "Does the intention of communicative planning theory to enable bottom-up civil society participation, and relax bureaucratic control of planning processes, inadvertently serve neoliberal aims?" If the communicative approach is highjacked by neoliberal entrepreneurial interests the prospects of a pluralist representation of interests fade. The argument rests on the assumption that neoliberalism has secured a hegemonic position, an assumption disputed in an article by Sager (2015): "When an ideology becomes hegemonic, it changes what is generally accepted as truth and common-sense knowledge. The relationship between knowledge and action will be affected, indicating that a new mode of planning might emerge … In the case of neo-liberal hegemony, public planning's archetypal subject areas – such as externalities, common goods provision, information deficiencies and equity concerns – will be handled by more market-oriented means. In the discussion about the alleged hegemony of neo-liberalism in urban planning, this article sides with the view that neo-liberalism is an important influence, but finds reason to question its hegemony".

3.2.5 *Uneasy Similarities of Communicative and Rational Planning*

One is entitled at this stage to wonder whether communicative planning does indeed differ from rational planning, with all its past sins and oversimplifications. This is a point referred to earlier in Chap. 2 on typologies. It seems that the present author is not the only one to speak of similarities between these two theoretical models. Davoudi (2018, 22) quotes M. Huxley and O. Yiftachel, who wrote: "The communicative planning field […] shares […] a tendency to see planning as a mainly procedural field of activity […]. There is the same sense of searching for the right decision-rules be they rational-comprehensive or rational-communicative, universal or local". Some commentators argue that these antagonistic models of planning can be used simultaneously: "[S]patial planning problems are primarily tackled using two concepts of rationality, classical analytical rationality on the one hand and communicative rationality on the other hand. These approaches are utilized to varying degrees depending on the situation at hand" (Diller et al. 2018). Innes and Booher (2015) also imply a sort of reconciliation between rational and communicative forms of planning, by focusing on the sensitive issue of "knowledge" and its acquisition: "The dichotomization of knowledge into scientific and lay categories disempowers both planners and community members, making planning goals difficult to achieve and keeping the community marginalized from the decisions that affect their lives. This opposition between hard 'scientific' data and the more qualitative 'soft' knowledge in the community is unnecessary and counterproductive. These knowledges can be integrated through communication power without making one trump the other. With communicative action among scientists, planners, and laymen, all can learn and take advantage of what the other does best, and the resulting knowledge can be both more accurate and more meaningful".

A synthetic disposition and a conciliatory tone is evident in the work of Yvonne Rydin. In an early paper, Rydin sought "to bridge some of the debates between modernist planning theorists and contemporary postmodern planning theory by arguing for the specific contribution of knowledge within planning while still seeing knowledge as socially constructed, multiple and constituted in the form of claims, open to contestation and recognition" (Rydin 2007, 66). Rydin's main contribution concerns the "actor /network theory" (ANT), a theme also developed by Boelens (see above in this chapter). Boelens exploits sociological theories about ANT, notably by Bruno Latour, and claims that ANT "starts off with actors and relations (or networks) – not only between each other but also between the human and non-human actors, for example, the specific characteristics and entities of the locality – in order to reassemble them in such a way as to become more innovative, enforceable and associative" (Boelens 2010, 36). In this context, "geography (and its application, planning) becomes the science, or skill, of the analysis (and/or planning) of heterogeneous associations or actor-networks in time and space. Spatial relations are reduced to network relations and spatial planning is understood as a process of network building, in which entities of various kinds are assembled in ways that allow networks to undertake certain functions" (op.cit., 37). The effect of networks is explained by Parker and Doak (2012, 49): "Researchers have recognised how networks can act to crumple or 'pleat and fold' space and time. By taking the effect of networks into consideration, planners can think about policies and plans for particular bounded areas – asking, for example, how discrete areas are connected and influenced by each other. This kind of analysis has been made possible primarily because of an increased awareness of the complex assemblage of relations that structure and shape places, and the socio-economic and environmental attributes affecting quality of life and sustainability, as key objectives of spatial planning".

Out of these premises, Boelens develops his own behavioural actor-relational view, which "demands a prominent role for a more neutral moderator and an open medium in which to sketch opportunities … [T]he approach is not about actors as such, in the broad sense of interactive planning (i.e., all affected parties), but about *leading actors*, who are primarily encountered in the world of human action" (op. cit., 41). It is here that Rydin objects: "[I]n considering the networks of human actors and non-human entities and how these operate, the key actant (to use the generic term) may be either human or non-human. However, Boelens quickly rejects this key assumption and focuses instead on human actors within networks. He sees the inclusion of non-human entities within networks as the last innovative feature of ANT rather than the first" (Rydin 2010, 265). Thus, Boelens "moves quickly on to a network perspective that is centred on human actors, with non-human entities – climate, environment, landscape, housing, stations, cars, budgets, materials, planning instruments inter alia – being described instead as 'significant *f*actors of importance' … In my view this is to miss the opportunity of considering how ANT could be directly relevant to planning theory" (op.cit., 266).

Rydin sketches her own approach, in which the key is "the identification of the appropriate actants. In my view this is achieved not by focusing on the material context for planning activity – the *f*actors – but rather on the specific socio-technical systems that planning practice seeks to influence. Planning practice is about trying to shape the social, economic and physical world within which we live. But it does so by engaging with specific systems" (ibid.). When, later, Rydin elaborated further the concept of ANT, and after pointing out that it was originally introduced for a better appreciation of how scientific knowledge is generated, she made it clear that "the central image at the heart of ANT is the network, an image that suggests linkages between different elements" (Rydin 2018, 303). At this point, Rydin opens a side perspective, which links the actor/network vista with that of associations and assemblages. "Within ANT the term 'associations' is commonly used to describe such relationships. Assemblage thinking focuses almost exclusively on such associations and how they are built …; this is the key point of contact between ANT and assemblage approaches" (ibid.). Wohl (2018) notes that "the notion of 'assemblage' is roughly equivalent to the concept of 'Emergence' in complex adaptive systems theory". This is an interesting additional justification for including ANT and assemblage theory in a section on the "uneasy similarities of communicative and rational planning". Given that both networking and systems theory are essential ingredients of the rational current, while the emphasis on actors is linked to the communicative current, one would not be off the mark if ANT and assemblage thinking are positioned in this interface.

Based on the work of Deleuze and Guattari, and on that of DeLanda, Van Wezemael believes that the conception of assemblages provides "a valuable basis to conceptualize governance networks, allowing for the heterogeneity of their 'parts'. Assemblage theory is based on the understanding that theories, which begin with 'actual' physically perceived systems, do not adequately explain the origins of those systems. It therefore subordinates the term 'real' to the terms 'virtual' and 'actual'. What we usually understand as the 'real world' in an assemblage perspective is merely the experience of the 'actual' state of final products: they are devoid of their virtual becoming" (Van Wezemael 2008, 166). He suggests, "that assemblage theory offers a basic ontogenetic perspective and a source of concepts in network governance. It builds on heterogeneity and non-linear relations, it deals with the unpredictability of processes; and it adequately conceptualizes relations of government agencies, economic organizations, social groups and so on as contingent, exterior relations" (op.cit., 180). He later expanded his analysis (Van Wezemael 2018) and confirmed the intention to bypass what he saw as the weaknesses of prior theories. For him, "an assemblage can conceptually and empirically be grasped along two axes. The first one refers to the processes of production (de-stabilization and re-stabilization); the second one refers to the question what it can do, which role it can play" (Van Wezemael 2018, 329). On the basis of assemblage theory, "planning theory can (1) avoid the essentialism that characterizes naïve realism and forms of rationalism (collaborative planner's critique toward rationalism), and simultaneously (2) remove one of the main objections that non-realists make against the postulation of an autonomous reality as it offers an integration of language and negotiation into a realist and materialist worldview" (op.cit., 328).

It was mentioned in Chap. 2 that Allmendinger (2017) refers to relativism and to reliance on exogenous social theories and gives as examples regime theory, actor-network theory, and assemblage theory, which, he suggests, "do not represent an approach to planning in the same way that the systems or collaborative approaches do". The author of this book has discussed regime and regulation theories elsewhere and argued that regime theory (Lauria 1997) provides theoretical support for the view that governance is not simply about social control, but rather aims at local objectives which can be better attained by empowering regimes constituted by partnerships and alliances between local government, the private sector, non-governmental organizations and social groups. In this sense, a fundamental goal of regime theory, e.g., at the level of a city, is a better understanding of the political dimension of local economic development. This is consistent with regulation theory which accepts that the capitalist system has shifted to a different mode of regulation, from Fordism to post-Fordism, and to a globalized economy. The decentralization and diffusion of this economy forces local administrations to adopt a more extrovert and entrepreneurial approach and cooperate with the private sector. The bottom line is that regulation theory was trying to respond to temporal and spatial changes of capitalism (Wassenhoven 2018, 269–270; see also Judge et al. 1995).

The task of finding a bridge between rational and collaborative theories and of reconciling various intruding theories was also undertaken by Davoudi (2015): "Instead of thinking about knowledge as having an instrumental place in the planning process (i.e., to inform action), it is more useful to think about planning as a process of knowing and learning". An unexpected case for rapprochement is found in the liberal philosophy of John Rawls (1995 and 1999), whose theory for "justice as fairness" had a great influence on communicative theory. "Rawls elaborates his theory through a process of seeking what he calls 'reflective equilibrium'. He is, I believe, reflecting on the experience of *injustice* in actual societies and trying to derive principles of justice in which these felt injustices could be avoided. Equilibrium is reached when our intuitive judgements about what justice could mean coincide with the principles which we put into words and sentences to express such judgement" (Low 2020, 192). Equilibrium is a critical term, because Rawls treads on a road of conciliation between modernism and postmodernism. "His political liberalism provides a pragmatic *normative rationale* for the sort of public planning which even planners who sound rather postmodernist ... often end up advocating, an approach to planning which is neither modernist nor postmodernist. It responds to the postmodernist critique of planning by eliminating the scientistic features of modernism, but retaining its positive liberal (humanistic) aspects" (Stein and Harper 2005, 166).

Perhaps this conciliation (shall we say "compromise"?) can be achieved through a "critical realist" approach, as Naess (2015, 1240–1241) seems to maintain, when he concentrates his attention on goal-oriented planning: "Distinct from incremental and some communicative planning approaches, where goal-oriented planning and instrumental rationality are typically regarded with great scepticism, critical realist planners would consider it possible to develop action alternatives likely to contribute positively to the achievement of a given goal". An accusation of lack of realism may be substantiated if communicative planning can be shown to ignore important procedures, parameters and tools. Palermo and Ponzini claim that "the 'collaborative turn' runs the risk of becoming an ideological simplification or even a fallacy". The approach, in their view, is devoid of thinking about the social and institutional conditions necessary for guaranteeing the expected results (Palermo and Ponzini 2010, 9 and 79). "Communicative planning has been widely criticized for having little to do with the official legal procedures and for low-quality spatial solutions. It has also been blamed to be an empty concept, referring to an action that in itself has no content" (Damurski and Oleksy 2018). But it was also blamed for ignoring the burdens imposed on citizens whose participation is trying to engage: "[M]uch existing theory does not fully acknowledge the demands it would make of these ideal citizen-subjects and has therefore sidestepped a significant challenge for the project of shaping more democratic planning practices … Drawing attention to the presence of a range of hidden costs, I have argued that they raise significant questions about the demands that are made of people in the name of democracy" (Inch 2015, 421–422).

Intelligent use of digital technologies might, although not certainly, reduce this burden and increase the efficiency of collaborative processes. "For too long, collaborative planning theory has been blind to what has been obvious in the literature on planning technology: that information technologies are sources of both insight and social oppression … Unlocking the insight and avoiding the oppression requires much more than 'complementing' the technical work of planners with new social practices …; it requires a rethinking of how planners use and manage technology. The role of planning technology must be recast from providing access to neutral information to facilitating a process of social inquiry" (Goodspeed 2016). Better access to neutral information is unfortunately a tall order. "When the possibility of conflicting interests is considered, then the meaning and implications of information becomes contested and information may be produced by different parties in a partisan way to support their case. Not only is society made up of conflicting interests, however, but also those interests have differential access to resources and degrees of control and influence. In this context information becomes a means of power and control, and it may even be systematically used to bolster and legitimate the position of the controlling elite" (Stephenson 2000, 108). This use and misuse of information may in itself become a major impediment of communicative planning. Major participation exercises have often provided the arena for powerful stakeholders who commission highly specialized experts to mobilize information and knowledge so that they can tip the balance of, e.g., a public inquiry, in their favour.

There is here a genuine dilemma; access to such information technologies may prove an impediment in the process of deliberation and consultation with deprived minorities, whose situated knowledge the collaborative planner is so keenly hoping to acquire. The demands made on people, to which Inch was alluding, may well involve the forbidding costs of access to technology. On the other hand, the issue of costs concerns not only the deprived, marginal groups, but also those who wish to manipulate the groups whose participation the planner wants to secure. These "transaction costs", as Sager suggests, may prove decisive. "The terminology of transaction cost theory is adopted from economics, and it is suggested that a core task of the critical and communicative planner is to apply cost-raising strategies against agents in the planning process who wield power in ways working against the public interest and to lower the transaction costs of deprived groups whose interests are easily ignored" (Sager 2006, 224).

3.2.6 Complementarities and Integration of Theories

The obstacles on which communicative approaches stumble are plenty but not unforeseen. This is not a reason for rejecting their opposition to technofix logics which for a long time have plagued the rational model. There may be a reason, however, for opening a dialogue about their complementarity, as suggested by those who detected similarities. Nigel Taylor (1999, 167–168) had made a plea in this direction, even though he recognized the existence of "irrelevant" theories. Archibugi, who had made reference to the "rigmarole" of theoretical debate (Archibugi 2008, 55), concedes "that 'planning science' needs (or would benefit from the attempt) to aim at a more integrative methodology. This may well resuscitate the moribund condition of the discipline, fractured as it stands (or falls) now by partial, disjointed, and fragmentary approaches that talk past each other instead of fostering real progress" (op.cit., 101).

Bitter theoretical disputes and unfounded conflictual stances can only perpetuate a proliferation of adversarial positions and a multiplication of explanations closely related with the indifference shown by practicing planners. As Dryzek and Dunleavey (2009, 300), somewhat ironically wrote, "'frames', 'perspectives', 'paradigms', 'discourses', and 'ideologies' are concepts that now pervade attempts to explain the content of public policy". "Today planning theory seems to have become a set of dividing discourses. People talk past one another. Blame, criticism, and incivility often crowd out scholarly dialogue and inquiry … Theorists belong to discourse communities which employ different languages and methods toward different ends. Students are often confused and frustrated, craving a way to make sense of the differences. While the brouhaha may have started as a war over turf and over which views will be dominant, the result today is that we, as theorists, have little ability to learn from our differences" (Innes and Booher 2015). This situation is very damaging to the image of theory, especially in the eyes of students and practitioners. Brooks is presenting a bleak situation: "Planning theory, I have written elsewhere, 'is a term that strikes terror in the hearts of many planners; it conjures up images of esoteric word games played by planning educators who have little knowledge of what practicing planners actually do'. Robert Beauregard paints an even bleaker picture, observing that planning theory 'is generally held in low esteem. … Within academia, planning theory is marginalized; within practice, it is virtually ignored'. This state of affairs is unfortunate, to say the least … [a] sound body of theory is an essential component of the planning profession …" (Brooks 2017, 21). Palermo and Ponzini (2010, 35) too do not mince their words: "[I]t seems that we might refer to the rise and crisis of planning theory. An ideological and scientific programme – or more than just one – failed. We can acknowledge this without regrets because other directions begin to open". Allmendinger and Haughton (2019) look for an answer in "opening-up planning" and in a process of meta-governance. They define the latter as "the processes for shaping governance mechanisms, setting out remit, rules, membership, accountability and evaluation criteria". As Dryzek and Dunleavey (2009, 140) wrote: "Governance was once synonymous with government, but the more recent usage of the term involves something a bit different: the production of collective outcomes (in the contest of public problems) that is not controlled by centralized authority". The concept of governance is interesting because, on the one hand, it favours a bottom-up approach, but on the other, it received a top-down blessing, as in the case of the European Union. "Multi-level governance can involve authority moving up from the national level as well as down", as in the EU (op.cit., 147). "The importance of governance and meta-governance in planning", Allmendinger and Haughton argue, "becomes clearer if planning is viewed as a series of strategy-making processes which aim to shape and coordinate the actions of others, for instance home builders and infrastructure providers. In governance terms, local plans are more likely to achieve their objectives if they are developed by working towards agreement across diverse actors, not least those with the capacity to facilitate or inhibit future development".

In view of this chaotic state of theory, the distinction that Baum makes between "espoused" theory and "theory-in-use" may be helpful in overcoming the confusion. "The distinction between 'espoused theory' and 'theory-in-use' helps understand the meanings of theorists' writings as part of their practice … 'Espoused theory' refers to public statements about how one should act, how others should act, or how one or others do act. 'Theory-in-use', by contrast, comprises principles inherent in action, whether or not the actor is aware of or intends them. People are conscious of their espoused theory but usually not of their theory-in-use. Moreover, the theories typically follow different logics. Public espousals are actions for an audience. Accordingly, people tend to describe their own action and to advise others on how to act in ways that imply they know how to act effectively. They are likely to present their actions as reasonable, high-minded, coherent, and successful. In reality, however, people often act inconsistently, selfishly, uncertainly, mistrustfully, and even unsuccessfully. Argyris and Schön believe these incongruities between espoused theories and theories-in-use reflect two intentions: one is to deceive others about one's goodness and efficacy: the other is to deceive oneself" (Baum 1996, 372).

3.3 Pragmatism

Before Baum's "theory-in-use" we had encountered, in the previous chapter, Hoch's pragmatist approach and Alexander's contingency model. Reference was made repeatedly to the notion of compromise, a theme on which we will concentrate in later chapters, using Greece as a case study. But, after a section on complementarities between theories and the wish of many theorists for greater integration, this author is convinced that we must return to pragmatism, as a unifying philosophy. Hoch will be here our companion. It has been made clear that, unlike a score of distinguished planning theorists, this author does not feel obliged to return to the roots of philosophy and of social science, simply because he does not feel capable to do so. Nevertheless, he was impressed by a definition found in the Encyclopaedia Britannica in a lemma on the American pragmatist philosopher John Dewey:

"Dewey's particular version of pragmatism, which he called 'instrumentalism,' is the view that knowledge results from the *discernment* of correlations between events, or processes of change". This is knowledge which gives "access to truth about society, as distinct from mere opinion … [T]he most central condition for the development of true knowledge is the process itself of free inquiry and experiment" (Low 2020, 23–24). Equally impressive is a comment on the views of the other great American pragmatist philosopher Richard Rorty, according to whom, "being objective or impartial negates the characteristic of man and it is no longer maintainable in the society … Similarly, Rorty believes that man ought to abandon any petitions for truth due to the fact that one can never have a proper understanding of the world. Therefore, man should never attempt to create a normative understanding of truth".[8] John Rawls, to whose "reflective equilibrium" I referred earlier, was equally loath of a claim of universal truth. "For Rawls there is no single transcendental truth. There are a number of 'truths', but there is a single *political* rule specifying the conditions under which people holding different 'truths' can cooperate for reciprocal advantage" (Low 2020, 195). It is apparently an open debate whether Rawls' liberalism belongs to a pragmatist tradition, but this is not relevant in the present discussion. For a pragmatist view, we can turn to Hoch's assistance: "Instead of conceiving cognition as mental modelling that represents the world, the pragmatist approach conceives cognition as actively using the world to inform and assess judgments about what to do" (Hoch 2019, 11). The relationship of pragmatist philosophy with action (i.e., "what to do") is stressed by Bolan: "Pragmatism, in general, has as its main thrust the notion of the human being as agent – as a *doer*. Human experience is viewed as an active 'transaction' between the human organism and its social and physical environment. The orientation of these transactions is contingent and anticipatory" (Bolan 2017, 156). "Simply stated", wrote Harrison (2001, 74), "pragmatism is a belief that the truth or value of an idea is to be found within its usefulness in practice and not in the accuracy of its relationship to some essential reality. Pragmatism therefore places focus on the consequence or outcome of ideas rather than on the connection of these ideas to an external or transcendental truth. It directs attention to learning through experience and by doing". Hoch recognizes that "pragmatism can seem anachronistic and uninspired", but refuses to be entangled in a conception of planning which strives to bridge the gap between knowledge and action: "The claim that we need a strong utopian vision to provide a compelling attachment to a desirable future place sets off alarms for the pragmatist who rejects that ideas compel consent" (sic) (Hoch 2019, 65). Utopian visions may divert attention from immediate actions. "Instead of conceiving the public interest as aggregate instrumental benefits, which might ensue from a policy or as a form of legal or moral right inspiring and directing policy compliance, the pragmatist focuses on the meaning of consequences or rules for practical judgments about what to do in specific situations" (Vidyarthi and Hoch 2018, 631). Vidyarthi (2019) adds: "Employing a pragmatist approach, I reconceived the disparate local initiatives as purposeful

[8] See https://ivypanda.com/essays

future-oriented efforts (or plans) and not just informal spontaneous reactions contesting formal planning efforts". A further problem is that a linear projection into the future can be received by different audiences in diverging ways. "For the pragmatist each audience creates its own utopia, scenario and plan within the historic contours of culture and place" (Hoch 2019, 72). This is an observation which could be made by a communicative planner, who would also probably agree with the next statement, which, however, could equally be acceptable to a rational planner: "The pragmatist does not believe we can craft a utopia, scenario or plan for a place without already imagining and deploying the active imagination we each use to conceive and compare options for practical attention and judgement" (op.cit., 78). In addition, rational and communicative planners might partly agree that "rational analysis may offer objectivity and precision, but like the judgment of the blind people, it sacrifices context and continuity. A pragmatic outlook embraces context and seeks continuity among diverse viewpoints. It avoids the separation between analysis and action. The Rationalist asks what we need know to assure our analysis is correct and certain. The pragmatist asks what we need to know to cope with the problems we face" (op.cit., 111). For a pragmatist, Ferreira (2018) remarks, "moving from an objectified understanding of reality and problems towards one where attention is given to how individuals conceptualise reality, is a transformational step. It replaces irreconcilable debates with pragmatic inquiries about why different individuals and organisations defend different arguments". Hoch keeps a careful balance, maintains ties with all the major currents of planning, and avoids dangerous divisions. "The pragmatists abandon the grand separation between theory and practice. The Rationalists insist upon this distinction believing that the power of ideas will trump the uncertainties we face. We need theory to guide practice. The pragmatists, like many current postmodern thinkers, worry that the quest for certainty becomes a power trip as those with little democratic sensibility use Rationality to subject others to purposes that masquerade as necessary and inevitable conditions" (Hoch 2019, 112). Planners, in this perspective, do not behave as scientists engaged in research. "[P]lanners can foster activities that encourage decisive deliberations among the diverse assortment of antagonists and allies. A pragmatic outlook supports public deliberations about a plan, not as civic window dressing for professional expertise, but a crucial source of practical comprehension useful for building consensus" (op.cit., 126). Planners are not charged with the task of imposing their rationalist beliefs. "Instead of trying to tame the contingency and uncertainty of the planning enterprise through a privileged Rational method and Professional expertise, we should try to comprehend and organize the different viewpoints and the ambiguity among them by adopting a pragmatic outlook" (op.cit., 128). Hoch, it is apparent, walks along a median separating line. Ambiguous causes, purposes and explanations co-exist and the planner must weave them in a fabric which is rationally woven, yet can be worn by the members of a multitude of audiences, singled out through a deliberative process. "The plan for a single uniform public should be recast as a complex combination of many proposals for the future coordinated through creative provisional experimentation" (op.cit., 153). Is this a futile, even if pragmatic, ambition? It may be seen as the agonizing effort of a tightrope walker, yet this is the effort of a pragmatist planner, faced with a tangle of contradictory views, situations, and beliefs, which can be alternately rigid and fluid, depending on shifting interests and/or established convictions. Hoch's pragmatism is, for the present author's purposes, a fitting introduction to compromise planning (see Chap. 9). For more on that the reader's patience is requested.

Epilogue

Writing this chapter was a dizzying adventure for the writer, which hopefully provides a useful picture for the benefit of the reader. The chapter included a large number of quotations from past theoretical work, which were often difficult to incorporate in a manner that would make a variety of complex concepts more lucid. The rational and communicative currents are the solid foundation of most past theoretical planning treatises, books and articles and cannot be easily summarized in a limited number of pages. Still, without some basic appreciation of their value, one cannot pretend to make a further contribution to planning theory. The geographical "concentration" of the origins of these theories should not escape our attention. The theorists mentioned in the chapter write on the basis of the experience of advanced industrial societies, which have their own distinctive histories, economic systems, social structures, governance and administrative practices, public-private interactions, spatial organization, and accumulated planning knowledge. The philosophical, economic, sociological and technical viewpoints that found their way into the subject matter and language of planning had a clear "western" character. This was bound to affect planning theory, which in its mainstream, as described in this chapter, was often critical, but remained on the whole within the limits of the capitalist system characterizing the industrial West. The radical current, to which we turn in the next chapter, openly challenges this orthodoxy. So does, an approach which looks at planning theory from a Southern perspective.

References[9]

Ackoff RL (1969) Systems, organizations, and interdisciplinary research. In: Emery FE (ed) Systems thinking. Penguin, Harmondsworth

Albrechts L (2002) The planning community reflects on enhancing public involvement – views from academics and reflective practitioners. PTP 3(3):331–347

Albrechts L (2003) Reconstructing decision-making: planning versus politics. PT 2(3):249–268

[9] **Note**: The titles of publications in Greek have been translated by the author. The indication "in Greek" appears at the end of the title. For acronyms used in the references, apart from journal titles, see at the end of the table of diagrams, figures and text-boxes at the beginning of the book.

Abbreviations of journal titles: AT Architektonika Themata (Gr); DEP Diethnis kai Evropaiki Politiki (Gr); EpKE Epitheorisi Koinonikon Erevnon (Gr); EPS European Planning Studies; PeD Perivallon kai Dikaio (Gr); PoPer Poli kai Perifereia (Gr); PPR Planning Practice and Research; PT Planning Theory; PTP Planning Theory and Practice; SyT Synchrona Themata (Gr); TeC Technika Chronika (Gr); UG Urban Geography; US Urban Studies; USc Urban Science. The letters Gr indicate a Greek journal. The titles of other journals are mentioned in full.

Albrechts L (2006) Bridge the gap: from spatial planning to strategic projects. EPS 14(10):1487–1500
Albrechts L (2008) Strategic spatial planning revisited: experiences from Europe. Keynote address at the 3rd Regional Development and Governance Symposium. Mersin, 27–28 November 2008 (Power Point presentation)
Albrechts L (2018) Strategic planning: ontological and epistemological challenges. In: Gunder M, Madanipour A, Watson V (eds) The Routledge handbook of planning theory. Routledge, London, pp 28–40
Albrechts L, Barbanente A, Monno V (2019) From stage-managed planning towards a more imaginative and inclusive strategic spatial planning. Environ. Plan. C Politics Space 37(8):1489–1506
Alden J, Morgan R (1974) Regional planning: a comprehensive view. Leonard Hill, Leighton Buzzard
Alexander ER (1992) Approaches to planning: introducing current planning theories, concepts and issues. Gordon and Breach, Luxembourg
Alexander ER (2002) The public interest in planning: from legitimation to substantive plan evaluation. PT 1(3):226–249
Alexander ER (2005) Institutional transformation and planning: from institutionalization theory to institutional design. PT 4(3):209–223
Allen J (1996) Our town: Foucault and knowledge-based politics in London. In: Mandelbaum SJ, Mazza L, Burchell RW (eds) Explorations in planning theory. Center for Urban Policy Research/Rutgers, New Brunswick, pp 328–344
Allmendinger P (2001) Planning in postmodern times. Routledge, London
Allmendinger P (2002) Planning theory. Palgrave, Basingstoke
Allmendinger P (2017) Planning theory, 3rd edn. Palgrave, Basingstoke
Allmendinger P, Haughton G (2019) Opening up planning? Planning reform in an era of 'open government'. PPR 34(4):438–453
Allmendinger P, Tewdwr-Jones M (eds) (2002) Planning futures: new directions for planning theory. Routledge, London
Altshuler A (1995) The goals of comprehensive planning. In: Stein JM (ed) Classic readings in urban planning. McGraw-Hill, New York, pp 81–105. (first published 1965)
Angyal A (1969) A logic of systems. In: Emery FE (ed) Systems thinking. Penguin, Harmondsworth, pp 17–29. (originally published 1941)
Archibugi F (2008) Planning theory: from the political debate to the methodological reconstruction. Springer, Milan
Arnstein SR (1995) A ladder of citizen participation. In: Stein JM (ed) Classic readings in urban planning. McGraw-Hill, New York, pp 358–375. (first published 1969)
Baeten G (2018) Neoliberal planning. In: Gunder M, Madanipour A, Watson V (eds) The Routledge handbook of planning theory. Routledge, London, pp 105–117
Banfield EC (1973) Ends and means in planning. In: Faludi A (ed) A reader in planning theory. Pergamon Press, Oxford, pp 139–149. (first published 1959)
Barca F (2009) An agenda for a reformed cohesion policy: a place-based approach to meeting European Union challenges and expectations. Independent report commissioned by the EU Commissioner for Regional Policy
Barrett S, Fudge C (eds) (1981) Policy and action. Methuen, London
Batty M, Hutchinson B (eds) (1983) Systems analysis in urban policy-making and planning. Plenum Press, New York
Baum H (1996) Practicing planning theory in a political world. In: Mandelbaum SJ, Mazza L, Burchell RW (eds) Explorations in planning theory. Center for Urban Policy Research/Rutgers, New Brunswick, pp 365–382. (revised version of 1988 publication)
Beauregard RA (1996a) Commentary – advocating pre-eminence: anthologies as politics. In: Mandelbaum SJ, Mazza L, Burchell RW (eds) Explorations in planning theory. Center for Urban Policy Research/Rutgers, New Brunswick, pp 105–110

Beauregard RA (1996b) Between modernity and postmodernity: the ambiguous position of U.S. planning. In: Campbell S, Fainstein S (eds) Readings in planning theory. Blackwell, Oxford, pp 213–233. (first published 1989)
Beauregard RA (2005) Introduction: institutional transformations. PT 4(3):203–207
Beauregard RA (2020) Advanced introduction to planning theory. Elgar, Cheltenham
Beckman N (1973) The planner as a bureaucrat. In: Faludi A (ed) A reader in planning theory. Pergamon Press, Oxford, pp 251–263. (first published 1964)
Benko G, Strohmacher U (eds) (1997) Space and social theory: interpreting modernity and postmodernity. Blackwell, Oxford
Berry BJL, Horton FE (1970) Geographic perspectives on urban systems. Prentice-Hall, Englewood Cliffs
Boelens L (2010) Theorizing practice and practising theory: outlines for an actor-relational-approach in planning. PT 9(1):28–62
Bolan RS (1973) Community decision behavior: the culture of planning. In: Faludi A (ed) A reader in planning theory. Pergamon Press, Oxford, pp 371–394. (first published 1969)
Bolan RS (1996) Planning and institutional design. In: Mandelbaum SJ, Mazza L, Burchell RW (eds) Explorations in planning theory. Center for Urban Policy Research/Rutgers, New Brunswick, pp 497–513
Bolan RS (2017) Urban planning's philosophical entanglements: the rugged, dialectical path from knowledge to action. Routledge, London
Bolan RS, Nuttall RL (1975) Urban planning and politics. Lexington Books/D.C. Health, Lexington, MA
Breheny M, Hooper A (eds) (1985) Rationality in planning. Pion Ltd, London
Bridge G, Watson S (eds) (2002) The Blackwell city reader. Blackwell, Oxford
Bridge G, Watson S (eds) (2003) A companion to the city. Blackwell, Oxford
Brindley T, Rydin Y, Stoker G (1996) Remaking planning: the politics of urban change. Routledge, London
Brooks MP (1996) Planning and political power: toward a strategy for coping. In: Mandelbaum SJ, Mazza L, Burchell RW (eds) Explorations in planning theory. Center for Urban Policy Research/Rutgers, New Brunswick, pp 116–133
Brooks MP (2017) Planning theory for practitioners. Routledge, London (first published 2002)
Bruton MJ (ed) (1974) The spirit and purpose of planning. Hutchinson, London
Bulkens M, Minca C, Muzaini H (2015) Storytelling as method in spatial planning. EPS 23(11):2310–2326
Burns T (2004) A practical theory of public planning: the Tavistock tradition and John Friend's strategic choice approach. PT 3(3):211–223
Byrne D (2003) Complexity theory and planning theory: a necessary encounter. PT 2(3):171–178
Camhis M (1979) Planning theory and philosophy. Tavistock Publications, London
Campbell H (2002) 'Evidence-based policy': the continuing search for effective policy processes. PTP 3(1):89–90
Campbell H (2012) "Planning ethics" and rediscovering the idea of planning. PT 11(4):379–399
Campbell S, Fainstein S (1996) Introduction: the structure and debates of planning theory. In: Campbell S, Fainstein S (eds) Readings in planning theory. Blackwell, Oxford, pp 1–14
Campbell H, Marshall R (1999) Ethical frameworks and planning theory. International Journal of Urban and Regional Research 23(3):464–478
Campbell H, Marshall R (2000) Public involvement and planning: looking beyond the one to the many. Int Plan Stud 5(3):321–344
Campbell H, Marshall R (2002) Utilitarianism's bad breath? A re-evaluation of the public interest justification for planning. PT 1(2):163–187
Carr EH (1964) What is history? Penguin Books, Harmondsworth
Castles FG, Murray DJ, Potter DC (eds) (1971) Decisions, organizations and society. Penguin, Harmondsworth
Chadwick G (1971) A systems view of planning. Pergamon Press, Oxford

Chambers R (1997) Whose reality counts? Putting the first last. ITDG Publishing, London
Chapin FS Jr (1965) Urban land use planning. University of Illinois Press, Urbana
Chapin FS Jr, Kaiser EJ (1979) Urban land use planning. University of Illinois Press, Urbana
Churchman CW, Ackoff RL, Arnoff EL (1957) Introduction to operations research. Wiley, New York
Cooke P (1983) Theories of planning and spatial development. Hutchinson, London
Cowan P (ed) (1973) The future of planning. Heinemann, London
Cowling TM, Steeley GC (1973) Sub-regional planning studies: an evaluation. Pergamon Press, Oxford
Cullingworth B, Nadin V (2006) Town and country planning in the UK. Routledge, London
Damurski Ł, Oleksy M (2018) Communicative and participatory paradigm in the European territorial policies – a discourse analysis. EPS 26(7):1471–1492
Dasgupta AK, Pearce DW (1972) Cost-benefit analysis: theory and practice. Macmillan, London
Davidoff P (1973) Advocacy and pluralism in planning. In: Faludi A (ed) A reader in planning theory. Pergamon Press, Oxford, pp 277–296. (first published 1965)
Davidoff P, Reiner TA (1973) A choice theory of planning. In Faludi A (ed) A reader in planning theory. Pergamon Press, Oxford, pp 11–39. (first published 1962)
Davies R, Hall P (eds) (1978) Issues in urban society. Penguin, Harmondsworth
Davoudi S (2015) Planning as practice of knowing. PT 14(3):316–331
Davoudi S (2018) Spatial planning: the promised land or rolled-out neoliberalism? In: Gunder M, Madanipour A, Watson V (eds) The Routledge handbook of planning theory. Routledge, London, pp 15–27
De Leo D, Forester J (2017) Reimagining planning: moving from reflective practice to deliberative practice – a first exploration in the Italian context. PTP 18(2):202–216
de Satgé R, Watson V (2018) Urban planning in the Global South: conflicting rationalities in contested urban space. Palgrave Macmillan and Springer Publishing, Cham
Dear MJ, Flusty S (eds) (2002) The spaces of postmodernity: readings in human geography. Blackwell, Oxford
Department for Communities and Local Government (2009) Multi-criteria analysis: a manual, London
Derycke PH (1979a) Economie et planification urbaines – Vol. I: L' espace urbain. Presses Universitaires de France, Paris
Derycke PH (1979b) Economie et planification urbaines – Vol. II: Théories et modèles. Presses Universitaires de France, Paris
DiGaetano A, Klemanski JS (1999) Power and city governance: comparative perspectives on urban development. University of Minnesota Press, Minneapolis
Diller C, Hoffmann A, Oberding S (2018) Rational versus communicative: towards an understanding of spatial planning methods in German planning practice. PPR 33(3):244–263
Dobrucká L (2016) Reframing planning theory in terms of five categories of questions. PT 15(2):145–161
Donnison D, Soto P (1980) The good city. Heinemann, London
Dror Y (1973) The planning process: a facet design. In: Faludi A (ed) A reader in planning theory. Pergamon Press, Oxford, pp 323–343. (first published 1963)
Dryzek JS, Dunleavey P (2009) Theories of the democratic state. Macmillan International/Red Globe Press, London
Etzioni A (1973) Mixed-scanning: a "third" approach to decision-making. In: Faludi A (ed) A reader in planning theory. Pergamon Press, Oxford, pp 217–229. (first published 1967)
Etzioni A (1975) A comparative analysis of complex organizations: on power, involvement and their correlates. The Free Press, New York
Fagence M (1977) Citizen participation in planning. Pergamon Press, Oxford
Fainstein SS (2002) The changing world economy and urban restructuring. In: Fainstein SS, Campbell S (eds) Readings in urban theory. Blackwell, Oxford, pp 111–123. (first published 1990)

Fainstein SS (2016) Spatial justice and planning. In: Fainstein SS, DeFilippis J (eds) Readings in planning theory. Wiley – Blackwell, Chichester, pp 258–272. (first published 2013)
Fainstein SS, Fainstein N (1996) City planning and political values: an updated view. In: Campbell S, Fainstein S (eds) Readings in planning theory. Blackwell, Oxford, pp 265–287
Faludi A (1973) Planning theory (1984 edn). Pergamon Press, Oxford
Faludi A (1982) Towards a combined paradigm of planning theory? In: Paris C (ed) Critical readings in planning theory. Pergamon Press, Oxford, pp 27–38. (originally published 1979)
Faludi A (1996) Rationality, critical rationalism, and planning doctrine. In: Mandelbaum SJ, Mazza L, Burchell RW (eds) Explorations in planning theory. Center for Urban Policy Research/ Rutgers, New Brunswick, pp 65–82
Faludi A (2000) The performance of spatial planning. PPR 15(4):299–318
Faludi A (2004) The impact of a planning philosophy. PT 3(3):225–236
Ferraro G (1996) Planning as creative interpretation. In: Mandelbaum SJ, Mazza L, Burchell RW (eds) Explorations in planning theory. Center for Urban Policy Research/Rutgers, New Brunswick, pp 312–327
Ferreira A (2018) Towards an integrative perspective: bringing Ken Wilber's philosophy to planning theory and practice. PTP 19(4):558–577
Flyvbjerg B (1998) Rationality and power: democracy in practice. The University of Chicago Press, Chicago/London
Foglesong RE (1996) Planning the capitalist city. In: Campbell S, Fainstein S (eds) Readings in planning theory. Blackwell, Oxford, pp 169–175. (first published 1986)
Forester J (1996a) The rationality of listening, emotional sensitivity, and moral vision. In: Mandelbaum SJ, Mazza L, Burchell RW (eds) Explorations in planning theory. Center for Urban Policy Research/Rutgers, New Brunswick, pp 204–224
Forester J (1996b) Argument, power, and passion in planning practice. In: Mandelbaum SJ, Mazza L, Burchell RW (eds) Explorations in planning theory. Center for Urban Policy Research/ Rutgers, New Brunswick, pp 241–262
Forester J (2000) The deliberative practitioner: encouraging participatory planning processes. The MIT Press, Cambridge, MA
Forester J (2013) On the theory and practice of critical pragmatism: deliberative practice and creative negotiations. PT 12(1):5–22
Forester J (2016) Daunting or inviting: 'context' as your working theory of practice. PTP 17(2):169–172
Friedmann J (1987) Planning in the public domain: from knowledge to action. Princeton University Press, Princeton
Friedmann J (1995) Planning in the public domain: discourse and praxis. In: Stein JM (ed) Classic readings in urban planning. McGraw-Hill, New York, pp 74–79. (first published 1989)
Friedmann J (1996) Two centuries of planning theory: an overview. In: Mandelbaum SJ, Mazza L, Burchell RW (eds) Explorations in planning theory. Center for Urban Policy Research/ Rutgers, New Brunswick, pp 10–29. (first published in Friedmann J (1987) Planning in the public domain: from knowledge to action. Princeton University Press, Princeton (NJ))
Friedmann J, Alonso W (eds) (1964) Regional development and planning. The MIT Press, Cambridge, MA
Friedmann J, Alonso W (eds) (1975) Regional policy: readings in theory and applications. The MIT Press, Cambridge, MA
Friend J (1996) Commentary: designing planning processes. In: Mandelbaum SJ, Mazza L, Burchell RW (eds) Explorations in planning theory. Center for Urban Policy Research/Rutgers, New Brunswick, pp 514–521
Friend JK, Jessop WN (1969) Local government and strategic choice. Tavistock Publications, London
Gabor D (1969) Open-ended planning. In: Jantsch E (ed) Perspectives of planning. Proceedings of the OECD working symposium on Long-Range Forecasting and Planning. Bellagio, Italy, 27 October – 2 November 1968, pp 326–347 (available on the web)

George P, Guglielmo R, Kayser B, Lacoste Y (1964) La géographie active. Presses Universitaires de France, Paris
Giaoutzi M, Nijkamp P (1993) Decision support models for regional sustainable development: an application of geographic information systems and evaluation models to the Greek Sporades Islands. Avebury, Aldershot
Giaoutzi M, Sapio B (2013) In search of foresight methodologies: riddle or necessity. In: Giaoutzi M, Sapio B (eds) Recent developments in foresight methodologies. Springer, New York, pp 3–9
Gillingwater D (1975) Regional planning and social change. Saxon House, Farnborough
Glasson J (1978) An introduction to regional planning: concepts, theory and practice, 2nd edn. Hutchinson, London
Godschalk DR (ed) (1974) Planning in America: learning from turbulence. American Institute of Planners, Washington, DC
Goodspeed R (2016) Digital knowledge technologies in planning practice: from black boxes to media for collaborative inquiry. PTP 17(4):577–600
Goodwin M (1999) Structure – agency. In: Cloke P, Crang P, Goodwin M (eds) Introducing human geographies. Arnold, London, pp 35–42
Greed C (1993) Introducing town planning. Longman, Harlow
Greed C (ed) (1999) Social town planning. Routledge, London
Greiving S et al (2008) A methodological concept for territorial impact assessment applied to three EU environmental policy elements. RuR1/2008: 36–51
Gunder M (2011) An introduction: psychoanalytical thought and planning theory. PT 10(4):299–300
Guyadeen D, Seasons M (2016) Plan evaluation: challenges and directions for future research. PPR 31(2):215–228
Habermas J (1971) Toward a rational society. Heinemann, London
Haggett P (1975) Geography: a modern synthesis, 2nd edn. Harper and Row, New York
Hall P (1967) The world cities. Weidenfeld and Nicolson, London
Hall P, Pfeiffer U (2000) Urban future 21: a global agenda for twenty-first century cities. E. and F.N. Spon, London
Hall P, Gracey H, Drewett R, Thomas R (1973a) The containment of urban England, Vol. I: Urban and metropolitan growth processes. PEP – George Allen and Unwin, London
Hall P, Gracey H, Drewett R, Thomas R (1973b) The containment of urban England, Vol. II: The planning system. PEP – George Allen and Unwin, London
Ham C, Hill M (1993) The policy process in the modern capitalist state. Harvester Wheatsheaf, New York
Harrison P (2001) Romance and tragedy in (post) modern planning: a pragmatist's perspective. International Planning Studies 6(1):69–88
Harvey D (1973) Social justice and the city. Edward Arnold, London
Harvey D (1996) On planning the ideology of planning. In: Campbell S, Fainstein S (eds) Readings in planning theory. Blackwell, Oxford, pp 176–197. (first published 1985)
Harvey D (2002) Social justice, postmodernism, and the city. In: Fainstein SS, Campbell S (eds) Readings in urban theory. Blackwell, Oxford, pp 386–402. (first published 1993)
Hauser PM, Schnore LF (eds) (1965) The study of urbanization. Wiley, New York
Hayward J, Watson M (eds) (1975) Planning, politics and public policy: the British, French and Italian experience. Cambridge University Press, Cambridge
Haywood I (ed) (1979) Implementation of urban plans. Organisation for Economic Co-operation and Development, Paris
Healey P (1992) Planning through debate: the communicative turn in planning theory. Town Plan Rev 63(2):143–162
Healey P (1996) The communicative work of development plans. In: Mandelbaum SJ, Mazza L, Burchell RW (eds) Explorations in planning theory. Center for Urban Policy Research/Rutgers, New Brunswick, pp 263–288. (first published 1993)
Healey P (1997) Collaborative planning. Macmillan, London
Healey P (2003) Collaborative planning in perspective. PT 2(2):101–123

Healey P (2005) On the project of "institutional transformation" in the planning field: commentary on the contributions. PT 4(3):301–310
Healey P (2010) Introduction: the transnational flow of knowledge and expertise in the planning field. In: Healey P, Upton R (eds) Crossing borders: international exchange and planning practices. Routledge, London, pp 1–25
Healey P (2016) The planning project. In: Fainstein SS, DeFilippis J (eds) Readings in planning theory. Wiley – Blackwell, Chichester, pp 139–155. (first published 2010)
Healey P (2017) Finding my way: a life of inquiry into planning, urban development processes and place governance. In: Haselsberger B (ed) Encounters in planning thought: 16 autobiographical essays from key thinkers in spatial planning. Routledge, New York/London, pp 107–125
Healey P, McDougall G, Thomas MJ (eds) (1982) Planning theory: prospects for the 1980s. Pergamon Press, Oxford
Healey P et al (2000) Editorial. PTP 1(1):7–10
Healy A (2002) Commentary: evidence-based policy – the latest form of inertia and control? PTP 3(1):97–98
Hickling A (1974) Managing decisions: the strategic choice approach. Mantec, Rugby
Hill M (1968) A goals-achievement matrix for evaluating alternative plans. J Am Inst Plan, January: 19–28
Hillier J (2003) Agon'izing over consensus: why Habermasian ideals cannot be 'real. PT 2(1):37–59
Hillier J (2005) Straddling the post-structuralist abyss: between transcendence and immanence? PT 4(3):271–299
Hillier J, Gunder M (2003) Planning fantasies? An exploration of a potential Lacanian framework for understanding development assessment planning. PT 2(3):225–248
Hoch C (1996) A pragmatic inquiry about planning and power. In: Mandelbaum SJ, Mazza L, Burchell RW (eds) Explorations in planning theory. Center for Urban Policy Research/Rutgers, New Brunswick, pp 30–44. (earlier version published in 1988)
Hoch C (2019) Pragmatic spatial planning: practical theory for professionals. Routledge, New York
Hodgson G (1984) The democratic economy: a new look at planning, markets and power. Penguin, Harmondsworth
Holden M, Scerri A (2015) Justification, compromise and test: developing a pragmatic sociology of critique to understand the outcomes of urban redevelopment. PT 14(4):360–383
Howe J, Langdon C (2002) Towards a reflexive planning theory. PT 1(3):209–225
Inch A (2015) Ordinary citizens and the political cultures of planning: in search of the subject of a new democratic ethos. PT 14(4):404–424
Innes JE (1995) Planning theory's emerging paradigm: communicative action and interactive practice. J Plan Edu Res 14(3):183–189
Innes JE, Booher DE (2015) A turning point for planning theory? Overcoming dividing discourses. PT 14(2):195–213
Jacobs J (1965) The death and life of great American cities. Penguin, Harmondsworth
Jacobs J (1969) The economy of cities. Penguin, Harmondsworth
Jantsch E (1969a) Integrative planning of technology. In: Jantsch E (ed) Perspectives of planning. Proceedings of the OECD working symposium on Long-Range Forecasting and Planning. Bellagio, Italy, 27 October – 2 November 1968, pp 179–200 (available on the web)
Jantsch E (ed) (1969b) Perspectives of planning. Proceedings of the OECD working symposium on long-range forecasting and planning. Bellagio, Italy, 27 October – 2 November 1968 (available on the web)
Judge D, Stoker G, Wolman H (eds) (1995) Theories of urban politics. Sage, London
Kaiser E, Godschalk D, Stuart Chapin F (1995) Urban land use planning. University of Illinois Press, Chicago
Kast FE, Rosenzweig JE (1972) The modern view: a systems approach. In: Beishon J, Peters G (eds) Systems behaviour. Harper and Row (for the Open University Press), London, pp 14–28
Klosterman RE (1985) Arguments for and against planning. Town Planning Review 56(1):5–20
Kriesis P (1970) Meta-planning with reference to physical planning, London

Krumholz N, Forester J (1990) Making equity planning work. Temple University Press, Philadelphia
Kunzmann KR (2017) Places matter. In: Haselsberger B (ed) Encounters in planning thought: 16 autobiographical essays from key thinkers in spatial planning. Routledge, New York/London, pp 202–221
Labasse J (1966) L'organisation de l'espace: Éléments de géographie volontaire. Hermann, Paris
Lagopoulos A-P (1973) Structural urbanism: the settlement as a system, Summary. TEE, Athens
Lai LWC (2016) "As planning is everything, it is good for something!" A Coasian economic taxonomy of modes of planning. PT 15(3):255–273
Lancaster LW (1968) Masters of political thought (vol. 3): Hegel to Dewey. George G. Harrap, London. (first published 1959)
Lauria M (ed) (1997) Reconstructing urban regime theory. Sage, Thousand Oaks (California)
Layard R (ed) (1972) Cost-benefit analysis. Penguin, Harmondsworth
Lean W (1969) Economics of land use planning – urban and regional. The Estates Gazette Ltd, London
Lefebvre H (1968) Le droit à la ville. Anthropos, Paris
LeGates RT, Stout F (eds) (1996) The city reader. Routledge, London
Leontidou L (2011) *Ageographitos Chora* [Geographically illiterate land]: Hellenic idols in the epistemological reflections of European Geography. Propobos, Athens. (in Greek)
Levin PH (1967) Toward decision-making rules for urban planners. J Town Plan Inst 53(10):437–442
Lichfield N (1956) Economics of planned development. The Estates Gazette Ltd, London
Lichfield N (1964) Cost-benefit analysis in plan evaluation. Town Plan Rev, July: 159–169
Lichfield N, Kettle P, Whitbread M (1975) Evaluation in the planning process. Pergamon Press, Oxford
Lieto L (2013) Place as a trading zone: a controversial path of innovation for planning theory and practice. In: Balducci A, Mäntysalo R (eds) Urban planning as a trading zone. Springer, Dordrecht, pp 143–157
Lindblom CE (1973) The science of "muddling through". In: Faludi A (ed) A reader in planning theory. Pergamon Press, Oxford, pp 151–169. (first published 1959)
Low N (2020) Being a planner in society: for people, planet, place. Edward Elgar, Cheltenham
Mäntysalo R, Bäcklund P (2018) The governance of planning: flexibly networked, yet institutionally grounded. In: Gunder M, Madanipour A, Watson V (eds) The Routledge handbook of planning theory. Routledge, London, pp 237–249
March A (2010) Practising theory: when theory affects urban planning. PT 9(2):108–125
Marcou G, Kistenmacher H, Clev H-G (1994) L'aménagement du territoire en France et en Allemagne. DATAR et La Documentation Française, Paris
Massé P (1965) Le plan ou l'anti-hasard. Idées nrf – Gallimard, Paris
Mattila H (2016) Can collaborative planning go beyond locally focused notions of the "public interest"? The potential of Habermas' concept of "generalizable interest" in pluralist and trans-scalar planning discourses. PT 15(4):344–365
McAuslan P (1980) The ideologies of planning law. Pergamon Press, London
McConnell S (1981) Theories for planning. Heinemann, London
McDougall G (1996) Commentary: the latitude of planners. In: Mandelbaum SJ, Mazza L, Burchell RW (eds) Explorations in planning theory. Center for Urban Policy Research/Rutgers, New Brunswick, pp 188–198
McLoughlin JB (1969) Urban and regional planning: a systems approach. Faber and Faber, London
Mintzberg H (1994) The rise and fall of strategic planning. Prentice Hall, Harlow
Mishan EJ (1972) Elements of cost-benefit analysis. Allen and Unwin, London
Mitchell GD (1968) A dictionary of sociology. Routledge and Kegan Paul, London
Morgan JR (1971) AIDA[10] – a technique for the management of design. Monograph No 2, The Institute for Operational Research, Tavistock Institute of Human Relations, London

[10] AIDA: Analysis of Interconnected Decision Areas.

Morley D, Proudfoot S, Burns T (eds) (1980) Making cities work: the dynamics of urban innovation. Croom Helm, London
Moroni S (2015) Complexity and the inherent limits of explanation and prediction: Urban codes for self-organising cities. PT 14(3):248–267
Moroni S (2018) The public interest. In: Gunder M, Madanipour A, Watson V (eds) The Routledge handbook of planning theory. Routledge, London, pp 69–80
Naess P (2015) Critical realism, urban planning and urban research. EPS 23(6):1228–1244
Needham B (2017) A renegade economist preaches good land-use planning. In: Haselsberger B (ed) Encounters in planning thought: 16 autobiographical essays from key thinkers in spatial planning. Routledge, New York/London, pp 165–183
Ozdemir E, Tasan-Kok T (2019) Planners' role in accommodating citizen disagreement: the case of Dutch urban planning. US 56(4):741–759
Pahl RE (1964) Urbs in rure. Weidenfeld and Nicolson, London
Pahl RE (ed) (1968) Readings in urban sociology. Pergamon Press, Oxford
Pahl RE (1970) Patterns of urban life. Longmans, London
Pahl RE (1975) Whose city? Penguin, Harmondsworth
Pahl R (1982) "Urban managerialism" reconsidered. In: Paris C (ed) Critical readings in planning theory. Pergamon Press, Oxford, pp 47–67. (originally published 1975)
Palermo PC, Ponzini D (2010) Spatial planning and urban development: critical perspectives. Springer, Dordrecht
Parfitt T (2002) The end of development: modernity, post-modernity and development. Pluto Press, London
Paris C (1982a) A critique of pure planning: introduction. In: Paris C (ed) Critical readings in planning theory. Pergamon Press, Oxford, pp 3–11
Paris C (1982b) The future of planning (theory?). In: Paris C (ed) Critical readings in planning theory. Pergamon Press, Oxford, pp 307–308
Parker G, Doak J (2012) Key concepts in planning. Sage, London
Perroux F (1968) Programming quantitative techniques. Papazisis, Athens. (in Greek) (translated from French original)
Perrow C (1970) Organizational analysis: a sociological view. Tavistock, London
Pile S, Brook C, Mooney G (eds) (1999) Unruly cities? Order/disorder. Routledge, London
Potts R (2020) Is a new 'Planning 3.0' paradigm emerging? Exploring the relationship between digital technologies and planning theory and practice. PTP 21(2):272–289
Pressman JL, Wildavsky A (1979) Implementation. University of California Press, Berkeley
Prest AR, Turvey R (1966) Cost-benefit analysis: a survey. In: Surveys of economic theory: resource allocation, vol III. Macmillan, London
Preteceille E (1982) Urban planning: the contradictions of capitalist urbanisation. In: Paris C (ed) Critical readings in planning theory. Pergamon Press, Oxford, pp 129–146. (originally published in French 1974)
Pugh DS (ed) (1971) Organization theory. Penguin, Harmondsworth
Ravetz A (1986) The government of space: town planning in modern society. Faber and Faber, London
Rawls J (1995) Justice as fairness. In: Stein JM (ed) Classic readings in urban planning. McGraw-Hill, New York, pp 63–73. (first published 1971)
Rawls J (1999) A theory of justice (Rev. edn). Oxford University Press, Oxford
Reissman L (1970) The urban process. The Free Press, New York
Rivero JJ (2017) Making post-truth planning great again: confronting alternative facts in a fractured democracy. PTP 18(3):490–493
Roy A (2010) Poverty truths: the politics of knowledge in the new global order of development. In: Healey P, Upton R (eds) Crossing borders: international exchange and planning practices. Routledge, London, pp 27–45
Rydin Y (1998) Urban and environmental planning in the UK. Macmillan, Basingstoke
Rydin Y (2007) Re-examining the role of knowledge within planning theory. PT 6(1):52–68

Rydin Y (2010) Actor-network theory and planning theory: a response to Boelens. PT 9(3):265–268
Rydin Y (2018) Actor network theory. In: Gunder M, Madanipour A, Watson V (eds) The Routledge handbook of planning theory. Routledge, London, pp 302–313
Safier M (1983) The passage to positive planning. Habitat Int 7(5/6):105–116
Sager T (2006) The logic of critical communicative planning: transaction cost alteration. PT 5(3):223–254
Sager T (2015) Ideological traces in plans for compact cities: is neo-liberalism hegemonic? PT 14(3):268–295
Sager T (2018) Communicative planning. In: Gunder M, Madanipour A, Watson V (eds) The Routledge handbook of planning theory. Routledge, London, pp 93–104
Sager T (2020) Populists and planners: 'we are the people. Who are you?'. PT 19(1):80–103
Salet W (2019) The making of the public. PT 18(2):260–264
Sandercock L (2000) When strangers become neighbours: managing cities of difference. PTP 1(1):13–30
Sandercock L (2002) Towards cosmopolis. Wiley, Chichester
Sanyal B (2002) Globalization, ethical compromise and planning theory. PT 1(2):116–123
Sassen S (1991) The global city: New York, London, Tokyo. Princeton University Press, Princeton
Sassen S (2002) Cities in a world economy. In: Fainstein SS, Campbell S (eds) Readings in urban theory. Blackwell, Oxford, pp 32–56. (first published 2000)
Schön D (1973) Beyond the stable state: public and private learning in a changing society. Penguin, Harmondsworth
Schön D (1983) The reflective practitioner. Basic Books, New York
Scott AJ (1980) The urban land nexus and the state. Pion, London
Seldon A, Pennance FG (1965) Everyman's dictionary of economics. Dent and Sons, London
Self P (1981) The future of urban planning. J R Soc 1981, No 5305: 30–38
Self P (1982) Planning the urban region. George Allen and Unwin, London
Short JR (1996) The urban order: an introduction to cities, culture and power. Blackwell, Cambridge, MA
Simmie J (1974) Citizens in conflict: the sociology of town planning. Hutchinson, London
Simmie J (1981) Power, property and corporatism: the political sociology of planning. Macmillan, London
Simmie J (1993) Planning at the crossroads. UCL Press, London
Smith PF (1974) The dynamics of urbanism. Hutchinson, London
Smith MP (1980) The city and social theory. Basil Blackwell, Oxford
Solesbury W (1974) Policy in urban planning. Pergamon Press, Oxford
Solesbury W (2002) The ascendancy of evidence. PTP 3(1):90–96
Soltys, J, Lendzion J (n.d.) Scenarios for the future in the regional and local territorial planning. https://www.imp.gda.pl/fileadmin/scitechfound/Publikacje/CetscenE.doc
Stein SM, Harper TL (2005) Rawls' 'Justice as fairness': a moral basis for contemporary planning theory. PT 4(2):147–172
Stephens GR, Wikstrom N (2000) Metropolitan government and governance. Oxford University Press, New York
Stephenson R (2000) Technically speaking: planning theory and the role of information in making planning policy. PTP 1(1):95–110
Stewart M (ed) (1972) The city: problems of planning. Penguin, Harmondsworth
Stojanović M, Mitković P, Mitković M (2014) The scenario method in urban planning. Facta Universitatis – Architecture and Civil Engineering series 12(1):81–95
Sturzaker J, Lord A (2018) Fear: an underexplored motivation for planners' behaviour? PPR 33(4):359–371
Tait M (2016) Planning and the public interest: still a relevant concept for planners? PT 15(4):335–343
Taylor N (1999) Urban planning theory since 1945. Sage, London

Thomas MJ (1982) The procedural planning theory of A. Faludi. In: Paris C (ed) Critical readings in planning theory. Pergamon Press, Oxford, pp 13–25. (originally published 1979)
Thorpe A (2017) Rethinking participation, rethinking planning. PTP 18(4):566–582
Throgmorton JA (1996) 'Impeaching' research: Planning as persuasive and constitutive discourse. In: Mandelbaum SJ, Mazza L, Burchell RW (eds) Explorations in planning theory. Center for Urban Policy Research/Rutgers, New Brunswick, pp 345–364. (revised version of 1993 publication)
Throgmorton JA (2003) Planning as persuasive storytelling in a global-scale web of relationships. PT 2(2):125–151
Van Hulst M (2012) Storytelling, a model *of* and a model *for* planning. PT 11(3):299–318
Van Wezemael J (2008) The contribution of assemblage theory and minor politics for democratic network governance. PT 7(2):165–185
Van Wezemael J (2018) Assemblage thinking in planning theory. In: Gunder M, Madanipour A, Watson V (eds) The Routledge handbook of planning theory. Routledge, London, pp 326–336
Vidyarthi S (2019) Charles Hoch: a pesky pragmatist. Plan Theory 19(4):445–451
Vidyarthi S, Hoch C (2018) Learning from groundwater: pragmatic compromise planning common goods. Environ Plan C Polit Space 36(4):629–648
Vigar G, Cowie P, Healey P (2020) Innovation in planning: creating and securing public value. EPS 28(3):521–540
Vogelij J (2015) Is planning theory really open for planning practice? PTP 16(1):128–132
Von Bertalanffy L (1969) The theory of open systems in physics and biology. In: Emery FE (ed) Systems thinking. Penguin, Harmondsworth
Von Bertalanffy L (1973) General system theory: foundations, development, applications. Penguin, Harmondsworth
Wannop U (1970) An objective strategy – the Coventry-Solihull-Warwickshire sub-regional study. J R Town Plann Inst, April
Wannop U (1995) The regional imperative: regional planning and governance in Britain, Europe and the United States. Jessica Kingsley, London
Wassenhoven L (1974) Planning techniques. In: Extension service teaching manual. Development Planning Unit, School of Environmental Studies, University College London (chapters written by Louis Wassenhoven)
Wassenhoven L (1978) Urban management: needs and lessons from international experience. Epitheorisis tis Topikis Aftodioikiseos (Central Union of Greek Municipalities and Communes), June 1978, No 6: 504–519 (in Greek)
Wassenhoven L (2002) The democratic quality of spatial planning and the contestation of the rational model. Aeichoros 1(1):30–49. (in Greek)
Wassenhoven L (2018) Hellinikon … passion: the development of the former Hellinikon airport, Athens. Sakkoulas, Athens. (in Greek)
Webber MM et al (1964) Explorations into urban structure. University of Pennsylvania Press, Philadelphia
Williams RH (ed) (1984) Planning in Europe: Urban and regional planning in the EEC. Allen and Unwin, London
Wilson A (1970) Entropy in urban and regional modelling. Routledge, London
Wilson BR (ed) (1977) Rationality. Blackwell, Oxford
Wohl S (2018) Tactical urbanism as a means of testing relational processes in space: a complex systems perspective. PT 17(4):472–493
Wright Mills C (1959) The sociological imagination. Oxford University Press, New York
Young RC (1966) Goals and goal-setting. J Am Inst Plan XXXII(2):76–85
Young IM (2016) Inclusion and democracy. In: Fainstein SS, DeFilippis J (eds) Readings in planning theory. Wiley – Blackwell, Chichester, pp 389–406. (first published 2000)
Zitcer A (2017) Planning as persuaded storytelling: the role of genre in planners' narratives. PTP 18(4):583–596

Chapter 4
Theoretical Challenges: The Radical Current and Southern Theory

Abstract The aim of this chapter is to present the theoretical constructs which challenge the universal validity of the mainstream rational and communicative currents. This challenge emanates from more radical, transformative, even subversive theories, of which the roots can be found both in the conditions of developed economies and in the world's periphery, often labelled as the Global South. It is the latter which is the prime focus because it is a catch-word representing the reaction against the transfer of spatial planning practices from the North and West to the South and East. The "periphery" encapsulates a variety of environments, from post-colonial to simply dependent. The chapter offers the opportunity to refer to essential concepts, like southern theory, informality, insurgency, justice, minority groups and movements, marginalization, becoming, agonism, alternative planning, militancy, activism and many others.

Keywords Radical · Transformative · Activism · Marxism · Critical theory · State · Insurgency · Global south · Ethics · Justice · Becoming · Less developed countries · Action planning · Informal · Agonism

Prologue

The passage from a planning "theory" dominated by the power of design and the authority of the designer to a theory based on process, expert reasoning, policy making, and rationalist choice was predicated on a transformation of planning as action. The passage from the rational comprehensive model to a conception of diversity, communication, collaboration, multiplicity of knowledges, and diverging conceptions of reality, also assumed a transformation of planning. Yet, the critics of these currents considered that both transformations had inadequately challenged social and political structures of power. A radical approach would have to endorse thorough social and political reform. In this chapter, the author attempts to provide an overview of challenging and critical theory, which reacts against the universality of dominant planning theory currents. Critical views appear both in developed and less

L. C. Wassenhoven, *Compromise Planning : A Theoretical Approach from a Distant Corner of Europe*, https://doi.org/10.1007/978-3-030-94331-8_4

developed countries to use an old terminology, but the causal factors are not the same. Our interest however lies more in the theoretical "revolt" which originates in the "South" and "East", although there exists in them an inevitable diversity. The main purpose is to argue that theorizing is place and country specific and cannot claim planetary validity.

4.1 The Radical Current

Keywords of the Radical Current
Ideas, concepts, preoccupations, interpretations, concerns, and socio-political conditions embedded in the radical current, which inspire radical planning theory and practice:

Transformation, otherness, subversion, insurgency, agonism, antagonism, revolt, minoritarian values, becoming, resistance, collective values, feminism, activism, challenge, systemic defiance, class interests, race, self--defense, power overthrow, ethnicity, radical politics, Marxism, freedom, oppression, unemployment, policing, poverty, exclusion, Global South, underdevelopment, colonialism, dependence, informality, refugees, periphery, dispossession, prison syndrome, unsettling, enclaves, terror, fear.

4.1.1 The Radical Turn

For many, a true radical approach was absent in mainstream theories. If planning was to adopt a radical agenda it would have to be genuinely transformative. Angotti traces the emergence of radical planning in the United States of America: "In the twentieth century a new trend in urban planning emerged … Let's call it *radical planning*. It broke out in the 1960s with roots in the Civil Rights Movement when some professional planners broke ranks with their profession and joined the movements to stop the displacement of communities of color from central cities that was promoted by 'urban renewal' (aka 'Negro removal') plans" (Angotti 2020, 6). The question is whether this is an agenda of planning as such or of a social movement aiming at overturning the political system. The author still remembers a planning simulation exercise, when he was a planning student in England in the late 60s and early 70s, where roles were distributed, i.e., those of planners, developers, municipal officials, stakeholders, activists etc. When a planning project was discussed and the point was made that it would never be sanctioned by the ruling political system, a radical Latin American fellow student exclaimed: "Well then, I shall have to change the system". Alexander refers to that period in the following comment: "The late 1960s and early 1970s have also seen a number of proposals for the involvement of planners in social change outside the governmental establishment or in active opposition to it. These schemes range from ideological reorientation through community self-regulation and self-help to communitarian planning philosophies. The trouble with these proposals, as some of the initiators have recognized, is that they can never be more than interstitial in society as we know it. The radical planners, once they have won the battle against the status quo, cannot avoid becoming part of the very institutions they have sworn to alter" (Alexander 1992, 104–105).

On the whole, planning theorists approach the radical turn by placing emphasis on the combative and activistic potential of planning to dislodge the structures of power. The power-society bipolar relationship is full of mutual interactions and this makes any discussion of the role of planning exceptionally tricky, because planning does not simply obey power, but also reinforces its foundations. "Explaining how the effects we label as 'power' are produced, instead of using 'power' as an in-itself all covering explanation of societal events, demands a conceptualization of power as the outcome of social processes rather than as a causal variable behind them" (Metzger et al. 2017, 204). Power holders and the status quo need the planners and their expertise for a variety of reasons, as N. and S. Fainstein explain: "Planning is necessary to the ruling class in order to facilitate accumulation and maintain social control in the face of class conflict. The modes by which urban planners assist accumulation include the development of physical infrastructure, land aggregation and development, containment of negative environmental externalities, and maintenance of land values. Social control is effected through a combination of containment tactics and the provision of social services. Urban planners are predominantly agents of the state. Even those planners who are not employed by the state are forced to work within the agenda established by the state" (Fainstein and Fainstein 1982, 148). This is the essence of the Marxist critique: "[M[arxism points to the severe limitations of the most typical forms of left-oriented planning activity. Radical advocacy, from a Marxist perspective, suffers from its cooptative tendencies, the negation of planning function and transformation of planners into political agitators (thereby raising questions about the special legitimacy of planners to lead the masses), and its inability to move beyond triumphs of veto and negation to orchestrated, positive, system-wide movement. Guerrillas in the bureaucracy are seen as weak and easily eliminated …" (op.cit., 168–169). The radical affinity with the Marxist interpretation has always been recognized in planning theory, because it "encourages awareness and understanding of existing social, social, economic and political conditions … Critical theory may lead to revolutionary praxis – and that is where its utility for urban planning theory can be questioned" (McConnell 1981, 25).

4.1.2 Planners, Activism and Transformative Planning

The argument of the Fainsteins, regarding the role of planners, is summarized, by Brooks. There are three possibilities for the planners: "First, work within the state to make it more humane. Find out what can be defended, and do so; open up repressive bureaucratic structures; share knowledge and expertise; de-professionalize; fight secrecy. This may entail functioning, for example, as 'a watchdog, a whistle-blower, a guerrilla in the bureaucracy, or as a monitor of communication flows who guards against the dissemination of false information'. Second, work outside the state, attempting to affect governmental policy as a social critic and activist. Third, develop alternative systems of production and distribution – symbolized in the 1970s, perhaps, by communes, and today by food cooperatives". Other authors, writes Brooks, were more inclined toward a fourth role: "planning for 'reconstruction of society', including 'the replacement of existing social institutions benefiting capital by new ones serving the interests of society at large'". Brooks concludes in a rather sarcastic manner: "How does one do this? Well … revolution comes to mind" (Brooks 2017, 40–41). A further alternative, one could add, can be followed if a sheltered position is found, even within the public sector, e.g., in universities. States of authoritarian inclinations may well prefer to keep radicals "encased" in an academic environment. As to working with social movements, of various descriptions, in a volunteer fashion, raises issues of identity, Beauregard remarks: "Contributing time, money, expertise, and one's body (e.g., at protest marches) is implied, but this hardly distinguishes planners from any other participant. I suspect that the actual tasks that radical planners would perform are not that much different than what is expected of an advocacy planner, that is, providing technical research and advice. Again, this maintains the expert-client relationship. Forming networks of opposition groups who can share resources and ideas, moreover, is not a skill unique to planners and certainly not their strength as the profession has been conceived" (Beauregard 2020, 52). Beauregard is rather underestimating the potential of radical planning action and the similarity with advocacy planning does not diminish the radical action potential. One can also add the possibility of offering services to political parties, which frequently set up think tanks for the formulation of political agendas. Theorists of planning or of other fields are often engaged as policy making advisors or suppliers of ideas. There is also room for taking up employment in administrations, e.g., local ones, which are openly opposing the central state.

Perhaps the most accurate view of radical planning, which will be discussed again in Chap. 10, although he uses the term "critical" planning, is that of Peter Marcuse: "I believe a new paradigm for progressive planning – I would describe it as a paradigm of critical planning but other names would do as well – is emerging from this ferment. As a slogan for good planning, it might be represented by: Analyze, Expose, Propose, Politicize" (Marcuse 2020, 33). The radical element in his conception lies particularly in "exposing", for which he borrows a phrase by Sandercock: "Making the invisible visible". For Marcuse, "exposing means communicating the results of the analysis, in comprehensible fashion, to the client, but also making the results public and converting the findings into a weapon in the struggle to achieve the desired goals … Exposing also means communicating to the group being served the full range of alternatives available, starting with the utopian and working backwards to the strategic and the tactical. It means exposing to them the realities of the plan and the planning process" (op.cit., 34). Of equal weight is naturally the element of politicizing, which "means squarely addressing issues of power of showing how the resolution of a particular issue may lead to conflict and may require challenging existing power relationships" (ibid.). Marcuse, an advocate of transformative planning (see below), places his analysis in the context of a democracy, where, although powerful interests may be dominant and discriminatory practices abound, the components of a pluralist regime and of the rule of law are nevertheless present. In this he follows the precedent of other advocates of radical politics, which perhaps explains the criticism of Beard, who remarks that "the literature on radical planning does not explain how citizens engage in radical planning for social transformation within authoritarian environments". She claims that her study "demonstrated that radical planning is possible in highly restrictive political environments, but that it is the outcome of a social learning process that over an extended period creates a powerful sense of collective agency and action" (Beard 2003, 29–30). The common denominator is that radical approaches refute the acceptability of "official" planning and call for an overthrow of practices which perpetuate social inequalities and power-driven choices. In this sense, radical theory is destabilizing or "unsettling", to use the title of an article by Barry et al. (2018): "Words like 'unsettle,' 'unsettlement,' and 'unsettling' are, therefore, evoked by scholars to disrupt, and to call to account the regressive ways in which structures, processes, and relations impede equity, justice, and ecological sustainability. The call to 'unsettle' … often signals an impulse to challenge the dominant forces that shape settlements, mobilities, knowledges, relations, and experiences. More generally, we might see 'unsettling' as a characteristic move of critical, post-structuralist analysis that seeks to expose the contingency underpinning all social arrangements".

Radical thought challenges the process of planning but only in its most extreme versions rejects the necessity of any planning at all. Instead, it looks for interstices in the planning system into which it can inject a different equity perspective. Total rejection, or simply "planning for not having a plan", in the words of Jean Hillier (2017), is not, ironically, the hallmark of market enthusiasts but rather of anarchist movements. Hillier's interest is in "possible relationships between anarchist practices and macro-structures, asking whether anarchisms / anarcho-syndicalisms might replace, reform or exist in dual power with existing systems. I conclude", Hillier writes, "that prefigurative experimental practices can offer potentialities for both exposing and prising open cracks in existing planning systems and also for doing planning differently … Anarchism and anarcho-syndicalism have always been apparent in human (and non-human) interactions. Self-organisation, mutual aid and co-ordination are currently lauded, and anarchism and anarcho-syndicalism are enjoying both theoretical and practical resurgence".

4.1.3 Ethics and Justice

Justice as a planning goal has already been discussed in connection with communicative theory. It acquires a different meaning when injustice is a constitutive and systemic part of the socio-political environment (Harvey 1973). The interplay between human rights and human capabilities can become a critical component of urban and regional planning, through the delineation of territorial entities (Basta 2016). Spatial inequalities and preferential branding of territorial entities, so frequently sanctioned by planning policy and land-use limitations, can be yet another source of injustice, often supported by claims of moral superiority among ruling groups. This is akin to what Baum (2011) describes, somewhat dramatically, as a "problem of evil". This could be a reason why planners, as Baum (2015) advocates elsewhere, should show greater emotional understanding and appreciation of motivation.

The acts of exposing the dark side of planning (Flyvbjerg 1996) and unsettling systems that perpetuate injustice and inequality, touch on sensitive ethical questions, involving "power and rationality; discipline and social control; and governmentality, which in this [planning] field tend to have a common emphasis on the way power subverts the overt progressive aims of planning" (Huxley 2018, 208). It might be argued that inequality of opportunity is not related to spatial planning, but the "hoary question", to use the expression of Susan Fainstein, is whether inequality of living conditions impedes equality of opportunity. Just planning, for just places, should aim, she argues, at equality, diversity, and democracy (Fainstein 2016, 263 and 268). A radical approach must not only expose the hidden contents of the planning process, but also give voice to disempowered communities, for the sake of all three aims. The difficulty, in the view of Harper and Stein, is the incommensurability of these communities: "Postmodernists, communitarians, feminists, and deep ecologists have made valuable and valid critiques of planning. They have pointed out that institutional planning processes frequently conceal the disempowerment of certain communities, that not all voices are heard. They have also made credible normative recommendations: planners are urged to empower voiceless communities by ensuring that their voices are heard". Harper and Stein focus "on one important premise shared by many postmodernists, communitarians, feminists, and deep ecologists. This premise may seem so obviously false that no one would pay attention to it … If not dealt with, however, it is likely to infect many future debates … The premise is that communities are incommensurable. That is, their stories, their discourses, and their languages are not understandable to others. The implication of this premise is that 'outside' critique is invalid and consensus-seeking dialogue is futile" (Harper and Stein 1996, 414–415). The task of a radical planner is therefore to make certain that all views are integrated in the process of analysis and then judged or measured by the same standards, to ensure commensurability.

The difficulty of eliciting genuine views is multiplied if we neglect that group members are not just static beings, but are rather in a state of "becoming", to which reference was made in Chap. 3. Group members find themselves in a flux caused either by external forces and/or as a result of past struggles, e.g., for emancipation and securing a place in a new "assemblage" of relationships. Jean Hillier, who follows the philosophy of Deleuze and Guattari, from which the concepts of becoming and assemblage (*agencement*) are borrowed, relates becoming with global contingent developments, growing uncertainty, and the constant production of difference: "Becoming … is not a phase between states of being through which something passes, but the dynamism of change itself …Becoming is an encounter between bodies and/or essences or forces that effects a transition of both" (Hillier 2018, 339). To capture the voice of individuals or groups in a state of becoming is immensely more difficult for the planner. The convenience of relying on past accounts of personal opinions and individualities is shattered by this fluid transition.

Hillier's thought had been made plain in an earlier article ("Plan(e) speaking: a multiplanar theory of spatial planning"). in which she declared from the start: "I develop a new theory of spatial planning". Using a couple of quotations from other authors, she stated that "Planners' representations of people and space have traditionally been regarded as value-free and objective. In their plans, strategic documents and decisions, planning practitioners have tended to assume that 'groups, individuals, race, attitudes, intentions, the state, societies and symbols exist' … 'out there' in a world which is 'more or less specific, clear, certain, definable and decided' … and that they can create accurate and valid representations of those objects in texts, maps and plans … Meaning cannot be 'given' as such, but is understood, often differently by different people (terms such as dense, village, park, sustainability and so on) … I regard planning and planners as *experiments* or *speculations* entangled in a series of contingent, networked relationships in circumstances which are both rigid (e.g., legally constrained) and flexible, where outcomes are volatile, where problems are not 'solved' once and for all but which, over the 'lifetime' of a strategic plan, are constantly recast by changing actors, situations and preferences, to be reformulated in new perspectives … I offer a Deleuzoguattarian-inspired view of spatial planning as experimental, speculative becoming; a creative reterritorialization which allows for the aleatory (chance) to occur … I argue for multiplanar spatial planning practice to encompass a broad trajectory of possible visions as background planes of immanence/consistency, together with more specific local/short-term plans and projects as foregrounded planes of transcendence/organization" (Hillier 2008, 25–27; also, Hillier 2005). Hillier's line of argument is not always easy to follow, but it is hoped that the above extracts provide a further understanding of the concept of becoming and summarize the key features of a radical approach which attempts to reconcile previous theoretical currents with the reality of ever-expanding uncertainty and contingency. It is interesting to note that, ultimately, she makes what seems like a concession: "I regard spatial practice as temporally engaged experimentation, improvisation or phronetic practical wisdom" (op.cit., 27). She is here invoking the ancient Greek notion of *phronesis* (Friedmann 1995, 74). To this reference to improvisation and practical wisdom we will return later in connection with compromise theory.

The balance between abstract judgements regarding social justice and practical wisdom is critical when planners speak vaguely about objectives like that of the "just city". Fighting gross inequalities and the violation of human rights are undisputed goals, but beyond that, references to just cities may end up by being an escape. Moroni (2020) touches on this danger. "In the fields of planning theory and human geography, there is a growing discussion of the *just city*. The impression is that in order to continue the discussion of the crucial issue of the just city, certain methodological considerations and precautions are necessary". Moroni focuses on "(a) (urban) institutions as the first subject of justice, (b) the incomplete overlap between social justice and distributive justice, and (c) the distinction between the concept and the conceptions of social justice". Such conceptions vary immensely among different groups in society and they may range from the protection of legally entrenched property rights to securing elementary rights for a decent life. A critical component of spatial planning is its relationships with human rights, in itself a vast subject of inquiry. "Human rights and spatial planning", according to Davy, "enjoy a relationship of mutual influence". The main reason is that "the right to life, the right to work, or the right to housing never are enjoyed in a spatial void, but always need to be contextualized with a certain place and realized in a spatial environment of many – and often conflicting – demands and claims to the use of space ('spatiality of rights')" (Davy 2014, 329–330). Working with conflict becomes a routine situation (Fisher et al. 2000). For radical planning, ethical issues arise particularly when the insurgent spirit has to accommodate place-related demands for equity and justice. McClymont (2019) sought a way "to link the creative conflict of agonistic/dissensus-based planning theory with ethical judgement by actively articulating the substantive content and meaning of terms such as 'justice' and 'social equity' rather than simply assuming agreement over these". Her intention is to "deepen the concept of planning as 'the art of situated ethical judgement' … by challenging the usual basis of moral judgements in planning practice". The problem is that, when it comes to planning theory, the theoretician's concern is to marry epistemological assumptions regarding knowledge with ethical principles. "[H]ow we act on or towards individuals, groups and objects depends not only upon our adopted epistemological standpoints but also upon our value-based judgements (or ethics), because ethical principles guide decision-making processes and actions. In sum, planning interventions necessitate an awareness of how we know (epistemology), some kind of action, as well as value-based judgements (or ethical principles)" (Winkler and Duminy 2016, 111–112).

4.1.4 Planning for Development

Long before the theoretical debates about the Global South, in the 1960s and 70s planning in the "less developed countries" was a highly developed subject of inquiry and theoretical analysis. Its central theme was a reaction against the import trade of western planning methods which were plainly irrelevant for the conditions of LDCs. This was a concern, which until the rise of Southern theory, was sadly ignored in a great deal of mainstream theoretical work. This section is an effort to remedy this omission and recognize the importance of this chapter of planning history. Radical planning, as we understand it today, was not the framework of these debates; rather, the discussion of planning in the LDCs was fueled by the phenomenon of rapid urbanization in Asian, African, and Latin American countries, explosive population growth, the weakness of the state apparatus, the unsuitability of western comprehensive and rigid planning methodologies, the informal nature of urban development, self-help housing initiatives, and the dramatic living conditions in cities. The situation was vividly pictured in a considerable literature of urban and development studies.[1] From our perspective the most important aspects of this concern for an appropriate model of planning were the rejection of the rational long-winded procedural planning, exemplified in western "master plan" approaches, the search for flexible solutions more suitable for the government structures of LDCs, and the corresponding allocation of roles to a variety of actors, including community and self-help housing movements. Importing planning ideas had ceased to be the standard recipe. "For many decades", Ananya Roy (2017, 13) remarks. "the canon of urban theory had remained primarily a theory of a Euro-American urbanism, a story of urban change in a handful of global cities: Chicago, New York, Los Angeles, all located in the global north". In the 1970s, at a time when McLoughlin's and Chadwick's systems approach and Faludi's procedural theory of planning had just seen the light of day, proposals appeared which tried to escape from the cumbersome trappings of rational, comprehensive planning. A certain similarity with the incremental model may have existed. Bendavid-Val (1975) and Waller (1975) advocated, respectively, a "themes-strategy-projects" approach and a "reduced" planning strategy. The latter, according to Waller, "is a systematized, iterative search process for key projects in the development of a region, as it were a middle way between comprehensive planning and the trial-and-error approach" (Waller 1975, 20). As to the themes-strategy-projects approach, Bendavid-Val (1975. 6) explained that it is "a logical three-step sequence, from the general to the specific". Planning promotes a programme "that spells out the approach to be taken with respect to each of the various aspects of the region's … activity. As such, it provides the specific framework for project recommendations". A project is defined as "any action that will bring about a change in the nature of quantity of production in the region, in the supporting system (including administration), in social welfare, or in the spatial framework" (op.cit., 7–9). In his contribution towards the development of an alternative planning style, what he calls "the sketch plan concept", Jakobson also insisted

---

[1] The list is purely indicative and includes mainly publications with an interest in urban studies and planning: Beier et al. 1975; Bendavid-Val and Waller 1975; Bourne et al. 1984; Breese 1966, 1969; Butler and Crooke 1973; Demko et al. 1970; Donaldson 1973; Drakakis-Smith 2000; Fanon 1967 (original 1961); Friedly 1974; Furtado 1971; Griffin and Enos 1970; Hardoy 1975; Hardoy and Satterthwaite 1981, 1989; Harris 1992; Harris and Fabricius 1996; Herbert 1979; Herbert and Van Huyck 1968; Johnson 1970; Jones 1975; King 2007 (original 1976); Lewis 1966; Linn 1979, 1983; Mabogunje et al. 1978; Pacione 1981a, b; Roberts 1978; Rondinelli 1975, 1979, 1983; Singer and Ansari 1977; Sjoberg 1960; Smith 1979; Stöhr 1980; Stretton 1978; Taylor and Williams 1982; Todaro 1971; Turner 1980b, 1976; Turner and Fichter 1972; Wakely and You 2001; Wakely et al. 1976; Waterston 1965; Weaver 1984; Wilber 1973; World Bank 1972.

on the provision of a framework for projects: "The purpose of sketch planning is to provide immediate development guidance for action programs and projects within a broad conceptual scenario/strategy framework. The concept is offered as a heuristic statement of plan formulation in complex, uncertain, and unstable situations. It is a plan formulation process that differs from the traditional rational-comprehensive process and leads to a plan whose contents also differ from much past work" (Jacobson 1980, 44).

The "project" had become the catchword for an approach which stresses "acting" instead of vague "planning". This was of course an exaggeration since "project planning" was a major task in itself, although heavily directed and targeted towards clearly defined objectives and with strict performance criteria. It was written by K.C. Sen that "the concept of project itself has undergone a metamorphosis since it has come to embrace a wide variety of technical assistance and investment activities, ranging from single-purpose infrastructure and commercial projects to multipurpose regional and social development projects" (Sen 1980, i). Colin Bruce explained project planning in terms of a two-way classification of planning. Spatial/institutional categories included national, regional, sub-regional, and local (district or urban) planning. Functional categories included macro-planning, sectoral planning, and project planning (Bruce 1980, 6). Rondinelli viewed sectoral analysis and planning as the primary link between development planning and individual project identification and selection (Rondinelli 1979, x). Although project planning was increasingly seen as an essential component of development planning, it would be a mistake to describe the latter as merely "planning by projects". The importance of the project is that it can be steered towards the achievement of precisely defined objectives and that it can be targeted on specific groups. Such is the case of the urban poor who were receiving special attention in international assistance programmes of the period (Beier et al. 1975, 381–386). Herbert saw as central tasks of project planning the identification of the urban poor as a target group and the measurement of impact of actions aimed at relieving their plight (Herbert 1979, 17–38). Both he and A. Turner offered detailed descriptions of the project planning process (op.cit., 36–38; Turner 1980a, 88–91). Sen (1980) and Bruce (1980) also discussed the project cycle from identification to monitoring and feedback, via appraisal and implementation. Of importance, however, is not the planning process which is internal to a project cycle, but the position of projects as instruments in development-oriented "action planning".

Action planning, as proposed in 1964 by Otto Koenigsberger (n.d.) *is* development planning in inspiration and ideology, but was also characterized by a distinct methodological approach, which contrasted with the rational systemic and comprehensive approach of goal-oriented planning, which had appeared in the 1950s. The reader is reminded that Banfield had proposed his "ends and means" methodology in 1959 (see Banfield 1973, first published in 1959). Koenigsberger, after a long career in India, had reached the conclusion that imported planning models were not producing the desired results. "The politics and pragmatics of place now seemed less tractable to the application of universalist, cross-cultural ethic, aesthetic and technique". In 1951, the year Koenigsberger quit India, he wrote that, "the longer I did planning work in India the firmer became my conviction that master plans and reports are not enough. It is necessary to create a live organization, preferably anchored in the structure of local government, which constantly deals with planning problems and keeps the basic conception alive'" (Windsor Liscombe 2006, 173; Wakely 1999; Aldhous et al. 1983). Koenigsberger had reached his conclusions after a vast amount of work especially on Indian new towns and neighbourhood design (Vidyarthi 2010, 83–90; Ward 2010, 56–57). For him, action planning (Koenigsberger 1964) was a style of planning, the aim of which would not be so much to control private initiative, like conventional land-use planning, but rather to guide public action. As explained by Michael Safier (n.d., 11–12), "the process of action planning can be considered as containing five interconnected operations: (a) the reconnaissance, a rapid and selective survey … to identify strategic issues and problems …; (b) the guiding concept, a broad 'structural' perspective plan …; (c) the action programmes, a connected series of major sector development strategies …; (d) the role casting, an administrative process to consolidate the link between planning and implementation by assigning appropriate roles in action planning …; (e) monitoring and feedback, an institutional arrangement to ensure constant and effective learning by doing…". Although certain elements in this process are reminiscent of systems planning, action planning departed from "master planning" by transforming the position of *the* plan in the whole sequence. Action planning shifts the emphasis to the stages of action programmes and role casting. This is not merely a question of shortening the duration of the planning process and of responding faster to the problems diagnosed in the reconnaissance, or "windscreen survey" as Koenigsberger had named it, although this was also part of the intention. More than this it was a question of focusing on key interventions and operations, of glaring priority, on the basis of guidelines drawn from national urban policy, and then of allocating responsibilities to agencies forming part of the entire network of potential actors.

Fifteen years after the publication of Koenigsberger's paper on action planning, the concept was reviewed by Colin Rosser (1979),[2] who had succeeded Koenigsberger as director of the Development Planning Unit, University College London. The interest in Rosser's reappraisal of the concept is due not only to the fact that he points out the ageing of this style of planning, but, on the contrary, because he reaffirms the disastrous persistence of over-ambitious master planning and underlines the importance of communication between planning theory (about planning) and planning (procedural) practice. Rosser exhibits a typical pragmatist vein, when he confirms his interest "in the essential relationship between planning and implementation – as opposed to planning for its own sake: wall-to-wall planning or yet-more-comprehensive planning or yet-more-elaborate internalised models of the planning process in the abstract – the methodological pedantry which seems to characterise much planning discussion". It was in this context that he revisited the concept of action planning: "I want to argue that the direction of planning thought implied by

[2]The author is grateful to Harry Dimitriou for rescuing a file of Colin Rosser's address.

this concept is right – even if the ideas involved are excessively elementary, if the analysis on which they are based is currently flawed, and if the concept does indeed require intensive re-examination and 'programmatic' development. The concept as it stands seems to me to do little more than emphasise the need for a more positive developmental planning style (as a reaction to the sterile tenets of master plan formulation – still widespread). While it is far more being a new and more useful urban planning paradigm, it has the general merit of concentrating attention on the view of planning as 'getting things done rather than on thinking things out.' This is a useful beginning, given well-nigh universal past experience with 'planning' – and, in my view, is worth preserving on these grounds alone, provided we see the concept as a beginning and not an end of thinking with regard to planning theory (for those of us specially interested in this field – and it is certainly not my view that we all should be) … For all its elementary nature, 'action planning' is a theory of the planning process (a procedural statement) rather than of any particular relevance to what planning is about". The study of planning and the practice of planning are "two different animals. However, it is surely necessary for the study of planning to be 'modelled' on the practice of planning if the gap between academic theory and professional practice is not to be unbridgeable" (see also Leys 1969).

4.1.5 Informal Planning

The view espoused in the action planning model was heavily influenced by the conditions of LDCs, or in recent jargon "Global South", above all the dominance of foreign planning models, the diffuse reality of action, and the multiplicity of actors operating in highly informal modes of urban development. Informality has characterized most cities of the world at some stage of their development. "Informal planning, although non-institutionalized, leads to planned outcomes that serve particular interests. The informal planning path runs parallel to or mixes with formal planning, shaping the context of formal planning practice" (Briassoulis 1997). In the past literature of urban studies, informal planning was associated with the so-called Third World. It is still dominant in the Global South, and remains a key element in Southern theory. As McFarlane points out, on the basis of a Mumbai study, it never ceased to be a form of practice, often side-by-side with formal urbanism, but these two processes have changing relations, which he identifies as "speculation, composition, and bricolage" (McFarlane 2012, 89). In colonies, European-inspired urban planning coexisted with indigenous informal quarters (Triantari 2005, 2008; Kostopoulos 2015). Leontidou refers to the concept of "spontaneity", as proposed by Gramsci. She had already used it in her study of the Mediterranean city (Leontidou 1990) to describe spontaneous urban development reproduced by popular movements: "This is mass democracy in action, where leadership (and planning), are difficult to discern. Further on, there is an interaction between *popular* city-building and the creation of intersubjective realities. Self-built environments reinforce anti-planning conceptions and related movements which undermine the hegemony of conscious leading groups, like urban planners and administrators. Spontaneous land and housing movements, though emerging within capitalism, can be distinguished from all manifestations of modernism in the city, i.e., from market/speculation on the one hand, which is integrated in and created by capitalism and its modernist culture, and reform/planning on the other" (Leontidou 1996, 189).

An interesting question, from our present perspective, is whether informality of settlement is associated with social phenomena such as patronage and clientelism. In an impressive statistical analysis, Deuskar (2019) identified "a statistical correlation between clientelism (the informal provision of benefits, including urban land and services, to the poor in contingent exchange for political support) and informal urban growth, across a globally representative sample of 200 cities". In particular, he found a strong correlation between clientelism and informal land subdivisions. Informality is mainly identified with urban growth and remains a key feature of the Global South. "Much of the urban growth of the twenty-first century will take place in the cities of the Global South. It is therefore tempting to make the case for a 'Southern urbanism', one characterized by 'urban informality as a way of life', the 'habitus of the dispossessed'. But as I have already argued, urban informality is not the ecology of the mega-slum; rather, it is a mode of the production of space and a practice of planning. Not surprisingly, the dynamics of urban informality vary greatly from context to context" (Roy 2016, 533). This is a process which owes its importance to the dominant importance of housing in the developing world. "Housing constitutes some 60–80 percent of the developed land of urban areas and in the order of 50–70 percent of the value of the fixed capital formation of towns and cities … It is fundamental to alleviating the impacts of urban poverty and to poverty reduction, to social change and cultural conservation. Thus, the production, maintenance and management of housing play fundamental roles in developing cities" (Wakely 2019).

Indeed, the dynamics of informality not only vary but are often prevalent in peri-urban, even rural, space. Acuto et al. (2019) "note how informality has frequently been perceived as the formal's 'other' implying a necessary 'othering' of informality that creates dualisms between formal and informal, a localised informal and a globalising formal, or an informal resistance and a formal neoliberal control". And they add: "Transcending dualisms in informality and making it a valuable locus of urban theorising is a mission that, in our view, emerges from a well-established tradition. Academic studies on the question of urban informality produced across the Global North and South have long shown the limitations of inevitably thinking the in/formal in dichotomous terms". The duality of "formal-informal" can take the form of "order-disorder" which is an arguably dangerous simplification. This is a point made by Leontidou (1996, 180), with reference to large Mediterranean cities, like Athens and Lisbon: "Uniformity and alienation are broken by colourful mosaics, the absence of a welfare state is moderated by family solidarity, segregation is fractured by vertical differentiation, contractual relationships fade with informal social networks, elite residences as well as emphasis on leisure and tourism light up the dark streets of the inner city, zoned space is dissolved into a patchwork of economic activity". Understanding this sense of informality is a matter of appreciation of the difference of place and of integration of planning work in a different agenda, which, in the present author's view, is both radical and pragmatist, but inevitably becomes highly political. Informality, in the guise of the "temporary", "transient", and "removable", can be transposed to an otherwise highly-regulated urban landscape. It is, in Webb's expression, a form of "tactical urbanism" (Webb 2018). Wohl (2018) takes this discussion even further, beyond the grassroots initiative level for undertaking action and realizing an enactment: "Less attention has been paid to the 'hows' of enactment: with tactical interventions situated as insertions within a pre-existing entanglement of relational forces that are subsequently altered and reconstituted. Furthermore, while some tactical projects are conceived as prototypes that might become permanent, little work has reflected upon how this prototyping might be executed in a more systematic manner and thereby leveraged as a tool for planners".

4.1.6 *Militant Planning for Change*

Informality, as opposed to rational order, has been embraced by a variety of social movements, the work of which is evident in the South and had a direct impact on Southern theory, as we shall see later. In this context, the transfer of planning processes from the North to the South is no longer a simple matter of effectiveness, if it ever was. The political role of planners is no longer disputed even by hard core rationalists. The mere fact that they are dealing with land, territory and space is sufficient reason to politicize their work. "Territory is more than merely land, but a rendering of the emergent concept of 'space' as a political category: owned, distributed, mapped, calculated, bordered, and controlled" (Elden 2007, 578). Planning in these conditions can become easily "hegemonic" and oppressive (Jabareen 2018). However, the political nature of the work of planners does not necessarily mean that they turn against power. "The activist and disruptor of power has a legitimate, indeed honourable place in pluralist democracy. In the French critical tradition, Jacques Rancière views disruption of political regimes and their ethical legitimations as the essence of democracy … The question, however, is whether 'activist' and 'disruptor' are roles for planners. There is a well-established argument in planning theory that answers the question with a resounding 'yes'" (Low 2020, 39). Activism, in many parts of the Global South, is associated, generally or particularly in cities, with contemporary geopolitics. "It was not until the formal decolonization of the European empires beginning in the late fifties and the emergence of a small but significant dissident intellectual culture in the sixties that the imperial heritage of modern geography came to be recognized and questioned" (Ó Tuathaigh 1996, 44). In this context, insurgency is naturally a far greater movement than that which concerns planning theory. "So-called 'insurgent' groups have been able to come together to prize apart the structural assumptions of liberalism and demand empowerment and an end to forms of oppression … However, 'insurgency' is not limited to groups that assume egalitarian reciprocity as the foundation of social justice and fair judgment. As we have seen in recent years, 'alt-right' insurgencies in Britain, America and some countries of continental Europe have united those marginalised and excluded from the onward march of capitalism. Their rejection of liberal values has not led them to the intellectual 'left' but to social conservatism, projected by neo-fascist strongmen as utopias of the past in which people knew their place" (Low 2020, 185).

In the countries of the Global South insurgency is manifested principally in purely political terms, of which issues of land and urban planning are subordinate, although political insurgency and counterinsurgency create by themselves rearrangements on the ground, as spaces of security and insecurity (Ingram and Dodds 2009). However, when geopolitical expediency on the side of great powers leads to a policy of counterinsurgency, the latter may well mobilize social policy as an instrument. In such cases "counterinsurgency doctrine seeks to constitute governable national 'societies' with distinct 'social realms' in which populations are managed by 'social policy' intervention and the expansion of 'social' forms of control" (Owens 2011, 140). It is claimed, with undoubted hypocrisy, that this intervention is implemented with due respect for local cultural values (Gregory 2008). It is at this juncture that spatial planning, as a form of control, becomes relevant. All these phenomena are not new in the history of cities, nor are they exclusive to some parts of the planet. If there is a nightmarish condition in our times, it is well presented by Graham who speaks of "cities under siege" (Graham 2010a) and then elaborates as follows: "Cities … provide much more than just the *backdrop* or *environment* for war and terror. Rather, their buildings, assets, institutions, industries, infrastructures, cultural diversities, and symbolic meanings have long actually *themselves* been the explicit target for a wide range of deliberate, orchestrated, attacks" (Graham (2010b, 166). Whatever the location of these cities, in the North or South, in the world core or the periphery, "globalization takes the periphery straight into the centre, the frontier between First and Third Worlds starts running through the middle of world cities" (Misselwitz and Weizman 2003).

The interaction between social mobilization for the sake of social justice in general and planning policy on the ground is far from simple, but is certainly real. This is how Beauregard presents this situation: "Particularly within the just-city and right-to-the-city literatures, a belief exists that a just planning is only possible when social movements bring public pressure on insensitive governments and exploitative corporations to act in just ways. Significant social change happens through opposition and resistance. Social justice cannot be found in governmental planning that is, at best, reformist. More basic change is needed and this is unlikely if planners work within the government or advocate for relatively isolated groups. Planners must, instead, become involved in anti-globalization movements, anti-racist coalitions, citywide anti-gentrification initiatives, and corporate boycotts. They must develop an insurgent planning that is 'counter-hegemonic, transgressive, and imaginative'" (Beauregard 2020, 51–52). This is no easy task for a typical planner in the state administration. "This dilemma arises from the relatively privileged position of the planner compared to marginalized communities. On the one hand, planners can decide to use their status, knowledge, and professional skills to the fullest. In this way, they effectively promote their view of social justice in the planning process and can assertively represent the interests of marginalized communities to administrators and other influential actors. However, when they wield power in this way, they run the risk of assuming positions of superiority during struggles for equality and side-lining segments of marginalized communities with different conceptions of what justice entails or how it should be achieved" (Uitermark and Nicholls 2017, 33). In Allmendinger's view insurgent and radical planning are one and the same thing (Allmendinger 2017, 285): "Insurgency broadly involves a reaction against prevailing approaches and planning regimes and action or resistance to a strategy outwith the offered or state-sanctioned channels of involvement" (ibid.). Taking her cue from a comment of John Friedmann on insurgency, the Good Society, and the struggle for "becoming human", Rangan adds: "The struggle had to arise from within – surge up, *insurge* – the world of social planning to make it more dialogic, comprehensible, and meaningful for all, without resorting to the principle of hierarchy and power that forms the underlying logic of the state" (Rangan 2017a, 2). Miraftab places insurgent planning squarely in the context of neoliberalism: "What does insurgency mean for the practice and pedagogy of radical planning? Legitimation is central to hegemonic relations of power … [W]e have discussed how neoliberalism seeks legitimation through governance that promotes political inclusion, but avoids translating it into redistributive equity. Rather, neoliberalism's structures of inclusion and participation contain citizens' collective action into sanctioned spaces of invited citizenship – for example, formal, decentralized state channels or a legitimated NGO sector that functions to replace social movements. This strategy is often complemented by a bifurcated conceptualization of civil society as authentic versus a criminalized ultra-left. In such a context, radical planning practices should be insurgent. To promote social transformation, insurgent planning has to disrupt the attempts of neoliberal governance to stabilize oppressive relationships through inclusion" (Miraftab 2016, 489; Miraftab 2009). For Miraftab, "insurgent planning is transgressive in time, place, and action", which means that "it transgresses false dichotomies, by public actions spanning formal/informal arenas of politics and invited/invented spaces of citizenship practice. It transgresses national boundaries by building transnational solidarities of marginalized people. It transgresses time bounds by seeking a historicized consciousness and promoting historical memory of present experiences. Being transgressive, insurgent planning is not Eurocentric in its theorization. It rather recognizes how the global core and the peripheries North and South might exist within each other" (op.cit., 494). As he explained later, a change in state-citizens relation has placed planning on a different footing: "The planning profession as institutionalized in the West has traditionally served to mediate state relations with citizens and the market. Conditions that have reconfigured the state's relationship with citizens and the market have placed the idea and the practice of insurgent citizenship high on the political agenda; those same conditions have also had important implications for planning thought and theorization (Miraftab 2018, 276).

We have already seen Peter Marcuse's view of "critical planning" as he outlined it in 2020. In fact, this view can be considered as a further elaborated version of "transformative planning", i.e., of another variant in the same radical planning conception: "[I] think the principled planner … can see, in almost anything he or she does, the outline of what more really needs doing, of what beyond the immediately feasible is the ultimately desirable, what principle would dictate really should be. That involves bringing out the hidden dimension of the alternatives underlying the one dimension of the actual. And then shaping the actual and realistic goal so that it points in the direction of the hidden dimension, the ultimately desirable, and makes them visible – puts them into play, even if they are not currently implementable. I would consider such an approach, combining doing what can be done now with raising what should be done in the future, transformative planning. I commend this approach. I call this 'transformative planning' and locate it … between the social/liberal and the critical/radical" (Marcuse 2017, 46). Radical planning and transformative change are also present in John Friedmann's writings. Such change "cannot be achieved through mediation, consensus-building and deliberation" (Porter 2017, 12). Rangan develops further this line of thought to throw light on the dilemmas of radical planning: "The mediations of radical planning put radical practice in a contradictory position, standing in one instance alongside civil society groups *in opposition* to state and market, and in the second instance as intermediary *between them* to bring about positive outcomes through transformative theory and practice. Radical planning, from this mediating position, must, on the one hand, work with civil society groups selectively to delink from the state and market and become self-reliant and self-empowered. On the other hand, it needs to deploy transformative theory in realistic terms to enable civil society groups to work with (and within) existing structures of state and market" (Rangan 2017b, 98).

One can detect in Marcuse's definition a shade of pragmatism, but we should not overlook the fact that transformative planning was proposed in the context not of the Global South, but in that of a great industrial nation, the USA. We can recall the comments by Angotti, which were quoted earlier. Those expressed by Kennedy refer equally to the American situation: "Transformative community planning is a way of working with communities across divisions. It is not based on the superficial pasting together of short-lived, issue specific conditions, but on transforming relations between groups. In this sense, it is participatory planning which empowers the community to act in its own interests" (Kennedy 2020, 11). This form of community planning has attracted interest and activist mobilization in other industrial countries, apart from the USA (Wates 2000), sometimes successful even within adverse conditions (Tulumello 2016). Mobilization and resistance are often associated with opposition to urban redevelopment and regeneration schemes which displace residents, who become "immovable subjects" by refusing to move, as Inch et al. (2017) phrase it. Mobilization may be small-scale, transitional, and community-focused and therefore mistakenly classified as of "minor" significance compared to large-scale planning initiatives. However, it involves major issues of politics, and hence of planning, because, underneath it, lie explosive issues of poverty or oppression. In politics, there are arrangements which can be, according to Deleuze, major (or "molar") and minor (Low 2020, 40). "[M]inor politics, in Deleuze's thinking is a realm of creation that will emerge from the 'cramped spaces' of oppression in majoritarian and nominally pluralist societies" (ibid.).

4.1.7 Movements: Feminism and "Othering"

Social movements fighting for the rights of women or of racial and ethnic minorities had a significant influence on the conception of planning, and far beyond it. This is not surprising considering that group rights have been advocated by a broad range of movements "based on gender, race, sexual orientation, disability, age, and social class" (Dryzek and Dunleavey 2009, 201). From the viewpoint of changing attitudes embedded in planning what matters is the changes provoked by such new perspectives and understandings in the scientific justification of planning. "[E]pistemological standpoints and accompanying actions shape – and are shaped by – ethical principles, and … *epistemologies of ethical action* undergo periodic shifts in accordance with ever-changing philosophical understandings of the world. Thus, contemporary ways of knowing and doing are, on the whole, grounded in poststructural, postcolonial, subaltern or feminist scholarships that value subjective and situated knowledge" (Winkler 2018, 89). The influence on planning was not simply due to injustices arising from the spatial and social organization of cities; it was extended to the field of planning theory as well, e.g., in the form of gendered theory. "Initially (from the early 1970s), feminist theorists' interest in epistemology was provoked by the apparent exclusion of women from the epistemology that has dominated the social sciences since the nineteenth century, that of positivism … History was written from the point of view of men" (Sandercock and Forsyth 1996, 471–472). "Fundamentally, feminist thought comprises an attack on male domination. Within the urban realm feminists have contended that men design cities to serve male needs" (Fainstein 1996, 456). The rise of social movements claiming the "right to the city" in southern Europe is extensively reviewed by Leontidou (2010). In studying professional plans in Australia, Jean Hillier wrote, "I discovered quickly that women are virtually invisible in these plans … They are the Other, embedded in a metanarrative that is inattentive to diversity but denies its own biases" (Hillier 1996, 289). In the view of these theorists, history must be rewritten from the perspective of marginalized minorities, a view which is rejected by Sanyal who warns against an "identity trap" and insists that the unity of the historical perspective will be thus lost to the detriment of our understanding of forces that produce inequalities (Sanyal 2010, 338). The importance of issues and conditions like diversity, gender, race, ethnicity, otherness, marginalization, South-North inequalities, migration, and asylum-seeking has permeated planning thought (Speak and Kumar 2018, 155–159). The relation of "self" and "other", the concept of "reflexivity", the syndrome of "othering" are crucial parameters for understanding human spatial behaviour. "To reflect on the self in relation to space and society has been seen as a key with which to open up new kinds of human geographies which relate to individuals more closely, and which individuals can relate to more closely" (Cloke 1999, 44).

Diversity, division and difference are not of course static. Particular conditions, especially in cities, create a fluid landscape in which difference is measured in comparison to norms which constantly change. "In relation to the city, difference and diversity arguably are constitutive of the very urban experience itself in all its complexity of relations and subjectivities; the city concentrates difference which in turn defines the urban as a (quasi) distinct realm" (Watson 2017a, 387). One is bound to remember the old motto that the city makes men (and women of course) free. At least, it should do so.

4.1.8 Agonism and Antagonism

It is clear that militant versions of planning are not the exclusive preserve of Global South theorizing. Chantal Mouffe's notions of agonism and antagonism are a good example. Low quotes from her writings: "*Antagonism* is a struggle between enemies, while *agonism* is a struggle between adversaries. We can therefore reformulate our problem by saying that envisaged from the perspective of 'agonistic pluralism' the aim of democratic politics is to transform *antagonism* into *agonism*. This requires providing channels through which collective passions will be given ways to express themselves over issues which, while allowing enough possibility for identification, will not construct the opponent as an enemy, but as an adversary" (Low 2020, 31; see also Dryzek and Dunleavey 2009, 196). Low adds a doubt: "How different 'agonism' is from American pluralism is questionable". Referring to Mouffe's concept of "The political", he repeats that Mouffe "is bringing back the necessity of conflict into 'the political' reconceives pluralism as 'agonism'" (op.cit., 231). Hillier's opposite view had been that "agonistic space … is a political space embracing legitimate and public contestation over access to resources … Its pluralism is axiological, recognizing the impossibility of ever adjudicating without contest and without residue between competing visions" (Hillier 2003, 42). Kühn (2020) points out that "some planning researchers have already declared agonism to be a new paradigm for planning theory". Legacy et al. (2019, 274) are worried about this dangerous opposition. They observed that theoretical papers "have questioned the restrictive nature of the communicative planning – agonistic planning binary by highlighting how such a polarisation may serve to destabilise the possibilities of participatory planning". Allmendinger resorts to a quotation from McClymont to clarify the concept of agonism: "Agonism is defined as irresolvable disagreement over political meanings and actions, though each party accepts the right of the other to express an opinion. It is a form of political engagement which acknowledges the permanence of conflict, viewing this as necessary for democratic politics to function, rather than as detrimental to it" (Allmendinger 2017, 212). The rise of agonistic theories is placed in the context of a period of post-politics, in which conventional and established forms of governance were questioned. Post-politics is a period variously described as an era, a condition, an arrangement, an imaginary, a managerial framework, a rhetoric, a governance technique, or a form of practice (Metzger 2018, 186–187).

Regardless of whether agonism is a version of post-political pluralism, Mouffe's analysis has been transferred to planning theory, the reason being, according to Pløger (2018, 264), that "planning is permanent conflict. Conflict is partly a matter of how planning institutions are supposed to act according to regulations, but also planning institutions' own ideas about power, consensus and democracy … Agonism is critical mode of thinking about conflict and conflictual consensus. Chantal Mouffe has been a key figure in planning theory in this endeavour …". Pløger (2004), had explained earlier that the relevance of the concept of agonism is due to the fact that "the key complex of problems in contemporary planning is how to work with 'strife'". The elevation of agonism to the status of a theory taking over from communicative planning is explained by Roskamm (2015): "Any planning theory needs a notion or an idea of how things are related, how society works, how cities work. In contemporary planning theory, the perhaps most popular version of such an idea is the collaborative, consensual approach of communicative theory … Mouffe's agonistic pluralism is used as an alternative framework in planning theory, and it seems to be in the process of replacing communicative and collaborative planning theory as the hegemonic paradigm in the field". Roskamm's point is that "Mouffe's proposed 'agonistic pluralism' has an internal and fundamental flaw and that the advocated 'taming of antagonism into agonism' is neither possible nor necessary". Yamamoto (2020) too quotes the view of McAuliffe and Rogers who dispute the ethical basis of agonistic theory. "According to them, this is because Mouffe makes a sharp distinction between subjective and ethical or moral values that are beyond reason and political/normative values that are more susceptible to rational debate … They claim that a marked distinction Mouffe subsequently makes between these values prohibits her from analysing, e.g., why people persist in fighting for particular radically distinct values (e.g., truth, justice), empirical conditions that might facilitate what Mouffe calls the 'democratic task' of agonistic pluralism, or the transformation from more ethically driven antagonistic forms of politics towards more productive normative value-driven agonistic ones".

4.2 Southern Theory

Southern theory is a major component of the radical current of ideas and influences which arose as the theorists of the South consolidated their position in the international academic scene and their access to the production of planning literature. Connell (2014), who has herself argued vigorously for a "southern theory", outlines the syndrome of "academic dependency" as follows: "There is a global division of labour, running through the history of modern science and still powerful today. The role of the periphery is to supply data, and later to apply knowledge in the form of technology and method. The role of the metropole, as well as producing data, is to collate and process data, producing theory (including methodology) and developing applications which are later exported to the periphery. Within this structure, Hountondji argues, the attitude of intellectuals in the periphery is one of 'extraversion', that is, being oriented to sources of authority outside their own society. This is very familiar in academic practice even in a rich peripheral country like Australia. We travel to Berkeley for advanced training, take our sabbatical in Cambridge, invite a Yale professor to give our keynote address, visit a Berlin laboratory, teach from US textbooks, read theory from Paris and try to publish our papers in *Nature* or the *American Economic Review*". To be noted that, as an English-speaking intellectual, Connell does not suffer from the additional handicap of academics whose native language is one spoken by a very limited potential audience. The reality of dependence that Connell stresses, albeit based on the experience of a far from "developing" country, is of prime interest for planning theory because it highlights the North-South divide and the dominant place of northern theory in southern conditions. "A central theme is that ideas deployed in much social theory and description originate in the north of the world and that ideas originating in the south are ignored in these hegemonic accounts" (Mabin 2017, 23).

Yiftachel outlines with great clarity the situation: "The main debates in planning theory during the last decade and a half have been commonly described as 'communicative', 'deliberative' or 'discursive', focusing on finding analytical and normative frameworks to understand and mobilize planners … Most theories emerging from the North-West have … concentrated on *planners rather than planning*, the latter standing for the broader arena of publicly guided transformation of space … The emphasis on planners and decision processes has left a particular void for those working in the diverse south-eastern settings … From south-eastern perspectives, the credibility of leading communicative theories is challenged by a constant mismatch with a wide range of south-eastern 'stubborn realities' (to invoke a useful Gramscian term), where liberalism is not a stable constitutional order, but at best a sectoral and mainly economic agenda; where property systems are fluid; inter-group conflicts over territory inform daily practices and result in the essentialization of 'deep' ethnic, caste and racial identities. As perceptively noted by Vanessa Watson …, this has resulted in the development of conflicting and often irreconcilable rationalities" (Yiftachel 2006, 212–213). Interestingly, Yiftachel turns to another crucial dimension of planning practice and theory, i.e., to the role of the state. His emphasis, naturally, is placed on conditions of ethnic inequality. "Theories of the state abound, and their impact on planning theory has been significant. However, these have centred on discussions of the state's (and planning's) role in facilitating capital accumulation and developing regimes of regulation, surveillance and developmentalism. Little has been said in the planning literature on a most fundamental function of the modern state – imposing ethno-national spatial control, often in conflict with groups holding counter territorial claims" (op.cit., 217). The implication from these grave omissions is that the education of planners and professional attitudes are deeply affected. "While planning theory may influence the work of practitioners far less than theorists hope, it nonetheless can and does, perhaps in uneven and unpredictable ways, shift both pedagogy and professional approaches. Hence when ideas about planning processes and outcomes (spatial models and urban forms), developed to address issues in other parts of the world, are inappropriately transported and applied in very different contexts in other parts of the world, it can directly affect lives and livelihoods. This sometimes happens in ways that cause mere inconvenience, but also can, and often does, impose unforeseen costs and hardship on those who are poor and vulnerable" (Watson 2017b, 99). From a Southern and radical viewpoint, Watson's remarks lead directly to the need of independent research or "engaged scholarship". In Winkler's words, "engaged scholarships allow for the coproduction of knowledge by explicitly desisting from establishing narrow strategies that converge on a 'correct' answer'. Instead, such scholarships involve multiple perspectives on how to tackle an identified problem" (Winkler 2017, 220).

The concept of the Global South and the notion of "southern" are fairly recent additions, at least in the planning literature. Roy argues "that is necessary to craft 'new geographies of theory', those that can draw upon 'the urban experience of the global South … I mean the global south as a 'concept-metaphor' that interrupts the 'flat world' conceits of globalization" (Roy 2017, 15). Global South is a "distinctive vantage point". De Satgé and Watson admit having borrowed the terms from Dados and Connell when they write: "Global South functions as more than a metaphor for underdevelopment. It references an entire history of colonialism, neo-imperialism, and differential economic and social change through which large inequalities in living standards, life expectancy and access to resources are maintained, and opens new possibilities in politics and social science" (de Satgé and Watson 2018, 30). It is obvious that these distinctive qualities appear in different shades and forms outside the North and that a Southern turn does not concern planning only. It is necessary to stress here that for a country outside the developed core of the world, being in the periphery and having been relegated to dependent status have relative scale values. Adopting foreign practices or copying inappropriate institutional arrangements may be the outcome of a wide gamut of situations, colonial, semi-colonial, or quasi-colonial. But, as far as planning theory and urbanism studies are concerned it is important to recognize the marks of their origins before giving them worldwide applicability. Southern planning theory does not aim at replacing its northern counterpart as a universal theoretical construct. "[A] southern theorising project can be viewed as developing critical perspectives on existing theory and practice, rather than attempting to establish theory (pseudo-universal theory) which claims to apply to all parts of the world" (op.cit., 24). Southern theory is not simply a matter of different voices and multiple subjectivities, which communicative theory had already emphasized. It has to do in addition, as Beauregard underlines, with "transnational relationships of two types. First, there are the global flows of finance, people, commodities, and ideas and their impact on cities and regions … Second, there is the development planning that originates in the global North (and West) – the metropole – and is practiced in the global South (and East) – the periphery. The critical premise is [that] … planning knowledge is sensitive to and shaped by context. As regards a global perspective, planning's ideas and its theory are primarily generated in affluent and powerful countries where they are supported by nation states, transnational corporations, elite universities, well-funded development agencies, and international financial bodies. The questions thus arise as to whether this North-West planning knowledge is sufficiently attentive to and appropriate for global relations as seen from other places and whether European and liberal democratic ways of thinking and acting are applicable in the global South-East" (Beauregard 2020, 40). Porter is right when she refers to "decolonization of planning theory" (Porter 2018, 175).

Based on his experience in South Africa, Harrison argued for an anti-realist approach: "I begin by reflecting on the impasse in planning theory, referring to the introspection within the field itself and to my personal experience as a planner in a state bureaucracy. I do give serious consideration to the possibility that 'the southwards turn' in planning may break the impasse but conclude that this will be difficult to achieve within the current frame of anti-realist ontology" (Harrison 2014, 66). "The anti-realist tendency initially provoked anxiety among planning theorists …, but, by the late 1990s, it was hardly contested". Harrison invokes Allmendinger to support his argument that "theories must be placed in their larger historical and social contexts" (op.cit., 67). In his view, "one of the most important developments in planning theory over the past half-decade or so has been a partial shift to intellectual vantage points in the global South (or South-East as Yiftachel puts it). Watson … pointed to the inadequacy of a body of planning theory produced almost exclusively in the global North, and by mid-decade, a handful of planning theorists were proactively and self-consciously taking the 'southwards turn'" (op. cit., 69). Indeed, Vanessa Watson has repeatedly blamed the transplant of northern systems to southern regions for the failure to recognize local conditions: "[T]he fact remains that in most of these regions the planning systems in place have been either inherited from previous colonial governments or have been adopted from Northern contexts to suit particular local political and ideological ends … The position taken here is that a significant gap has opened up between increasingly technomanagerial and marketised systems of government administration, service provision and planning (including, frequently, older forms of planning) and the every-day lives of a marginalised and impoverished urban population …" (Watson 2016b, 541; see also Watson 2009). This leads to a renewed conception of southern planning theory: "A view of planning from the global south argues for contextualized and historicized grounded research which also recognizes the location of any place and process in a system of global relations" (Watson 2017b, 105). The author's intention in this book is to do precisely that, not necessarily out of a radical inclination, but rather out of theoretical honesty.

The transfer of knowledge and what is known as "best practice" from developed to developing countries has been a standard routine in planning. However, this routine may ignore the particularities of places where an instrument is used on the strength of its success in a different context. "Many planning ideas energetically promoted and 'carried' by strong advocates fail to make much progress when they arrive in a new context. This can be because, locally, key people were looking for something different, or because of tensions and struggles over whether to adopt a particular approach, or because the arena in which the exogenous planning idea was promoted was not connected to stakeholders whose backing was essential for an idea to work" (Healey 2010, 9). Lieto (2015) warns against the uncritical adoption of planning ideas: "In the current debate on cross-national planning, scholars maintain that planners need good justifications for an idea already experienced elsewhere to be able to implement it in a specific context. They should know where and how the idea came about, and why it succeeded in its 'origin site' … The underlying assumption is that ideas cannot be treated as templates deprived of their biases and history. When they land in a destination, they will inevitably go through a complex process of translation and become different from the 'original version'. What travels is not just 'ideas', unless we use this term in its generic, conversational meaning. Nor should we consider that what travels is a truth statement which reflects empirical realities in a mostly unmediated way". As Beza (2016) warns, "in a planning and design sense best practice encompasses a transfer of expert knowledge developed in one setting to address particular issues which, through recognition and interest, is then applied in another setting to achieve the same desired improvement. Context and social appropriateness when applying best practice in the new setting, however, are areas of concern … and may largely be overlooked by the designer acting as prime transfer agent …". Even the efforts of international agencies proved inadequate to secure successful development planning, as Archibugi (2008, 3) admits, for a variety of reasons linked to the relationships of North and South. S.V. Ward (2010, 48) has developed a typology, within which the principal distinction "lay between 'borrowing', in which the inward flow of planning knowledge was shaped by indigenous agency, and 'imposition', in which it was shaped by external actors". The problem of transferring "best practices" by well-intentioned consultants was not limited to developing countries. Hillier refers to Australia, on the basis of work by Brian McLoughlin, who "suggests that such problems are exacerbated by the extensive use of consultants. Consultants' reports depend on briefs, which have tended to think in specific and linear 'silos', such as demographic analysis, retail forecasts and so on, not encouraging relational thinking. Moreover, consultants present their reports at the end of their contract and 'walk away', generally leaving issues of relationality, implementation and so on unaddressed" (Hillier 2013, 34). It should be recognized, however, that the transfer of "best practices", incidentally a favourite in European Union policy (Fischer 1999), is not necessarily bound to fail, as even studies in the Global South admit (Dávila 2017). In certain conditions imported practices may be useful not to say welcome, as Sorensen explained: "Ideas from elsewhere can be useful because they arrive with ready-made expert analysis, and/or moral authority that enables policy entrepreneurs to frame credible, legitimate solutions to pressing local problems. The borrowed legitimacy of imported planning ideas may be especially important for smaller actors, who are politically and economically weaker, and who lack the ability either to impose ideas or solutions on other actors or even to pursue them autonomously without permission or aid. For such weaker actors, the endorsement, credibility, and imaginaries of proponents elsewhere may be key to being taken seriously. This is likely to be particularly true for advocates of improved urban conditions and greater sustainability and livability, who are so often weaker politically and economically than those pursuing profit through urban development or redevelopment. Planning ideas, then, can be deployed as strategic resources by actors engaged in struggles over the control of urban development and change processes" (Sorensen 2010, 135). Sorensen is probably over-optimistic because even the most impeccable credentials of a best practice cannot guarantee its success. In plenty of cases best practices either fail, or remain unused and are quickly forgotten (see Text-boxes 9.1 and 9.2, in Chap. 9).

Vanessa Watson's warning regarding the systems of government turns the light to the fundamental role of the state and to government failures in the conditions of the Global South, especially in urban regions where the inadequacy of practices borrowed from the North is mostly felt. Roy (2018, 153) argues "that a theory of the state must be advanced in relation to a theory of the city". He insists on "the question of historical conjuncture" and on the historical conditions under which the urban becomes "the terrain of government and the territory of politics". Roy is right in asserting that the state is a critical parameter in planning theory, as it will be discussed again in Chap. 7, but we must not overlook the web of forces that constitute the "state" in its semi-formal sense. The state's impact in practice is expressed through its established, and often little discussed, norms of action. These are embedded in institutions and regulations which evolve into a stable framework that is taken for granted independently of individual actors. "While recognizing systemic power and the pervasive impacts of powerful actors, institutions such as norms and shared understandings are usually conceived as operating beyond the scale of individual actors or groups" (Sorensen 2018, 252). This unorthodox nature of the state apparatus is increasingly reflected in Global South planning theory, as Vidyarthi (2018) emphasizes in connection to India: "Scholarship in this vein recognizes that city planning is not an exclusive prerogative of the state and its statutory agencies, or capitalists and their clever consultants, but city planning work on the actual ground is in fact a much broader enterprise involving a diverse range of urban actors (homeowners, squatters, developers, politicians, etc.) making different kinds of plans (informal, tacit, spontaneous, incremental, formal, and more … I find that India's contemporary planning practices constitute a dynamic arena characterized by many existing and emergent players pursuing diverse aims and purposes. This is a significant shift, which stands out in even sharper relief when seen against the rather staid and bureaucratic model of expert-oriented and state-centered planning adopted in the post-independence period". In a way, Vidyarthi reiterates here the observations regarding "informality".

4.3 Neoliberalism

The dissemination of best practices has been a frequent occurrence in urban and regional planning, affecting legislation, rules, instruments, procedures and site design. But what is relevant in the context of this analysis is the modes of planning promoted in the current neoliberal and globalization period. "Neo-liberalism is a keyword for understanding the urban condition, the re-thinking of social democracy, and the regulatory regimes of our time. Neo-liberalism mobilises urban space as an arena for market-oriented economic growth and elite consumption practices, and in so doing it transforms the politico-economic setting in which public plans and projects are implemented" (Sager 2011, 149). Neoliberalism is increasingly associated with mega-projects, mega-events, large scale leisure facilities and *nouveau-riche* developments capable of exciting the fantasy of the consumer, by encouraging an illusion of cosmopolitanism, urban imaginings and an enchanted city (Sandercock 2016, 410). "The speeded-up inter-referencing of city visions across the globe, where images of Dubai and Shanghai are impacting on the London skyline much as they are on the emerging profiles of Lagos and Kigali … speaks to a growing planning mono-culture in which even older justifications for planning as serving the public good now seem quaintly romantic" (Watson 2016a, 663).

This a form of social engineering of the high modernism variety. "[I]t is the ideology par excellence of the bureaucrats, intelligentsia, technicians, planners, and engineers … This vision of a great future is often in sharp contrast to the disorder, misery, and unseemly scramble for petty advantage that the elites very likely see in their daily foreground" (Scott 2016, 83). The economic crisis which started in 2008 made the contrast all the more painful. "The current economic downturn and the rise of austerity-driven public policies have further sharpened these inherent contradictions between ideals of sociocratic discovery and modernist urban transformations" (Savini et al. 2015, 297). "Planning for fantasy" is not a new slogan at least in the developed North-West (Ward 1983), but theme parks, shopping malls, sports venues, spectacular architectural designs, tourist resorts, casinos, dizzying skyscrapers etc., and more generally the "McDonaldization of society", have a very modern brand (Hannigan 2002). Disney World is a "utopia of leisure": "At Disneyland one is constantly poised in a condition of becoming, always someplace that is 'like' someplace else" (Sorkin 2002, 342). Such symbols are identified with successful, creative and smart cities, capable of attracting entrepreneurial talent. They represent a new mode of capitalist organization, what Rossi (2019) calls "platform capitalism", capable to ensure "the extraction of value from society and life itself", and, at the same time give a new meaning to urbanism which, if spread worldwide, can be a force of urban creativity. Large scale sports venues have a similar orientation, as "over-the-top, intensely corporate-dominated visions", as Williamson (2020) describes the Rio de Janeiro Olympics. It is interesting that this neoliberal form of planning can be associated, as Gunder claims, with the distortive exploitation of communicative ideas by neoliberal forces: "During the third quarter of the last century, the rational scientific management of space reflected the one-dimensional positivism of instrumental planning posited by Faludi's planning theory. In more recent years, the reintroduction of values into planning via 'ideologically freed' communicative planning theory facilitated its very hegemonic capture by the neoliberal supporting state. 'Nodal points' of planning concern emerged as unquestioned planning deficiencies requiring resolution: global competiveness, sustainable development and 'appropriate' urban design that facilitated the attraction of talent to globally ranked world cities, all became topics of collaborative planning discourse that sought to promise fantasies of harmony, security and above all – enjoyment – within the cities and populations for which planning provides both hope and discipline" (Gunder 2010, 308–309). Many observers have associated the onslaught of neoliberalism and ostentatious projects with the financial crisis of 2008 and attributed the reaction of activist planners to the crisis itself. Kunzmann has responded that "as a rule, activist planning, predominantly in bigger metropolitan cities, is responding to neoliberal challenges and economic recession, not to the financial crisis as such" (Kunzmann 2016, 1315–1316). The signs of cosmopolitan city-views and mega-projects may be few in countries outside the orbit of advanced economies, but they gradually emerge in these contexts too (see Text-box 8.11).

Epilogue

In this chapter we had the opportunity to glimpse at the variety of alternative conceptions of planning which dispute the universality of mainstream planning theories. They do not yet represent a cohesive body of knowledge, but they have a common motivation, which is mainly manifested in the difference of their vantage points. They look at planning from the perspective of less advantaged, marginalized, peripheral, or dependent situations. They do not claim global validity, because, if they did, they would deny the *raison d'être* of the situations they explain and represent, their identity as "outsiders", of being on the margin, or at the edge of the North-West, of being in the South and East, often in a metaphorical sense. These theoretical alternatives essentially advocate a multiplicity of planning theories, of which the unifying theme is hard to single out. But there is a global theme which might unite them, and this is no other than their joint concern. The fate of their common world, the planet of which they are all part. Can the global environment, the developments that threaten it, and its sustainable future provide a unifying framework? Can the acceleration of climate change provide a common platform, which all theories will share? If the answer is positive, it will have implications not only for the practice of planning, but also for its theoretical explanation. This is what the author had in mind when he referred in Chap. 2 to the climate current, which is the object of the next chapter.

References[3]

Acuto M, Dinardi C, Marx C (2019) Transcending (in)formal urbanism. US 56(3):475–487

Aldhous W et al (eds) (1983) Otto Koenigsberger festschrift: action planning and responsive design. Habitat Int 7:5/6

---

[3] **Note**: The titles of publications in Greek have been translated by the author. The indication "in Greek" appears at the end of the title. For acronyms used in the references, apart from journal titles, see at the end of the table of diagrams, figures and text-boxes at the beginning of the book.

Abbreviations of journal titles: AT Architektonika Themata (Gr); DEP Diethnis kai Evropaiki Politiki (Gr); EpKE Epitheorisi Koinonikon Erevnon (Gr); EPS European Planning Studies; PeD Perivallon kai Dikaio (Gr); PoPer Poli kai Perifereia (Gr); PPR Planning Practice and Research; PT Planning Theory; PTP Planning Theory and Practice; SyT Synchrona Themata (Gr); TeC Technika Chronika (Gr); UG Urban Geography; US Urban Studies; USc Urban Science. The letters Gr indicate a Greek journal. The titles of other journals are mentioned in full.

Alexander ER (1992) Approaches to planning: introducing current planning theories, concepts and issues. Gordon and Breach, Luxembourg
Allmendinger P (2017) Planning theory, 3rd edn. Palgrave, Basingstoke
Angotti T (2020) Introduction to transformative planning. In: Angotti T (ed) (2020) transformative planning: radical alternatives to neoliberal urbanism. Black Rose Books, Montréal, pp 4–10
Archibugi F (2008) Planning theory: from the political debate to the methodological reconstruction. Springer Verlag Italia, Milan
Banfield EC (1973) Ends and means in planning. In: Faludi A (ed) A reader in planning theory. Pergamon Press, Oxford, pp 139–149. (first published 1959)
Barry J et al (2018) Unsettling planning theory. PT 17(3):418–438
Basta C (2016) From justice *in* planning toward planning *for* justice: a capability approach. PT 15(2):190–212
Baum H (2011) Planning and the problem of evil. PT 10(2):103–123
Baum H (2015) Planning with half a mind: why planners resist emotion. PTP 16(4):498–516
Beard VA (2003) Learning radical planning: the power of collective action. PT 2(1):13–35
Beauregard RA (2020) Advanced introduction to planning theory. Elgar, Cheltenham
Beier G et al (1975) The task ahead for the cities of the developing countries, World Bank Staff Working Paper No 209. World Bank, Washington, DC
Bendavid-Val A (1975) The themes-strategy-projects approach to planning for regional development. In: Bendavid-Val A, Waller PP (eds) (1975) action-oriented approach to regional development planning. Praeger, New York, pp 3–17
Bendavid-Val A, Waller PP (eds) (1975) Action-oriented approach to regional development planning
Beza BB (2016) The role of deliberative planning in translating best practice into good practice: from placeless-ness to placemaking. PTP 17(2):244–263
Bourne LS, Sinclair R, Dziewoński K (eds) (1984) Urbanization and settlement systems. Oxford University Press, Oxford
Breese G (1966) Urbanization in newly developing countries. Prentice-Hall, Englewood Cliffs
Breese G (ed) (1969) The city in newly developing countries: readings on urbanism and urbanization. Prentice-Hall, Englewood Cliffs
Briassoulis H (1997) How the others plan: exploring the shape and forms of informal planning. J Plan Educ Res 17:105–117
Brooks MP (2017) Planning theory for practitioners. Routledge, London. (first published 2002)
Bruce CMF (1980) The stages of project planning: an introduction to project planning, EDI training materials / course note series. International Bank for Reconstruction and Development (World Bank), Washington DC
Butler J, Crooke P (1973) Urbanisation. Angus and Robertson, London
Cloke P (1999) Self – other. In: Cloke P, Crang P, Goodwin M (eds) Introducing human geographies. Arnold, London, pp 43–53
Connell R (2014) Using southern theory: decolonizing social thought in theory, research and application. PT 13(2):210–223
Dávila JD (2017) Urban fragmentation, "good governance" and the emergence of the competitive city. In: Parnell S, Oldfield S (eds) The Routledge handbook on cities of the global south. Routledge, London. (first published 2014), pp 474–486
Davy B (2014) Spatial planning and human rights. PT 13(4):329–330
de Satgé R, Watson V (2018) Urban planning in the global south: conflicting rationalities in contested urban space. Palgrave Macmillan/Springer, Cham
Demko GJ, Rose HM, Schnell GA (eds) (1970) Population geography: a reader. McGraw-Hill, New York
Deuskar C (2019) Informal urbanisation and clientelism: measuring the global relationship. US 57(12):2473–2490
Donaldson P (1973) Worlds apart. Penguin, Harmondsworth
Drakakis-Smith D (2000) Third world cities. Routledge, London

Dryzek JS, Dunleavey P (2009) Theories of the democratic state. Macmillan International/Red Globe Press, London
Elden S (2007) Governmentality, calculation, territory. Environ Plan D Soc Space 25(3):562–580
Fainstein SS (1996) Planning in a different voice. In: Campbell S, Fainstein S (eds) Readings in planning theory. Blackwell, Oxford, pp 456-460 (first published 1992)
Fainstein SS (2016) Spatial justice and planning. In: Fainstein SS, DeFilippis J (eds) Readings in planning theory. Wiley – Blackwell, Chichester, pp 258–272. (first published 2013)
Fainstein SS (2018) Urban planning and social justice. In: Gunder M, Madanipour A, Watson V (eds) The Routledge handbook of planning theory. Routledge, London, pp 130–142
Fainstein NI, Fainstein SS (1982) New debates in urban planning: the impact of Marxist theory within the United States. In: Paris C (ed) Critical readings in planning theory. Pergamon Press, Oxford, pp 147–173. (originally published 1979)
Fanon F (1967) The wretched of the earth. Penguin, Harmondsworth. (French original 1961)
Fischer H (1999) Good practice in view of problems of sustainability. Paper presented at the international conference held in Athens in May 1999 on "Sustainable Development and Spatial Planning in the European Union: Prospects for the twenty-first century in the European Union, its member states, the Balkans and the Black Sea countries. National Technical University of Athens
Fisher S et al (2000) Working with conflict: skills and strategies for action. ZED Books, London
Flyvbjerg B (1996) Commentary – the dark side of planning: rationality and '*Realrationalität*'. In: Mandelbaum SJ, Mazza L, Burchell RW (eds) Explorations in planning theory. Center for Urban Policy Research/Rutgers, New Brunswick, pp 383–394
Friedly PH (1974) National policy responses to urban growth. Saxon House/Lexington Books, Westmead/Hants
Friedmann J (1995) Planning in the public domain: discourse and praxis. In: Stein JM (ed) Classic readings in urban planning. McGraw-Hill, New York, pp 74–79. (first published 1989)
Furtado C (1971) Development and underdevelopment. University of California Press, Berkeley
Graham S (2010a) Cities under siege: the new military urbanism. Verso, London
Graham S (2010b) Postmortem city: towards an urban geopolitics. City 8(2):165–196
Gregory D (2008) 'The rush to the intimate': counterinsurgency and the cultural turn. Radic Philos 150:8–23
Griffin KB, Enos JL (1970) Planning development. Addison – Wesley, London
Gunder M (2010) Planning as the ideology of (neoliberal) space. PT 9(4):298–314
Hannigan J (2002) Fantasy city: pleasure and profit in the postmodern metropolis. In: Fainstein SS, Campbell S (eds) Readings in urban theory. Blackwell, Oxford, pp 305–324. (first published 1998)
Hardoy JE (ed) (1975) Urbanization in Latin America. Anchor Books, Garden City
Hardoy JE, Satterthwaite D (1981) Shelter: need and response. Wiley, Chichester
Hardoy JE, Satterthwaite D (1989) Squatter citizen: life in the urban third world. Earthscan, London
Harper TL, Stein SM (1996) Postmodernist planning theory: the incommensurability premise. In: Mandelbaum SJ, Mazza L, Burchell RW (eds) Explorations in planning theory. Center for Urban Policy Research/Rutgers, New Brunswick, pp 414–429
Harris N (ed) (1992) Cities in the 1990s: the challenge for developing countries. UCL Press, London
Harris N, Fabricius I (eds) (1996) Cities and structural adjustment. UCL Press, London
Harrison P (2014) Making planning theory real. PT 13(1):65–81
Harvey D (1973) Social justice and the city. Edward Arnold, London
Healey P (2010) Introduction: the transnational flow of knowledge and expertise in the planning field. In: Healey P, Upton R (eds) Crossing borders: international exchange and planning practices. Routledge, London, pp 1–25
Herbert JD (1979) Urban development in the third world: policy guidelines. Praeger, New York
Herbert JD, Van Huyck AP (eds) (1968) Urban planning in the developing countries. Frederick A Praeger, New York

Hillier J (1996) Deconstructing the discourse of planning. In: Mandelbaum SJ, Mazza L, Burchell RW (eds) Explorations in planning theory. Center for Urban Policy Research/Rutgers, New Brunswick, pp 289–298
Hillier J (2003) Agon'izing over consensus: why Habermasian ideals cannot be 'real. PT 2(1):37–59
Hillier J (2005) Straddling the post-structuralist abyss: between transcendence and immanence? PT 4(3):271–299
Hillier J (2008) Plan(e) speaking: a multiplanar theory of spatial planning. PT 7(1):24–50
Hillier J (2013) On relationality and uncertainty. Plan Rev 49(3):32–39
Hillier J (2017) On planning for not having a plan? PTP 18(4):668–675
Hillier J (2018) Lines of becoming. In: Gunder M, Madanipour A, Watson V (eds) The Routledge handbook of planning theory. Routledge, London, pp 337–350
Huxley M (2018) Countering "the dark side" of planning: power, governmentality, counter-conduct. In: Gunder M, Madanipour A, Watson V (eds) The Routledge handbook of planning theory. Routledge, London, pp 207–220
Inch A et al (2017) Planning in the face of immovable subjects: a dialogue about resistance to development forces. PTP 18(3):469–488
Ingram A, Dodds K (2009) Chapter 1: spaces of security and insecurity: geographies of the war on terror. In: Ingram A, Dodds K (eds) Spaces of security and insecurity: geographies of the war on terror. Ashgate, Farnham
Jabareen Y (2018) Hegemonic planning and marginalizing people. In: Gunder M, Madanipour A, Watson V (eds) The Routledge handbook of planning theory. Routledge, London, pp 289–302
Jacobson L (1980) The sketch plan concept: a method for programming action, Regional planning and area development project, international studies and programs. University of Wisconsin, Madison
Johnson EAJ (1970) The organization of space in developing countries. Harvard University Press, Cambridge, MA
Jones R (ed) (1975) Essays on world urbanization. George Philip and Son, London
Kennedy M (2020) Transformative planning for community development. In: Angotti T (ed) Transformative planning: radical alternatives to neoliberal urbanism. Black Rose Books, Montréal, pp 11–24
King AD (2007) Colonial urban development: culture, social power and environment. Routledge, London. (first published 1976)
Koenigsberger O (1964) Action planning. Architectural Association Journal, May
Koenigsberger O (n.d.) Action planning. In: Mumtaz B (ed) Readings in action planning – Vol.I, DPU working paper no 13. Development Planning Unit, University College London, London, pp 2–9. (originally published 1964)
Kostopoulos D (2015) The production of space in the Dodecanese Islands during the period of Italian occupation: the case of Leros. Doctoral thesis. National Technical University of Athens (in Greek)
Kühn M (2020) Agonistic planning theory revisited: the planner's role in dealing with conflict. PT 20(2):143–156
Kunzmann KR (2016) Crisis and urban planning? A commentary. EPS 24(7):1313–1318
Legacy C et al (2019) Beyond the post-political: exploring the relational and situated dynamics of consensus and conflict in planning. PT 18(3):273–281
Leontidou L (1990) The Mediterranean city in transition. Cambridge University Press, Cambridge
Leontidou L (1996) Alternatives to modernism in (southern) urban theory: exploring in-between spaces. Int J Urban Reg Res 20(2):178–195
Leontidou L (2010) Urban social movements in 'weak' civil societies: the right to the City and cosmopolitan activism in southern Europe. US 47(6):1179–1203
Lewis WA (1966) Development planning. George Allen and Unwin, London
Leys C (1969) A new conception of planning? Paper for a conference. The Institute of Development Studies, University of Sussex

Lieto L (2015) Cross-border mythologies: the problem with traveling planning ideas. PT 14(2):115–129

Linn JF (1979) Policies for efficient and equitable growth of cities in developing countries, World Bank staff working paper no 342. World Bank, Washington DC

Linn JF (1983) Cities in the developing world. Oxford University Press, New York

Low N (2020) Being a planner in society: for people, planet, place. Edward Elgar, Cheltenham

Mabin A (2017) Grounding southern city theory in time and place. In: Parnell S, Oldfield S (eds) The Routledge handbook on cities of the global south. Routledge, London. (first published 2014), pp 21–36

Mabogunje AL, Hardoy JE, Misra RP (1978) Shelter provision in developing countries: the influence of standards and criteria. Wiley, Chichester

Marcuse P (2017) From utopian and realistic to transformative planning. In: Haselsberger B (ed) Encounters in planning thought: 16 autobiographical essays from key thinkers in spatial planning. Routledge, New York/London, pp 35–50

Marcuse P (2020) Changing times, changing planning: critical planning today. In: Angotti T (ed) Transformative planning: radical alternatives to neoliberal urbanism. Black Rose Books, Montréal, pp 30–35. (first published 2010)

McClymont K (2019) Articulating virtue: planning ethics within and beyond post politics. PT 18(3):282–299

McConnell S (1981) Theories for planning. Heinemann, London

McFarlane C (2012) Rethinking informality: politics, crisis, and the city. PTP 13(1):89–108

Metzger J (2018) Postpolitics and planning. In: Gunder M, Madanipour A, Watson V (eds) The Routledge handbook of planning theory. Routledge, London, pp 180–193

Metzger J et al (2017) 'Power' is that which remains to be explained: dispelling the ominous dark matter of critical planning studies. PT 16(2):203–222

Miraftab F (2009) Insurgent planning: situating radical planning in the global south. PT 8(1):32–50

Miraftab F (2016) Insurgent planning. In: Fainstein SS, DeFilippis J (eds) Readings in planning theory. Wiley – Blackwell, Chichester, pp 480–498. (first published 2009)

Miraftab F (2018) Insurgent practices and decolonization of future(s). In: Gunder M, Madanipour A, Watson V (eds) The Routledge handbook of planning theory. Routledge, London, pp 276–288

Misselwitz P, Weizman E (2003) Military operations as urban planning. Mute [www.metamute.org]

Moroni S (2020) The just city. Three background issues: institutional justice and spatial justice, social justice and distributive justice, concept of justice and conceptions of justice. PT 19(3):251–267

Ó Tuathaigh G (1996) Critical geopolitics: the politics of writing global space. University of Minnesota Press, Minneapolis

Owens P (2011) From Bismarck to Petraeus: the question of the social and the social question in counterinsurgency. Eur J Int Rel 19(1):139–161

Pacione M (ed) (1981a) Urban problems and planning in the developed world. Croom Helm, London

Pacione M (ed) (1981b) Problems and planning in third world cities. Croom Helm, London

Pløger J (2004) Strife: urban planning and agonism. PT 3(1):71–92

Pløger J (2018) Conflict and agonism. In: Gunder M, Madanipour A, Watson V (eds) The Routledge handbook of planning theory. Routledge, London, pp 264–275

Porter L (2017) "Resistance is never wasted": reflections on Friedmann and hope. In: Rangan H et al (eds) Insurgencies and revolutions: reflections on John Friedmann's contributions to planning theory and practice, RTPI library series. Routledge/The Royal Town Planning Institute, New York/London, pp 7–15

Porter L (2018) Postcolonial consequences and new meanings. In: Gunder M, Madanipour A, Watson V (eds) The Routledge handbook of planning theory. Routledge, London, pp 167–179

Rangan H (2017a) Introduction. In: Rangan H et al (eds) Insurgencies and revolutions: reflections on John Friedmann's contributions to planning theory and practice, RTPI library series. Routledge/The Royal Town Planning Institute, New York/London, pp 1–3

Rangan H (2017b) Are social enterprises a radical planning challenge to neoliberal development? In: Rangan H et al (eds) Insurgencies and revolutions: reflections on John Friedmann's contributions to planning theory and practice, RTPI library series. Routledge/The Royal Town Planning Institute, New York/London, pp 95–104
Roberts B (1978) Cities of peasants: the political economy of urbanization in the third world. Edward Arnold, London
Rondinelli DA (1975) Urban and regional development planning. Cornell University Press, Ithaca
Rondinelli DA (ed) (1979) Project planning and implementation in developing countries: a bibliography. US Agency of International Development, Washington, DC
Rondinelli DA (1983) Development projects as policy experiments. Methuen, London
Roskamm N (2015) On the other side of "agonism": "the enemy," the "outside," and the role of antagonism. PT 14(4):384–403
Rosser C (1979) Knowledge and action: the problem of progressive synthesis in planning studies. Paper presented to the Development Planning Unit, University College London, 19 June
Rossi U (2019) Fake friends: the illusionist revision of Western urbanology at the time of platform capitalism. US 57(5):1105–1117
Roy A (2016) Urban informality: the production of space and practice of planning. In: Fainstein SS, DeFilippis J (eds) Readings in planning theory. Wiley – Blackwell, Chichester, pp 524–539
Roy A (2017) Worldling the south: toward a post-colonial urban theory. In: Parnell S, Oldfield S (eds) The Routledge handbook on cities of the global south. Routledge, London. (first published 2014), pp 9–20
Roy A (2018) The grassroots of planning: poor people's movements, political society, and the question of rights. In: Gunder M, Madanipour A, Watson V (eds) The Routledge handbook of planning theory. Routledge, London, pp 143–154
Safier M (n.d.) An action planning approach to possible patterns and solutions for accelerated urbanization. In: Mumtaz B (ed) Readings in action planning – Vol.I, DPU working paper no 13. Development Planning Unit, University College London, London, pp 10–23. (originally published 1974)
Sager T (2011) Neo-liberal urban planning policies: a literature survey 1990–2010. Prog Plan 2011(76):147–199
Sandercock L (2016) Towards a cosmopolitan urbanism. In: Fainstein SS, DeFilippis J (eds) Readings in planning theory. Wiley – Blackwell, Chichester, pp 407–426. (first published 2009)
Sandercock L, Forsyth A (1996) Feminist theory and planning theory: the epistemological linkages. In: Campbell S, Fainstein S (eds) Readings in planning theory. Blackwell, Oxford, pp 471–474. (first published 1992)
Sanyal B (2010) Similarity or differences: what to emphasize now for effective planning practice. In: Healey P, Upton R (eds) Crossing borders: international exchange and planning practices. Routledge, London, pp 329–350
Savini F, Majoor S, Salet W (2015) Dilemmas of planning: intervention, regulation, and investment. PT 14(3):296–315
Scott JC (2016) Authoritarian high modernism. In: Fainstein SS, DeFilippis J (eds) Readings in planning theory. Wiley – Blackwell, Chichester, pp 75–93. (first published 1998)
Sen KC (1980) The project cycle, Regional planning and area development project, international studies and programs. University of Wisconsin, Madison
Singer H, Ansari J (1977) Rich and poor countries. George Allen and Unwin, London
Sjoberg G (1960) The preindustrial city: past and present. The Free Press, New York
Smith DM (1979) Where the grass is greener: living in an unequal world. Penguin, Harmondsworth
Sorensen A (2010) Urban sustainability and compact cities ideas in Japan: the diffusion, transformation and deployment of planning concepts. In: Healey P, Upton R (eds) Crossing borders: international exchange and planning practices. Routledge, London, pp 117–140
Sorensen A (2018) New institutionalism and planning theory. In: Gunder M, Madanipour A, Watson V (eds) The Routledge handbook of planning theory. Routledge, London, pp 250–263

Sorkin M (2002) See you in Disneyland. In: Fainstein SS, Campbell S (eds) Readings in urban theory. Blackwell, Oxford, pp 335–353. (first published 1992)

Speak S, Kumar A (2018) The dilemmas of diversity: gender, race and ethnicity in planning theory. In: Gunder M, Madanipour A, Watson V (eds) The Routledge handbook of planning theory. Routledge, London, pp 155–166

Stöhr WB (1980) Development from below: the bottom-up and periphery-inward development paradigm, Discussion paper 6. Interdisziplinäres Institut für Raumordnung Wirtschaftsuniversität Wien, Wien

Stretton H (1978) Urban planning in rich and poor countries. Oxford University Press, Oxford

Taylor JL, Williams DG (eds) (1982) Urban planning practice in developing countries. Pergamon Press, Oxford

Todaro MP (1971) Development planning: Models and methods. Oxford University Press, Nairobi

Triantari M (2005) European town planning of modernization and development in the Arab world: the case of Michel Ecochard in Morocco (1946–1952). Doctoral thesis. National Technical University of Athens (in Greek)

Triantari M (2008) Modernization “away from home” and the conflicts of cultures: French urbanisms and the charter of Athens from the opposite coast of the Mediterranean. Ergon IV, Athens. (in Greek)

Tulumello S (2016) Reconsidering neoliberal urban planning in times of crisis: urban regeneration policy in a ‘dense’ space in Lisbon. UG 37(1):117–140

Turner JFC (1976) Housing by people. Marion Boyars, London

Turner A (1980a) Project planning. In: Turner A (ed) The cities of the poor: settlement planning in developing countries. Croom Helm, London, pp 65–92

Turner A (ed) (1980b) The cities of the poor: settlement planning in developing countries. Croom Helm, London

Turner JFC, Fichter R (eds) (1972) Freedom to build. Macmillan, New York

Uitermark J, Nicholls W (2017) Planning for social justice: strategies, dilemmas, tradeoffs. PT 16(1):32–50

Vidyarthi S (2010) Reimagining the American neighborhood unit for India. In: Healey P, Upton R (eds) Crossing borders: international exchange and planning practices. Routledge, London, pp 73–93

Vidyarthi S (2018) Spatial plans in post-liberalization India: Who’s making the plans for fast-growing Indian urban regions? J Urban Aff 43(8):1063–1080. (published online)

Wakely P (1999) Otto Koenigsberger obituary: cities of light from slums or darkness. The Guardian, 26 January

Wakely P (2019) Development planning unit record in housing 1971–2021, DPU working paper no 201. Development Planning Unit, University College London, London

Wakely P, You N (eds) (2001) Implementing the habitat agenda: in search of urban sustainability. Development Planning Unit, University College London, London

Wakely PI, Schmetzer H, Mumtaz BK (1976) Urban housing strategies: education and realization. Pitman, London

Waller PP (1975) The reduced planning approach for regional development programs in lagging areas. In: Bendavid-Val A, Waller PP (eds) Action-oriented approach to regional development planning. Praeger, New York, pp 18–28

Ward C (1983) Planning for fantasy. New Soc 63(1062)

Ward SV (2010) Transnational planners in a postcolonial world. In: Healey P, Upton R (eds) (2010) crossing borders: international exchange and planning practices. Routledge, London, pp 47–72

Waterston A (1965) Development planning: lessons of experience. A World Bank Publication/ Johns Hopkins University Press, Baltimore

Wates N (ed) (2000) The community planning handbook. Earthscan, London

Watson V (2009) Seeing from the south: refocusing urban planning on the globe’s central urban issues. US 46(11):2259–2275. (republished as Watson 2016b)

Watson V (2016a) Planning mono-culture or planning difference? PTP 17(4):663–667

Watson V (2016b) Seeing from the south. In: Fainstein SS, DeFilippis J (eds) Readings in planning theory. Wiley – Blackwell, Chichester, pp 540–560. (first published 2009)

Watson S (2017a) Spaces of difference: challenging urban divisions from the north to the south. In: Parnell S, Oldfield S (eds) The Routledge handbook on cities of the global south. Routledge, London. (first published 2014), pp 385–395

Watson V (2017b) Learning planning from the south: ideas from new urban frontiers. In: Parnell S, Oldfield S (eds) The Routledge handbook on cities of the global south. Routledge, London, pp 98–108. (first published 2014)

Weaver C (1984) Regional development and the local community: planning, politics and social context. John Wiley, Chichester

Webb D (2018) Tactical urbanism: delineating a critical praxis. PTP 19(1):58–73

Wilber CK (ed) (1973) The political economy of development and underdevelopment. Random House, New York

Williamson T (2020) Rio's real v. unmet Olympic legacies: what they tell us about the future of cities. In: Angotti T (ed) Transformative planning: radical alternatives to neoliberal urbanism. Black Rose Books, Montréal, pp 141–149. (first published 2016)

Windsor Liscombe R (2006) In-dependence: Otto Koenigsberger and modernist urban resettlement in India. Plan Perspect 21(2):157–178

Winkler T (2017) The "radical" planning practice of teaching, learning, and doing in the informal settlement of Langrug, South Africa. In: Rangan H et al (eds) Insurgencies and revolutions: reflections on John Friedmann's contributions to planning theory and practice, RTPI library series. Routledge/The Royal Town Planning Institute, New York/London, pp 219–228

Winkler T (2018) Rethinking scholarship on planning ethics. In: Gunder M, Madanipour A, Watson V (eds) The Routledge handbook of planning theory. Routledge, London, pp 81–92

Winkler T, Duminy J (2016) Planning to change the world? Questioning the normative ethics of planning theories. PT 15(2):111–129

Wohl S (2018) Tactical urbanism as a means of testing relational processes in space: a complex systems perspective. PT 17(4):472–493

World Bank (1972) Urbanization: sector working paper. World Bank, Washington, DC

Yamamoto A (2020) From value to meaning: exploring the ethical basis of Chantal Mouffe's agonistic pluralism. PT 19(2):237–241

Yiftachel O (2006) Re-engaging planning theory? Towards 'South-Eastern' perspectives. PT 5(3):211–222

Chapter 5
The "Climate" Current: Environmental Concerns in the Anthropocene Age

Abstract This chapter is dedicated to an emerging theoretical current, motivated by the phenomenon of climate change, which seems to unite planning theory around a theme of global importance. The scientific base of this approach is broad and encompasses physics, biology and ecology, systems analysis, economics, but also the sociology of inequality and disadvantage. This multiplicity of origins is made clear in planning approaches which are designed to tackle natural hazards, risk and vulnerability, not only through an exclusive environmental ideology, but rather in a social context of the effects of the Anthropocene age. Tools like resilience planning, with all its pros and cons, are proposed to address these effects. The new theoretical foundations include philosophical and sociological sources, but also a sense of pragmatism and of understanding of the individual and the particular, the isolated person, the social group, and the world we all live in. Humanism, reciprocity and mutual respect are stressed as essential values.

Keywords Climate · Sustainability · Global heating · Risk · Resilience · Vulnerability · Natural disasters · Ecology · Anthropocene · Systems · Dependency · Being · Worldling · Terrestrial · Biosphere

Prologue

The threat of climate change, with all its obvious and latent manifestations, is now monopolizing the attention of the policy-makers, planners and the public at large. From a theoretical perspective it seems to contain a promise of uniting planning theory under a common banner. There are already theoretical treatises of great weight which point to this direction, e.g., those of Low (2020), Rees (2018) and Roy (2017) to which we return later. In this chapter, we start with a statement which could be described as the reaction of the common man. A section follows in order to discuss the precedent of the dialogue on sustainable development and the first concerns caused by the overheating of the planet. Planning for the mitigation of natural disasters has been an old theme developed by a distinct scientific community, which gradually gave emphasis to parameters like risk, vulnerability, exposure,

L. C. Wassenhoven, *Compromise Planning : A Theoretical Approach from a Distant Corner of Europe*, https://doi.org/10.1007/978-3-030-94331-8_5

coping capacity and resilience. These concepts became increasingly topical in the present circumstances of climate change, but also acquired a significance which is not narrowly environmental. Their social and economic content is now emphasized in a context of inequality, poverty and disadvantage, which links it with the radical approaches to planning. The consequences of climate change have both a local and global dimension, which a number of theorists now place in the framework of the Anthropocene age and of the position of the individual, the Self, in the world.

Keywords of the Climate Current

Ideas concepts preoccupations interpretations concerns and socio-political conditions embedded in the "climate" current which inspire ecological – climate planning theory and practice:

Ecologism, belonging, natural values, perception, protection, trend reversal, adaptation and adaptability, restoration, coexistence, circular economy, socio-eco-systems, balanced action, being in nature, technical adjustment, equilibrium, life values, self-organization, earth care, corrective action, practice re-evaluation, ecological cooperation, risk, resilience, climate change, natural disasters, vulnerability, mitigation, redundancy, Anthropocene heritage, spaceship earth, sustainability, maritime space, biosphere, pollution, species extinction, landscape disfigurement, biodiversity loss, economic repercussions, resource exhaustion, solidarity, preservation, energy exchange, unity, contamination, earth science progress, responsibility, resourcefulness.

5.1 Climate Change and Planning in Simple Language

The current of planning theory, which the author described as a set of ideas related to climate change, is bound to be multi-faceted and complex. This is because it has historical, biological, philosophical, and global geopolitical roots, while, at the same time, touches more pedestrian issues addressed by urban and regional planning. Its very complexity makes it inaccessible to those not immersed in the fields, mostly of the physical sciences, which explain its rise to fame. However, the recent frequency of the devastating effects of climate change are gradually hammering in the public conscience the realization that something is going disastrously wrong. People in a variety of places start to realize that, maybe, what is happening in their locality, next door, even on their own land, may be contributing to a global threat. They become aware that their village or neighbourhood is threatened not just by an unexpected or isolated disaster but by the effects of the accumulated impact of climate change which are fully predictable. Using the example of Greece, which will be first discussed at the end of this chapter, an attempt will be made to throw light on this interplay. Indeed, in Part II of the book, which is dedicated to the history, the state, and the planning system of the country, we inevitably reach the point of wondering what the next stage will be. How will history evolve in the twenty-first century under the impact of climate change? How the state and its administration will cope, when it is already under pressure to accommodate conflicting demands for economic development, environmental protection, and climate adaptation? What will be the implications for the planning system? This chapter has, as a result, a twin purpose. On the one hand it has to present the theoretical contributions which sketch the content of the "climate current", occasionally in a convoluted language, by no means rare in all planning theory literature. On the other, the last section of the chapter functions as an introduction to Part II, when the lofty theoretical speech will cede its place to an analysis of a real example. Needless to add, this chapter is not written from the perspective and with the knowledge of a climatologist or an expert in the natural environment, but of a planner who is anxious to understand how urban and regional planning will have to adapt to the challenge of climate change.

Before entering the review of theories which constitute an emerging corpus of planning theory of the climate current, we need to approach the subject in a way that "popularizes" it and opens the way to a more elaborate presentation of the theoretical corpus. In Chap. 2 on planning theory typologies, reference was made to a current of ideas included in what was called the "climate current". It is a current which is taking over from an environmental movement and related literature. In the same chapter the author discussed the first quasi-theoretical stirrings which accompanied the urban design tradition of the late nineteenth and early twentieth centuries. Landscape protection and the reconciliation of the city with nature were from those days a primary concern of urban planning, after the traumatic experiences of the industrial revolution, which disfigured both city and rural landscapes. The ruins that two world wars left behind, the pressures for reconstruction, explosive population growth, and the over-heating of the planet brought to public attention the dangers for "spaceship earth". Thus, in the first postwar decades, a community of experts and non-experts came to realize the intimate web that connects the environment, the landscape, geographical space, and urban and regional planning. Ian McHarg (1971) presented in a masterly way a method for approaching the protection and use of the landscape, not simply for the purpose of respecting landscape beauty but also for the preservation of ecological systems and natural habitats. Such pioneering analyses were a most important step. The second step is made in the 1980s, when an international community becomes conscious of the need for sustainable development. After the year 2000, the escalating climate change and the magnitude of its effects leads to global mobilization. This is the third and more critical step. Every step is accompanied by a change of perceptions in the field of urban and regional planning. At the beginning, it was a question of order in the organization of space, to secure more humane cities, protect the environment, restore the balance between artificial and natural environment, and fight pollution. Later on, a new perception appears that corrective actions are not sufficient and that a new strategy is necessary to invent a new future, compatible with sustainable development and with an ecosystem approach. In the third step, almost in a state of panic, spatial planning is called upon to assist in the reversal of, not simply adaptation to, climate change, and to reinstall critical economic and social processes on a new base. This is not the exclusive task of spatial planning, but, it too, has an important role to perform. Its practice has to embrace urgently notions like climate change response, prevention, mitigation of vulnerability, adaptive governance and resilience, not only in environmental, but also social and economic terms.

All the evidence suggests that the phenomena generated or exacerbated by climate change have convinced the international community, in spite of short-sighted objections, that it finds itself in a situation of emergency. People everywhere are aware of rising temperatures, heat waves, gas emissions, heat islands, ecological footprints, forest fires, increased rainfall, floods, and rising sea levels. They gradually, albeit slowly, realize that, as a result, the landscape and the geographical space around them, in the cities and the countryside, are changing or could change in the immediate future, and that this change may be accelerated dramatically. They are also beginning to understand that this is a two-way influence. It is not only that climate change transforms the earth we live on, hence our ability to use this earth in the future. The reverse also happens, i.e., the way we use the earth and determine land uses causes or reinforces climate change. Land use and climate change is a hen and egg relationship. We all live in a geological era, that the specialists call Anthropocene, in which humans shape the state of our planet. It is an age which has started, depending on the scientific opinion, several millennia or only two or three centuries ago, after the industrial revolution.

The crucial question is what alters geographical space or brings about land use changes. The reasons can be multiple, for instance natural disasters, the actions of individuals and productive units, wars and conflicts, mass migrations, or geopolitical factors. Changes, however, are also due to the initiatives and policies of modern states, at central, regional, or local level, and to land regulation, imposed or encouraged through urban and regional planning. Modern states are equipped with a statutory and legally-backed planning system and a hierarchy of plans, which regulate space and permitted activities, from the national scale down to that of villages and neighbourhoods. Spatial plans can either provide guidance or include legally binding rules, which allow, forbid or impose compulsory uses. The state prescribes land use classes, like housing, exclusive or accompanied by social activities, commercial areas, industrial zones, transport infrastructures, energy production installations, protection zones etc., from which the planner can select. The planning authorities can also determine quantitative limitations, e.g., density or intensity of use. This is perhaps an oversimplified outline, which, nevertheless, shows the impact which land use can have on the environment and on the mechanisms which contribute to or speed up climate change.

Land use is not the only thing that matters. It is also the broader organization of urban space and countryside that affects the relationship between humans and their environment. The so-called ecological footprint is an expression of the geographical extent of the influence of human and social activity, given that, e.g., a city diffuses its wastes and gas emissions over a large area, and, simultaneously, imports foodstuffs and energy from distant locations. Heat islands are formed over cities. Disasters like forest fires, common in many countries in the present climate change conditions, like those that the wider region of Athens experienced in 2021, produce a cloud of smoke and material particles over urban areas, threatening public health. Another sensitive geographical zone, of great interest for Greece, is the coasts. Their vulnerability is not only due to their being sensitive eco-systems, but also to the potential, sometimes already observable, threat of sea level rise and to rising temperatures, with their long-term implications for tourism. Sadly, the state itself has encouraged development in coastal and mountain zones, through special regulations which in the long run severely damage natural landscapes. In all these situations, and in several other cases, the role of the state and of interstate organizations, such as the European Union, is of prime importance. The state, as in the Greek example, adopts a range of policies, of which spatial planning is only one, which influence these interactions. The state's actions are affected by their political nature and the networks that link the state with citizens and the electorate. This produces an immediate effect on spatial planning, as it will become clear in the Greek example, and the building processes – often illegal – in areas outside the boundaries of statutory plans. Social awareness, public conscience, dissemination of information, transparency, and consultation become urgent priorities. Ethical issues and individual responsibility acquire a renewed significance. Humans are asked to put aside their character as economic and atomistic actors and adopt a wise and prudent attitude. They are of course aware that large business interests have the main responsibility, but they should not hide behind this excuse to eschew their own.

5.2 Environment, Sustainability, and First Steps Towards a Climate Current Theory

The earlier reference, in the previous chapter, to creative and smart cities in relation to the imagery of a fantasy world was not accidental because the aura of new urbanism ideology was shrewdly associated with environment-friendly and sustainable choices. This was a clever move in that it enabled the promoters of this ideology to claim that this was a planning style that contributed to the fight against climate change. In other words they highjacked the precepts of passive architecture, energy-saving planning, low environmental impact development to make them the flag of their planning orientation. But such tactics and the distortion of environmentally-inspired ideas to suit the priorities of business interests should not deter us from paying attention to a rising wave of concerns which constitute the background of what has been called the "climate" current in planning theory (see Chap. 2). It has been mentioned already that an extensive environment-conscious literature had prepared the "sustainability" turn, long before the famous Brundtland Report (1987) of the United Nations and the ensuing flood of international agencies' reports, government documents, or academic studies on environmental policy and sustainable development.[1] "'Sustainable planning' has been associated with sustainable-, eco-green-, or low-carbon- urbanism; and smart-, circular- or healthy-cities, among others" (Turcu 2018). Duty and care, Turcu points out, were always present in sustainability ideology. Sustainable development ideology, sadly used sometimes as an opportunistic political slogan, was not immune to radical criticism. Sustainability was often identified with maintenance of the status quo, or, according to Sorensen to business as usual. The trouble is, he points out, that "the ways in which urban sustainability issues are perceived and the solutions that make sense in different contexts are strongly influenced by the available and effective policy levers, and by past patterns of institutional development. These factors create both capacities and preferences among relevant actors. As sustainable cities ideas are so varied, and possible policy approaches so diverse, it is not surprising to find that different actors will interpret the imperative to promote sustainability differently, and will adopt those aspects of the idea that best fit their own situation" (Sorensen 2010, 119 and 133).

It is true that sustainability ideas and policies frequently had a "pick and choose" character. These ideas no longer seem adequate to address climate change. Long and Rice (2018) claim that there is "an ongoing paradigm shift in urban planning focused on the development of 'climate-friendly' and 'climate-resilient' cities, which we argue is reframing the rhetoric of sustainable urbanism as the dominant policy narrative among the world's major cities. We call this shift climate urbanism, and it can be characterised by new policies, programmes, and development initiatives …". The inevitable question is about the reasons of the failure of the sustainability paradigm. Marcuse answers as follows: "In practice 'sustainability' had its origins in the environmental movement and in most usage is heavily focused on ecological concerns.[2] But why, given limited resources and limited power to bring about change, are efforts thus focused? I would suggest that the environmental movement is a multi-class, if not upper- and middle-class movement, in its leadership, financing, and politics. While the environmental justice movement is making a substantial contribution to both social justice and environmental protection, the environmental movement as a whole often proclaims itself to be above party, above controversy, seeking solutions for which everyone will benefit, and to which no one can object" (Marcuse 2020, 105–106).

As with sustainable development, equally rich was the literature on the threat of climate change, years before the famous Stern Review (Stern 2007). I still remember a speech by the American Senator Patrick Moynihan at an international conference held in Rotterdam in 1970, which I had attended as a graduate student of the Architectural Association School of Architecture. Moynihan described in dramatic terms the melting of ice in the Arctic Circle and its consequences (European Cultural

[1] See Adams 1999; Allen and You 2002; Burgess et al. 1997; Franck and Brownstone 1992; ICLEI 1996; Laskaris 1996a, b; Nadin 2001; Panayotou 1998; Riddell 2004; Soubbotina 2004; Strange and Bayley 2008; among many-many others.

[2] In Greek, "sustainable" is usually translated as *viosimos*, which means viable, although many sustainability advocates prefer the word *aeiforos*, which is the standard term used by foresters.

Foundation 1971). We have now reached the point of no return. As the well-known environmentalist David Attenborough put it very recently: "We know in detail what is happening to our planet, and we know many of the things we need to do during this decade … Tackling climate change is now as much a political and communications challenge as it is a scientific or technological one … We have the skills to address it in time, all we need is the global will to do so".[3] Attenborough is rightly addressing his plea to the leaders of powerful nations. "If there is to be effective action on global heating it will, in the first instance, be through the agency of national states" (Low 2020, 249). What is, however, important for our interests here is the responsibilities of spatial planning, intricately linked as they are with the functioning of states. Radical movements were bound to seize the initiative. "Modern urban planning is only a century old but now faces extinction along with the urbanized world that fostered global climate change. Unless planning is radically transformed and develops serious alternatives to neoliberal urbanism and disaster capitalism it will be irrelevant in this century. Nonetheless, planning alone is not enough. The last century of urban planning was dominated by white European males, who helped to rationalize the hegemony of capitalism rooted in the global North. Planning ignored gaping inequalities of race, class, and gender while promoting unbridled growth and environmental injustices. Just and equitable planning is urgently needed now more than ever. Will activist planners be part of such a transformation and advocates for real, democratic long-term planning?" (Angotti 2020a, 4).

The challenge for urban and regional planning is how it will shoulder its share of duties in the effort to reverse not just the effects, but also the causes of climate change. As Bolan (2019, 161) wrote, "a person skilled in urban planning needs full recognition of climate and environmental impacts". The challenge for planning theory is how the ideas inherent in what is described here as "climate current" will be incorporated in cohesive theory. This is beginning to happen although it may take time. The difficulty lies in the reappreciation by planning theorists of the scientific sources which in the last decades have inspired their task. They will have, once again, to turn to natural science and systems theory, without relinquishing the tenets of social science. The environmental input is quickly increasing its presence in planning methodology and practice, e.g., in the form of the eco-system approach (Shepherd 2004; Kidd et al. 2011a, b, c) and environmental or territorial impact models (Glasson and Therivel 2019; Nosek 2019). The conceptual and knowledge load of urban planning is growing exponentially as the city is no longer figured solely as a place of production, habitation, and leisure. The return of "nature" as an essential function places new demands on planning: "Attempts are currently being made to 're-nature' cities to support local and global ecosystems, increase human well-being and address environmental issues such as climate change" (Duvall et al. 2018). Not only planning practice must prove worthy of the task, but planning theory must follow.

---

[3] Message by David Attenborough to G7 leaders (The Independent, 13 June 2021).

5.3 Risks and Resilience

The difficulties that present-day dominant planning currents, such as the communicative current, will face, are convincingly outlined by Ash Amin. Apart from the difficulty of communicating with diverse groups and stakeholders, the communicative approach may be found wanting for totally different reasons, that of ignoring materiality and technical expertise, as Amin points out. Amin casts doubts on the deliberative school of planning, not because he disagrees, as he says (perhaps out of politeness), but because it fails to come to grips with the extremely complex reality of the modern city. This is not due only to the growing technological complexity and global processes, but also because it does not address the issues of risk, resilience, threats, pandemics etc., all imbued with organizational and technical complexities. Amin explains his argument as follows: "Although I am sympathetic to the arguments and proposals of the deliberative tradition and also believe that the relationally constituted city requires a negotiated approach …, two aspects of its thinking strike as problematic regarding the urgency to act decisively in an uncertain world that generates grave hazards and risks. The first concerns what counts as a stakeholder and intermediary in urban life. My claim is that the deliberative tradition makes light of non-human and non-cognitive inputs, an omission that not only overstates the potential of inter-human deliberation but also limits thinking on how the materiality of cities – brick, stone, metal, wires, software, and physical space – is implicated in the regulation of uncertainty. The second concerns the scepticism of the deliberative tradition towards expert judgment and programmatic intervention, which raises the important question of how to respond effectively to serious hazards and risks without recourse to the authoritarian excesses of the knowing tradition. Is it possible to act with authority in an uncertain urban environment without compromising stakeholder involvement and the mobilization of diverse knowledge?" (Amin 2016, 158–159).

In other words, how is planning to respond to the challenge of a "risk society", a concept made famous through the work of Ulrich Beck (2004), and in a host of scholarly works (Wisner et al. 2007; Hewitt 1997; Sapountzaki 2007). The hazards and risks to which Amin refers are obvious to the citizens in a world tortured by disasters like fires, hurricanes, floods, sea level rising, heat waves, or drought. They impose an urgent need for increased resilience, which in itself has become an important subject in planning studies (Othengrafen and Serraos 2017), although it has been an important concept in disaster research for some time. The response to climate change, or indeed to the existence of natural hazards, has been associated with the pursuit of resilience, especially in cities. "The concept of the *resilient city*, a framework developed by supra-national bodies such as the United Nations Office for Disaster Risk Reduction (UNISDR) in the early twenty-first century, is based on the idea of a *risk society* … It imagines the city as an entity capable of withstanding and rebounding from disruptive natural and human threats and challenges, such as economic crises, disease pandemics, or terror attacks" (Hatuka et al. 2018). According to Sapountzaki, who provides a most illuminating analysis of hazards, risk, vulnerability, exposure, and resilience, the usual view is that "resilience indicates reactions and counter-actions against risks, abrupt changes, external pressures or adversities aiming at survival or preservation of the pre-existing structure of the exposed entity (a region, a city, an urban area, a community, a social group, a household etc)" (Sapountzaki 2019a, b). She has also pointed out that disaster risk reduction and climate change adaptation are still at pains to find their common ground. We shall return to her views which incorporate a much wider spectrum of resilience planning aims. Resilience is a goal not only to resist the impact of natural disasters but also that of economic crises (Giannakis and Bruggeman 2017; Kakderi and Tasopoulou 2017). An expanded targeting is also present in a paper by Mehmood: "Resilience is not just about economy and environment but also society and culture. It does not merely refer to readiness to the surprise of isolated occurrences but also refers to long-term strategies to mitigate and adapt to socio-economic as well as environmental challenges. In a world of limited resources, resilience thinking can help integrate the issues of social, economic and environmental well-being by strategically navigating the policy and planning to proactively create, assume and shape change" (Mehmood 2016, 416).

Resilience is rapidly making its way into planning theory, because "resilience thinking constitutes an alternative approach. 'Planning for resilience' can find a home in planning theory as an analysis of the external dynamics that accelerate urban economic, social and spatial vulnerability and as an approach that helps to link social and economic processes with ecological processes, calling for a reconsideration of the 'substance' of planning so as to enhance capacity to deal with slow and sudden changes of different forms. This can occur within a process that focuses on 'building a self-organisation capacity' alongside a change in the value system that can overcome the unequal power relations" (Eraydin 2013, 19). Wilkinson emphasized the unity of social-ecological resilience, the "disproportional detrimental impact cities have on the global environment", and the importance of "adaptive cycles". "*Adaptability* to change is a key focus for governing for social-ecological resilience in complex adaptive systems facing irreducible uncertainty. Social-ecological resilience offers the ideal process of adaptive co-management ... Human and natural systems are conceptualized as truly interlinked and interdependent systems and are thus defined as one system, a social-ecological system, with the separation between human and natural systems being a human construct that had immense impact in shaping our world views" (Wilkinson 2012, 153).

M. Scott includes an even greater breadth of argument in favour of enhancing resilience: "[I]n a time of volatility and uncertainty, perhaps greater emphasis should be given to coping with crisis and managing risks; in other words, developing urban and regional resilience in the face of external social, political, or environmental stresses and disturbances" (Scott 2012, 3). Blečić and Cecchini (2020) discuss resilience but in comparison to the concepts of robustness and antifragility. "To assess if something is fragile, robust, resilient or antifragile means to examine its possible responses to stressors, perturbations and volatility, and to place those responses somewhere along a harm-gain dimension. For planning, the usefulness of the conceptual triad fragile-robust/resilient-antifragile goes beyond its analytical relevance and offers also some operational guidance we want to explore ... … Rather than aspiring to obtain hard predictions, we can, and cannot but, settle for something we could call soft predictions, where the point is not to accurately and precisely predict what will happen, when and where, as a consequence of what planning action; rather the point is to examine what are the system's possible responses to perturbations, volatility, even low-probability events, in other words to detect its fragility, robustness, resilience or antifragility, and to explore what makes it such".

The problem of effective resilience is that it is not just a technical problem. Once again, a radical perspective is bound to differ from engineering approaches or simple participatory exercises. "Urban planners all over the world sell technological fixes and best practices, instant recipes for green infrastructure, sustainable communities, citizen participation and – yes – social justice. They advance formulas for urban density and diversity and the cosmopolitan society, and sell popular brands like Smart Growth and the New Urbanism. Local sustainability and resiliency plans *seem* good because they attempt to go beyond the fashionable fixes and narrow reductionist science, towards a holistic, ecological approach to the city. However, unless local plans place the fundamental inequalities across the land at the *center* of their work, they will instead turn our cities into fortified enclaves for the privileged while the rest of the planet faces the catastrophic effects of climate change. Therefore, the challenge is not just adaptation or resilience but *climate justice*, which means asking the question, *resilience for who*?" (Angotti 2020b, 28). Consistently with a radical view of planning, Angotti rejects the whole "package" of conventional planning as a recipe for fighting climate change: "The thinking behind resiliency planning is in line with mainstream planning, not a fundamental departure from it. Resiliency is about protecting the basic structures of the failed system, solving problems through innovation while salvaging the basic systems of economic and political power. This reactionary impulse is imbedded in the history of urban planning" (Angotti 2020c, 109). Angotti in fact implies that resilience planning is presented as an excuse for resorting to further control. "Resilience", writes Neocleous (2013), "is nothing if not an apprehension of the future, but a future imagined as disaster and then, more importantly, recovery from the disaster. In this task resilience plays heavily on its origins in systems thinking, explicitly linking security with urban planning, civil contingency measures, public health, financial institutions, corporate risk and the environment in a way that had previously been incredibly hard for the state to do". In Kaika's view, "this is a mature and opportune moment to pay attention to socio-environmental innovations and methods forged not out of social consensus, but out of social dissensus (e.g., out of wide-spreading practices of dissent)" (Kaika 2017). The real challenge however is not to denounce scientific approaches as a cover of hegemonic structures, but rather to instill social values into the scientific and/or engineering debate, perhaps in what Tironi (2015) calls "mode of technification". The experience of the Covid-19 pandemic and the scientific-political-ethical controversy that accompanied it is valuable. What has been labelled as "citizen science" has gained ground.[4] The coronavirus, from 2020 onwards, has already generated a debate on its implications for planning (e.g., Jon 2020; Ramos 2020; European Commission 2022).

Notwithstanding the radical critique of resilience, we still need the concepts on which to base a relevant theoretical approach. To do that we must be fully aware of the broader conditions which constrain the task. It has been remarked already that a planning theory embedded in the climate current, will be obliged to revisit the tradition of systems thinking, which was discarded in the flurry of sociologically-influenced communicative and radical approaches. There are two troublesome realities if this "revisiting" is indeed feasible. The first is whether "turbulence", an old anxiety in planning, and uncertainty are manageable. De Roo (2018) reminds us vividly of the existence of turbulence, discontinuous change, non-linearity, circular causalities, butterfly effects, chaos, complexity, and bifurcations. For Skrimizea et al. (2019, 123), "it … becomes clear that many issues, planning aims to address, can be considered to some degree as problems of complex adaptive systems (CAS). Indeed, complexity has been useful at conceptualizing a variety of phenomena relevant to planning". The second trouble is the growing confusion and fuzziness arising from the globality of world systems. Environmental change and world-wide political priorities give birth to unexpected associations. Placing climate change and its roots and causes on a global scale of world systems produces a mix of radical and science-based planning that, at first sight, may seem curious. Yet, this is exactly what Gotts proposes in a paper which "compares two ambitious conceptual structures. The first is the understanding of social-ecological systems developed around the term 'resilience' and more recently the term 'panarchy' … The second is Wallerstein's 'world-systems' approach to analysing hierarchical relationships between societies within global capitalism as developed and applied across a broader historical range by Chase-Dunn and others. The two structures have important common features, notably their multiscale explanatory framework, links with ideas concerning complex systems, and interest in cyclical phenomena" (Gotts 2007). "Panarchy",[5] Gotts adds, "refers … to the framework for conceptualizing the type of coupled human-environment systems … This framework may be divided into two parts …, 'the resilience conceptual framework' and 'the adaptive cycle metaphor'".

As explained in a web site, "no system can be understood or managed by focusing on it at a single scale. All systems … exist and function at multiple scales of space, time and social organization, and the interactions across scales are fundamentally important in determining the dynamics of the system at any particular focal scale. This interacting set of hierarchically structured scales has been termed a 'panarchy' … The panarchy framework connects adaptive cycles in a nested hierarchy. There are potentially multiple connections between phases of the adaptive

[4] Television interview of Alexis Papazoglou (21 March 2021).

[5] The title is derived from the name of the ancient Greek god Pan. In itself, this suggests unpredictability and randomness.

cycle at one level and phases at another level" (Resilience Alliance n.d.). The existence of cycles of adaptation is a process akin to that of co-evolution, i.e., "the entwined evolution of two or more systems or entities, whereby changes in one affect changes in the other. What can co-evolve will be different depending on discipline and theory: systems, species, networks, organizations, societies, perspectives" (Van Assche et al. 2018, 221). As mentioned, Gotts relies on Immanuel Wallerstein's concept of "world systems" (Wallerstein 2004) and on the analysis of Chase-Dunn (1989), on global formations. "For Wallerstein, 'a world-system is a social system, one that has boundaries, structures, member groups, rules of legitimation, and coherence. Its life is made up of the conflicting forces which hold it together by tension and tear it apart as each group seeks eternally to remould it to its advantage. It has the characteristics of an organism, in that it has a life-span over which its characteristics change in some respects and remain stable in others… Life within it is largely self-contained, and the dynamics of its development are largely internal'" (Martinez-Vela 2001). What I find revealing, is that according to Martinez-Vela, "world-system theory is in many ways an adaptation of dependency theory … Wallerstein draws heavily from dependency theory, a neo-Marxist explanation of development processes, popular in the developing world, and among whose figures is Fernando Henrique Cardoso, a Brazilian. Dependency theory focuses on understanding the 'periphery' by looking at core-periphery relations, and it has flourished in peripheral regions like Latin America. It is from a dependency theory perspective that many contemporary critiques to global capitalism come from". It is as if, suddenly, seemingly different theoretical currents converge: Systemic change, resilience, climate change, dependency, global inequality. Dependency, of a colonial, semi-colonial, or simply power-wielding variety comes into play. Systemic change is shown to be associated with world-wide networks of political power and domination, i.e., of dependency structures, which impact on the socio-ecological imbalances, that climate change is all about. Dependency is a theme to which we will inevitably return in the chapters on Greece, because of its value as an explanatory variable in Greek social affairs.

What has to be addressed in this chapter is the implications of ecological thinking, climate change, resilience, risk mitigation, and the Anthropocene heritage for planning, of which the priorities are rapidly changing. From the moment the environment started being "conceptualized as a political issue", things were no longer the same (Dryzek and Dunleavy 2009, 243). The repercussions are far-reaching. "Urban problems of today and the future extend far beyond architecture, engineering and land-use zoning. Climate problems, created historically and increasingly by growing urbanization since the nineteenth century, now need top-priority academic focus … With global warming, urban planning has taken on a new and encouraging focus, the concept of *resiliency*. Resiliency implies the ability to absorb the effects of a completely unpredictable and devastating event while being able to survive and restore viability of the victimized communities. The increasing frequency of disastrous events around the globe in recent years has seen urban officials increasingly involved in planning for emergency preparedness. These events (hurricanes, tornadoes, forest fires, and flooding) stimulate debate about their cause, as to whether it stems from global circumstances. Whatever the cause, urban officials realize such events pose a significant challenge" (Bolan 2019, 194 and 196). Perhaps Bolan fails here to make the necessary distinction between pro-active and re-active resilience, either to avoid the effects of an eventual hazard, or to painlessly absorb these effects when they occur.

As indicated earlier, the situation is far more complicated than that which Bolan refers to in the USA, but the fact remains that the challenge for planners must be seen in a new perspective. For this we can turn to views regarding the search for resilience and the place of planning in a global environment. Sapountzaki touches the real dilemma of resilience planning: "Spatial planning turns a blind eye to the vulnerability of cities and regions, while searching for resilience (i.e., the strong points of communities versus risks) ... Spatial planning should deal with vulnerability reduction also, the only way to spatial, environmental and vulnerability justice. Resilience, cannot replace the public policies for risk and vulnerability reduction. The best thing it can achieve is adaptation of only some agents / structures / actors at the expense of others" (Sapountzaki 2019b). The risks against which resilience-enhancement is conceived as the ultimate weapon are not an independent variable: "Risks and disasters are not exogenous to development. Not only disasters jeopardize development gains, but there are specific development paths producing risks and paving the way for disasters. Examples include the rapid urbanization and unplanned spatial development; poverty and social inequalities in terms of access to the resources; financial and socio-political crises and armed conflicts; privatization of public services and reduction of the state's capacity to provide social safety nets; expansion of the real estate market in hazardous or sensitive ecosystems and unbounded building construction activity; unwise exploitation of resources (e.g. water, land) and pressures on limited reserves; even conceptualizations of quality of life in terms of dominating nature with the help of technology, i.e. Western world visions" (Sapountzaki et al. 2022).

This is a radical critique calling for a departure from established practices. Sapountzaki et al. attempt "to unveil the complex relationship between risk and development on one side, and risk mitigation and development planning on the other: what risks, by whom, for whom? How are risks (re)generated, transferred and (re)distributed through development and development planning? How the latter can become a risk mitigation process for those more at risk, even at the expense of development gains?" Among other conclusions, the same authors are pointing out that "according to the Anthropocene, as well as the climate change and systemic thinking, hazard, vulnerability and risks interact within development processes, reinforcing or reducing one another". They propose "principles and guidelines for risk-development planning, to become an attainable, purposeful, socially acceptable, effective and efficient project. These guidelines / recommendations refer to the vision, objectives and content, spatio-temporal scope, institutional background, methodological approach, tools and decision-making processes of the risk mitigation – development planning venture". The approach of Sapountzaki et al. is at the core of what was described in this book as the climate current in planning theory. Nicholas Low, in a real *tour de force*, takes over (Low 2020, 208–263).

5.4 Being in the Anthropocene Age

Although Low's treatment of theory has a definite environmental bend, his analysis is synthetic *par excellence*. He draws on the philosophical, political, ecological and sociological theories of Martin Heidegger, John Rawls, Bruno Latour, Chantal Mouffe, Michael Bookchin, Aldo Leopold, David Pepper, John Dewey, John Dryzek and many others. He blends all the planning theory streams and expands his comments on planning to all levels. He is a radical in his social justice and ecology ideology, but a rationalist in his view of planning work. He has an eye on the local, but keeps a global perspective. In the present author's view, he is a pragmatist of the age of climate change. Often, he mobilizes complex concepts in his analysis: Rawls's "reflexive equilibrium", "bare persons", and "fairness", "Self" and "I" (Heidegger's *Dasein*), "land ethic", "biospherical egalitarianism", nature's vulnerability, climate emergency, "extinction rebellion", eco-feminism, "eco-centric" ethic, "globalization minus" and the "terrestrial" (Latour), Terra, antagonism and agonism (Mouffe), "discursive designs" (Dryzek), Anthropocene, empathy and solicitude. Low, however, keeps his feet on the ground. This is most evident in his unwavering confidence in global governance, human action, consent (rather than consensus), state initiative, and planning. "Many instances of environmental injustice are … a particular category of social injustice: unfair distribution of environmental harms amongst humans … [U]tilitarian economics conceives of what Rawls called 'bare persons'. We can go a little further and say that the utilitarian Self is conceived as a sort of homunculus, a bounded Self, living inside one's body, isolated, except for sense impressions from the external social and physical environment, and responding to those impressions by generating preferences related only to the internal Self … It is exactly that idea of the Self (what I *am*) that Heidegger … sought to demolish, reconstructing instead a conception of the Self that is built in constant engagement with the world around it. That 'I', called *Dasein* (Being-there), replaces the subjective-objective distinction, and thus the 'bare person', with the holistic idea of being-in-the-world, caring about the world, and so building the 'I' in interaction with the world. Indeed, as Heidegger says, 'there is no such thing as the side-by-side-ness of an entity called *Dasein* with another entity called the 'world''" (Low 2020, 209).

Being-in-the-world is a concept more or less adapted by Roy when she speaks of "worldling". She borrows the term from Radhakrishnan, who saw worldling as "a perennial process of a lived and immanent contingency", but, following Heidegger, she views worldling as "a way of finding a privileged place, of being at home, of crafting the art of being global" (Roy 2017, 17). The next important step in Low's account is his discussion of ethics and ecological justice. He quotes Aldo Leopold, who wrote that "the land ethic simply enlarges the boundaries of the community to include soils, waters, plants, and animals, or collectively the land'" (Low 2020, 212). "Ecological justice … governs our instruments of collective action and must eventuate in rules for their conduct and the protection of rights. Some eco-philosophers have seemed to demand no discrimination among species, at the extreme, a 'biospherical egalitarianism'. In this view all nature has an equal right with humans" (op.cit., 213). Low continues, following David Pepper: "As we develop the means to learn about nature's vulnerability we may also change our self-perception to one of mutual dependency. But this mutual dependency has to work three ways. Human persons are dependent on other humans, human society is dependent on non-human nature, non-human nature is dependent on human persons and society. The enlarged Self thus conceived is one in which the reality of the person-society-nature relationship is fully realized and in which reifications and fetishes such as the isolated 'individual' and 'Nature' are done away with … Problematic human relationships with nature today do not lie within the person, or even with the person's relationship with nature, but rather with the social systems, which are systems of domination that shape personal conduct" (op.cit., 215–216).

It would appear that the "human subject" is lost and marginalized in a situation of declining humanistic concerns, at least in the view of Good et al. (2017, 291), who "value humanism that identifies humans … as the ultimate source of politico-ethical decisions". Low seeks support in the views of Bruno Latour: "To reach an accommodation between the politics of human society and the politics of nature, Latour … argues, requires a fundamental realignment of the imagined structural fissures in society. His central premise is 'that the climate question is at the heart of all geopolitical issues and that it is directly tied to questions of injustice and inequality'" (Low 2020, 220). Latour uses the term "globalization minus", which Low interprets as "globalization of bare persons", in reaction to which people in various nations "withdraw into stockaded national economic interests" (op.cit., 221). Another of Latour's concepts is the "Terrestrial" which is a synthesis of the Global and the Local: "The Terrestrial pole is the new vector towards which humanity needs to move" (op.cit., 223). According to Low, "*Terrestrial*, as Latour means it, is the sense of belonging to the Earth, *Terra*, and caring about what is happening there … The *Terra*, as Latour makes very clear, is the 'Critical Zone' in which life takes place on the planet … The Critical Zone is more commonly called the biosphere. Now our occupancy of *Terra* (understood as biosphere) is under threat. The extinctions of the Anthropocene are now feared to include our human species because, like other life forms, we are ecologically dependent on our biospheric ecology" (op.cit., 224–225). Low expands on the views of Chantal Mouffe on antagonism and agonism, which were mentioned earlier, and on the democratic paradox, to overcome which "she posits *agonistic* democracy as a process of managing ineradicable conflict and the same time harnessing the power of the state" (op.cit., 232). He recalls John Dryzek's answer to the democratic paradox, which consists of "discursive designs" to be developed in "public spheres", to be created in the "interstices of the power structure" (ibid.).

The Anthropocene is at the centre of the analysis of William Rees, who shows quite clearly the systems logic which should prevail in the priorities of tackling climate change. Fifty years after the work of McLoughlin and Chadwick on the systems approach to planning, systems thinking returns with a vengeance. Rees starts by stressing that "the cumulative effect of human activities (including meticulously *planned* activities) is eroding the biophysical structure and life-support functions of the ecosphere" (Rees 2018, 53). Rees asks, "what are some of the biophysical and economic principles that should help shape urban land-use, resource and development planning theory in the twenty first century?" He replies, that "my starting premises in addressing this question are that: (1) the biophysical foundations of planning, particularly pertaining to urban ecosystems, are seriously underdeveloped; (2) the prevailing neoliberal framing of economic planning is dangerously misguided; (3) resolving the ecological crisis requires reasserting the common good and the public interest over selfish individualism and corporate privilege; (4) the 'disjointed incrementalism' characterizing typical development planning is woefully inadequate for the Anthropocene" (op.cit., 54). Rees mobilizes the full arsenal of biology, ecology and systems analysis terminology to explain the systemic and dissipative structure of cities and the built environment, but also to stress the importance of the cities' hinterland. "Clearly planners must begin to acknowledge in practical terms that arable land, forested land and other ecosystems are a more important functional component of urban ecosystems than are the lifeless roads, parkades and building lots that currently command professional attention. Planning for the Anthropocene should strive to re-conceive cities – or better, urban regions – as self-renewing, regenerative human ecosystems" (op.cit., 57). For Rees, it is apparent "that a steady-state economy functioning on available biocapacity is an appropriate economic model for our proposed system of increasingly self-reliant bioregional city-states" (op,cit., 62).

Can we hope that such actions can be promoted in each country separately, or even locally, or is it essential that climate change must be tackled first and foremost through international coordination? At the global level, Low speaks of a terrestrial constitution: "The social conflicts generated by climate heating and an unjust and self-destructive global economy cannot be resolved by retreat behind national boundaries" (Low 2020, 243). Sanyal (2010, 345–346) had a similar perspective: "As planners we need to acknowledge that we live in one world, not three or four, and our lives across the fragile globe are more interconnected than ever before. There are huge overlaps in the concerns, aspirations, and expectations of people around the world, and it is important to appreciate such overlaps without ignoring the specificities of each setting … [T]he purpose of refocusing on similarities over differences is not to strengthen the current ideological dominance of the West – just the opposite". This is probably the reason behind his call for "territories of inclusion" (Sanyal 2017, 21).

Nicholas Low does not limit his arguments to the global level; simultaneously, at all levels, he calls for a re-evaluation of "the art of planning, *public* planning … The investigation of city planning in the last 50 years has taught us much about society, or in the words of Patrick Geddes: *Place*, *Work*, *Folk* – or as I want to re-express it, 'People, Planet and Place'. The problem, though, is that public planning has been devalued. Revaluing planning means revaluing a kind of society that values planning" (Low 2020, 246). We could add that the very concepts planning uses need reformulation. E.g., territorial capital, as Tóth (2015) argues, should be divested of its exclusive economic emphasis and embrace the environment. Low insists on the things that matter: biosphere, state, work, poverty, inequality, free markets, democracy, planning, and "our being". Being highly selective, we can quote here some of his comments on the state and on planning. "If there is to be effective action on global heating it will, in the first instance, be through the agency of national states. By 'the state' I do not mean those elements essential to democracy – executive, legislative and judiciary – but the part that devises policy options for political choice and carries out their implementation. This part is variously described as 'the public service', the 'civil service' or pejoratively 'the bureaucracy' or 'the state apparatus'" (Low 2020, 249). The present author hastens to add that the very fact that environmental effects defy conventional state and regional boundaries and that their complexity requires cooperative action at the supra-state level must make us cautious in any discussion about the ability of the individual states' capacity to cope with this challenge (Dryzek and Dunleavy 2009, 254–268). The requirements Low is talking about are "necessary for effective *public planning* to take place. Faludi's model of planning is not so far off the mark … Of course, there are all sorts of barriers and difficulties to be overcome …" (Low 2020, 251). And elsewhere: "If capitalism … is to survive and adapt yet again, as I believe it will, public planning must be revalued … [I]t is not just the climate for which we need to plan, but also the biosphere emergency, the pollution emergency, the emergency of cities, and social emergency of poverty and inequality. Human occupation of the biosphere (in 'the Anthropocene') is destroying ecologies that support the habitats of the immense diversity of species. Mega-cities around the world are experiencing toxic air pollution endangering health and even lives. The crushing of the unemployed into poverty is making the transition to the zero-carbon economy dangerous to society by creating a desperate underclass scarcely able to survive, let alone prosper in the period between losing an old carbon-based job and finding a new zero carbon one. Similarly, jobs that are destroying habitat and endangering species … will require similar transition planning. Planning of cities is needed to mitigate global heating, adapt to a hotter world, abolish air and water pollution, provide for safe mobility …". For all that, planning must be public and based on facts and truth. In addition, it "must be for a specified 'place'" (op.cit., 260–261). The very success of the human race now threatens its existence. "Planning to improve the conditions of life of our fellow humans (social justice) and planning to include the welfare of our fellow animals (ecological justice), requires a choice to respond to our capacity for empathy over aggression … Empathy remains a choice for us – kindness, respect, recognition, tolerance, civility, 'care' in the sense of Heidegger's 'solicitude'" (op. cit., 262).

Can we remain optimistic? Are we as individuals prepared for the necessary sacrifices? Ridley and Low express doubts: "At the center of all environmentalism lies a problem whether to appeal to the heart or to the head – whether to urge people to make sacrifices on behalf of the planet or to accept that they will not, and instead rig the economic choices so that they find it rational to be environmentalist … Those who do recognize the problem often conclude that their appeals should not be made to self-interest but rather should be couched in terms of sacrifice, selflessness, or, increasingly, moral shame. We believe they are wrong. Our evidence comes from a surprising convergence of ideas in two disciplines that are normally on very different tracks: economics and biology" (Ridley and Low 1996, 199). In a discussion about the ever-present problem of the "tragedy of the commons", they argue "that the environmental movement has set itself an unnecessary obstacle by largely ignoring the fact that human beings are motivated by self-interest rather than collective interests. But that does not mean that the collective interest is unobtainable. Examples from biology and economics show that there are all sorts of ways to make the individual interest concordant with the collective – so long as we recognize the need to" (op.cit., 201). We must heed their advice about what it is to be rational and the convergence of economics and biology. This is a pragmatic observation which planning should not ignore. But this requires a revalued public planning, as Low suggests, and practical, yet committed, planners.

5.5 A Greek Example

It is not easy to predict what a "revalued public planning" will be in particular national contexts, e.g., in the Greek case to which is dedicated Part II of this book. Climate change will no doubt be a critical development. This does not mean however that there have not been in the past isolated research initiatives to promote planning which would be climate and energy conscious. One such initiative was a project carrying the title "Energy Systems and Design of Communities", in which a Greek team worked alongside teams from the United States, Germany and Italy (Wassenhoven 1980, 1982). Such efforts, at the planning level, came to nothing, in spite of the interest in passive architecture.

The threat of climate change is already clearly taken into account in Greek regional spatial plans (see Chap. 8), and in the official brief of 2021 for the new generation of Local Town Plans which will cover the whole country. It is stated in the guidelines that the new plans will have to promote the prevention of, and adaptation to, climate change, as well as the resilience to, and safety from, natural and manmade disasters.[6] The need to incorporate climate change adaptation in spatial plans and to undertake supporting studies had been stressed very early (Wassenhoven et al. 2009). In 2009 – and for a period of 6 years – the words "climate change" were added to the title of the Ministry of the Environment, which became Ministry of the Environment and Climate Change. A simple symbolic gesture! However, at the end of August 2021, after a series of disastrous forest fires, the creation of a new Ministry of Civil Protection and Climate Change was announced. In fact, the word "change" was immediately replaced by "crisis". Giannitsis has rightly emphasized that the policy for the environment and climate change is not just another policy, as those for education, industry, agriculture, space, or energy. It is a cross-cutting horizontal

[6] Greek Government Gazette, No 3545 B, 3 August 2021.

policy and a necessary ingredient of all policies.[7] The hazards to which Greece will be exposed because of climate change have been repeatedly documented (Doussi 2017, 17). Fires and floods in the recent years have caused enormous destruction. Over one hundred people died in the fire at Mati in Attica in 2018, catastrophic floods hit Euboea and Thessaly in 2020, forest fires caused the loss of hundreds of thousands of acres in Euboea and Attica in 2021. Recurring heat waves, apart for providing favourable conditions for fires, have clear effects on urban living and economic activity, especially tourism (Wassenhoven 2016). In 2016, Greece incorporated in her legislation the international convention for climate change and in the same year adopted a new strategy on adaptation. Climate change will affect the Greek maritime environment, which is important to take into account in view of developments in the field of maritime spatial planning (see Chap. 8). This intimate liaison has been often emphasized (Wassenhoven 2017, 2018; Kyvelou and Papadaki 2016; Kyvelou 2016) and it is obvious in the case of the coastal zone and maritime protected areas (Papageorgiou 2016; Vasilopoulou and Krasanakis 2016).

It has been claimed that the necessary reforms for climate change adaptation may be less interventionist than those already implemented because of the Covid-19 pandemic, but they will have to involve more systemic change in the direction of different standards of production and consumption (EKPAA 2020). The claim of "less interventionist reforms" probably implies that measures to tackle the pandemic were taken within a very short period of time and in conditions of the utmost pressure. This claim may be an exaggeration and can divert attention from the fact that climate change is a relatively gradual, if predictable, process. It is preferable to argue that a climate change policy, particularly in dense urban areas, is not unrelated with a policy for long-term prevention of diseases. We can only imagine the implications for human health of sharp fluctuations of temperature levels and of gas emissions, through the increased risks for infections. Besides, "many of the root causes of climate change also increase the risk of pandemics".[8] Regardless of the scientific validity of such concerns, as Kartalis reminds us, according to an opinion poll among thousands of citizens in 14 countries, the great majority estimated that climate change is a graver threat than coronavirus.[9] It is a healthy sign in the public debate in Greece, that well-argued reports appear on the environment and sustainable development. In a recent report, a Greek Green Deal was presented with operational plans, for a number of fields, from agriculture to energy, and from mineral extraction to urban development. In the last case a plan is advocated for the transformation of a large number of centres into "smart cities", through an Operation of Urban Restructuring of the twenty-first century. The tools to be used are the new Local Town Plans (see Chap. 8), the plans of integrated urban interventions in cities with a population of over 50,000, local renewal projects, re-use of abandoned

[7] Prof. Tassos Giannitsis (former Minister of Employment), Political dimensions of climate change. Newspaper Kathimerini, 7 October 2018.

[8] See "Coronavirus and climate change", in https://www.hsph.harvard.edu/

[9] Prof. K. Kartalis, Climate change in the post-coronavirus epoch. diaNEOsis, June 2020.

buildings, digital technologies, and the national digital register of spatial data (diaNEOsis 2020). The subject of this section will be taken up again in the section on the environment and natural disasters of Chap. 8.

Epilogue

The climate current, as introduced in Chap. 2, is the fourth major current of planning theory, after the rational, communicative and radical currents. In this chapter the author endeavoured to explain its significance and its role in uniting older theories, from which no doubt it borrows concepts, within an environmental, egalitarian, humanist, but also pragmatist, perspective. This new route has been already well developed in path-breaking theoretical works, as the reader must have appreciated. The bridges with both rationalism and collaborative thinking which the new direction promises to build are indeed impressive. What is also striking is that the new approach attempts to overcome narrow perceptions developed in particular advanced societies and place them in a global context, the hallmark of the Anthropocene age. However, it does not deny the importance of place, history, social structure, and individuality. Together, with the previous chapters on the rational, communicative, radical, and Southern perspectives, this chapter prepared us to focus on a particular case, the example of Greece. In Part II of the book, the author will have the arduous task of placing all this theoretical search in the realities of a single country, by studying the relevance of history, the role of the state, and the structure of a national planning system.

References[10]

Adams WM (1999) Sustainability. In: Cloke P, Crang P, Goodwin M (eds) Introducing human geographies. Arnold, London, pp 125–132

Allen A, You N (eds) (2002) Sustainable urbanisation. London, Development Planning Unit, University College London

Amin A (2016) Urban planning in an uncertain world. In: Fainstein SS, DeFilippis J (eds) Readings in planning theory. Wiley – Blackwell, Chichester, pp 156–168 (first published 2011)

Angotti T (2020a) Introduction to transformative planning. In: Angotti T (ed) Transformative planning: radical alternatives to neoliberal urbanism. Black Rose Books, Montréal, pp 4–10

Angotti T (2020b) Advocacy, planning and land: how climate justice changes everything. In: Angotti T (ed) Transformative planning: Radical alternatives to neoliberal urbanism. Black Rose Books, Montréal, pp 25–29 (first published 2016)

---

[10] **Note**: The titles of publications in Greek have been translated by the author. The indication "in Greek" appears at the end of the title. For acronyms used in the references, apart from journal titles, see at the end of the table of diagrams, figures and text-boxes at the beginning of the book.

Abbreviations of journal titles: AT Architektonika Themata (Gr); DEP Diethnis kai Evropaiki Politiki (Gr); EpKE Epitheorisi Koinonikon Erevnon (Gr); EPS European Planning Studies; PeD Perivallon kai Dikaio (Gr); PoPer Poli kai Perifereia (Gr); PPR Planning Practice and Research; PT Planning Theory; PTP Planning Theory and Practice; SyT Synchrona Themata (Gr); TeC Technika Chronika (Gr); UG Urban Geography; US Urban Studies; USc Urban Science. The letters Gr indicate a Greek journal. The titles of other journals are mentioned in full.

Angotti T (2020c) Resilience is not enough: think seven generations, act now for climate justice. In: Angotti T (ed) Transformative planning: radical alternatives to neoliberal urbanism. Black Rose Books, Montréal, pp 107–113

Beck U (2004) Risk society: towards a new modernity. Sage, London (German original Risikogesellschaft: Auf dem Weg in eine andere Moderne, 1986)

Blečić I, Cecchini A (2020) Antifragile planning. PT 19(2):172–192

Bolan R (2019) Urban planning in planet earth's tragedy of the commons. iUniverse, Bloomington

Brundtland Report (Report of the World Commission on Environment and Development) (1987) Our Common Future (available in the web)

Burgess R, Carmona M, Kolstee T (eds) (1997) The challenge of sustainable cities. Zed Books, London

Chase-Dunn C (1989) Global formation: structures of the world-economy. Basil Blackwell, Oxford

de Roo G (2018) Spatial planning and the complexity of turbulent, open environments: about purposeful interventions in a world of non-linear change. In: Gunder M, Madanipour A, Watson V (eds) The Routledge handbook of planning theory. Routledge, London, pp 314–325

diaNEOsis (2020) A Greek Green Deal – How green development can become the answer to the environmental crisis. Report by Prof. J. Maniatis. Athens (in Greek)

Doussi E (2017) Climate change. Papadopoulos, Athens (in Greek)

Dryzek JS, Dunleavey P (2009) Theories of the democratic state. Macmillan International / Red Globe Press, London

Duvall P, Lennon M, Scott M (2018) The 'natures' of planning: evolving conceptualizations of nature as expressed in urban planning theory and practice. EPS 26(3):480–501

EKPAA (2020) Environment and health 2019. National Centre for the Environment and Sustainable Development, Athens (in Greek)

Eraydin A (2013) "Resilience thinking" for planning. In: Eraydin A, Tasan-Kok T (eds) Resilience thinking in urban planning. Springer, pp 17–37

European Commission (2022) Cohesion in Europe towards 2050: eighth report on economic, social and territorial cohesion publications Office of the European Union, Luxembourg

European Cultural Foundation (1971) Citizen and city in the year 2000. Kluwer, Deventer

Franck I, Brownstone D (1992) The green encyclopedia. Prentice Hall, New York

Giannakis E, Bruggeman A (2017) Determinants of regional resilience to economic crisis: a European perspective. EPS 25(8):1394–1415

Glasson J, Therivel R (2019) Introduction to environmental impact assessment. Routledge, London

Good RM et al (2017) Confronting the challenge of humanist planning/Towards a humanist planning/A humanist perspective on knowledge for planning: implications for theory, research, and practice/To learn to plan, write stories/Three practices of humanism and critical pragmatism/Humanism or beyond? PTP 18(2):291–319

Gotts NM (2007) Resilience, panarchy, and world-systems analysis. Ecology and Society 12(1):24

Hatuka T et al (2018) The political premises of contemporary urban concepts: the global city, the sustainable city, the resilient city, the creative city, and the smart city. PTP 19(2):160–179

Hewitt K (1997) Regions of risk: a geographical introduction to disasters. Addison Wesley/ Longman, Harlow

ICLEI (1996) The Local Agenda 21 planning guide: an introduction of sustainable development planning. International Council for Local Environmental Initiatives, Toronto

Jon I (2020) A manifesto for planning after the coronavirus: towards planning of care. PT 19(3):329–345

Kaika M (2017) 'Don't call me resilient again!': the New Urban Agenda as immunology … or … what happens when communities refuse to be vaccinated with 'smart cities' and indicators. Environment and Urbanization 29(1):89–102

Kakderi C, Tasopoulou A (2017) Regional economic resilience: the role of national and regional policies. EPS 25(8):1435–1453

Kidd S et al (2011a) The ecosystem approach and planning and management of the marine environment. In: Kidd S, Plater A, Frid C (eds) The ecosystem approach to marine planning and management. Earthscan, London, pp 1–33

Kidd S et al (2011b) Developing the human dimension of the ecosystem approach: connecting to spatial planning for the land. In: Kidd S, Plater A, Frid C (eds) The Ecosystem Approach to marine planning and management. Earthscan, London, pp 34–67

Kidd S et al (2011c) The ecosystem approach to marine planning and management – The road ahead. In: Kidd S, Plater A, Frid C. The Ecosystem Approach to marine planning and management. Earthscan, London, pp 205–219

Kyvelou S (2016) Maritime spatial planning under the lens of geo-philosophy, geography and geopolitics: thoughts on spatial planning of “mare liberum” and of maritime spatial zones. In: Kyvelou S (ed) Maritime spatial issues: the maritime dimension of territorial cohesion – maritime spatial planning – sustainable blue development. Kritiki, Athens, pp 37–62. (in Greek)

Kyvelou S, Papadaki M (2016) Maritime spatial planning in practice: experiences, practical instruments, and evaluation of eco-systemic services. In: Kyvelou S (ed) Maritime spatial issues: the maritime dimension of territorial cohesion – maritime spatial planning – sustainable blue development. Kritiki, Athens, pp 606-617 (in Greek)

Laskaris C (1996a) The concept of sustainable development. In: Laskaris C (ed) Sustainable development: theoretical approaches to a critical concept. EU Petra II Programme. Papasotiriou, Athens, pp 13–86 (in Greek)

Laskaris C (ed) (1996b) Sustainable Development: theoretical approaches of a crucial notion. EU Petra II Programme. Papasotiriou, Athens

Long J, Rice JL (2018) From sustainable urbanism to climate urbanism. US 56(5):992–1008

Low N (2020) Being a planner in society: for people, planet, place. Edward Elgar, Cheltenham

Marcuse P (2020) Sustainability is not enough. In: Angotti T (ed) Transformative planning: radical alternatives to neoliberal urbanism. Black Rose Books, Montréal, pp 104–106 (first published 2004)

Martinez-Vela CA (2001) World systems theory https://web.mit.edu/esd.83/www/notebook/WorldSystem.pdf · PDF file

McHarg I (1971) Design with nature. Doubleday, New York

Mehmood A (2016) Of resilient places: planning for urban resilience. EPS 24(2):407–419

Nadin V (2001) Sustainability from a national spatial planning perspective. In: OECD Towards a new role for spatial planning. A collection of essays. OECD, Paris, pp 77–94

Neocleous M (2013) Resisting resilience. Radical Philosophy 178:1–7

Nosek Š (2019) Territorial Impact Assessment – European context and the case of Czechia. AUC Geographica 54(2):117–128

Othengrafen F, Serraos K (eds) (2017) Urban resilience, climate change and adaptation: coping with heat islands in the dense urban area of Athens, Greece. Leibniz Universität (Hannover), Institut für Umweltplanung, National Technical University of Athens

Panayotou T (1998) Instruments of change: motivating and financing sustainable development. Earthscan, London

Papageorgiou M (2016) Spatial planning and protected maritime areas in Greece. In: Kyvelou S (ed) Maritime spatial issues: the maritime dimension of territorial cohesion – maritime spatial planning – sustainable blue development. Kritiki, Athens, pp 148–158 (in Greek)

Ramos SJ (2020) COVID-19 and planning history: a space oddity. Planning Perspectives 35(4):579–581

Rees WE (2018) Planning in the Anthropocene. In: Gunder M, Madanipour A, Watson V (eds) The Routledge handbook of planning theory. Routledge, London, pp 53–66

Resilience Alliance (n.d.) Panarchy https://www.resalliance.org/panarchy

Riddell R (2004) Sustainable urban planning. Blackwell, Oxford

Ridley M, Low BS (1996) Can selfishness save the environment? In: Campbell S, Fainstein S (eds) Readings in planning theory. Blackwell, Oxford, pp 198–212 (first published 1993)

Roy A (2017) Worldling the South: toward a post-colonial urban theory. In: Parnell S , Oldfield S (eds) The Routledge Handbook on cities of the Global South. Routledge, London (first published 2014), pp 9–20

Sanyal B (2010) Similarity or differences: what to emphasize now for effective planning practice. In: Healey P, Upton R (eds) Crossing borders: international exchange and planning practices. Routledge, London, pp 329–350

Sanyal B (2017) Territoriality: which way now? In: Rangan H et al (eds) Insurgencies and revolutions: reflections on John Friedmann's contributions to planning theory and practice. Routledge, New York, and The Royal Town Planning Institute (RTPI Library Series), London, pp 16–23

Sapountzaki K (2019a) Less vulnerable, more adaptive societies: can spatial planning make a contribution? In: Serraos C, Melissas D (eds) Natural disasters and spatial policies. Sakkoulas, Athens, pp 49–62 (in Greek)

Sapountzaki K (2019b) Less vulnerable. more resilient territories – What can be a contribution of spatial planning? Lecture at Université Savoie Mont-Blanc, 3–5 (November)

Sapountzaki K (ed) (2007) Tomorrow in danger: natural and technological disasters in Europe and Greece. Gutenberg, Athens (in Greek)

Sapountzaki K et al (2022) Chapter 12: A risk-based approach to development planning. In: Eslamian S, Eslamian F (eds) Handbook of disaster risk reduction for resilience (HD3R). Vol.4. Springer

Scott M (2012) Planning in the face of crisis. PTP 13(1):3–6

Shepherd G (2004) The ecosystem approach: five steps to implementation. IUCN, Gland/Cambridge, UK

Skrimizea E, Haniotou H, Parra C (2019) On the 'complexity turn' in planning: an adaptive rationale to navigate spaces and times of uncertainty. PT 18(1):122–142

Sorensen A (2010) Urban sustainability and compact cities ideas in Japan: the diffusion, transformation and deployment of planning concepts. In: Healey P, Upton R (eds) Crossing borders: international exchange and planning practices. Routledge, London, pp 117–140

Soubbotina TP (2004) Beyond Economic Growth: an introduction to sustainable development, 2nd edn. The World Bank, Washington, DC

Stern N (2007) The economics of climate change – The Stern Review. Cambridge University Press, Cambridge, UK

Strange T, Bayley A (2008) Sustainable development. OECD, Paris

Tironi M (2015) Modes of technification: expertise, urban controversies, and the radicalness of radical planning. PT 14(1):70–89

Toth BI (2015) Territorial capital: theory, empirics and critical remarks. EPS 23(7):1327–1344

Turcu C (2018) Responsibility for sustainable development in Europe: what does it mean for planning theory and practice? PTP 19(3):385–404

Van Assche K, Beunen R, Duineveld M (2018) Co-evolutionary planning theory: evolutionary governance theory and its relatives. In: Gunder M, Madanipour A, Watson V. The Routledge handbook of planning theory. Routledge, London, pp 221–233

Vasilopoulou V, Krasanakis V (2016) Concepts and approaches to maritime spatial planning and pilot application in the Adriatic-Ionian macro-region. In: Kyvelou S (ed) Maritime spatial issues: the maritime dimension of territorial cohesion – maritime spatial planning – sustainable blue development. Kritiki, Athens, pp 297–311. (in Greek)

Wallerstein I (2004) World-systems analysis: an introduction. Duke University Press, Durham (NC)

Wassenhoven L (1980) The design of new settlements. In Energy Systems and Design of Communities (EN.SY.DE.CO.). Report. In: Thymio Papayannnis and Associates (1980–82) Energy systems and design of communities. Ministry of Coordination and National Energy Council, Athens

Wassenhoven L (1982) Energy boom towns. Analysis of case studies. In: Thymio Papayannnis and Associates (1980–82) Energy systems and design of communities. Ministry of Coordination and National Energy Council, Athens (in cooperation with Simos Yannas)

Wassenhoven L (2016) The systemic character of heat waves and long-term mitigation: the case of Athens. Geografies 27 (Spring): 89–102 (in Greek)

Wassenhoven L (2017) Maritime spatial planning: Europe and Greece. Crete University Press, Herakleion (in Greek)

Wassenhoven L (2018) Maritime spatial planning in the context of European spatial planning. In: Serraos C, Melissas D (eds) Maritime spatial planning. Sakkoulas, Athens, pp 13–37 (in Greek)

Wassenhoven L (coordinator), Doussi E, Koutalakis Ch, Lalos A, Housianakou M (2009) Climate change, environmental vulnerability, European challenges. In: ISTAME Green development: consultation documents. Athens (general coordinator: J. Maniatis), pp 3–54 (in Greek)

Wilkinson K (2012) Social-ecological resilience: insights and issues for planning theory. PT 11(2):148–169

Wisner B, Blaikie P, Cannon T, Davis I (2007) At risk: natural hazards, people's vulnerability, and disasters. Routledge, London

Part II
Greece as a Case Study

Chapter 6
Greece: On the Edge of North and South – A Historical Perspective

Abstract This chapter is an essential introduction to the presentation of a "Greek case study", which occupies Part II of the book. The author's look at history privileges a series of crucial parameters: Building of a nation state, irredentism and the dream of the Great Idea, gradual accretions to the national territory, north-south division, national schisms and civil wars, defeats and bankruptcies, foreign control and dependence, land question, distribution of land, agrarian question, land property fragmentation, modernization v. tradition, West v. East, state power v. local power structures, Hellenism of the diaspora, local elite classes, clientelism, land use, Asia Minor disaster and refugee problem, settlement of refugees and town planning, migration, urban-rural dichotomy, dominant position of Athens, post-Second World War reconstruction, economic development priorities, regional inequalities, and recent economic crisis.

Keywords Modern Greece · History · Geographical expansion · Spatial structure · Dependence · Crises · Social rifts · Land

Prologue

This chapter opens Part II of the book which is dedicated to a national case study, that of Greece. History is here the subject, with the following Chaps. 7 and 8 addressing in more detail the issues of the national state and the spatial planning system. During the early years of the twentieth century, when planning in western countries was being developed as practice and discipline, social and spatial structures in Greece were totally different. Greece was only half of what it is today. The interwar period was dominated by the exchange of populations triggered by the Balkan Wars, the First World War, the annexation of northern Greece and the influx of refugees after the 1922 Asia Minor military disaster. Urban planning inevitably focused on the settlement of incoming refugees. The events of the 1940s (Second World War, German occupation, civil war) caused an inflow of migrants into Greater Athens and a wave of reconstruction, unruly urban housing development, unplanned urban sprawl, and illegal building in peri-urban zones and the countryside. In the

L. C. Wassenhoven, *Compromise Planning : A Theoretical Approach from a Distant Corner of Europe*, https://doi.org/10.1007/978-3-030-94331-8_6

1950s the government's attention, with considerable foreign aid, focused on national and regional development, with scant attention to spatial aspects. After the interim period of a shameful military dictatorship and the accession to European Union membership, normal democratic conditions were restored, with considerable economic and political stability, succeeded in the 2010s by a severe financial crisis. Overall, in spite of political upheavals, domestic dissent, social rifts, and perpetual external dependence, Greece had managed to establish a successful "western-type" democracy, but traditional elements still survived intact in its social fabric.

This chapter is not, and could not be, given that the author is not a historian, a short history of modern Greece. The purpose here is to isolate elements which continue to mark the character of spatial planning and to emphasize the importance of historical heritage. In previous chapters the author underlined the differences of historical circumstances and their impact on the evolution of planning. Southern theory, as stressed several times, blames imported planning practice and theory for failing to take into account the history of countries and places where foreign planning practice is being implemented. The history of modern Greece starts in 1821, the year when the independence war, to overthrow the yoke of the Ottoman Empire, began in the region of the Peloponnese. As these lines are written, the country celebrates the 200th anniversary of the national revolution. In the course of these two centuries, a broad range of developments and upheavals left their imprint on the space which is now Greek territory. The aim of the author is to shed light on them and to focus on how and why they affected the problems which urban and regional planning, when it was finally established in a cohesive and institutionalized way, had to solve. The history of modern Greece, not surprisingly, has been the subject of endless historical studies, mainly in Greek. These were frequently multi-volume histories, sometimes of a particular political shade, nationalistic, Marxist, or liberal. Many were openly partisan in their approach or focused on the key individual protagonists – heads of state, prime ministers, or revolutionary leaders. In the recent past, given the progress of history as an academic field, history-writing has become more objective than it was in the past when political passions discouraged detachment and objectivity. Whether this is a right or wrong impression is irrelevant, because for sheer practical reasons the author could not possibly cover all existing historical studies and had to rely in his overview on a shorter list of history publications, mainly of the recent past.[1] Exceptions are made for particular events and developments, which are specifically mentioned (Fig. 6.1).

[1] See in particular: Beaton 2020; Campbell and Sherrard 1968; Clogg 2021; Close 2002; Dertilis 2016, 2019; Diamantopoulos 2017–2018; Gallant 2017; Hadjivasileiou 2015; Kalyvas 2015; Kostis 2018; Mavrogordatos 2016, 2017; Rafaelidis 1993; Svoronos 1972, 1976; Vakalopoulos 2005; Veremis and Koliopoulos 2013; Voulgaris 2013; Woodhouse 1968, 1991. In the text the author refers to many more. Use is made in this chapter of material from Wassenhoven 1980, 2022. Sources that were also used include Beckinsale and Beckinsale (1975), Eleftheroudakis Encyclopaedia (1929), Encyclopaedia Papyrus – Larousse – Britannica (2006), Great Greek Encyclopaedia (1928), Heurtley et al. (1967), Lamprianidis (2001), Markezinis (1966 and 1973), Roberts (1996), Stamatelatos and Vamva-Stamatelatou (2006), Vournas (1974 and 1983), and Stavrianos (1958 and 1974).

Fig. 6.1 The 25 largest urban centres of Greece. (Note: List of cities from the site of the Greek National Statistical Service. Map drawn by the author)

6.1 Independence and Dependence, Nationalism, and Foreign Powers

Almost the first century in the life of independent Greece, from the beginning of the independence war in 1821 against the Ottoman Empire, is a story of efforts to create o modern state, of irredentist struggles, of the constant pursuit of the Great Idea (*Megali Idea*), i.e., of the liberation of territories still occupied by the Turks and their annexation in free Greece, of total dependence on European Great Powers, and of financial bankruptcies. The concepts of "nation" and "state" are constantly in a precarious balance. Greece in not of course unique among countries that faced the challenge of building a state. As Dryzek and Dunleavy (2009, 198–199) remark, "the history of many states is characterized by the often coercive efforts of state-builders to craft an identity – usually national, though sometimes religious – to accompany their political project. While such efforts lie deep in the past of some established liberal democracies, to the point the nation is taken for granted, in other states they are a recent memory or in some cases still in progress". National identity and state building were constantly important parameters in Greece, where "all foreign relations during the nineteenth century and the early twentieth century were in one way or another viewed through the lens of the Megali Idea. Moreover, both the very identity of the nation and the development of the state became bound up in this ideology. Identity, nationalism and irredentism, then, formed an essential core that influenced almost every aspect of the development of Greece during the nineteenth century. The same, however, cannot be said about Greeks outside the kingdom. There were multiple and competing identity narratives to the Athenocentric one … In sum, the National Question was answered differently depending on which Greek community was doing the responding. The Megali Idea was the one adopted in Greece" (Gallant 2017, 99). The notion of "national space" was subordinated to the goal of the Great Idea, which, arguably, was used by the ruling classes to consolidate their hegemony, in the Gramscian sense.

Hellenic nationalism started growing at the time when the Byzantine Empire was already in decline, roughly from the eleventh century onwards, when the long struggle of central imperial power against the rise of feudalism lost its impetus and the local lords began to have the upper hand. The contribution to the process of decline of the Franks, who captured Constantinople (now Istanbul) in 1204, long before the Ottomans finished off the Empire in mid-fifteenth century, was decisive and never to be forgotten. Earlier than that, towards the end of the eleventh century, the Empire was in effect under an imperialist quasi-colonial domination, not of the colonial empires of much later centuries, but of the Venetians and Genoese, who secured independent quarters in Constantinople as the British were to do centuries later in Peking. Historical circumstances were naturally vastly different, but there are similarities, e.g., in the feeling of "dependence", a term which of course did not exist at the time. Educated people in Greek lands of the Byzantine Empire were most certainly aware of the extent to which their fate depended on external forces. Under the Ottoman occupation, the trend towards a form of feudalization, certainly totally different from that of the West European Middle Ages, continued, as a class of Greek notables, the *kodjabashis*, dominated the supposedly self-governed Greek communities. In the years of Ottoman decline, economic imperialist penetration from the West gradually undermined Ottoman rule and the trading agreements, or capitulations, between European countries and the Ottomans facilitated the rise of a class of Greek merchants, working for the Europeans and enjoying special privileges. The pressure of this class, established in communities (*paroikies*) within major European or Near Eastern cities, played an important role in stimulating nationalist feelings, taking of course into account the virtual absence of a genuine bourgeois class in the Greek lands. This international Greek diaspora was the seed of the heterochthonous (foreign-resident) element of the political elite of independent Greece, which clashed with the autochthonous, native element of local notables (Clogg 2021, 47; Vakalopoulos 2005, 19). It is important to realize that the concepts of Greek "nation", let alone of a Greek independent "state", diverged radically, depending on the position of an individual in Greek communities, in London,

Odessa, Trieste, or in the territories of present-day Greece. "Before the war of liberation, there really was no Greek nation, let alone people possessing a fully formed national consciousness. There was an awareness, either acute or vague depending on a person's station in life, of being a member of the Millet-i Rum, of being a Christian (Orthodox), and of speaking the Romaic language, that is, Greek. But such awareness did not constitute a consciousness of being nationally Greek. Instead, peoples' primary identity was rooted in their local village community ..." (Gallant 2017, 84).

In 1828 Count Capodistrias (Capo d' Istria), a nobleman of Corfu, foreign minister of the Russian Tzar and one of the most prominent European diplomats of his day, was appointed Governor of Greece and settled first in Aegina, then in Nafplion, first Greek capitals. By introducing a western conception of centralized state, Capodistrias buried for good the local "states" of the provincial primates, as Alivizatos (2015, 36) remarks. His effort to establish a modern state, admittedly of autocratic colour, to subdue the local notables and revolutionary chieftains, to introduce a Europeanized system of governance, and to secure unity led to his assassination in 1831 in Nafplion. It was an ominous event and an instructive lesson. Within a short period of time the work of Capodistrias fell to ruins and the country plunged in complete anarchy (Kostis 2018, 175). "The internal condition of the country was now pitiful", Woodhouse writes. "No progress had been made with Capodistria's systematic plans for developing a modern state. On the contrary, things were now worse than under Turkish rule..." (Woodhouse 1991, 161). The years of the national revolution itself, but also the following decades witness constant struggles between, on the one hand, the innovators and modernizers inspired by West European prototypes, mostly Greeks of the diaspora, and, on the other, the caste of local notables and primates who were reluctant to relinquish their powers. After the assassination of Capodistrias their privileges remained intact and they continued to oppose the modernizing efforts of the *heterochthones*, with their "Frankish" customs and attire. The result was a compromise which produced a long-lasting, uneasy coexistence of modern state and traditional society (Vatikiotis 1974, 6) and prevented, or delayed, the emergence of a modern national bourgeoisie. Compromise was to become a strategy of securing a delicate balance of interests.

National space was fragmented, still ruled by this caste of notables and the clergy, and made up of traditional communities. In fact, Capodistrias and other similarly-minded politicians in later stages have been accused by ultra-nationalist writers (Tsakonas 1970, 1971), who idealized the lost world of Greek communities, of undermining local communal values. These values were a mixed inheritance of customs and religious beliefs, but also of oriental traditions, after centuries of Ottoman rule. Kontogiorgis (2020, 88–90, 138–139), in spite of his idealistic – nationalist approach, believes that Capodistrias respected traditional communities (*koina*), which operated as local federations or civitates, and integrated them in a hierarchical administrative system. In contrast to Greek intellectuals of the diaspora who contributed to the construction of a new nationhood, like Adamantios Korais, Capodistrias was following the "Greek road", while Korais chose the "western road" (op.cit., 137). When, in 1999, the number of municipalities and communes based on traditional communities was reduced drastically under a local government reform bearing the symbolic name "Capodistrias", there have been voices in the Council of State itself claiming that the reform was unconstitutional because of the historical inviolability of traditional administrative organization (Alivizatos 2011, 73) (Fig. 6.2).

These divisions apart, Greece, ever since the independence war, remained in perpetual debt and constantly dependent on foreign loans. The ability to create an administrative system capable of uniting the national territory was effectively curtailed. The models available to the Greeks, to imitate and install in the new governmental machine, were inevitably of foreign origin. The war against the Ottomans had been won with the decisive military intervention of foreign countries, the "Protecting Powers" (Great Britain, France, Russia). The influence of the Great Powers and their role in tilting the balance in favour of the revolutionaries and in securing the creation of a new state out of the body of the Ottoman Empire meant that decisions in Greece were made under their constant supervision. This included any effort to embark on new irredentist adventures, but also internal affairs of minor importance if they were considered as affecting the interests of citizens or firms of

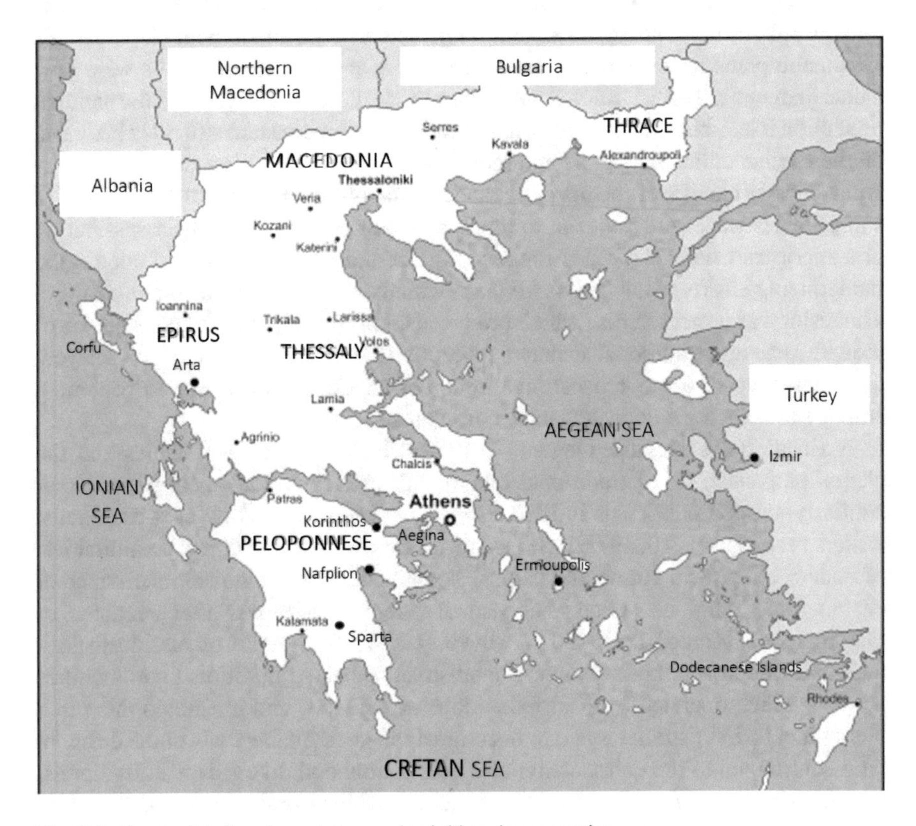

Fig. 6.2 Geographical regions, towns and neighbouring countries

the Great Powers.[2] Naturally, the regime of the new state could not be an exception to this rule. After the assassination of Capodistrias, the London Protocol of 1832 which fixed the frontier line between independent Greece and the Ottoman Empire was, as expected, signed by the Protecting Powers and the Ottoman Empire. It created a version of a protectorate (Kontogiorgis 2020, 162). The territory of the new state was but a fraction of today's Greece. Soon afterwards, the allies installed in the Kingdom of Greece a new royal regime under the Bavarian Prince Otto. The Greek capital was transferred to Athens, then a mere village, because of its ancient glamour.

King Otto brought with him a Bavarian administration, which started a new effort to organize and modernize the new state, always of course on the basis of western prototypes. He was also accompanied by a Bavarian army. Provincial administration was organized around prefectures, copied from the Napoleonic model. During an initial stage of a regency, while Otto was still a minor, a member of the regency triumvirate was a professor of law from Munich. An attempt was made, without success, to introduce a land cadastre. "The Bavarians in Otto's entourage were bent on equipping the new state with institutions fitting for an embryonic European state" (Clogg 2021, 281). Such initiatives were resisted by local centres of power which found an excuse for their reactions in the absolutist nature of the regime. The Bavarian regime became known as *xenocratia* (foreign rule) (Kontogiorgis 2020, 167). In 1843 King Otto was forced to accept a more democratic constitution. The Europeanization of the country, if there was any, was most evident in Athens, where a town plan of neoclassical conception was produced, but only partly implemented because of landowning interests. The social and economic distance between, on one hand, rural Greece, and, on the other, Athens and the scattered minor urban centres remained enormous. Popular dissatisfaction was growing, while the irredentist ideology of Otto, who espoused the Great Idea, caused the grave concern of the Protecting Powers. In 1862 Otto was forced to leave the country. A new constitution was voted in 1864 by a sovereign constitutional assembly establishing a reigned democracy in place of the previous constitutional monarchy. The new model was much closer to European countries like Belgium or Denmark (Alivizatos 2011, 117–118). A Danish prince was selected by Britain, France and Russia to succeed Otto as King George. The national assembly endorsed the choice, although the Greeks' favourite for the throne was Prince Alfred, a son of Queen Victoria (op.cit., 111). George I reigned until his assassination in 1913.

[2] The most famous such affair, in 1850, was that of Don Pacifico, a Portuguese consul and merchant, but British subject. His conflict with the Greek government over a matter of compensation, when his house was set on fire during an anti-Semitic riot, led to a British intervention and the blockade of Piraeus by a naval squadron. It was on this occasion that Lord Palmerston, the British foreign secretary, said, in a speech to parliament, that "a British subject, in whatever land he may be, shall feel confident that the watchful eye and the strong arm of England will protect him against injustice and wrong" (Encyclopedia Britannica). The affair remained in history as either the Pacifico affair or the "Parkerika", Parker being the name of the British flotilla commander (Gallant 2017, 70; Clogg 2021, 53).

6.2 National Lands, First Development Efforts, Bankruptcies and Defeats

As mentioned earlier, the nineteenth century was dominated by irredentist claims, which ultimately led to a traumatic military defeat by the Turks in 1897. An equally important source of domestic troubles, which directly touches spatial organization, was the problem of land distribution, or, more broadly of the agrarian question (Tsoulfidis 2015, 115–125). The so-called National Lands, formerly the property of the Ottoman State, had become Greek public property after independence. Land reform, ownership of agricultural land, and the distribution of the National Lands was a major problem and stayed so for almost a century (Dertilis 2019, 98–101, 164–167, 323–329, 560–564). The claims of landless peasants and erstwhile independence fighters caused serious political problems for the Greek government, for which these lands were essential as a guarantee in the process of contracting much-needed bank loans. The prospect of their distribution brought to the surface social inequalities between wealthy primates and poor peasants. "The national estates formed the major potential resource for the economic regeneration of liberated Greece, but they also became a focal point for the rivalries and power struggles among contending groups and factions" (McGrew 1985, 64). Soon after liberation the problem of national lands assumed unmanageable proportions. "The primates, being the most powerful, helped themselves to the choicest Turkish estates. Thus, the new state was born with a serious land problem" (Stavrianos 1958, 296). The state owned at least one third of cultivated land, an estimate raised to two thirds by some commentators (Vergopoulos 1975, 104–115; Sakellaropoulos 1991, 65–82). One quarter was owned by the Greek Church, which left only a pitiful percentage in the hands of private individuals. For several decades after independence only a small amount of land was distributed to farmers (ibid.; Svoronos 1976, 79 and 102). To be noted that, at the time of the 1861 census, almost 52% of the national population lived in settlements of less than 1000 inhabitants. The percentage rises to almost two thirds if the settlements of between 1001–2000 are included. 74% of adult men were agriculturalists (Gallant 2017, 127–128). A land reform undertaken in 1871 was of questionable success because of the extremely small size of the land lots and hence their competitive disadvantage in comparison to large estates which remained intact (Vournas 1974, 446–447). Many of these estates had been acquired by Greeks from the original Turkish owners. The problem of large estates took on a new significance, when the region of Thessaly was annexed by Greece in the 1880s, because of the existence there of large formerly Turkish *chifliks*. The whole situation amounted to a serious social problem, which caused a peasant rebellion and a bloody conflict in 1910, in Kileler of Thessaly, and lasted until the agrarian reform of the 1930s. The fragmentation of rural land gradually became the precondition of one of the thorniest problems of spatial planning after the Second World War.

The incorporation into the social fabric of impoverished peasants was of vital importance for the construction of the Greek state and for reviving a sense of nationhood. Forging a national identity remained a goal throughout the years after independence, as indeed it had been the dream of expatriate intellectuals even in the eighteenth century. A new vision had to be crafted with elements drawn from the past, a difficult task, which was by no means exclusive to Greece, since other nations too had to go through such a process. For a long time since then, the "local" and the imported "Western" elements have been facing each other in a variety of conflicts between "orientalism" and "Europeanism", backwardness and progress, with various social groups taking sides as the circumstances demanded (Sotiropoulos 2019, 48). When this opposition was still raging, in mid-nineteenth century, guidance for the future had to be found in the historical past (Veremis et al. 2018, 243). Linking Hellenism with its Ancient Greek or Byzantine past was the task of distinguished historians, particularly Konstantinos Paparrigopoulos in the 1860s (Dertilis 2019, 242; Clogg 2021, 2). Nationalism, embedded in a "romantic narrative", served as a funnel in which social tensions disappeared (Sotiropoulos 2019, 64, 179). "Finally, through the conscious modification of popular culture, the state and the intelligentsia facilitated the internalization by the masses of this invented Greek identity. This new vision of Hellenism was literally inscribed on to the landscape. In rebuilding the towns that had been destroyed during the Revolution, city plans were designed specifically to make the new urban areas look Western. Built on a grid system and adorned with public buildings and houses in the neoclassical style, the new towns were to give concrete form to the new Hellenic national identity by obliterating all traces of the Ottoman past and reviving the memories of the Classical past. With a similar purpose in mind, the names of hundreds of rural villages were changed from 'Slavic' or 'non-Hellenic' names to names that were 'totally' Greek" (Gallant 2017, 93).

Text-Box 6.1: Agricultural Reform: The Drainage of Lake Copais

Lake Copais in Boeotia, located 110 kms from Athens, had been first drained in prehistoric times. Modern drainage projects in the nineteenth century are closely associated with the effort to modernize the Greek state and to develop its economy. In their final phase, they were a centrepiece of the policies of the progressive Tricoupis administration. They must be seen in the wider context of rural reform and repeated attempts to assist the agricultural economy. Banks, Greek financiers, and two foreign engineering firms from France and England participated in successive drainage projects. Soil engineering problems caused successive failures, until the final deal with an English company and the drainage of about 24,000 hectares and the inauguration of the project in 1892. The drainage of Copais lake became a symbol of modernization and development. New cultivations were introduced and although most of the land remained in English possession employment opportunities were created, a side benefit being the elimination of malaria. In 1953 all land was compulsorily acquired and distributed to landless peasants, with the management being assumed by a public company. The Copais development changed the face of the area. The township of Aliartos, where the headquarters and rural produce processing plants of the English Copais company were located, practically owes its existence to the project. But apart from the regional impact of the drainage scheme, the plan itself, albeit not an urban plan, is representative of a kind of allotment parceling and land distribution plan, which of course can be frequently found in other countries. In Greece, it was a pioneering land use scheme (see Fig. 6.3).

The last quarter of the nineteenth century was a period of mixed developments. Under the leadership of Charilaos Tricoupis, who, interestingly, was first elected to parliament as representative of the Greek community of London, the country embarked in a period of modernization and strengthened democratic, constitutional and parliamentary processes. Tricoupis, a staunch believer in British-style parliamentary institutions (Alivizatos 2015, 41), initiated a programme of road and railway construction of great importance for the unification of national space, although the alignment of railways was possibly the choice of foreign creditors (Close 2002, 3). "Tricoupis believed that the state needed to be strengthened politically and economically before it could contemplate engaging in irredentist adventures. He therefore sought to establish the country's international creditworthiness, to encourage incipient industrialisation, to improve communications through railway construction and the building of the Corinth canal,[3] and to modernise the army and navy" (Clogg 2021, 63; see also Vakalopoulos 2005, 285–292; Tsoulfidis 2015, 202; Woodhouse 1968, 173).

In the meantime, the Ionian Islands had been annexed in 1864 and southern Epirus (Arta province) and Thessaly were to follow in the 1880s. "The annexation of Thessaly, the second extension of Greece's borders and, like the first, the cession

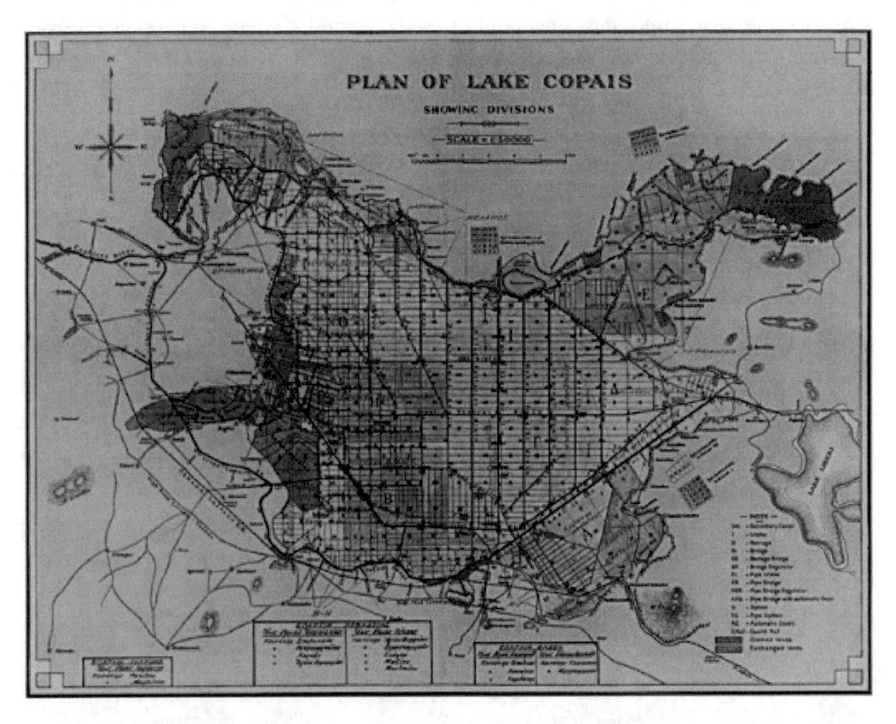

Fig. 6.3 The English plan of agricultural land parcels in the drained lake of Copais. (See Text-Box 6.1. Source: orchomenos.gr.)

[3] Built across the isthmus separating Mainland Greece from the Peloponnese.

of the Ionian Islands in 1864, the result not of irredentist agitation but of Great Power mediation, brought the frontier to the borders of Macedonia" (Clogg 2021, 67). However, the state of public finances caused the 1893 bankruptcy, and, a few years later, persistent nationalistic aspirations, exploited by an adventurous new government, resulted in the disastrous 1897 military conflict with Turkey (op.cit., 69). At every turn, the subordination of the country to the policies of the Great Powers was more than evident, particularly when Greek ports were blockaded to force the Greek government into submission. The British, French and Russian ambassadors were the effective rulers in Athens, each trying to neutralize the influence of the other. This was of course a reality ever since independence. In the British ambassador's view in 1841, "a really independent Greece is an absurdity. Greece is either Russian or English and since she must not be Russian she must be English" (Stavrianos 1958, 292; see also Kordatos 1973, 40, and Korizis 1975, 111). The Greek economy and the country's political stability were totally dependent on the support and on the economic interests of the Protecting Powers. The transfer of a term like "dependence" from the condition of post-Second World War developing world to that of Greece in the nineteenth century may seem too bold but, in the author's view, justified: "By dependence", Dos Santos suggested, "we mean a situation in which the economy of certain countries is conditioned by the development and expansion of another economy to which the former is subjected" (Dos Santos 1973, 109). Dependence, in his view, is a "conditioning situation". Kühn (2015) defines it in broader terms: "Regarding concepts that try to explain peripheries in terms of political power, there is the long tradition of 'Dependency theory'. This is a school of thought that attempts to explain the 'underdevelopment' of regions and countries within the capitalist world economy. Emerging in Latin America during the 1960s, these theories have argued that there is a power asymmetry between developed and underdeveloped states. 'Underdevelopment' is explained with reference to the neo-colonial 'domination' of the periphery by capitalistic metropolis". Dependence in the case of Greece did not involve colonial domination, but was a permanent condition, even though it was not accompanied by military occupation (Sotiropoulos 2019, 41–42). The country was however constantly under the threat of gunboat diplomacy, a reality which reached a peak during the First World War.

6.3 Social Classes, Greeks of the Diaspora, Emerging Elite and Cities

After the 1893 bankruptcy, Greece had to suffer the humiliating imposition of foreign control, exercised by an International Financial Commission to which half of the state's revenues were directly assigned.[4] The plague of public finances continued to force the state into dependence on foreign and expatriate capital, with little

[4] A street in central Athens still bears the name of Edward Law, chairman of the commission.

effect for industrial development. Foreign imperialism played a major part in the rise of the bourgeoisie during the last years of the nineteenth century, through the flourishing of what Svoronos (1972, 60) describes as "peripheral Hellenisms", the Greek expatriates of Russia and the Near East, who dominated the economy of the new state and conferred to it an importance which it could not otherwise acquire, given its small size (Markezinis 1966, v.1, 295). Big international lenders and public loan managers, Dertilis (2019, 460) observes, were in a position to always adjust in time their investment strategy on the basis of sound predictions, which was simply impossible for simple bond holders in Greece. According to Svoronos, it was Istanbul and not Athens which was the economic capital of the Greeks, although Athens became their cultural centre. Dependence on external centres and domestic resistance to it are frequent themes in historical studies, what D.P. Sotiropoulos (2019, 50) calls *Svoronean* historiography, but it is no doubt true that the territory of Greece, not for the first time in her history, had to rely on external space. The central character of peripheral Hellenism, which amounted, as Dakin put it, to a commercial empire, created even before a nation-state was built, was the Greek colonist, the merchant-financier of the Ottoman Empire and of European capitals and ports (Dakin 1972, 266). From his perspective, the Greek merchant hoped that Greece would move eventually from her peripheral position to the metropolitan centre and control the business life of the strategic centres of Istanbul, Athens, Izmir, Thessaloniki, Alexandria, and urban centres in Russia. Subsequent events – the Asia Minor catastrophe of 1922, the communist revolution in Russia, and, much later, the rise of nationalism in Egypt – put an end to such hopes.

Both the weak presence of middle classes in independent Greece and the strong influence of the bourgeoisie of the Greek diaspora contributed to the confusion and distortion of Greek social stratification which Svoronos (1976, 13–20) considers as a central theme of modern Greek history. Even when, with considerable delay, the Greek middle classes began to take shape, the weight of their most vigorous component, which resided out of the country, created a visible lack of identity and an outward orientation. Socio-political life, the practice of government, and international attitudes were strongly influenced by expatriate Hellenism and its domestic tentacles. Gradually, the powerful land owners or merchants saw their influence diminish because of the rise of bankers and financiers (Kostis 2018, 508). The ruling class inside Greece – the "big bourgeoisie" – was satisfied with a centralized and bureaucratic form of government, the immediate control of which was in the hands of a middle bourgeoisie, for which the state and a complicated system of political patronage was a mode of survival. Not only did this "rule by intermediary" create a class of poorly educated white collars and professionals, amenable to effective external control, but it also perpetuated conditions of favouritism and corruption, behind the façade of a modern, efficient state machinery. "Given the rudimentary development of the economy, the state assumed a disproportionate importance as a source of employment, and the proportion of bureaucrats to citizens was far higher than in western Europe" (Clogg 2021, 59) Political manipulation and foreign control thus became closely linked with a top-heavy administration and with centralization, evident in the structure of the urban system, while the growth of a dependent middle class, living in the shadow of their rich local and overseas relatives, encouraged the pursuit of quick and easy benefits and profits, a short-sighted and narrow-minded perspective, and a parasitic mentality. Political parties – for a while called the English, French, or Russian parties – competed in demonstrating their subservience to foreign powers, while an ideology of *compradorism*, in Moskof's language, limited the ability of urban classes to effectively unite and lead the country, thus subordinating the country to the international market and maintaining a city-countryside barrier (Moskof 1972, 151). These parties were loose assemblages of autochthonous and heterochthonous representatives, not substantially different from the model defined by Dryzek and Dunleavy (2009, 68): "Until the late nineteenth century parties had mainly operated as small cadres of national political elites in parliament, sustained by a network of established regional and local elites (in Europe often aristocrats or the local wealthy). These 'notables' delivered the votes of people in their areas, in return for influence with national elites and an opportunity to create clientelist networks of patronage controlling titles, government jobs, publicly funded services and economic concessions". If we remove the reference to aristocrats the definition suits perfectly the Greek situation. However, the absence of an aristocratic nobility is not a negligible feature, because it differentiates Greece from other European nation states.

In terms of development, the Greek territory was fraught with inequalities, especially between cities and rural areas. To a large extent this explains the introverted mentality of the Greek family, particularly in villages, and the constant effort to include powerful external allies in the family fold. "Peasants coped with the reality of inequalities in a number of ways. One was to make useful outsiders insiders through the practice of fictive kinship, blood brotherhood and godparenthood, for example. Reliance on men who shared the same blood, even metaphorically, did not diminish a man's reputation. Finally, men formed strategic alliances with those from the higher strata of society in which they willingly adopted the subservient position in order to obtain material benefit and protection. These bonds between a client and his patron provided the weaker partner with an important insurance policy in the unsure environment of rural Greece. Because the dominant party came from a higher social class and because the client benefited from this arrangement, becoming a client did not diminish a man's stature among his peers. The key here was that the client obtained goods and services that increased his family's fortunes, and so it enhanced his reputation as a man who takes care of his own. Thus, there was often intense competition between peasant men to forge a bond with a powerful figure" (Gallant 2017, 147). The roots of patronage and clientelism were deep in rural Greece and were to influence political life for a long time to come (Campbell 1964). Family loyalty remains a respected value (Close 2002, 215). In addition, family priorities and clientelism determine the relations between cities and rural society. "Patronage had originally developed as a kind of defence mechanism against the harshness, and particularly the arbitrariness, of the Ottoman system of government. There was a need for patrons and protectors to mediate with the Ottoman authorities and to mitigate the capriciousness of the judicial system. Many Greeks regarded the impositions of the new state as scarcely less oppressive than those of the Ottomans, and values and attitudes shaped under Ottoman rule persisted into the independence period. Patronage, indeed, proved wholly compatible with the formal institutions of parliamentary democracy. The local *kommatarkhis* or political boss simply took over the role of the Ottoman *ağa*. Until modern times a parliamentary deputy has seen it not only as an obligation but as the indispensable precondition of political survival to secure favours for his voters" (Clogg 2021, 61) (see also Chap. 7).

The growth of cities in the nineteenth century was modest and depended on administrative expansion and internal trade for foodstuffs and handicraft products, because demand for industrial goods, originating in a small social group, was satisfied by imports. The genuine Greek bourgeoisie, the "adventurous merchants", operated from centres on foreign soil (Filias 1974, 125; Campbell and Sherrard 1968, 93), while the marginal class which grew up in Greek cities, "cities of silence" according to Filias (op.cit., 124–132), produced what he considered spurious or false "urbanification" and an urban cretinism. Before the massive inroads of expatriate financial interests, the domestic landed oligarchy kept control of banking and shipping, and of course of the export trade of crops. Small maritime ports gradually lost their trade and gradually only Piraeus and Patras, and temporarily Ermoupolis in the Cyclades Islands, survived (ibid.; Markezinis 1966, v.1, 278–280). Foreign and expatriate capital chose to associate with local interests and did not take the lead for the development of a manufacturing industry, with the exception of a metallurgy plant in Attica, on the site of the ancient Greek Lavrion mines. Cities grew without an industrial base, while the spread of capitalist relations in the countryside accelerated urbanization, migration to Athens, and after the financial collapse of the 1890s out-migration to America and Australia. This does not mean that the rural areas were cut-off and totally isolated. Greece was a small country and even her mountainous areas, most of which, in the north, remained in the Ottoman Empire throughout the nineteenth century, had settlements maintaining cross-country communication and, in rare cases, even contacts with foreign countries. As Gallant points out, there has been an impression "that rural men and women were born, lived and died without ever transcending the narrow confines of their village and its surrounding fields. The reality was that the Greek countryside was alive with movement. When anthropologists began writing about the remote villages, they had studied during the 1950s and 1960s, they either implicitly or explicitly reproduced this assessment. The view of the Greek village as isolated in space and frozen in time has thus had a long history. But, at best, it is misleading and, at worst, inaccurate. For one thing … villages were enmeshed in very extensive networks of settlements that spread over fairly wide regions" (Gallant 2017, 139).

What is true, however, is that the network of towns was still weak. The absence of domestic centres of some dynamic autonomy meant that all determinants of centrality favoured Athens. The capital was the gateway of all external influence, the seat of the king, the mediator of international influence, and the point of encounter between the "foreign factor" – one more popular expression –, the Greeks of the diaspora, and the local oligarchy. Thus, the centralizing forces were strengthened by functions, not only located *in* Athens, but also mediated *through* Athens, a point of vital significance, certainly not lost to the average citizen, since it made the abstraction of outside influence a reality, because of the pressures that domestic interests could bring to bear on office-holders through extra-political manipulators, such as courtiers, foreign officials, and Greek expatriates. As years passed, the influence of foreign ambassadors or creditors became subtler than in Otto's days (Svoronos 1969, 21–22), but otherwise the role of the palace as a locus of power, and increasingly as antagonist of the rising bourgeois classes, was as clear as ever, with the "royal clan" (Meynaud 1965, 335–338) remaining the unchallenged representative of British interests. The ensemble of these actors, it could be claimed, formed a power elite, consistent with a typical definition, "controlling economic, social and political life through its expertise, ownership of wealth and property, social status, intelligence, and economic and political guile" (Dryzek and Dunleavy 2009, 58). The local Greek elite, in the view of Kontogiorgis (2020, 171–172), had fully succumbed to foreign protection and had distanced itself from the social collective identity, even from the Hellenism of foreign lands and from historical roots. Clientelism was one of the symptoms of this condition. Athens, the habitat of this elite, though also full of semi-parasitic clerks, harboured an administration of "unbridled centralism" (Moskof 1972, 153) and, in Moskof's dramatic style, "the compradorist city … was dominant in Greek space" (op.cit., 164). Nevertheless, it must not escape our attention that the Athenian bourgeoisie – at least its most progressive component – was at the same time creating a new culture of westernized orientation. Typical example is the creation of universities, the Athens Conservatoire and its orchestra (1881), the Academy of Athens (1886), the National Archaeological Museum of Athens (1889), or, a little later, the Lyceum Club of Greek Women (1911), product of an embryonic feminist movement. The National and Capodistrian University of Athens and the National Technical University of Athens (or National *Metsovion*[5] Polytechnic) had been established in 1837. It was in that year that a "school" for architectural training appeared for the first time with a royal decree.

Social developments in the nineteenth century are the origin not only of the role that the capital, link with and window to the world of western metropolises and Greek communities, assumed in Greek society, but also of the city's later social and economic structures. This is not to say that today's organization is a direct product of a linear and inexorable evolution from past periods. But at least some of the conditions that were laid down in those days proved to be a fertile ground and a propitious habitat for developments extending after mid-twentieth century. Similarly, Greek attitudes to the "foreign factor" and to outside influence are the result of a long formative process in which nineteenth century events were an essential part. The end of the period from independence to the First World War was additionally marked by a major political and social schism, the infamous *dichasmos*, which was to be the most lasting of a series of rifts that had preceded it or were to follow. The *ethnikos dichasmos* (national schism) during the Great War "split the country into

[5]Thus named to honour the national benefactors born in the town of Metsovon in Epirus, who funded its creation. Many of the public buildings of this period, in neoclassical or byzantine style, were built thanks to donations by expatriate Greek businessmen.

two rival, and at times warring, camps. Not for the first and certainly not for the last time Greece was to be riven with domestic dissension at a time of grave international threat" (Clogg 2021, 83).

6.4 The Asia Minor Disaster and the Influx of Refugees

The 1897 financial bankruptcy brought to the boil the unrest of liberal middle classes, disillusioned by the ruling establishment, and of the peasants of Thessaly, still living in semi-feudal conditions (Kordatos 1958, 182–194). Young officers, sympathetic to the slogans of a rising liberal movement, revolted in 1909 and brought to power a liberal government under Eleftherios Venizelos, a brilliant Cretan politician, without, however, challenging the monarchy (Dakin 1972, 181–186 and 311; Markezinis 1966 v.3, 82 and 114–117). Like Tricoupis before him, Venizelos was a keen observer of institutional deficiencies and capable to delve into their real causes (Alivizatos 2015, 65). He was also the initiator of an ambitious undertaking to modernize Greek society and establish a national bourgeois state (Sotiropoulos 2019, 15–16). In this task, Venizelos relied on a new elite class of army officers, administrators, intellectuals, and technocrats (op.cit., 68, 72–85). Under his leadership, Greece fought the two Balkan Wars of 1912–1913 and occupied the remainder of Epirus and Macedonia, including the important cities of Thessaloniki, Ioannina, and Kavala, before entering the "Great War" in 1917 on the side of the Anglo-French Entente Cordiale and against the Triple Alliance of Germany, Austria-Hungary, and Italy. The island of Crete had been annexed in 1913. The First World War was accompanied in 1916 by a vicious clash between the Royalists and the Liberals – between King Constantine I and Venizelos – over the decision whether to side with the British and the French or the Germans. This caused a dramatic political and social rift, a national cleavage, which marked Greek history not only in social terms, but also territorially, as it produced the feeling in the north of the country of an unequal domination by the "State of Athens". The royal government ruled in Athens, while the Liberals set up a rival government in Thessaloniki. Two states existed for a while side-by-side, the State of Athens and the State of Thessaloniki (Diamantopoulos 2017–2018, v.1, 71). "Greece now had two competing governments", Gallant writes and then adds: "The rupture within the political system also took on a geographical dimension and inaugurated a division between 'Old' Greece to the south and 'New' Greece in the north. The significance of this north-south division would only increase in the decades ahead" (Gallant 2017, 184). Repeated splits, or "fault lines" in Beaton's expression, became a perpetual syndrome lurking at every historical turn. "Shifting tectonic plates" produced disastrous results at various points in Greek history (Beaton 2020, 224 and 385). We shall return, in Chaps. 7 and 8, to the role of the liberals in building a new state and in introducing town planning.

On the question of Greece's involvement in the Great War, the allied powers of Britain and France imposed their will through military intervention in Athens and ousted King Constantine (Clogg 2021, 89). A series of spatial divisions inherited from the early years of the new state of Greece reached their climax in these events. Exploiting the support of their allies, which proved half-hearted and of short duration, the Greeks occupied Izmir (Smyrna) and came close to capturing Istanbul. The national dream of a Greece of "two continents and five seas" was within their grasp, but was never realized. A foolish and ill-judged military campaign towards Ankara, launched by a new government, ended in an unprecedented defeat. In a sense, this was the last episode of the so-called Eastern Question, and also the final blow to the Great Idea. It was a war fought, perhaps unconsciously, to serve foreign interests, since, in Churchill's words, it was in fact Britain that was fighting by proxy (Pentzopoulos 1962, 44). The Asia Minor disaster turned the tables against Greece. Arguably, Kontogiorgis (2020, 194) claims, this was the predictable result of following to the end the "Western road". Not surprisingly, Greece lost some of the lands it had temporarily captured in the last phase of the war, including Eastern Thrace, and had secured with the Treaty of Sèvres in 1920 (Vakalopoulos 2005, 418 and 426). The Treaty of Lausanne of 1923 gave Greece its present form, with the exception of the Dodecanese Islands, which were annexed in 1947 (Kasperson 1966). New territories were more than double the old lands of the country thus posing a challenge comparable to that of the founding of the state a century earlier (Sotiropoulos 2019, 68). By far the most tragic outcome of the military defeat was a colossal movement of refugees who poured into Greece, following an agreement for the exchange of minorities (op.cit., 427–447). It was a development which scarred the social, ideological and economic life of the country for decades. The Great Idea had been shattered and so was the morale of the Greek people. In the aftermath of the disaster, a military uprising overthrew in 1922 the government deemed responsible for the catastrophe. The country's political and military leadership was court-martialled and executed. The task of the refugee settlement once again forced Greece into familiar arrangements for the supervision of the country's finances and of the servicing of the inevitable loans.

As mentioned, an exchange of populations between Turkey and Greece had been agreed in 1923. However, almost 850,000 refugees of Greek origin had already arrived in Greece earlier, after the First Balkan War of 1912. Another 250,000 were now added (Campbell and Sherrard 1968, 128–129). Within an extremely short period of time, the population of Greece rose by 18% (Leontidou 2001, 151). When, in 1928, a new census was taken, about 1.2 m. refugees were recorded, amounting to about 20% of the total population, with approximately half a million settling in rural areas (Sandis 1973, 13), although the precise numbers are impossible to fix with certainty (Mavrogordatos 2017, 157). There are alternative estimates raising the total number to 1,5 m.[6] The simultaneous departure of about half a million Turks left vacant land mainly in Macedonia, in the north of Greece, which was distributed

[6] Mavrogordatos uses estimates provided by Pentzopoulos (1962) and Eddy (1931). A further source of information is Mears (1929). According to national census data, in approximate numbers, the population of Greece was 2.6 m. in 1907. 5.5 m. in 1920, i.e., including annexed territories which were returned to Turkey after the Lausanne treaty, and 6.2 m. in 1928. The loss of these

to Greek refugees (Veremis and Koliopoulos 2013, 196–198). The settlement of refugees in agricultural occupations, although in their great majority they were not farmers before coming to Greece, was planned by the Refugee Settlement Committee (RSC) of the League of Nations. As Veremis and Koliopoulos explain, about half of land lots and dwellings distributed to the refugees had been abandoned by departing Turks and Bulgarians. Another half came from the implementation of land reform and the expropriation of landed estates (*chifliks*). Houses were built either by the RSC or by the Greek state, and were usually concentrated in independent townships and villages, given that the refugee settlement concerned groups of people of common local origin and not individual families. The aim was to reconstitute as much as possible the communities of origin (see also Chap. 8).

From the point of view of geographical transformation, the changes were revolutionary. Although planning was in the hands of the RSC, the direct and indirect role of the state, as Leontidou (2001, 203) remarks, was crucial, because it built houses, enacted ad hoc legislation, and created new institutions. But it was the colonization of the countryside which had the main spatial impact. According to Leontidou, it had a decisive effect on urbanization patterns, given the construction of a large number of remote and isolated rural settlements, mainly in Northern Greece, which accommodated 80% of the settlements, thus causing a ruralization process, against all urbanization trends (op.cit., 156–157). Within a decade from the second Balkan War of 1913 the ethnic profile of Macedonia had been completely altered (Vakalopoulos 1988, 219–224). Beaton underlines that "the country was transformed – not just demographically, but socially, politically, economically, even in its physical appearance. Shanty-towns spang up all around the cities of Athens, Piraeus and Salonica" (Beaton 2020, 234–235). And further: "Even after the distinction between the state and the wider nation had been all but erased, the centralized state would struggle to impose homogeneity upon its diverse subjects. When something approaching homogeneity did come, it would be due not to the Greek state but to its total collapse during the 1940s, to the horrors of a world war and a brutal occupation" (op.cit., 238).

Text-Box 6.2: Refugee Settlements

The traumatic experience of the settlement of Greek refugees from the Black Sea and, especially, from Asia Minor, after the 1922 military debacle, is being discussed in all chapters of Part II of this book. A large number of Greek and foreign architect-planners were employed in the interwar period to design new settlements in rural areas, especially in the north of the country, but also in and around large urban centres. Their work was a valuable inheritance for Greek town planning, concentrated of course in the design of townships, villages, suburbs and neighbourhood quarters. This chapter of Greek planning has been extensively studied in the academic, historical and geographical literature (see, e.g., Kafkoula 1990; Leontidou 2001, 2011; Katsapis 2002). It is best described by some examples of the plans themselves (see Fig. 6.4 and Chap. 8).

Piraeus (Drapetsona) and Athens (Nea Filadelfeia)

Sources: Angathi / espivblogs.net; Kyramargiou et al 2015.

Macedonia (Semaltos / Nea Doirani, Kato Tzoumagia, Doxato)

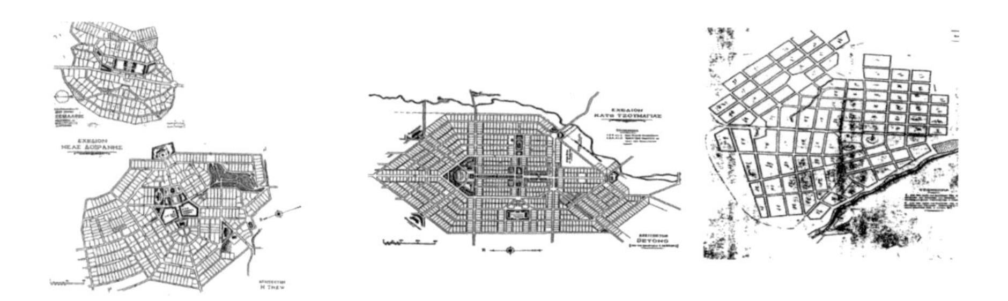

Source: Kafkoula 1990

Fig. 6.4 Plans of refugee settlements (see Text-Box 6.2)
Piraeus (Drapetsona) and Athens (Nea Filadelfeia). (Sources: Angathi/espivblogs.net; Kyramargiou et al. 2015)
Macedonia (Semaltos/Nea Doirani, Kato Tzoumagia, Doxato). (Source: Kafkoula 1990)

By 1938, almost three quarters of a million refugees had been settled, from whom 90% in Macedonia and Thrace, in the north and northeast. These figures demonstrate the magnitude of the settlement effort, its geographical concentration, but also the relative speed of the undertaking. The settlement of the refugees was the most administratively and technically demanding task undertaken by the state in the interwar years (Sotiropoulos 2019, 82–88). However, in the urban centres it proved relatively slow and was concentrated in Athens, Piraeus, and Thessaloniki (Salonica). Whether in rural areas or in cities, the refugees were settled in small villages, compact housing estates, or neighbourhoods, always under the pressure and urgency of the circumstances (Mavrogordatos 2017, 159–160 and 172). The Athens area received 125,000 refugees, settled in estates which in due course became incorporated in the urban fabric and became well known neighbourhoods (Biris 1995, 286). In many cases these neighbourhoods were totally informal precincts with wooden shacks (Leontidou 2001, Vaiou 2019). These hasty urban extensions, in several cities, sometimes created an enclosed space and caused problems of servicing and regulated expansion in the urban periphery (Papaioannou 1971, 264). In the case of Athens, the mushrooming of new neighbourhoods was the reason for the drafting of the first master plan of the city (Giannakourou 2003) (see also Chap. 8). Typical of the social antitheses of the period was the planning and construction in Athens, in parallel with the refugee estates, of private high-quality suburbs (Voivonda et al. 1977, 137; Kafkoula 1990; see also Text-Box 8.1).

To cover the cost of the refugee settlement, the Greek government contracted an international bank loan of 9 m. pounds sterling, guaranteed by the RSC, under the aegis of the League of Nations (Dafnis 1955, 365–366; Pentzopoulos 1962, 82–91). Half a million hectares of agricultural land were placed at the disposal of RSC for settlement of refugees (Dafnis 1955, 268). As mentioned earlier, large estates were compulsorily acquired to secure additional agricultural land. It is worth mentioning that the urgent needs of the settlement led to the acceleration of a land reform which had become a major challenge for the Greek state in the previous century. A series of special laws were enacted in the 1920s and 30s to complete this long-standing reform (Sakellaropoulos 1991, 117–119). A side-effect was the creation of an agricultural sector of smallholders and family enterprises in constant indebtedness, hence easily controlled through credit facilities, mainly handled by the state-owned Agricultural Bank. The peasants were regularly exploited by middlemen and this condition of insecurity fuelled migration, but did not surprisingly produce a serious political peasant movement, although at least a section of the peasantry supported the communist insurgents in the 1940s. In the 1930s the popularity of the Liberal Party, which had introduced a series of reforms, was a factor that kept most of the peasants away from radical political affiliations. As a result, the refugee settlers of Northern Greece, although outside the established networks of political patronage (Legg 1969, 210), presented no challenge to the Greeks of the south, an important fact for the continued power of the "core" of the country, dominated by the liberal establishment. This made party divisions largely irrelevant (Tsoucalas 1969, 44).

Text-Box 6.3: Land Distribution and Settlement of Landless Peasants

Town planning in the interwar years is mostly associated with the settlement of refugees, before and mainly after the 1922 Asia Minor disaster. However, it is also linked with land distribution to both refugees and native landless peasants. The question of national lands, which came to the possession of the Greek state after its independence, and their distribution was a constant source of trouble throughout the nineteenth century and a great part of the twentieth. Apart from that, when Thessaly with its fertile plains was annexed by Greece in 1981, the fate of rural workers became an explosive issue. These surfs were

(continued)

Text-Box 6.3 (continued)

employed originally in Ottoman *chifliks* and later in the estates of rich Greek financiers who hastened to buy land from the departing Turks. Greek governments vacillated for long between the option of retaining very productive estates and thus pleasing the owners, who were also lenders of the state, and the option of expropriating the land and distributing it to impoverished peasants. After bloody riots in the early twentieth century, the new liberal government paved the way for a decisive step finally taken by the revolutionary government of 1923. Land was distributed and settlements were created, one of which was the Palamas township in the Karditsa area of western Thessaly. The township plan is essentially a land allocation scheme of rural housing and cultivation allotments, on a grid pattern. It was surrounded by a narrow belt of common land. Palamas is located on the site of a Neolithic settlement and as a village had existed for about three centuries. It is a prosperous rural community, which had recently serious flooding problems. A modern plan was produced in the mid-1980s, in the context of the so-called town reconstruction operation, in which the author of the present book had participated (see Fig. 6.5).

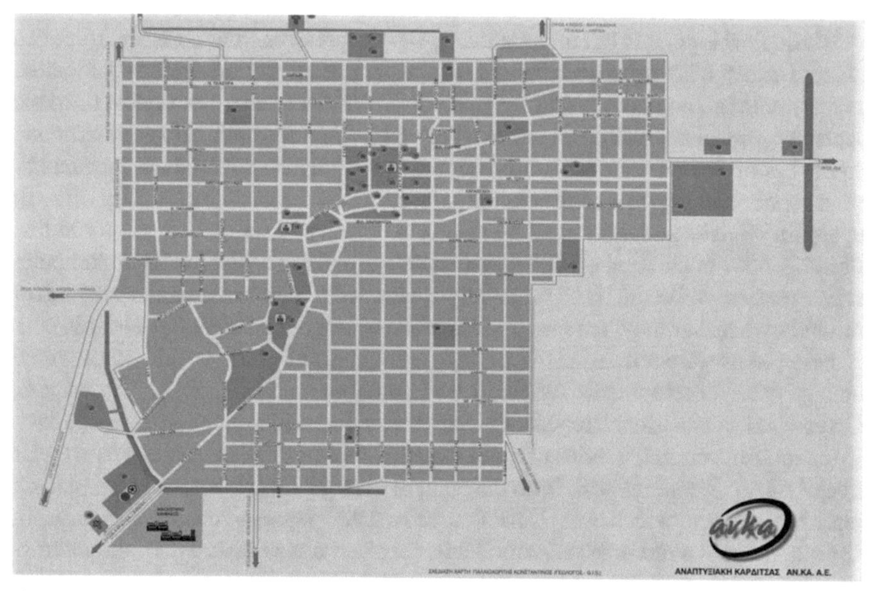

Source: Karditsa Development Company.

Fig. 6.5 Plan of Palamas township (see Text-Box 6.3). (Source: Karditsa Development Company)

Land distribution not only created a pattern of very small land properties, which was to have serious consequences in the long run when these properties turned into unlicensed building lots, but also acted as a counterweight to political and governance concentration (Leontidou 2001, 55). Since then, as a result of settlement policy and, more generally of land reform, small property size became a standard pattern both in cities and rural areas, giving rise to informal construction and economic activity (Mantouvalou et al. 1995). Another side effect of the inflow of refugees was the increase, on one hand, of cheap labour, and, on the other, of better educated potential entrepreneurs. These factors, to which can be added the savings of some of the early refugees, contributed to an unexpected growth of industrial production in urban centres (RIIA 1970, 122–123), and, generally, to a regeneration of the economy (Vakalopoulos 2005, 433; see also Sotiropoulos 2019, 39). In addition, industrial development was facilitated by the imposition of emigration quotas in the USA (Polyzos 1973, 39–40; Nikolinakos 1974a, b; Pangalos 1974), the expansion of the national territory in the Balkan Wars, hence of domestic markets, the devaluation and stabilization of the national currency (Dafnis 1955, 364–369), and the increased supply of money (Nikolinakos 1976, 50–59).

Conditions attached to refugee settlement loans and foreign capital infiltrating the trade of export crops, mainly tobacco, greatly increased the external control of the economy (ibid.), with the industrial sector growing in importance, along with the relative weight of the working class and the power of trade unions (Jecchinis 1967, 49–57). But, from the days of the 1936 dictatorship, trade unionism remained for a long time under the tight control of the state.

These developments left an indelible spatial impression, mainly in the so-called "New Lands" of Northern Greece. It was a parameter that no regional or spatial policy could ignore in the following years. To the refugee settlement were added the repercussions of the political schism of the 1910s and the split between north and south. Leontidou refers to the north-south division but also throws light on an additional split, which appeared during the refugee settlement and proved very difficult to repair. Divisions were created between the territory of refugee quarters and that of indigenous populations. The refugees tended to displace the locals in particular urban centres, while the latter frequently succeeded to discourage the settlement of refugees whenever they had the power to do so (Leontidou 2001, 161 and 164). As if these events were not sufficient to create a troublesome political situation, marked among other developments by the growth of a communist political movement, Greece had to experience repeated dictatorships and military coups. Nevertheless, in the interim periods, a series of modernization measures were implemented under liberal rule. Of great interest, from the point of view of the present book, was a town planning reform, which was legislated by a 1923 legislative decree (Chap. 8). Greece, already overburdened with debts, received a new blow when the international economic crisis which had hit the USA and Europe struck the Greek economy in the early 1930s. The country had to suspend debt payments and was practically bankrupt, once again (Kostis 2018, 635). Dependence on foreign capital was one of the reasons why Greece could not possibly escape the impact of the economic crisis, which quickly brought the country to a state of default (RIIA 1970, 62). Industrial concentration in Athens now became quite noticeable (Sweet-Escott 1954, 127), the provincial industry being less resilient, and so did the concentration of business ownership titles. The dominant position of the capital was reinforced in every respect.

6.5 Second World War, German Occupation, and Civil War

A right-wing dictatorship was imposed in 1936 under General Metaxas. It was under his leadership that the country entered the Second World War in 1940. His regime collapsed in front of the advancing Germans in 1941 and left a vacuum during the critical years of occupation when the country had a government of collaborators in Athens and a royal government in exile in Egypt. The vacuum was exploited by the National Liberation Front (EAM), offshoot of the Communist Party and the most important resistance group, which introduced new forms of local administration in rural areas. The occupied areas themselves were divided into three zones, depending on the occupying power, German, Italian, or Bulgarian (Tsoulfidis 2015, 322). This multiple division of governing forces claiming to control the state was unique and even more complex than the division of the country in 1916 between a royalist and a liberal government (Alivizatos 2011, 23). The 1936 dictatorship had cultivated an almost metaphysical nationalistic mentality (Sotiropoulos 2019, 94) and left an inheritance of extreme centralization and social dissent (Vatikiotis 1974, 30–31; Sweet-Escott 1954, 8–9), in short "a legacy of hatred and distrust" (RIIA 1945, 60). During the German occupation, the occupying forces controlled the cities and the lines of communication, but resistance organizations, in fact EAM, controlled four fifths of the total territory (Woodhouse 1968, 243; Jelavich and Jelavich 1965, 118). Extreme poverty and starvation were the destiny of large sections of the country, particularly the cities which were at the mercy of black marketeers (Tsoulfidis 2015, 328). The government in exile had virtually no contact with this reality (McNeill 1947, 95–96), which led to the existence of "two Greeces" (Tsoucalas 1969, 59). Large zones of the country functioned without control from Athens for the first time since independence, a historic event and a measure of the ferocity of the civil war that followed, a conflict won by Athens and lost by the rural periphery (Vatikiotis 1974, 11–12).

After the end of the German occupation in 1944, EAM clashed with the British-supported government. The December events (the *Dekemvriana*) of that year were but the first phase of the Civil War that lasted until 1949. The 1940s turned out to be "the most tragic and bloodiest decade" of the country's history (Kalyvas and Marantzidis 2015, 497). For long periods the Greek territory was divided into areas controlled by the official government and those under communist control. The new right-left schism was to prove much deadlier than that of 1916. During the 1940s, Greece endured massive destruction, famine, an internecine war, new social and political schisms, territorial division, forced population movements, and new forms of dependence, after the Truman Doctrine. To counter the threat of communist forces, the government, under American guidance, adopted compulsory urbanization policies (Stavrianos 1974, 38–40; Kousoulas 1965, 259). Many forced migrants or insurgent servicemen never returned to their villages, war having proved once more a most catalytic process of contact with urban living. Although it is impossible to provide totally accurate figures, the Civil War may have caused the displacement of 0,7 m. persons, i.e., almost 10% of the country's population in 1940 (Kayser 1968, 36), or up to 1,0 m. (Athenian 1972, 34), and greatly speeded up the process of urbanization. Similar was the effect of physical devastation which ruined the economy. During the war with Italy and Germany 2000 villages, 25% of the building stock, one third of forests, a large number of ports and bridges, vehicles, ships etc. were destroyed (Sweet-Escott 1954, 94–96). The number of destroyed settlements is raised in other estimates to 3700. 18% of the population was homeless (Kalfa 2019, 19). About 1 m. people, throughout the country, were left destitute and inflation virtually eliminated the official currency as a means of exchange (ibid,; Candilis 1968, 15–83). In the late forties, with two thirds of the population on state aid, agriculture in ruins (Coombs and Coppock 1947), and some regions in a condition of starvation, it was estimated that the loss of national wealth was equal to two prewar years' national income (Recovery Organization 1949, 28).

Although a detailed account of the developments of the interwar period, of the German occupation and of the Civil War is beyond the intentions of this book, it must not escape our attention that they are the source of problems that had to be tackled by the national development policy of the years 1945–1960. The study of these problems had begun in the days of the German occupation in a government service, under the direction of Konstantinos Doxiadis, who was to become later a planner of international reputation. This service, which recorded the damages of war, was in fact a department of postwar reconstruction, which, to cover up its real aim, was named "Office of Regional and Town Planning Studies and Research" (Kyrtsis 2006, 342; see also Voivonda et al. 1977, 138). The destruction, initially underestimated by the country's war allies, although it was the worst in Greek history (Candilis 1968, 11–19), was presented after liberation in a report, which, *inter alia*, is interesting as an image of the spatial structure of the country in 1940 (Kathimerini 2014). Doxiadis, later minister in the Ministry of Reconstruction, wrote that the entire settlement system of the country was radically altered in a way that it could no longer support the social and economic needs of Greece. Areas that had to be decongested were now congested more than ever, while areas which had to be populated had lost a large share of their population. The countryside was abandoned, northern Greece was being depopulated, and internal migration to cities was intensified (Doxiadis 1946, 1947). The situation was eloquently described in a famous report, written in 1947, by Paul Porter, Chief of the US Economic Cooperation Administration Mission to Greece (diaNEOsis 2017; see also Sotiropoulos 2019, 114–122)). Porter gives a dramatic account of damages in buildings and infrastructures, of the devastation of agricultural land, of the paralysis of manufacturing industry, of debilitating inflation. and of the incapacitated administration. The country, he concluded, was in a state of chaos. The civil service was overinflated, demoralized, underpaid, and frequently subjected to unequal treatment

(Kalyvas 2015, 102). Administration, according to a later report by the OECD expert Langrod (1965), was in tatters, with the result that the state, with its actions, produced the exact opposite result of the one it anticipated (Makrydimitris 1999, 52, 100).

Porter provides a vivid impression of this state of affairs: "Bitter internal strife and the rapid turnover of the governments have created a climate of insecurity and instability that has prevented any rational planning ... No government has been able to develop an effective economic policy and to inaugurate necessary controls. Those controls which have been attempted have failed as a result of various causes, among which is the lack of effective government machinery for impartial administration. Partly because of these factors, private capital instead of devoting itself to reconstruction and development has been preoccupied with schemes to hedge its risks, outside the Greek economy". It was only with the support of the Americans that the official Greek government won the Civil War. The United States undoubtedly rescued the crippled economy through massive aid, but also determined the subsequent political developments, characterized by solidly conservative governments, and dictated national policy. Whether foreign aid sustained a genuine development has remained a much-debated issue.

The Civil War was to continue the process of destruction. According to Kalyvas and Marantzidis (2015, 499) more than 1700 villages were devastated and hundreds of thousands of people were forced to flee to the towns. They point out that the internal conflict acted as a catalyst for the transformation of a rural country to an urban one. The ravages of successive wars had enormous spatial impacts, and so did internal migration mainly to large urban centres. The population depletion of the provinces continued for a long period (Valaoras 1974). The population of Greater Athens, as an approximate percentage of the national population, rose from 15% in 1940, to 18% in 1951, 22% in 1961, and 29% in 1971 (Wassenhoven 1980, 365–366). The number of internal migrants settling in Athens in the years 1946–1960 exceeded 0,4 m. (National Statistical Service 1964, 21). About three quarters of migrants who had settled in Athens until 1971 had moved there directly from their villages, i.e., without an intermediate settlement, e.g., in a smaller centre (Sandis 1973, 85). The smaller the place of origin, the greater was the probability of migration. When the migrant chose to migrate from his village to a foreign country, he/she did so without an intermediate stage (Tsaousis 1971, 106, 113–114, 142; see also Patellis 1980). Rural exodus became a much-studied phenomenon (Kayser et al. 1971; Pepelasis 1962). As observed by Leontidou (2001, 160–161), the exodus, although it had its roots in the interwar period, reached a peak in the 1960s. Excessive land fragmentation, agricultural production for self-consumption only, and exploitation by middlemen, added to the devastation of war, forced the typical villager to sell his land for a pittance and migrate. The reasons of migration became increasingly economic. Rural land fragmentation was a phenomenon observed in 1947 by Paul Porter too, to whom reference was made earlier: "Land reform ... was not a major problem. Except for one or two large farms, nearly all of the land was held by small holders".[7] As Patellis (1978, 94) was later to add, the peasant becomes conscious of the fact that in his village nothing is going to change. He raises his children with one aim only, to sell his land at whatever price and migrate. The flight to the cities, especially Athens, ceases to be simply an economic matter or a question of "unequal exchange" between industrial and agricultural production (Kavouriaris 1974, 45). It becomes now deeply cultural, reflecting a loss of self-confidence (Mouzelis and Attalides 1971, 187). In a typical case of a mountainous rural province (Voion in West Macedonia) the flight of the population took the dimensions of a total collapse. In the period 1951–1981, it lost one quarter of its population, but 70% of this loss occurred in the decade of the 1960s (Kaldis and Laskaris 1990, 406).

6.6 Development and Reconstruction

Reference was made already to the American control of the ailing economy, during the Civil War and after the announcement of the Truman Doctrine in 1947. But, even before responsibility for Greek affairs was handed over to the Americans, foreign control was tight, since the British were the effective governors (Sweet-Escott 1954, 38 and 102; Candilis 1968, 33). Their control extended well beyond economic affairs. In the words of an American official, Greece was reduced to client state status (Rousseas 1967, 74–75). The British were effectively charting the entire post-war political system (Legg 1969, 72). American control was even tighter, either as daily intervention or as indirect guidance (Rousseas 1967, 87; see also Munkman 1958, 143–146, Campbell and Sherrard 1968, 307–313, Psilos 1968, 38). It is superfluous to add the control of the army (O'Ballance 1966, 153–177). Foreign dependence, mediated through Athens, had a strong centralizing effect and contributed greatly to the diffusion of imported values and customs, of which the showroom was the capital, where economic activity was concentrated. The lack of any economic planning on the side of the Greek government, in the first phase of the aid programme, facilitated this development (Psilos 1968, 35–36). Planning improved too late, when development aid was drastically cut in 1951, as a result of the Korean war (op.cit., 45; Sweet-Escott 1954, 113). In the postwar years, the first priority was industrialization and infrastructures. As Hadjijoseph (1994, 23) emphasized, the reconstruction debate of 1945–1953 centred, almost exclusively, on the potential and mode of industrial development. The main goal of successive reconstruction plans was the creation of industrial and, even more so, energy generation infrastructure. This does not imply that there had not been disagreements. An ideologically loaded dialogue concerned the priority of agricultural versus industrial development, although in practice emphasis was first given to the stabilization of public

[7] National Archives/Harry S. Truman Library and Museum.
Paul R. Porter Oral History Interview | Harry S. Truman (trumanlibrary.gov)

finances and the fight against gigantic inflation, a choice that caused extreme criticism (op.cit., 31; see also Alexander 1964). Political divisions between right and left and fanaticism were the order of the day. The state was totally identified with the winners and the administration suffered because of political persecution (Sotiropoulos 2019, 136). The national schism of the Civil War was dominant. Its traumatic and horrific memories were to linger for a long time. A comment made by Gregory and Pred (2007, 2), although referring to the conditions of a totally different context, is nevertheless pertinent. They stress "the need to be sensitive to the fractured histories of violence, predation, and dispossession – as material fact, as lived experience, and as resonant memory – that erupt so vividly time and time again in our own present". They do erupt occasionally in Greece, even today.

This situation in Greece affected the dialogue on the importance of industrialization and on the need for national economic planning (Thomadakis 1994, 38). The battle between the left-socialist and the right-liberal positions was marked by two important documents, a 1947 book on the priority of heavy industry, by D. Batsis (1977), who was executed in 1952 as a traitor (Rafaelidis 1993, 320–321), and the report of K. Varvaresos (1952), an economics professor and banker, on the economic problem of Greece (Sotiropoulos 2019, 75, 119, 201–203). Batsis's advocacy of industrial development was echoed by other important figures of the period, like A. Angelopoulos and X, Zolotas (Angelopoulos 1974; Bank of Greece 2005; Zolotas 1961; Ioannidis et al. 1994; see also Gregorogiannis 1975). Varvaresos on the other hand advocated an emphasis on agriculture, the promotion of consumer goods industry, and the encouragement of low-income housing construction. One of his arguments was the shortage of domestic industrial capital (see also Psilos 1964). Ultimately, the development strategy adopted by the Greek governments of the period was a mixture of the reproduction of the existing domestic production process and of the socio-political power system that represented it (Sakkas 1994, 64–65). In fact, this was a typical example of Greek-style compromise. Housing construction did indeed increase exponentially and so did unlicensed building (Marmaras and Tsilenis 1994, 603; see also Korres et al. 2011). The irony is that in spite of the intensity of the debate on industrial v. agricultural development, industrial growth did take place, while important steps were made in the 1950s for the development of tourism. The conflict symbolized by the views of Batsis and Varvaresos caused an unfortunate polarization of opinion, because of the alleged communist influence in the first case, and of capitalist interests in the second. Not to be forgotten of course that the postwar situation was more the product of international conflicts than of internal social dynamics (Svoronos 1969, 34). But on the domestic front too, as G. Voulgaris (2013, 41–42) has stressed, the divided society bequeathed by the Civil War was ruled by a twin legal-institutional system, of which the main feature was the co-existence of the official Constitution and a parallel "para-constitution", i.e., a system of extraordinary measures against the "internal enemy" of the political left. For Svoronos (1976, 145) the 1950s were characterized by the privileges accorded to foreign capital investment which bred a parasitic economy of neo-colonial type, a view echoed by Kremmydas (1994, 16). Yet, already by 1950, as Kostis (2018, 754–759) points out, all the economic indicators had returned to prewar levels, although they did not justify enthusiasm. Yet, the performance of the economy in the first postwar decades was impressive by international standards, so that there were references to a Greek economic miracle (Tsoulfidis 2015, 365–374; see also Babanasis 1995). Thanks to the currency stabilization policy of 1953, or because economic growth started from a very low level, the rate of growth was exceptionally high and growth continued unabated for two consecutive decades. The years 1948–1960 served as a pre-industrialization phase, during which fundamental infrastructures were built and social policies were first promoted (Chiotis and Louri 1984, 201), while, at the same time, social structure and geographical organization changed, and unequal regional development intensified (Andreadis 1966, 162). From the mid-1950s onwards, economic growth is associated with a stable rule of conservative political forces, mainly under the leadership of Konstantinos Karamanlis (Voulgaris 2013, 464–465). According to the World Bank Atlas of 1977, GNP per capita grew during the period 1960–1975 by 6.6% p.a., a record surpassed by only seven countries in the world of which four were major oil-producing countries and the others were Singapore, Portugal and South Korea. According to data compiled by Tsoulfidis (2015, 376) the growth rate of GNP for the entire period 1950–1973 exceeds that of all OECD countries and is second only to that of Japan. Hadjivasileiou (2015, 371–372) claims that the developments of 1955–1963 are one of the most important – if not *the* most important – economic and social reform of modern Greece. They represent, in his opinion, not just the transition from underdevelopment to development; they are a decisive step to move the country from the periphery of an indistinct "Orient" to the heart of the West European system. The state and a technocratic elite class had the key role in this transformation (Sotiropoulos 2019, 17, 123–127).

One of the signs of this "westernization" effect is the formulation of multi-annual development plans, a practice followed in several states after the Second World War, now adopted by Greece too (Exarchos 1992, 210). In the period 1947–1966 at least six five-year plans were drafted, mostly of a rather casual character and with plenty of problems of internal cohesion, vagueness, and unrealistic objectives (Sakkas 1994, 59 and 68–72; Psilos 1968, 71). Liodakis agrees with this assessment and remarks that the first serious plan was produced in 1959, under the Karamanlis administration, for the period 1960–1964 (Liodakis 1994, 80–81). It is a period in which, in spite of the policy to attract foreign investment, the logic of a state-managed economy seems to prevail (Veremis and Koliopoulos 2013, 368). Central planning gains ground, a development which sounds odd, given the debate of the late 1940s and early 1950s. In a way it is a vindication of the radical approach, at the expense of the market-led policy. The overall logic of the plan was that balanced development would be the outcome of national and sectoral growth, through the stimulation of agriculture, industry, mineral extraction, and tourism. Special emphasis was placed on strong regional growth poles (MinCoord 1960, 96–97; Kafkalas 1981, 17; Hoffman 1972, 105), a policy which was to return repeatedly in later years. Partly with the possible exception of the 1960–1964 plan, the national plans of this period proved very weak instruments of regional, let alone spatial, policy and were poorly implemented (Angelidis 2000. 156). National planning was given a new boost with the rise to power of a political "centre" coalition in 1963, and the goal of regional development received increased emphasis in the five-year plan for 1966–1970 (Sakkas 1994, 71), the implementation of which was suspended by the military coup of 1967 and the seven-year dictatorship that followed. The military regime produced its own national plans, but this was mere window-dressing, with plenty of repetitions from previous plans and an enhanced emphasis on tourism and the attraction of foreign investment (Pesmazoglu 1972, 77; Les Temps Modernes 1969; Papaspiliopoulos 1969; Clogg and Yannopoulos 1972). In the period of the junta the economy depended on tourism and the building sector (Hadjivasileiou 2015, 521). If there were policies in the period after 1950 which had a clear spatial impact, these were the construction of transport infrastructure, the support to the building sector, and industrial incentives, although the latter often simply encouraged industrial development in the catchment area of Athens, just on the edge of the capital (Skagiannis 1994, 129; Argyris 1986, 37). The developments of the 1960s to the end of the dictatorship in 1974 are succinctly summarized by Kafkalas (1984, 70–71), who explains that they transformed the socio-economic structure of the country, contributed to capital accumulation, and modernized the production system. However, the social cost was very high, as entire regions witnessed the collapse of their demographic structure and degraded quarters multiplied in large cities, especially in the areas of illegal construction. Foreign control dominated key sectors of the economy. According to Lagopoulos (1992, 3), the decades from 1950 to the end of the 1970s bear the stamp of monopoly capitalism, an undoubted flaw of the development strategy which perpetuated dependency, in the words of Close (2002, 53–55). It is true that large scale investments are made, e.g., in bauxite mining, aluminum processing, oil refining, shipyards, steel industry, and fertilizer production (Poulopoulos 1972a, 21 and 1972b, 177; Hadjivasileiou 2015, 356–358; Nikolinakos 1976, 75–77; Blanc 1965, 41–43; Babanasis and Soulas 1976, 92–97; Meynaud 1966, 428–462; Kostis 1999). The attraction of foreign investment remained a central choice of the conservative governments (Meynaud 1966, 257; Rizas 2015, 59–61). Nevertheless, such a monopoly capitalistic boom coexisted with the total control over the economy by the state, which owned key nationalized industries, e.g., telecommunications and electricity production. Spatial organization changed dramatically, because the countryside was being abandoned and migrants, apart from those leaving for foreign countries (mainly Germany and Belgium), moved to Athens, and, to a lesser extent, Thessaloniki. A new social stratification emerged and the urban middle classes were reinforced (Kostis 2018, 763–764).

The scientific debate on the consequences of these developments, primarily the overgrowth of Athens, is not going to end. From a Marxist angle, the capital will continue to be blamed for being the locus of capitalist reproduction of labour (Sagias and Spourdalakis 1992). The fact that the majority of commentators condemn the growth of the capital is not, in the author's view, a final verdict. In an intensely globalized economy, in which urban centres spearhead economic growth, the possibilities offered by one or two vigorous urban economic centres are not to be neglected. The author has discussed in the past the regular references to the "hydrocephalous", "parasitic" and "gigantic" growth of Athens, and remarked that what really matters is not the sheer volume of the capital city but rather the economic activities that it harbours, the mechanisms of their control, their speculative character, and their anti-developmental nature, which may find allies eager to perpetuate a parasitic economy (Wassenhoven 1978). It is true that a parasitic or shadow or hidden economy thrived in the "informal sector" – and still does – but the economic size and performance of Athens should be judged in the framework of an enlarged geographical space. It can even be argued that a productive and economically healthy Athens can be the trump card of the Greek economy. However, regarding the period discussed at this point, condemnation has been almost unanimous. E.g., Kostis (2018, 764–765) makes the point that the development of Greece rested on intense regional inequality, with most industrial potential concentrated along the Athens-Thessaloniki corridor. Athens grew out of all proportion in relation to the population of the country. As he stresses, an equivalent asymmetry is evident in the national economy. Athens is not just the government seat but the country's main economic pole of which the continued growth leads to the stagnation or decline or entire regions, especially of small towns (ibid.).

The economic boom of the 1960s takes a downturn when the international oil crisis of the early 1970s struck the Greek economy which plunges in a prolonged crisis, which affected all sectors. Public finances were badly hit, as the balance of state expenditure and revenues deteriorated steadily (Lamprinidis and Pakos 1990, 175). The regional problem (Papageorgiou 1972, 1973) and regional development policy could not remain unaffected, as explained by Kafkalas (1990, 305). GNP and investment were on the decline, tight-handed authoritarian policies could no longer be imposed easily as in the past, life quality problems become acute mainly in declining regions, but also in large cities, while the centralization of decision making proves to be a source of the regional problem. A demographic study which covered the entire period 1928–1971 had shown the extent of population concentration on the eve of the economic crisis. The only region which had a positive population balance was the one which included Greater Athens. Rural population had fallen from about 61% of the national total to 35% (Demetras et al. 1973, 15–20), while urban population was steadily growing (Sotiropoulos 2019, 130, 138–139). This was the situation when the military dictatorship collapsed in 1974, mainly because of its disastrous initiatives in Cyprus and the economic downturn. The accession agreement signed in 1961 between the European Economic Community and Greece might have secured a more favourable external environment for the country, but it had been "frozen" after the imposition of the 1967 dictatorship (Hadjivasileiou 2015, 365; Kostis 2018, 761).

6.7 Stability, Affluence, and Renewed Crisis

The restoration of democracy in 1974 inaugurates a long period of stability, during which, with brief exceptions, the parliamentary system operates mostly with a one-party rule. For some, the fall of the dictatorship had a wider significance: "Liberal democracy as a universal model did not really take off until the mid-1970s, when Spain, Portugal and Greece removed their dictatorships" (Dryzek and Dunleavy 2009, 24). The smoothness of the transition to normal democratic conditions is in itself impressive. Within months of the collapse of the regime of the colonels a plebiscite put an end to the regime of the reigned democracy and installed a "kingless", presidential democracy (Alivizatos 2011, 495). A conservative government of the New Democracy party, under, as in the late 1950s, Konstantinos Karamanlis, who returned from exile in France (Clogg 2021, 287), was succeeded in 1981 by that of the Panhellenic Socialist Movement (PASOK) under Andreas Papandreou. The 1975 Constitution was a decisive break towards the consolidation of a liberal democracy (Sotiropoulos 2019, 161). It was drafted and voted in parliament in a very short period to make certain that the country would return to normality (Hadjivasileiou 2015, 50; Venizelos 1986; Papademetriou and Sotirelis 2001; Papademetriou 1993; Babalioutas 2018, 60–62). It was imbued with the conviction that the state has a broad range of economic and social responsibilities in the framework of a mixed economy, which ought to secure both economic efficiency and social justice (Voulgaris 2013, 186). The state coordinates economic activity in the country, in order to protect the "general interest". The Constitution's critics imply that it confirmed the dominance of large interests in the means of production and the privileges of shipping capital and foreign monopoly business firms (Kaltsonis 2017, 255–256). Its effect on the recognition of spatial planning, as practice and theory, in fact as a "science", was unprecedented. Article 24 of the Constitution refers to "spatial restructuring",[8] wrongly translated as "master plan" in the official translation of the Greek Parliament: "The master plan of the country, and the arrangement, development, urbanisation and expansion of towns and residential areas in general, shall be under the regulatory authority and the control of the State, in the aim of serving the functionality and the development of settlements and of securing the best possible living conditions. The relevant technical choices and considerations are conducted according to the rules of *science*. The compilation of a national cadastre constitutes an obligation of the State" (my italics).[9] No one thought at the time, Alivizatos (2011, 501) ponders, that this article was anything more than wishful thinking. After 1975, the role of spatial (urban and regional) planning was entirely different, not only because of the stipulations of the Constitution, but also because of the case law emanating from the supreme administrative court (Council of State) and of legislation enacted on the basis of the Constitution. Spatial planning was now on a new plane. The legal interpretation of planning law blossomed in the academic literature (Christofilopoulos 1988; Skouris 1991; Rozos 1994; Melissas 2002, 2007, 2019; Giannakourou 2012, 2019; Karatsolis 2020).

From 1975 to 1993, new laws were enacted on urban development, regional spatial planning, the environment, regional administration, and forest protection

[8] In Greek planning law and in professional practice, use is made of the noun *chorotaxia* (i.e., space ordering, after the German term *Raumordnung*) and of the adjective *chorotaxikos*, which the author translates as spatial.

[9] The Constitution (hellenicparliament.gr). See also Constitution of Greece 2013.

(Loukakis 2010, 102–104; see also Burgel and Demathas 2001), although the first of those concerned with urban and regional planning, voted soon after the new Constitution, remained unimplemented. This, however, was not the case with a major town planning reform in 1983 (see Chap. 8). When the socialist party ascended to power in 1981 (Kostis 2018, 809) the political environment was more favourable to regional development, the interventionist role of the state, and citizen participation. The achievement of an equilibrium was attempted between top-down and bottom-up approaches, which had inherent difficulties. Regional development planning was radically altered (Carter 1989, 208). This was evident in the five-year plan for 1983–1987 and in a more active dialogue with the regions, which aimed to demystify the meaning of regional development, which had remained a distorted concept for the majority of the people (Giannousis 1983, 51). Important initiatives were undertaken for the transfer of powers to local and regional governments, but an even more important innovation had been introduced earlier, when a separate ministry for Spatial Development, Settlement and the Environment was created in 1980 (Vamvakopoulos 1994; Hadjivasileiou 2015, 533; Angelidis 2000, 217; Rozos 1994, 59). A national council for regional planning and the environment had been set up in 1976. Under the socialists, a law was enacted to strengthen the now elected prefectoral authorities and 13 regional authorities, as arms of the central government (Babalioutas 2018, 216–243). Devolution of power explains the abandonment of the planning legislation of the previous conservative government (Giannakourou 1994, 29). The problem was that at the same time the priorities of regional policy had to undergo a radical change and shift from a nationwide regional development strategy to implementable policies focused locally in priority problem areas (Andrikopoulou 1994, 347; Papadaskalopoulos 2001, 238). The weaknesses or inadequacies of development plans were as a rule blamed on implementation failures or technical planning problems. From the perspective of a radical critique this was considered as an alibi, which overlooked the functioning of the capitalist system (Vaiou-Hadjimichalis and Hadjimichalis 1979, 70).

The critical event which opened the 1980 decade was the treaty of integration in the European Economic Community (later European Union) which was signed in 1979 by the conservative government and came into force in 1981 (Kostis 2018, 803; Mousis 2003; Stevens 1999). Prime Minister Karamanlis had clearly declared that "Greece belongs to the West" (Clogg 2021, 176). In spite of its initial rhetoric, the socialist party which won that year's election accepted EU membership as a fait accompli (op.cit., 186), although, initially its slogan had been that "Greece belongs to the Greeks". By and large, the Greeks had welcomed accession to the Union with pleasure. "An unspoken assumption underlying the enthusiasm of many Greeks for Europe was that membership would somehow place the seal of legitimation on their country's somewhat uncertain European identity: after all they habitually spoke of travelling to Europe as though Greece did not form part of the same cultural entity" (op.cit., 174). This attitude did not prevent many critics to complain later that Greece was ruled from Brussels through a displacement of power to a supra governmental organization: "The past presumption has been that even when states make such moves, they do so voluntarily and retain the ability to withdraw if they choose.

Globalization theorists argue that the enmeshing of states within many different encompassing international and trans-national processes is now so extensive that this presumption no longer holds … In the European Union a unique form of quasi-federalism has developed between the member states (which are no longer termed 'nation states') and the EU's powerful central institutions" (Dryzek and Dunleavy 2009, 5 and 28; see also Ohmae 1996; Moulaert 2002; Moulaert and Cabaret 2006). In Greece, albeit grudgingly, EU membership was welcomed as a political, economic, and social protective shield. In due course the EU was instrumental in a process of modernization, the most tangible evidence being new transport and digital economy infrastructures (Sotiropoulos 2019, 143, 154). Greece had to accept the Union's legal *acquis* and recognize the supremacy of European legislation over Greek law, including environmental law (Alivizatos 2011, 508, 515).

The symptoms of the oil crisis of the 1970s and inflation continued unabated and fears were expressed that the financial deficit would soon be out of control (Economist 1993; Komnenos 1993, 201; OECD 1979). Throughout the years of the Cold War, Greece was in an awkward geopolitical and territorial position with its northern border regions, practically cut off from its neighbouring international hinterland, a situation which was still prevalent in the 1980s before the fall of the Berlin Wall in 1989. In spite of bilateral relations with her northern neighbours, the effects of this situation caused regional development problems because international space was not accessible as in other European Union countries (Wassenhoven 1996, 48). The enlargement of the European Union to include Bulgaria changed this situation later.

The problems inherited by the post-dictatorship democratic governments are analytically presented in the national and regional development plans of 1981–1985 and then 1983–1987, where the emphasis on the regions and on spatial planning is rising (Kafkalas 1990; Angelidis 2000; Andrikopoulou 1994; CPER 1980; YPETHO 1985). The problem of specific "problem areas" and "poverty pockets", as opposed to that of entire regions, was gradually acknowledged as crucial (Tsakloglou 1995; Korres and Rigas 2000). Tourism was a steady hope for the national economy, but in those days was still concentrated in the Athens region (Attica), with the exception of some islands (Komilis 1986, 1987). By 1990, the percentage of urbanization had reached 90% in Attica, where most of the secondary and tertiary sectors were concentrated, as opposed to 60% in the country as a whole (Petrakos 1996, 100–111). Partly because of the inflow of funds from the European Union, partly because of the return to power of a liberal government in 1990 reliance on the market and foreign investment was again on the increase (Getimis and Economou 1992). The influence of European Union regional development policies, through a series of programmes launched by the EU Structural Funds, was now the main factor determining national policies. The EU Support Frameworks replaced the old national five-year plans (Giannakourou 2000, 2008; Angelidis 2004; Wassenhoven 1996). The influence of the EU was now paramount in other policy fields too, e.g., transportation (European Commission 2018; Dionelis and Giaoutzi 2008; Dionelis et al. 2008) or the environment. The environmental dimension was increasingly influential, not only because Greece had to incorporate in her legal framework the relative

EU directives, but also because it had ratified a series of international conventions (Siouti 2011).

Development orientation did not essentially change when the socialists regained control of parliament in 1993, particularly during the "modernization" period 1996–2004 under the premiership of Konstantinos Simitis. Greece joined the European Monetary Union (EMU or euro-zone) and large projects were undertaken, although some had started earlier (Athens Metro, new Athens international airport, a bridge connecting mainland Greece with the Peloponnese (near Patras), Egnatia Motorway across the northern provinces of the country etc). Modernization was evident in other fields too, touching nationalistic and religious feelings. "The Simitis government's commitment to political, economic and social modernisation was not in doubt. The kinds of obstacles that it faced, however, were strikingly illustrated by the way in which, in the summer of 2000, the issue of whether identity cards, compulsory for all citizens, should give the religious affiliation of the holder gave rise to fierce emotions and massive demonstrations" (Clogg 2021, 234). It was eventually decided that they should not mention religion. Greece hosted the 2004 Olympic Games, but just before the games were opened the conservative party was back in government which it controlled until 2009 (Voulgaris 2013, 140–152). Real estate values were rising and the building sector was once again the leading sector of the economy. A sense of real or illusory prosperity prevailed, although globalization and the prospects of further European integration caused some anxiety, even before the 2008 economic crisis which landed in Greece a year later in the most painful way (ibid.). Until that time, since roughly 1990, the economy was growing rapidly, but this was mainly due to increased consumption and the growth of non-productive sectors, while industry and agriculture were shrinking (Tsoulfidis 2015, 425). Inequalities were still evident and critics spoke of a "two-thirds" society (Pakos 2000). The crisis may not have been a real surprise. "The great crisis of 2009 is the latest of these boom-bust bailout crises. Greece's decision to join the EMU was a classic case of overreaching" (Kalyvas 2015, 203). Greece suffered once more a typical socio-political schism and entered a period of great turbulence and great social unrest. Social troubles had erupted even before the economic crisis was officially recognized by a new government of the socialist party. The riots in Athens of December 2008 were a violent and early warning. This explosion, wrote Kostis (2018, 838–841) was an expression of discontent and dissatisfaction with political practice and vacuous rhetoric. One cannot link directly this turmoil with spatial planning and policy, but it certainly left a feeling of uncertainty and a pressure for urgent and hasty decisions which had their impact on urban and regional planning, especially when the financial crisis erupted (see Chap. 8).

The years of the financial crisis (Kostis 2018, 856–901), are a period during which, in the expression of Drettakis (2012, 61), Greece was painfully and totally defeated by its public debt. The debt had started growing threateningly ever since 1981, but grew out of all control in the 2000 decade. "Few observers of the Greek scene in the year 2000 could have predicted the dire situation in which the country would find itself within less than a decade. Greece in the fifty years since the end of the civil war, a conflict described by one contemporary observer as a war of the poor against the very poor, had transformed itself from an impoverished, battle-scarred backwater, heavily dependent on US Marshall Aid for its very survival, into one of the twenty-eight most developed countries in the world" (Clogg 2021, 272). The bitter taste that remained is that it was the failures of state management which caused the wrath of citizens, against the governing class and the state institutions. These failures then justify the – no doubt populist – view that the profligacy of the state is due to the effort of politicians to feed a dysfunctional, overgrown public administration, to perpetuate conditions of dependency, and cover national overborrowing under a veil of secrecy (Loulis 2019, 263). The assessment provided by D.P. Sotiropoulos (2019, 173) is that the crisis exposed all the vulnerability of the state. It was a fractured corporatist state which had surrendered to rent-seeking interests, with reformist tendencies but also internal resistance forces, in a constant but unfinished Europeanization, with a development process based on financial deficits, and a social policy favouring the insiders and not the genuinely disadvantaged. In contrast to the crises of 1893 and 1932 which had occurred in totally different circumstances, Sotiropoulos maintains, the 2010 financial collapse is the crisis of a state colonized by private interest coalitions and ruled by an impotent and self-preserving political elite. The "curse" of dependence was returning, although this time it raised the spectre of another challenge, that of a better incorporation in the web of globalized capitalism (op.cit., 176).

During the crisis the economy shrank dramatically and GDP declined sharply (Gallant 2017, 458–463). During economic crises of the past, the economy recovered relatively quickly, but the effects of the new crisis continued unabated for several years (Tsoulfidis 2015, 419) and are still present, prolonged, as it were, by the Covid-19 pandemic. A Memorandum of Understanding to secure an international loan was signed in 2010 by Greece and her creditors, popularly known as Troika (International Monetary Fund, European Commission, and European Central Bank). A second memorandum replaced the first in 2012 (Pappas 2015, 137 and 142; Kostis 2018, 853, 874 and 885). The painful experience of the pressures is further discussed in Chap. 8. The memoranda demanded the support of entrepreneurship, the fight against bureaucracy, and the elimination of time-consuming administrative procedures (Kalyvas 2015, 171), a demand that clearly touched planning processes. After the second memorandum the country entered again a phase of rebellion and social troubles. Kostis (2018, 888–891) distinguishes the crisis period into three phases, but it was after the 2012 memorandum that the split "memorandum v, counter-memorandum" acquires dimensions of a social rift. In 2015 the political stage changed when the Coalition of the Radical Left (SYRIZA) came to power and was on the verge of making a decision to abandon the European Union. Eventually, the left-wing government signed a new bailout agreement and accepted the Troika's terms (Clogg 2021, 263). Kostis disagrees with the claim that the coming to power of SYRIZA signified the end of the period of *metapolitefsi* (change of political system), after the fall of the military junta in 1974. On the contrary, he argues, this was a new effort of the political system to survive. Subsequent developments show that he was right. There is no doubt, adds Kalyvas (2015, 189), that the country became more vulnerable to populist or nationalistic preaching, the latter exemplified by the rise of an extreme right-wing party. The country was going through a new phase in a long series of socio-political schisms and bankruptcies, thankfully not of open civil war. New historical analyses which appeared on the bookshop shelves bear eloquent titles: "Seven wars, four civil conflicts, seven bankruptcies: 1821–2016" (Dertilis 2016) or "Ten plus one decades of political divisions" (Diamantopoulos 2017–2018).

"Once again", writes Beaton (2020, 385), "the tectonic plates were shifting ... [A] new political alignment had suddenly emerged. Some commentators have defined the new split as being between those for and against acceptance of the Memoranda. Another sees it as 'ethnocentrism' versus 'Europeanism'. Either way, the new fault line transcends and replaces the traditional face-off between right and left. Its existence can be traced all the way back to the internal conflict during the Revolution of the 1820s, between rival versions of liberty: on the one side an ideal of absolute self-sufficiency, on the other integration into a Western-dominated world order. Once again, Greeks found themselves forced to choose". The 2019 elections, when a liberal (neoliberal for others) version of the conservative New Democracy party returned to power, proved that the *metapolitefsi* was alive and well. If anything, the Covid-19 pandemic of 2020–2021 strengthened the ties with the international community. One more political schism seems to have been overcome and kept within the normal workings of a parliamentary democracy. The country still has of course to face critical challenges, e.g., internal migration (Rovolis and Tragaki 2008), emigration of young talent, incorporation of economic migrants, reception of political refugees, and the question of Exclusive Economic Zones in its maritime space and oil-drilling rights in the sea bed, an issue which goes back to the 1970s (Clogg 2021, 173, Wassenhoven 2017, 168–183, and 2018). Climate change is also attracting a long overdue attention, following a series of flood and forest fire disasters, but so is also the policy for clean sources of energy. The energy sector has been for several decades dependent on lignite coal production (Nikolinakos n.d.; Papagiannakis 1976), now abandoned in line with European Union policy. 2021 has added new forest mega-fires and an international fuel price crisis. As to the perpetual emergence of schisms, if there is a division which deserves our attention, is the "rationality v. irrationality" rift – we could call it "modernism v. tradition" – which constantly resurfaces, and does again in the period of the Covid-19 pandemic and of the related vaccination campaign. The example may be considered as a jocular exaggeration, but it represents a deeper division. The question is, is this rift really new and is it limited to Greece? In terms of political divisions, a debate has opened on whether we are witnessing a new separation line which defies the time-honoured left-right configuration, a claim which is not of course shared by all observers.

Epilogue

In this chapter, the reader was exposed to an account of the history of modern Greece, which was inevitably incomplete and very personal. It was not an account that would be written by a proper historian worthy of his/her name. As in a previous book by the same author, it was "a planner's look at history", a planner who, under the influence of a Southern perspective, is trying to isolate the historical developments which formed the context of subsequent urban and regional planning. There have been plenty of such developments which were summarized in the prologue of the chapter. In the author's opinion this historical presentation fully supports the view that planning theory cannot possibly be history-independent, a-historical or of universal validity. A further analysis of the Greek state and spatial planning system, which is attempted in the next two chapters, will lend additional support to this thesis.

References[10]

Alexander AP (1964) Greek industrialists. Center of Planning and Economic Research, Athens

Alivizatos N (2011) The constitution and its enemies in modern Greek history, 1800–2010. Polis, Athens (in Greek)

Alivizatos N (2015) Pragmatists, demagogues and dreamers: politicians, intellectuals and the challenge of power. Polis, Athens (in Greek)

Andreadis S (1966) Regional development and the structure of the urban network in Greece. Koinoniologiki Skepsi 1966(2):153–175. (in Greek)

Andrikopoulou E (1994) Regional policy in the 1980s and prospects. In: Getimis P, Kafkalas G, Maravegias (eds) Urban and regional development: theory, analysis and policy. Themelio, Athens, pp 347–383. (in Greek)

Angelidis M (2000) Spatial regional planning and sustainable development. Symmetria, Athens. (in Greek)

Angelidis M (2004) European Union policies for spatial development. University Press NTUA, Athens. (in Greek)

Angelopoulos A (1974) Economics: Essays and studies 1946–1967. 2 volumes. Papazisis, Athens (in Greek)

Argyris T (1986) Regional economic policy in Greece: 1950–81. Adelfoi Kyriakidi, Thessaloniki. (in Greek)

Athenian (pseudonym) (1972) Inside the colonels' Greece. Chatto and Windus, London

Babalioutas L (2018) The contemporary institutional framework of Greek public administration. Vol. A. Sakkoulas, Athens. (in Greek)

Babanasis S (1995) Economic development and social impacts in Greece during the first postwar period 1945–1967. In: Sakis Karagiorgas Foundation (ed) Greek society in the first postwar period 1945–1967. Vol. B. Conference proceedings, Athens, pp 37–59. (in Greek)

Babanasis S, Soulas K (1976) Greece in the periphery of developed countries. Themelio, Athens. (in Greek)

Bank of Greece (ed) (2005) Xenophon Zolotas: a memory and honour event, Athens. (in Greek)

Batsis D (1977) Heavy industry in Greece. Kedros A (ed) 3rd edn. (first published 1947) (in Greek)

---

[10] **Note**: The titles of publications in Greek have been translated by the author. The indication "in Greek" appears at the end of the title. For acronyms used in the references, apart from journal titles, see at the end of the table of diagrams, figures and text-boxes at the beginning of the book.

Abbreviations of journal titles: AT Architektonika Themata (Gr); DEP Diethnis kai Evropaiki Politiki (Gr); EpKE Epitheorisi Koinonikon Erevnon (Gr); EPS European Planning Studies; PeD Perivallon kai Dikaio (Gr); PoPer Poli kai Perifereia (Gr); PPR Planning Practice and Research; PT Planning Theory; PTP Planning Theory and Practice; SyT Synchrona Themata (Gr); TeC Technika Chronika (Gr); UG Urban Geography; US Urban Studies; USc Urban Science. The letters Gr indicate a Greek journal. The titles of other journals are mentioned in full.

Beaton R (2020) Greece: biography of a modern nation. Penguin Random House
Beckinsale M, Beckinsale R (1975) Southern Europe: a systematic geographical study. Holmes and Meier, New York
Biris C (1995) Athens: From the 19th to the 20th century. Melissa, Athens (first published 1966)
Blanc A (1965) L'économie des Balkans. Presses Universitaires de France, Paris
Burgel G, Demathas Z (eds) (2001) Greece in the face of the 3rd millennium: space, economy, society in the last 40 years. Panteion University, Athens (bilingual)
Campbell J (1964) Honour, family and patronage. Clarendon Press, Oxford
Campbell J, Sherrard P (1968) Modern Greece. Ernest Benn, London
Candilis WO (1968) The economy of Greece 1944–66. Frederick A. Praeger, New York
Carter FW (1989) Post-war regional economic development: a comparison between Bulgaria and Greece. In: Dimitriadis EP, Yerolympos KA (eds) Space and history: urban, architectural and regional space. Prodeedings of the Skopelos symposium. Aristotle University of Thessaloniki, Thessaloniki. (papers in English and Greek), pp 203–210
Chiotis GP, Louri HD (1984) The Greek post-war experience on regional development planning and policy within the south European area of the enlarged EEC. In: Koutsopoulos KC, Nijkamp P (eds) Regional development in the Mediterranean. Phebus Editions, Athens, pp 201–215
Christofilopoulos DG (1988) The new legal framework of town planning. P. Sakkoulas, Athens
Clogg R (2021) A concise history of Greece. Cambridge University Press, Cambridge
Clogg R, Yannopoulos G (eds) (1972) Greece under military rule. Secker and Warburg, London
Close D (2002) Greece since 1945: politics, economy and society. Longman/Pearson Education, London
Constitution of Greece (2013) As revised with the Resolution of 27 May 2008 of the 8th Constitutional Assembly of the Parliament of Greece. Athens (in Greek)
Coombs CA, Coppock JO (1947) Agriculture and food in Greece. Operational analysis papers no. 19, United Nations Relief and Rehabilitation Administration, European Regional Office
CPER (1980) Regional Development Plan 1981–1985. CPER and Ministry of Coordination, Athens. (in Greek)
Dafnis G (1955) Greece between two wars, 1923–1940: Vol. A ("Venizelism" rules). Icarus, Athens
Dakin D (1972) The unification of Greece 1770–1923. Ernest Benn, London
Demetras E, Papadakis M, Siampos G (1973) Developments and prospects of the population of Greece, 1920–1985. National Centre of social research and Ministry of Shipping. Transport and Communications, Athens. (in Greek)
Dertilis G (2016) Seven wars, four civil conflicts, seven bankruptcies: 1821–2016. Polis, Athens. (in Greek)
Dertilis G (2019) History of modern and contemporary Greece 1750–2015, Revised edn. Crete University Press, Herakleion. (in Greek)
Diamantopoulos Th (2017–2018) Ten plus one decades of political divisions, vol 1–7. Thessaloniki, Epikentron (in Greek)
diaNEOsis (Organization of Research and Analysis) (2017) 70 years after the Porter Report (in Greek) (with attachment of the original report in English). https://www.dianeosis.org/2017/05/ekthesi_porter_70_xronia/
Dionelis C, Giaoutzi M (2008) The enlargement of the European Union and the emerging new TEN transport patterns. In: Giaoutzi M, Nijkamp P (eds) Network strategies in Europe: developing the future for transport and ICT. Ashgate, Aldershot, pp 119–132
Dionelis C, Giaoutzi M, Mourmouris J (2008) The network versus the corridor concept in transportation planning. In: Giaoutzi M, Nijkamp P (eds) Network strategies in Europe: developing the future for transport and ICT. Ashgate, Aldershot, pp 151–168
Dos Santos T (1973) The structure of dependence. In: Wilber CK (ed) The political economy of development and underdevelopment. Random House, New York, pp 109–117. (originally published 1970)
Doxiadis C (1946) Economic policy for the reconstruction of the country's settlements. No. 3 – Publications of the Ministry of Reconstruction, Athens (in Greek)

Doxiadis C (1947) Housing policy for the reconstruction of the country with a 20-year plan. No. 7 – Publications of the Ministry of Reconstruction, Athens (in Greek)
Drettakis M (2012) Crisis hour for Greece. En Plo, Athens. (in Greek)
Dryzek JS, Dunleavey P (2009) Theories of the democratic state. Macmillan International/Red Globe Press, London
Economist, The (1993) Last chance, Sisyphus: a survey of Greece. Special report. May 22nd, 1993
Eddy CB (1931) Greece and the Greek refugees. Allen and Unwin, London
Eleftheroudakis Encyclopaedia (1929) Athens (in Greek)
Encyclopaedia Papyrus – Larousse – Britannica (2006) (in Greek)
European Commission (2018) Transport in the European Union: current trends and issues. Brussels
Exarchos GS (1992) Pathways of development. Gavrielidis, Athens. (in Greek)
Filias V (1974) Society and power in Greece: I. The false urbanization 1800–1864. V. Makryonitis, Athens (in Greek)
Gallant TW (2017). Modern Greece: from independence to the present. Bloomsbury Academic, London (first published 2001)
Getimis P, Economou D (1992) New geographical inequalities and spatial policies in Greece. Topos 1992(4):3–44. (in Greek)
Giannakourou G (1994) The new spatial policy framework in the 1990s: institutional re-arrangements and uncertainties. Topos 1994(8):15–40
Giannakourou G (2000) Spatial policy in an enlarged Europe. In: Andrikopoulou E, Kafkalas G (eds) The new European space: enlargement and the geography of European development. Themelio, Athens, pp 401–424. (in Greek)
Giannakourou G (2003) Planning of metropolitan areas: institutions and policies. In: Getimis P, Kafkalas G (eds) Metropolitan governance: international experience and Greek reality. Panteion University, Athens, pp 63–82. Also published in Kafkalas G (ed) (2004) (in Greek)
Giannakourou G (2008) Regional spatial planning in the European Union. Papazisis, Athens. (in Greek)
Giannakourou G (2012) The statutory framework of city planning in Greece: historic transformations and modern demands. In: Economou D, Petrakos G (eds) The development of Greek cities: interdisciplinary approaches of urban analysis and policy. Thessaly University Press, Volos. (first published 1999), pp 457–480 (in Greek)
Giannakourou G (2019) Regional-spatial and urban planning law. Nomiki Vivliothiki, Athens. (in Greek)
Giannousis G (1983) Geographical distribution of the parameters of the Greek regional problem. TeC (Technika Chronika) 1983(9–12):29–51. (in Greek)
Great Greek Encyclopaedia (1928) 2nd edition. "O Foinix" Publishing, Athens (in Greek)
Gregorogiannis A (1975) Foreign capital in Greece. Grammi, Athens. (in Greek)
Gregory D, Pred A (2007) Introduction. In: Gregory D, Pred A (eds) Violent geographies. Fear, terror and political violence. Routledge, New York
Hadjijoseph C (1994) The reconstruction period 1945–1953 as a key moment in modern Greek and European history. In: Sakis Karagiorgas Foundation (ed) Greek society in the first postwar period 1945–1967. Conference proceedings. Vol. A, Athens, pp 23–33. (in Greek)
Hadjivasileiou E (2015) Greek liberalism: the radical current, 1932–1979. Patakis, Athens. (in Greek)
Heurtley WA, Darby HC, Crawley, Woodhouse CM (1967) A short history of Greece from early times to 1964. Cambridge University Press, London
Hoffman GW (1972) Regional development strategy in Southeast Europe. Praeger, New York
Ioannidis S, Kalogirou G, Lymperaki A (1994) The demand for development in the journal *Nea Oikonomia* 1946–1967. In: Sakis Karagiorgas Foundation (1994–95) Greek society in the first postwar period 1945–1967. Conference proceedings. Vol. A. Athens, pp 335–359 (in Greek)
Jecchinis C (1967) Trade unionism in Greece: a study in political paternalism. Roosevelt University, Chicago
Jelavich C, Jelavich B (1965) The Balkans. Prentice-Hall, Englewood Cliffs

Kafkalas G (1981) The regional organization of the Greek economy, 1948–74. PoPer 1981(2):7–38. (in Greek)
Kafkalas G (1984) Regional organization of industry. Doctoral thesis. Aristotle University of Thessaloniki (in Greek)
Kafkalas G (1990) Regional development and regional policy in the 1980s: crisis or transition to a new era? In: Psycharis Y (ed) The functions of the state in a period of crisis: theory and Greek experience. Conference proceedings. Sakis Karagiorgas Foundation, Athens, pp 302–309. (in Greek)
Kafkoula K (1990) The garden city idea in interwar Greek town planning. Doctoral thesis, Aristotle University of Thessaloniki. National Archive of Doctoral Theses (didaktorika.gr)
Kaldis P, Laskaris C (1990) The necessity of an interdisciplinary investigation of development potential and resources of selected mountain and backward rural areas in Greece. In: NTUA (ed) The interdisciplinary approach to development. Proceedings of an interuniversity interdisciplinary conference. Papazisis, Athens, pp 403–411. (in Greek)
Kalfa C (2019) Self-help shelter now!: the invisible side of American assistance to Greece. Futura, Athens. (in Greek)
Kaltsonis D (2017) Constitutional history of Greece 1821–2001, K.Ps.M. Editions, Athens (in Greek)
Kalyvas S (2015) Modern Greece. Oxford University Press, New York
Kalyvas S, Marantzidis N (2015) Civil conflict passions. Metaichmio, Athens. (in Greek)
Karatsolis C (2020) Introduction to town planning law in Greece and Cyprus: planning the city, constructing the Civitas. Nomiki Vivliothiki, Athens. (in Greek)
Kasperson RE (1966) The Dodecanese: diversity and unity in island politics. The University of Chicago, Chicago
Kathimerini (2014) The sacrifices of Greece in World War II. 2 volumes. Kathimerines Ekdoseis (Newspaper Kathimerini), Athens. (in Greek)
Katsapis K (2002) Refugee settlement in interwar Greece. Encyclopaedia of the Hellenic world, Asia Minor. Great online encyclopaedia of Asia Minor (fhw.gr)
Kavouriaris E (1974) Some thoughts on the causes and consequences of migration. In: Nikolinakos M (ed) Economic development and migration in Greece. Kalvos, Athens, pp 24–62. (in Greek)
Kayser B (1968) Human geography of Greece. EKKE, Athens (French original: Géographie humaine de la Grèce) (Greek translation)
Kayser B, Pechoux P-Y, Sivignon M (1971) Exode rural et attraction urbaine en Grèce. Centre National de Recherches Sociales (EKKE), Athènes
Komilis P (1986) Spatial analysis of tourism. Scientific studies. CPER, Athens. (in Greek)
Komilis P (1987) The spatial structure and growth of tourism in relation to the physical planning process: the case of Greece. PhD thesis. University of Strathclyde, Glasgow
Komnenos N (1993) Innovative growth in the peripheral regions: some implications for Greece. In: Getimis P, Kafkalas G (eds) Urban and regional development in the new Europe. Topos – Special series, Athens, pp 193–206. (in Greek)
Kontogiorgis G (2020) Hellenism and Helladic state: two centuries of quarrel, 1821–2021. Poiotita, Athens. (in Greek)
Kordatos GK (1958) History of modern Greece 1900–1924. 20th Century Publications, Athens. (in Greek)
Kordatos GK (1973) The interventions of the English in Greece. Epikairotita, Athens. (in Greek)
Korizis C (1975) The authoritarian regime, 1967–1974. Gutenberg, Athens. (in Greek)
Korres G, Rigas K (2000) Income distribution and its impact on Greek regional economic development. In: Pakos T (ed) The "two-thirds" society: dimensions of the modern social problem. Panteion University, Athens, pp 239–254. (in Greek)
Korres G, Kourliouros E, Marmaras E (2011) The impact of the construction industry in the economic and regional development of Greece. In: Panteion University (ed) Volume in honour of Prof. P. Loukakis. Gutenberg, Athens, pp 456–473. (in Greek)
Kostis C (1999) The myth of the foreigner or Pechiney in Greece. Alexandreia, Athens. (in Greek)

Kostis C (2018) "The spoiled children of history": the formation of the modern Greek state, 18th–21st centuries. Patakis, Athens. (in Greek)
Kousoulas DG (1965) Revolution and defeat: the story of the Greek Communist Party. Oxford University Press, London
Kremmydas V (1994) Greece in 1945–1967: The historical context. In: Sakis Karagiorgas Foundation (1994–95) Greek society in the first postwar period 1945–1967. Conference proceedings. Vol. A. Athens, pp 15–19 (in Greek)
Kühn M (2015) Peripheralization: theoretical concepts explaining socio-spatial inequalities. EPS (European Planning Studies) 23(2):367–378
Kyramargiou E, Prentou P, Christoforaki K (2015) Social and spatial atlas of refugee Piraeus. In: Maloutas T, Spyrellis S (eds) . Social atlas of Athens. www.athenssocialatlas.gr
Kyrtsis A (ed) (2006) Konstantinos A. Doxiadis: Texts, plans, settlements. Icarus, Athens (in Greek)
Lagopoulos AP (1992) The programming and planning system in Greece and urban planning theory. Teaching manual, Thessaloniki. (in Greek)
Lamprianidis L (2001) Economic geography: elements of theory and empirical examples. Patakis, Athens. (in Greek)
Lamprinidis MI, Pakos T (1990) The Greek fiscal crisis. In: Psycharis Y (ed) The functions of the state in a period of crisis: theory and Greek experience. Conference proceedings. Sakis Karagiorgas Foundation, Athens, pp 175–193. (in Greek)
Langrod G (1965) Reorganisation of public administration in Greece. OECD, Paris
Legg KR (1969) Politics in modern Greece. Stanford University Press, Stanford
Leontidou L (2001) Cities of silence: Working class colonization of Athens and Piraeus, 1909–1940. ETVA Cultural Technological Foundation, Athens (first published 1989) (in Greek)
Leontidou L (2011) Ageographitos Chora [Geographically illiterate land]: Hellenic idols in the epistemological reflections of European geography. Propobos, Athens (in Greek)
Les Temps Modernes (directeur Jean-Paul Sartre) (1969) Aujourd'hui la Grèce…: Dossier. Les Temps Modernes, 1969: No 276bis, Paris
Liodakis G (1994) Sociological theoretical foundations and economic programming practices in Greece in the first postwar period. In: Sakis Karagiorgas Foundation (1994–95) Greek society in the first postwar period 1945–1967. Conference proceedings. Vol. A. Athens, pp 77–94 (in Greek)
Loukakis P (2010) Critical examination of the completion processes of spatial planning and its implementation in contemporary Greece. In: Demetriadis EP, Kafkalas G, Tsoukala K (eds) The *logos* of the *polis*. Volume in honour of Prof. A.Ph. Lagopoulos. University Studio Press, Thessaloniki, pp 101–111. (in Greek)
Loulis C (2019) The survival of Greece through a succession of miracles: 200 years of Greek history. Psychogios, Athens. (in Greek)
Makrydimitris A (1999) Administration and society: public administration in Greece. Themelio, Athens. (in Greek)
Mantouvalou M, Mavridou M, Vaiou D (1995) Informal activities and micro-landownership vs planning in Greece: Local specificities in a unifying Europe. In Aegean, Seminars of the (1995) Geographies of integration, geographies of inequality in Europe after Maastricht. 1993 Syros seminar. Athens and Thessaloniki, pp 176–192
Markezinis S (1966) Political history of modern Greece 1828–1964. 4 volumes. Papyrus, Athens (in Greek)
Markezinis S (1973) Political history of modern Greece: Contemporary Greece. 4 volumes. Papyrus, Athens (in Greek)
Marmaras M, Tsilenis S (1994) Postwar policy of town plan expansion in Athens as a result of the absence of principles of planning and urban space formation. In: Sakis Karagiorgas Foundation (1994–95) Greek society in the first postwar period 1945–1967. Conference proceedings. Vol. A. Athens, pp 599–618 (in Greek)
Mavrogordatos G (2016) 1915: the national schism. Patakis, Athens. (in Greek)
Mavrogordatos G (2017) After 1922: the prolongation of the schism. Patakis, Athens. (in Greek)

McGrew WW (1985) Land and revolution in modern Greece 1800–1881. The Kent State University Press, Kent Ohio
McNeill WH (1947) The Greek dilemma: War and aftermath. Victor Gollancz (Left Book Club Edition), London
Mears EG (1929) Greece today: the aftermath of the refugee impact. Stanford University Press, Stanford
Melissas D (2002) Fundamental issues of regional spatial planning. Ant.N. Sakkoulas, Athens. (in Greek)
Melissas D (2007) Land uses and the General Town Plan (GPS). Sakkoulas, Athens. (in Greek)
Melissas D (2019) TPS and EPS plans. Sakkoulas, Athens. (in Greek)
Meynaud J (1965) Les forces politiques en Grèce. Études de Science Politique, Lausanne
Meynaud J (1966) Political forces in Greece. Byron, Athens (in Greek; translated from the French original)
MinCoord (1960) 5-year economic development plan 1960–1964. ETyp (National Press), Athens. (in Greek)
Moskof K (1972) National and social conscience in Greece 1830–1909: The ideology of compradorist space, Thessaloniki
Moulaert F (2002) Globalization and integrated area development in European cities. Oxford University Press, Oxford
Moulaert F, Cabaret K (2006) Planning, networks and power relations: is democratic planning under capitalism possible? Plan Theory 5(1):51–70
Mousis N (2003) Europan union: legislation, economy, policies, 10th edn. Papazisis, Athens. (in Greek)
Mouzelis N, Attalides M (1971) Greece. In: Scotford Archer M, Giner S (eds) Contemporary Europe: class, status and power. Weidenfeld and Nicolson, London, pp 162–197
Munkman CA (1958) American aid to Greece: a report on the first ten years. Praeger, New York
National Statistical Service (1964) In-migration to the Athens capital region. ESYE, Athens. (in Greek)
Nikolinakos M (1974a) Capitalism and migration. Papazisis, Athens. (in Greek)
Nikolinakos M (ed) (1974b) Economic development and migration in Greece. Kalvos, Athens. (in Greek)
Nikolinakos M (1976) Studies on Greek capitalism. Nea Synora, Athens. (in Greek)
Nikolinakos M (ed) (n.d.) The energy sector of the Greek economy. Nea Synora, Athens. (in Greek)
O'Ballance E (1966) The Greek Civil War 1944–1949. Faber and Faber, London
OECD (1979) Economic surveys: Greece, Paris
Ohmae K (1996) The end of the nation state: the rise of regional economies. HarperCollins, London
Pakos T (ed) (2000) The "two-thirds" society: dimensions of the modern social problem. Panteion University, Athens. (in Greek)
Pangalos T (1974) Some hypotheses for the study of the problem of Greek workers' migration to West Europe from Greek peripheral regions. In: Nikolinakos M (ed) Economic development and migration in Greece. Kalvos, Athens, pp 63–77. (in Greek)
Papadaskalopoulos A (2001) New trends in regional planning. In: Burgel G, Demathas Z (eds) Greece in the face of the 3rd millennium: space, economy, society in the last 40 years. Panteion University, Athens, pp 235–238. (in Greek)
Papademetriou G (1993) The principle of the social state in the political regime transition period of Greece (1974–1991). In: Sakis Karagiorgas Foundation (ed) Dimensions of present-day social policy. Conference proceedings. Athens, pp 131–140. (in Greek)
Papademetriou G, Sotirelis G (eds) (2001) The constitution of Greece, 3rd edn. Kastaniotis, Athens. (in Greek)
Papageorgiou C (1972) The regional problem of Greece. EpKE (Epitheorisi Koinonikon Erevnon) 1972(13) (in Greek)
Papageorgiou CL (1973) Regional employment in Greece. Volumes I and II. EKKE, Athens

Papagiannakis L (1976) The Greek energy sector and industrialization process. In: Nikolinakos M (ed) The energy sector of the Greek economy. Nea Synora, Athens, pp 87–121. (in Greek)
Papaioannou JG (1971) Evaluation of recent trends in metropolitan planning in Greece. In: The mastery of urban growth. Report of the International Colloquium, Brussels, 2–4 December 1969. Meus en Ruimte, V.Z.W. (M+R International). Brussels, pp 263–293
Papaspiliopoulos S (1969) Structures socio-politiques et développement économique en Grèce. In: Les Temps Modernes (directeur Jean-Paul Sartre) (1969) Aujourd'hui la Grèce…: Dossier. Les Temps Modernes, 1969, No 276bis, Paris, pp 37–66
Pappas T (2015) Populism and crisis in Greece. Icarus, Athens. (in Greek)
Patellis G (1978) Elements of Greek rural space. TeC (Technika Chronika) 1978(8):82–105. (in Greek)
Patellis G (1980) Level of living, quality of life and problem areas. In: TEE (ed) Spatial planning and development. Conference on development – 2nd pre-conference one-day meeting. Athens, pp 28–32. (in Greek)
Pentzopoulos D (1962) The Balkan exchange of minorities and its impact upon Greece. Social Science Centre, Athens/Mouton/Paris
Pepelasis A (1962) Surplus labour in Greek agriculture 1953–1960. CPER, Athens
Pesmazoglu J (1972) The Greek economy since 1967. In: Clogg R, Yannopoulos G (eds) Greece under military rule. Secker and Warburg, London, pp 75–108
Petrakos G (1996) The new geography of the Balkans: cross-border cooperation between Albania, Bulgaria and Greece. University of Thessaly, Volos
Polyzos N (1973) Interactions between industrial and population developments. In: NTUA (ed) Industrial development in space, Lecture series. NTUA, Athens, pp 31–47. (in Greek)
Poulopoulos SK (1972a) Economic geography of Greece – general part: the entire environment and geo-sociological correlations of the Greek economy. P. Sakkoulas, Thessaloniki. (in Greek)
Poulopoulos SK (1972b) The socioeconomic system of Greece and its effects on the shaping of national space. Thessaloniki (reprint) (in Greek)
Psilos D (1964) Capital market in Greece. CPER, Athens
Psilos D (1968) Postwar economic problems in Greece. In: Committee for Economic Development. Economic Development Issues: Greece, Israel, Taiwan, Thailand. Praeger, New York
Rafaelidis V (1993) (comical-tragic) history of the modern Greek state 1830–1974. Ekdoseis tou Eikostou Protou, Athens. (in Greek)
Recovery Organization (1949) The survival of the Greek nation – Part B: The plan. Reconstruction Studies No 2b, Greek Recovery Programme Co-ordinating Office, Ministry of Co-ordination, Athens
RIIA (Royal Institute of International Affairs) (1945) The Balkans, with Hungary. RIIA Information Department, Information Notes No 8, London
RIIA (Royal Institute of International Affairs) (1970) The Balkan states: I. economic. Oxford University Press, London. (A reprint of the Johnson Reprint Corporation, New York; originally published 1936)
Rizas S (2015) Konstantinos Karamanlis. Kathimerini, Athens. (in Greek)
Roberts JM (1996) A history of Europe. Helicon, Oxford
Rousseas SW (1967) The death of a democracy: Greece and the American conscience. Grove Press, New York
Rovolis A, Tragaki A (2008) The regional dimension of migration in Greece: spatial patterns and causal factors. In: Coccossis H, Psycharis Y (eds) Regional analysis and policy: the Greek experience. Physica-Verlag, Heidelberg, pp 99–117
Rozos N (1994) Legal problems of regional spatial planning. Ant.N. Sakkoulas, Athens. (in Greek)
Sagias I, Spourdalakis M (1992) Collective consumption, state, social reproduction. In: Maloutas T, Economou D (eds) Social structure and urban organization in Athens. Paratiritis, Thessaloniki, pp 33–66. (in Greek)
Sakellaropoulos T (1991) Institutional transformation and economic development: state and economy in Greece 1830–1922. Exantas, Athens. (in Greek)

Sakkas D (1994) The development programmes of the period 1947–1966 and their relationship with indicative planning. In: Sakis Karagiorgas Foundation (ed) Greek society in the first postwar period 1945–1967. Vol. A. Conference proceedings, Athens, pp 59–76. (in Greek)

Sandis EE (1973) Refugees and economic migrants in greater Athens: a social survey. National Centre of Social Research, Athens

Siouti G (2011) Manual of environmental law, 2nd edn. Sakkoulas, Athens. (in Greek)

Skagiannis P (1994) The role of infrastructures in the capital accumulation regimes of the first postwar periods in Greece. In: Sakis Karagiorgas Foundation (ed) Greek society in the first postwar period 1945–1967. Conference proceedings. Vol. A, Athens, pp 115–132. (in Greek)

Skouris V (1991) Spatial regional and town planning law. Sakkoulas, Thessaloniki. (in Greek)

Sotiropoulos DP (2019) Phases and contradictions of the Greek state in the 20th century, 1910–2001. Estia, Athens. (in Greek)

Stamatelatos M, Vamva-Stamatelatou F (2006) Single-volume geographical dictionary of Greece. Ermis, Athens. (in Greek)

Stavrianos LS (1958) The Balkans since 1453. Holt, Rinehart and Winston, New York

Stavrianos LS (1974) Greece in revolutionary period: forty years of struggle. Kalvos, Athens. (in Greek)

Stevens A (1999) The institutions of the European Union towards the new millennium. In: Dyker DA (ed) The European economy. Addison Wesley Longman, Harlow, pp 64–81

Svoronos N (1969) Esquisse de l'évolution sociale et politique en Grèce. In: Les Temps Modernes (directeur Jean-Paul Sartre) (1969) Aujourd'hui la Grèce…: Dossier. Les Temps Modernes, 1969: No 276bis, Paris, pp 7–36

Svoronos N (1972) Histoire de la Grèce moderne. Presses Universitaires de France, Paris (1ère édition 1953)

Svoronos N (1976) A review of modern Greek history. Themelio, Athens. (in Greek)

Sweet-Escott B (1954) Greece: a political and economic survey, 1939–1953. Royal Institute of International Affairs, London

Thomadakis S (1994) Reconstruction impasses and economic institutions of the postwar state. In: Sakis Karagiorgas Foundation (ed) Greek society in the first postwar period 1945–1967. Conference proceedings. Vol. A, Athens, pp 34–40. (in Greek)

Tsakloglou P (1995) Inequality in the EC: how much between and how much within regions? In: Aegean, seminars of the (1995) Geographies of integration, geographies of inequality in Europe after Maastricht. 1993 Syros seminar, Athens/Thessaloniki, pp 96–98

Tsakonas D (1970) Problems of Hellenicity: sociology and philosophy of modern Hellenism. Athens (in Greek)

Tsakonas D (1971) Introduction to the new Hellenism. Athens (in Greek)

Tsaousis DG (1971) Morphology of modern Greek society. Gutenberg, Athens. (in Greek)

Tsoucalas C (1969) The Greek tragedy. Penguin, Harmondsworth

Tsoulfidis L (2015) Economic history of Greece. University of Macedonia Press, Thessaloniki. (in Greek)

Vaiou D (2019) … in the popular neighbourhoods of Athens with Lila Leontidou. In: Afouxenidis A et al (eds) Geographies in an era of fluidity: critical essays on space, society and culture in honour of Lila Leontidou. Propobos, Athens, pp 269–274. (in Greek)

Vaiou-Hadjimichalis D, Hadjimichalis C (1979) Regional development and industrialization: monopoly investment in Pylos. Exantas, Athens. (in Greek)

Vakalopoulos KA (1988) Single-volume history of Macedonia. Adelfoi Kyriakidi, Thessaloniki. (in Greek)

Vakalopoulos KA (2005) History of Greece: from the birth of new Hellenism (1204) to modern Greece, Ant. Stamoulis, Athens. (in Greek)

Valaoras VG (1974) Urban-rural population dynamics of Greece 1950–1995. A preliminary report. ESYE and CPER, Athens

Vamvakopoulos V (1994) General competencies and urban development programmes of YPECHODE. In: Laskaris C (ed) National and community policies and programmes for the development of urban centres. European Social Fund and NTUA, Athens, pp 65–78. (in Greek)
Varvaresos K (1952) Report on the economic problem of Greece. Athens (in Greek)
Vatikiotis PJ (1974) Greece: a political essay, Washington papers no 22. Sage, London
Venizelos E (ed) (1986) The constitution of Greece 1975–1986. Paratiritis, Thessaloniki. (in Greek)
Veremis T, Koliopoulos I (2013) Modern Greece: a history since 1821. Patakis, Athens. (in Greek)
Veremis T, Koliopoulos I, Michailidis I (2018) 1821: the creation of a nation-state. Metaichmio, Athens. (in Greek)
Vergopoulos C (1975) The agrarian question in Greece – the social integration of agriculture. Exantas, Athens. (in Greek)
Voivonda A, Kizilou V, Kloutsinioti R, Kontaratos S, Pyrgiotis Y (1977) Space regulation in Greece: a brief historical account. AT (Architektonika Themata) 1977(11):130–151. (in Greek)
Voulgaris G (2013) Post-regime transition Greece 1974–2009. Polis, Athens. (in Greek)
Vournas T (1974) History of modern Greece: from the 1821 revolution to the 1909 Goudi rebellion. Tolidis, Athens. (in Greek)
Vournas T (1983) History of modern Greece: From the first years after the civil war to the military coup d' état of the colonels (21 April 1967). Tolidis, Athens (in Greek)
Wassenhoven L (1978) Athens: Is it really suffering from gigantism? Ta Nea 11-1-1978 (in Greek)
Wassenhoven L (1980) The settlement system and socio-economic formation: The case of Greece. PhD thesis. London School of Economics and Political Science, University of London, London
Wassenhoven L (1996) Regional and spatial policy: The European strategy of Greece. Pyrforos, May–June 1996: 42–50 (NTUA journal). Originally presented at a TEE conference (Dec. 1993) on "Greece in Europe: Spatial and regional policy towards the year 2000" (in Greek)
Wassenhoven L (2017) Maritime spatial planning: Europe and Greece. Crete University Press, Herakleion. (in Greek)
Wassenhoven L (2018) Maritime spatial planning in the context of European spatial planning. In: Serraos C, Melissas D (eds) Maritime spatial planning. Sakkoulas, Athens, pp 13–37. (in Greek)
Wassenhoven L (2022) Putting our country in order: a history of spatial regional planning in Greece after the Second World War. Kritiki, Athens. (in Greek)
Woodhouse CM (1968) The story of modern Greece. Faber and Faber, London
Woodhouse CM (1991) Modern Greece: a short history. Faber and Faber, London
YPETHO (1985) Economic and social development programme 1983–1987. Preliminary report and final recommendations, Athens. (in Greek)
Zolotas X (1961) Regional programming and economic development. Bank of Greece, Athens. (in Greek)

Chapter 7
The State as a Crucial Parameter for the Interpretation of Planning

Abstract The purpose of this chapter is to study the Greek state as a determining factor in developing a theory of spatial planning. The formation of the modern Greek state is dotted with parameters which explain subsequent developments of the planning system. Of central importance are the relationships between state and citizenry, foreign control and dependence, profligacy and corruption, the tradition of clientelist politics, the influence of these policies on land use and spatial planning, erratic legislation often tailor-made to suit favouritism priorities, unintelligible legal acts, administrative inefficiencies, and oversized administration. The influence of populism, in its personalized or partisan varieties, is also discussed. The state is looked at by the citizens, alternately, as oppressive, or as a guarantor of family and group interests. Reactions against it can be justified, especially when the state is seen as a threat to the cultural or natural heritage, when the state is patently unjust or when it has systematically tolerated illegality, as in the case of land use and building construction.

Keywords Formation · Role and size of the state · Administration · Legislation · Patronage and political clientelism · Populism · Global South · State-citizen relations · Dependence

Prologue

The aim of this chapter is to explore a key parameter of the author's theoretical model of compromise planning. This parameter is the state. In analyzing planning in a particular country, in this case Greece, the country's geographical and state entity must be viewed as a spatial structure, which has a distinctive organization arrived at via a particular historical process, in which its geopolitical position played an important role. The structure of the state and administrative organization are basic ingredients. Wars, civil conflicts, political divisions and failures, bankruptcies, the gradual construction of a nation state, foreign interference, all frequently leading to "national cleavages" (*dichasmoi*), had their role in shaping interactions between the state and the citizens, who always look at the state both as an enemy

L. C. Wassenhoven, *Compromise Planning : A Theoretical Approach from a Distant Corner of Europe*, https://doi.org/10.1007/978-3-030-94331-8_7

and as a father-figure. The state regularly acted as a mediator but, in order to maintain its internal legitimation, was prepared to foster special preferential links with the electorate. This has been the source of a widespread network of patronage and political clientelism, a phenomenon, not unknown in all Global South, which breeds specific government practices in social policy, including planning and environmental regulation. Such a background is a fertile soil for interactive processes between political power, economic interests, the administrative machinery, local groups, and, above all, the mass of middle- and low-income, land-owning class. The dependence of the mass of voters on real estate interests, hence on land use regulations and prohibitions, makes spatial planning a sensitive political issue.

7.1 The Formation of the Nation State

"What is invariably taken to characterize the modern state is the simultaneous combination of, on the one hand, its claim to act as a public power responsible for the governance of a tightly delineated geographical territory and, on the other, its separation from those in whose name it claims to govern. The modern state is, then, an institutional complex claiming sovereignty for itself as the supreme political authority within a defined territory for whose governance it is responsible" (Hay and Lister 2006, 5; see also Jessop 2002). Dryzek and Dunleavy (2009, 8–11) speak of two theoretical perspectives on state formation: "One perspective is framed top-down in terms of the behaviour of rulers and competition between states, while the other focuses on the bottom-up emergence of states from nations … The top-down view emphasizes the behaviour and skill of ruling political elites (in the past often generals, monarchs or aristocrats) as the key determinant of a state's survival and growth in the context of competition with other states or proto-states … The bottom-up view sees the emergence of states as a process of securing a progressively better fit between political structures and underlying pattern of human societies, each shaped by a common language, culture, religion, ethnicity or historical experience … This bottom-up view is epitomized by those who see nations as the obvious and most secure basis for states, such that 'the state' is shorthand for 'nation state', and not saying the 'nation' part underlines its taken-for-granted character". I believe that, unlike other nation states, the new Greek nation state was not the result of an initiative of its ruler or monarch keen to increase the internal cohesion of his territory and to protect it from external threat; nor was it the outcome of a domestic revolution against a sovereign of the same race and national identity, to put an end to absolutism and establish a democracy. It emerged instead through an independence war against a foreign conqueror and ruler to put an end to subjection and servitude, which had lasted for centuries.

This process of state building had two crucial distinguishing features. The first was that the initiative for the revolt originated in Greek communities of European and Near Eastern lands outside the territory where the war was fought and ultimately won. The initiators were Greeks of the diaspora who cherished the dream of a Greek independent state, or, possibly, had a vested interest in its creation. The primates of local communities in the areas which became the theatre of war were often ambivalent in their attitudes and, when they joined the revolution, were reluctant to give up the powers they enjoyed under Ottoman rule. These autochthons were soon to become enemies of the outsiders, the heterochthons, who brought to the new state western ideas and practices. Violent internal strife broke out even while the independence war was being fought.[1]

The second distinguishing feature was also very important. The great powers of the day (England, France, Russia) were pursuing their own strategies in the Balkans and the Eastern Mediterranean, with respect to the Eastern Question and the Ottoman Empire, the "sick man of Europe". Their involvement was decisive in tipping the balance of the Greek war of independence in favour of the insurgents. It was equally decisive in the creation of the new state, which was to adopt institutions and organizational standards of their own choice, above all the monarchical regime. The new state was from the start modelled on prototypes of the liking both of the "protecting powers" and of the elite of the international Greek communities. The examples of the royal regime, the Napoleonic prefectures and central government organization are eloquent testimony of this influence, resented of course by local provincial notables. Greece had no local nobility like other kingdoms of the early nineteenth century and no indigenous bourgeois class. Still, the state, as an "institutional ensemble" has to be understood as providing "a context within which political actors are seen to be embedded and with respect to which they must be situated" (Hay and Lister 2006, 10). The new ruling class which was to provide the social framework of the new Greek state recruited gradually its members from large estate owners, shipowners, courtiers, senior army personnel, budding politicians, and incoming merchants of the diaspora. Their children were soon being educated in foreign schools and universities thus creating a new stratum in the bourgeoisie of doctors, engineers, and lawyers that would soon offer their services to the state apparatus and its supporting institutions. But the state bureaucracy itself was manned by lower class individuals, often poorly educated and totally dependent for their employment on the rising ruling class. Thus, the great mass of the population looked at the state basically as a provider of employment, but also for other vital concessions like tax exemptions, land grants, public works, and other land-related actions, of long-term spatial consequence. The roots of political clientelism can be traced to those early years, although they had an Ottoman ancestry.

The "national" foundation of the new state as a political formation was far from clear, in spite of the paramount importance of religion and language as a unifying factor. The conditions of the long centuries of foreign occupation that preceded independence had not left a heritage of institutional structures which could be identified as "national". The institutions and practices imported from the West after the new state was formed were of alien extraction. It is here that a sizeable component of the diaspora had focused its attention and cultivated the ideology of a return to

[1] See S. Ramfos, The civil wars (A and B). Newspaper Kathimerini, 19 September and 10 October 2021.

the Greek glorious ancient past. The neoclassical and romantic movements in the West lent their support. Kontogiorgis (2020, 25–28, 58–59, 63, 71–72, 163) condemns the notion that it is the state that creates the nation, which implies that preexisting Hellenism was simply a Greek-speaking entity with no collective identity. In his opinion, the view that it was only after the independence war that a nation emerges, through the constitution of a state, is wrong and tantamount to an acceptance that Hellenism, as an "anthropocentric world system", suffered a historical defeat. Kontogiorgis returns to an age-old ecumene and an ecumenical cosmopolis to support his assertion that the state defeated the nation, which had its roots in the late years of the Byzantine Empire. This bold incrimination of the state is difficult to substantiate but the claim that in other respects in failed is not baseless. Its inefficient and anti-developmental administration is considered the main culprit (Makrydimitris 1999, 8).

The nineteenth century was the period in which the nation state was turned into a reality. It was not from the beginning democratic or liberal, but was able to become so through a long series of traumatic experiences of economic crises, wars, foreign tutelage, and injuries to national pride. The Greek state, D.P. Sotiropoulos (2019, 14–15, 21, 27, 34–35) remarks, followed its own modernization trajectories, adapting, more or less, foreign models, sometimes on its own volition and with its own reform plan, on other occasions after external pressures and coercion of protecting powers, lenders or partners. He adds that, in spite of pessimistic exclamations like "we shall never become Europe", the construction and development of the Greek nation state remains a remarkable achievement. The state incorporated all the norms of European political modernism in its Constitution and its political institutions. The assessment of the road towards modernity, Sotiropoulos correctly explains, depends on the vantage point of the observer, from the past to the future or vice-versa. Looked at from the point of departure, the evolution of the state is impressive. When observed from a point in time when the recent economic crisis brought to the surface all the usual pathologies, the assessment is less enthusiastic. What is clear, however, is that Greek society and the Greek state never evolved into a replica of the social and political models of countries from which it borrowed new institutions and government tools, with which it managed its internal affairs. The background of the mutual bonds between state and people was always there, and although it too changed, it did so much more slowly than its institutional surface. However, the difficulty, Sotiropoulos adds, is to abandon the simplistic theoretical construct of modernization v. traditionalism at war with each other and develop what he calls "reverse syncretism", i.e., a synthesis of contradictions between opposing ideas and practices within the local and international environment (op.cit., 57). If we accept that Greek constitutional history is a useful yardstick for assessing the evolution of the state, then we can borrow and adapt for our purposes the view of Alivizatos (2011, 673) that this evolution was a balancing act of modernist experimentation in a pre-modern environment.

The key role of the state and the decisive importance of its functions have been already made amply clear (Fioravantes 1990; Samaras 1985, 1990) in Chap. 6 on the history of modern Greece, and they will be studied again in Chap. 8 on the Greek planning system. Without their understanding we cannot conceive the elements of a convincing planning theory. It is this realization that permeates Southern theory and the radical current of planning, presented in Chap. 4. Time and time again the state was the channel through which foreign influence changed the full range of administrative practices, including urban and regional planning. This was the case from the moment the revolutionaries of the 1820s started drawing the constitutional map of the new Greek nation state they were aspiring to. But in the recent past the central place of urban and regional planning in the social policy of the state adds a new dimension. In a review of the role of the social state and its place in the 1975 Constitution, Papademetriou (1993, 133) includes among its main social clauses the compulsory property acquisition for reasons of public benefit, the exploitation of the sources of national wealth, the spatial reorganization of the country, urban regeneration, and town development. There is a clear transition, he underlines, from an emphasis on the rule of law to an emphasis on the social state. This, however, presumes the capacity of the administration to produce not just rules, but clear, well-argued policies aimed at the realization of goals and strategic objectives and not simply at the implementation of regulations. The Greek system repeatedly failed to adjust to these organizational requirements, a failure evident even at the level of the macro-region of Athens. It failed further, until the very recent past, to establish a connection with innovations in organizational development (Makrydimitris 1999, 52, 111, 185, 414). The current extraordinary situation of the fight against the coronavirus pandemic seems to have accelerated this adaptation (Fig. 7.1).

The forces that modified the position of state-controlled planning were many and often had external origins. From the moment the Greek state acquires the character of a capitalist state it must support capital accumulation, hence assist the "creation of social and economic conditions conducive to capitalist enterprise (such as securing private property rights, enforcing laws of contract and maintaining a predictable money supply)" (Dryzek and Dunleavy 2009, 95). It is this Marxist interpretation of the process which is offered by Kotzias (1995, 131), yet, in Greece, it is the state itself which assumes the role of promoting capitalist relations of production (Tsoulfidis 2015, 105). Kotzias claims that the internationalization of Greek society, migration and the adjustment of capitalist institutions propelled within the system of political institutions relatively advanced forms of social organization, due, depending on the political opinion, to growing westernization, or state monopolistic practices, or corporatist structures. Some analysts interpreted the evolution of the state as a result of foreign intervention and of conditions of dependence. Historically, there is no doubt that the room of manoeuvre of the Greek state was constrained by external factors; but then, as Hay and Lister (2006, 12) point out, "to understand the capacity for governmental autonomy is … to assess the extent of the institutional, structural and strategic legacy inherited from the past". This historical perspective is the approach of historical institutionalism which concentrates "on the origins and development of the state and its constituent parts, which it explains by the (often unintended) outcomes of purposeful choices and historically unique initial conditions in a 'logic of path-dependence'" (Schmidt 2006, 99 and 104). This historically-grounded interpretation might explain certain characteristics of a new state which seem, at first sight, unjustified, like the size of the Greek state. Path-dependence,

Fig. 7.1 Present division of Greece into 13 regions administered by elected authorities

D.P. Sotiropoulos (2019, 29) agrees, was one of the main factors in the evolution of the state, the others being institutional culture, power mechanisms, conjuncture-related popular demands, outside pressures, and the role of elite classes. There is in the relevant literature an almost universal assumption that the state was not consistent with the Greek economic and social formation and was excessively large. The theoretical "core-periphery" model was frequently invoked as an explanation by particular political forces, along with the bipolar concept "establishment-people" (Veremis and Koliopoulos 2013, 371–372). The justified conviction of dependence and of continuous meddling of the "foreign factor" in Greek affairs tended to become an ingredient of Greek identity (Kalyvas 2015, 8).

7.2 The Size of the State

The claims about an oversized state and an overmanned public sector are closely associated with the problem of state-client relations and the employment of political clients in the civil service. These deficiencies have been often attributed to the excessive centralization of the state and the concentration of powers in the central state, e.g., for planning, as opposed to a devolution to local government. Popular dissatisfaction with the state administration was always present and as a result the rationalization of government was a steady political promise, in the form of "structural reforms" or of "re-building the state" (Gallant 2017; Pappas 2015). The public sector can be approached in an alternative way, not in terms of the size of government but in terms of the services it offers and the cost to the taxpayer. Therefore, Ladi and Katsikas (2017, 98) argue, what really matters is efficiency and quality which need to be improved. D,P. Sotiropoulos (2019, 193–194) agrees that focusing on the size of the administration is an erroneous approach. E.g., in the 1950s, its top-heavy structure was more of a problem than total size (op.cit., 135). To note, that in the conditions of the civil war the appointment of migrants displaced from the country's provinces had increased disproportionately the number of government employees (Makrydimitris 1999, 104). Regardless of the disagreements regarding the size of the public sector, resentment against the government and public administration is always present in public debate and is regularly linked with the repeated failures and economic bankruptcies of the state. The "boom and bust" and "rise and fall" cycles, which Kalyvas (2015, 13 and 195) stressed in his analysis, have been mentioned already. The responsibility of the state, from which the citizen expects almost everything, was always linked subconsciously with the structure and objectives of the administration. In spite of expectations, society never ceases to harbour suspicions and to view the state as hostile, rapacious and oppressive.

The earlier reference to corporatism is a reminder that, as in every human organization, a problem in the state too is the priority that functionaries attach to their own interests, either as individuals or as persons of particular political affiliation. The impression that the administrative machinery is more inclined to protect the rights and privileges of its own members rather than to serve the public interest is widespread. "Government officials make their living from public services. Especially in senior ranks, their welfare is often dependent on the level of the budgets that their department or office receives. Market liberals assume that all bureaucrats try to maximize their budgets. More money means more promotion opportunities, a greater number of other officials that each tier in the hierarchy can supervise, and increases to the bureau's status and prestige" (Dryzek and Dunleavy 2009, 115). Regardless of the prejudice concealed in such "market liberal" views, it is true that government policies are not always motivated by the desire to improve the efficiency of the administration. They are related both with the attitude of the civil servants which is dependent on the support they get from the political personnel and with the wish of governments to keep the independence of the administration under control. We should remember that public choice theorists, who assume that "individuals are self-interested utility maximizers", argue "that the notion of the state acting as the guardian of the public interest is a myth, as state actors (as all individuals) rationally pursue their self-interest" (Hay and Lister 2006, 16). These theorists were taking issue with welfare economists who believed "that self-interest had its limits. For in extolling the possibilities of state intervention to correct the market failures, they assumed that those working for the state, would act as benevolent guardians of the public interest" (Hindmoor 2006, 85). More will be said about the public interest in Chap. 10, but we can continue, for the time being, with the weaknesses of the Greek state.

Kontiadis (2009, 235–238) makes the point that all too often the failure of reforms in Greece is due to resistance from the side of the administration, either at the stage of planning or at that of its implementation. A study of the pathology of the administration, in his opinion, shows clearly that public bureaucracy not only is overstaffed, inefficient and prone to patronage, failing to address the problems of partisanship and corruption; it proved also unable to deal with its functional defaults, low output quality, inadequate performance control, and poor modernization. The administration was consolidated during the postwar period of reconstruction and its reorganization proved a hard nut to crack (Makrydimitris 1994). A series of reports have been produced over the years with proposals for the improvement of the administration (Sotiropoulos 2019, 144–145). Closely related to the size of the public sector is the centralization of the state, which is perceived as the real source of the problem,[2] although not everyone shares this opinion. The roots of centralization can be unearthed in the first years of the life of the Kingdom of Greece, when the concepts of state and nation are being shaped (Beaton 2020, 114). Even today the power of regional, let alone local, authorities is severely circumscribed (Babalioutas 2018, 19; Babalioutas 2019; see also Chlepas 1999, 2003; Flogaitis 1989; Pappas 1995). E.g., after repeated planning reform bills, the approval of a local town plan for a municipality, even for part of it (municipal department), is made by presidential decree, because, in accordance with the Constitution, the Council of State has ruled that this process is not a "local affair". It is possible to argue that after 2019 centralization has become intensified because of the increasing role of the prime minister's office in running government affairs. This development has of course to be seen as part and parcel of the economic crisis which reached Greece in 2010 and was evident in other counties too. Holgersen made a general comment, not of course related to Greece only: "Apparently, since the 2007/2008 crisis, we have seen a new role of the state. It is not only back on stage but has undoubtedly also taken a lead role. It has also become obvious that the capitalist market economy was not – and can never be – a self-regulating process. This has led to an increasing focus on the state within academia, with claims of the so-called 'return' of the state, but profounder analysis claims that ideas such as the state had withdrawn from the economy was a 'neoliberal ideological myth' … The call for the 'return of planning' as a necessary step for dealing with the ecological crisis … strikes similar chords" (Holgersen 2015). In Greece, the state has never withdrawn from the economy, in spite of a neoliberal claim that the role of the private sector is being enhanced. Spatial planning initiatives are both evidence of state-led activity and of neoliberal bolstering of private investment. As put by Inch (2018), "given the status of planning as a primarily state-based activity, these initiatives are clearly related to broader transformations in the rationalities that shape governmental activity. Indeed, the proper limits and purposes of state activity have been central concerns of neoliberal governmentalities …".

---

[2] G. Prevelakis, Centralization as a national danger. Newspaper Kathimerini, 22 October 2017.

The commentary by Kontiadis on the role of the administration should be made more explicit, because it is dependent on the particular sector of public services. However, when viewed from the perspective of spatial policy, he is correct in his conclusion that central government should strengthen its role as staff headquarters coordinating strategic policy making (Kontiadis 2009, 240). Admittedly, this direction was not absent in the field of official spatial planning as practiced in the Ministry of the Environment (under its frequently changed title). The problem was rather the multiplicity of decision-making responsibilities within the overall government structure, i.e., *across* ministries. The trend towards improved coordination was encouraged by the European Union, both through its support frameworks and its supply of information and experiences to government cadres. It can be argued that the process of support programmes debased the role of the national legislature, but the top echelons of the administration acquired increased capacity to steer these programmes (Voulgaris 2013, 209). Ideally this could assist the autonomy of the civil service, if it wasn't for the habit of ministers to rely only on a cortège of trusted advisors. Centralization does not automatically guarantee strategic coordination and does not disappear if a certain amount of de-concentration is achieved, for instance through the creation of regional secretariats under centrally-appointed coordinators, as indeed it has happened in Greece. Besides, there are recent examples of re-concentration or responsibilities, e.g., of forest management. At the same time there has been progress at the top of the government towards a reinforcement of its strategic headquarters role. But strategic coordination and centralization are two totally different concepts and can coexist. Much-needed improvements in the prime ministerial headquarters to strengthen its coordinative capacity have been introduced, but they are of a nature that distinguishes them from considerations affecting centralization of decision-making at the level of central ministries. At this level, centralization, as already mentioned and will be repeated later, carries a historical burden closely associated with traditional politics, patronage and partisan priorities. In plain language, "modernization" can be a fairly smooth process in particular areas of the operation of the state, but runs against the historical current in others.

7.3 Legislation

The literature on Greek law is enormous and cannot be reviewed here, although a publication in English deserves mentioning (Kerameus and Kozyris 1993). Critical is the role of the executive which is initiating the processes of law making. The standard mode of action at the level of the executive is much more difficult to undergo a radical change and this inevitably affects the legislative function and by implication the legislature. It is true that, "modern states regulate social activities using a system of laws, and a constitution to control the activities of government institutions themselves" (Dryzek and Dunleavy 2009, 5). The tantalizing question concerns the way of regulating social activities and of the shared role of the executive and the legislature. In Greece, centralization and the mighty omnipresence of the central state impose an equivalent legal armoury and a proliferation of laws, with the legislature invited to vote even minute details, for instance building coefficients in particular development projects. We shall return to this phenomenon in Chap. 8 on the planning system, but what must be stressed at this point is the inevitable multiplication of laws and statutes and their doubtful quality, when they are ultimately published in the government gazette, where thousands of pages and innumerable amendments remain inaccessible to the average citizen, often even for the specialist. In common parlance this multiplication is described as *polynomia* and bad laws as *kakonomia*. The phenomenon is attributed to the polarization in political life, but was intensified in conditions of economic crisis (Sotiropoulos and Christopoulos 2017, 53–54; see also Sotiropoulos 2019, 165). *Polynomia*, the same authors believe, is congenital in public administration, which is of the opinion that the sheer volume of regulations justifies its mission. A second most influential reason is the culture of "legalism", which prevents political action beyond the limits of issuing administrative acts. A third reason is the desire of every minister to link his/her name with a legislative initiative. All these symptoms are familiar to those working in the fields of spatial planning and environmental protection. Laws are voted, only to be amended by the next government or the next minister in the same government, thus perpetuating a climate of legal uncertainty and insecurity.[3] Public debate is usually limited to "what has been voted in parliament" or what was the ruling of the Council of State. This is where government work is judged. The distance between law making and implementation is made apparent years later.

Text-Box 7.1: Law Making

The drafting of laws is a very claustrophobic affair, even when the task at hand is to remodel the entire planning system and the hierarchy of spatial plans. However, this is an exceptional case. From the time of the 1975 Greek Constitution, which elevated spatial planning to hitherto unknown heights of prestige, until the turmoil of the economic crisis of the 2010s, there had been only three occasions of laws which placed planning on a new footing, in the late 1970s, in 1983, and at the end of the 1990s. On the contrary, from 2014 to the present, three laws were enacted which practically rewrote the system from scratch, although the main structure remained the same. The drafting of these laws is usually the job of a small circle of trusted advisors, a provisional staff group, the invited *éminences grises* in a minister's office who depart when the minister is replaced or resigns. In exceptional cases the planner-minister who had been educated abroad brought his personal knowledge to the play. We must not forget that the minister who authored the important 1983 planning law was himself an architect-planner trained in the United

(continued)

[3] D. Karavellas (Director General WWF Hellas), Out-of-plan construction, unreasonable future. Newspaper Kathimerini, 6 June 2020.

Text-Box 7.1 (continued)

States. The deputy minister and professor of planning who wrote the 2020 planning law has studied in France. Permanent senior civil servants often complain that in the process of law drafting they are not consulted. The conditions at the peak of the economic crisis, which led to the 2014 and 2016 planning laws, were somewhat different because of the external pressure of the so-called Troika of the lending institutions, i.e., the International Monetary Fund, the European Commission, and the European Central Bank (see Chap. 6). An extended consultative group was formed to discuss a planning reform aiming at a more flexible, adaptable, and efficient planning system. The embarrassing – not to say humiliating – element in this process was the feeling that a sort of Damocles' Sword hung over the heads of the participants, which was no other than the danger of the Greek government failing to secure the necessary loans unless law and policy reforms were implemented in a variety of fields, not just in planning. However, a draft law was eventually agreed and the 2014 planning act was voted in parliament. It was scrapped of course by a new government and the 2016 act took its place, but the differences between these acts were minor.

The process of law making, even in the case of basic acts, has its dark side of secrecy, although when the bill is fully written an obligatory and extremely short consultation period follows before it is discussed in parliament. However, the situation is far worse with bills which amend a large number of clauses of existing legislation and tax the ability even of experts to comprehend what is changed and what not, and for what reason. To make matters worse, the amending provisions are as a rule added at the end of totally irrelevant bills. An amendment of great practical importance may thus be found in the penultimate article of a bill for the protection of domestic animals, carefully concealed amid hundreds of pages. Deputies in parliament rightly protest that they did not have the time even to have a glance at the whole bill, especially when so-called ministerial amendments are added at the close of the parliamentary debate minutes before a final vote is taken. The trouble is that the protesting deputies are, as a rule, only the members of opposition parties. A crucial decision on the permitted land use or building conditions of a specific plan or project may hang in the balance. This is the definition of compromise planning in all its majesty.

The public debate is predominantly formalistic, legalistic and prejudiced and stays far from essential problems (Mouzelis 1978, 134). Accidental circumstances shift the dialogue away from basic choices, a frequent occurrence in the spatial and environmental field. In spite of laudable exceptions this situation affects the daily routine of public administration. The claim that the endless regulations will in due course descend the hierarchical ladder and become the responsibility of the appropriate level of government is negated, in spite of assertions that the subsidiarity principle is respected. As Giannakourou (1994, 19–27) remarked a long time ago, the inability to integrate this principle in the centralized system of government underscores the difficulty of convergence with European traditions of state management and the perpetuation of traditional state practices.

Text-Box 7.2: Code of Planning Legislation

The code of planning legislation of 1999 is a densely-printed, 300-page document, which was not subsequently updated and thus quickly rendered useless. For several years successive laws stipulated that a new code should be produced. It was not until 2020 that an expert commission of jurists and planners was charged with this task. In early 2022 it was continuing its deliberations under the chairmanship of a former president of the Council of State.

On several occasions, the commission encountered cases of particular clauses of laws or decrees which were unconstitutional, contradictory, or unclear. In certain cases, it submitted to the Ministry of the Environment and Energy a recommendation that obscure clauses should be corrected through additional legislation; in others, it recommended that certain sections be excluded from codification. The work of the commission included exclusively statutes of urban and regional planning which were approved by law or presidential decree at the initiative of the competent Minister. A problem that arises as a result is that environmental legislation initiated by the same minister or legislation initiated by other ministries (e.g., industry, agriculture, or tourism) remain outside its responsibility, although it often contains urban and regional planning regulations.

The codification process can also throw light on glaring examples of derogations and interest-serving law making. Although the code is limited to legislation of nationwide application to the exclusion of local statutes and administrative bylaws, the ultimate intention of certain law clauses is easy to detect. An interesting example is the provision in a recent law that private land plots in out-of-plan areas (see also Text-Box 8.2 on out-of-plan areas). which acquired their present form before a specified date (like the publication of a previous law) can enjoy preferential treatment, e.g., regarding their minimum size and "buildability". The only precondition is that a building permit application was submitted before the new law. The loophole is big enough for all to see. Such an application might have been submitted on the eve of the final voting in parliament. Here again we have the ghost of compromise looming large.

In spite of all these concerns, the existence of a single planning legislation code is of utmost importance. It will make life easier for all potential users, planners, lawyers, administrators, and judges. The obvious necessity is that future legislation will be invalid unless the code is first accordingly modified and updated. If one recalls the haste and last-minute drafting with which planning law amendments attached to irrelevant bills (say, about child care, refuse collection, or the ratification of a European Union directive) are voted in late-night parliamentary sessions, one is justified to remain pessimistic.

These governance traditions conceal both the confusing complexity of state functions and the dominance of clientelist and populist logics. Fighting against centralization and the overweight central mechanisms, as well as devolution of powers to local self-governments, seem a fully justified step. Many accept the view that past reforms encouraging the transfer of powers proved limited. Yet, apart from the appearance of challenges, like climate change, which may require central action, those supporters of devolution who observe the results in local practice cannot but admit weaknesses. The worst ills of the central state, like clientelism or corruption, tend to be reproduced at the local level where old traditions are stronger.[4] Corruption has to be placed in the particular historical and social context of antagonism between client groups (Serafetinidou 2005). It does not necessarily involve private benefits, as it may relate to wider family and/or political interests (Vavouras and Manolas 2005). Corruption of course is a most sensitive issue, which, although not confined to the realm of planning, reminds us of the "dark side" of planning which was mentioned in Chap. 4. What is intriguing is that it has found its way into a most penetrating and literate published analysis of the civil service, written with a scathing humour (Avgoustos 2003). The author, writing obviously under a pseudonym, had the specialist knowledge to dig deep into the processes of production of professional studies commissioned to private consultants by public authorities and to expose how these studies are carried out, checked, and approved. He even focuses on the case of studies disguised as academic research projects (op.cit., 226–232). A variant of corruption is the pursuit of status-enhancing through budget maximization and power accumulation in a bureaucracy which was mentioned earlier. "Budget-maximizing bureaucrats can expect support from the private interests served by their agency, and from legislators who want to attract government spending to their own constituencies … Cosy three-way relationships can develop between the government department, the alleged oversight committee in the legislature, and rent-seeking private interests. In the USA and Japan these are called 'iron triangles'"(Dryzek and Dunleavy 2009, 118). Such relationships can of course be claimed as justifiable in the name of higher administrative efficiency and thus provoke further centralization.

7.4 State, Populism and Patronage

On several occasions in this book, political clientelism and populist politics were seen as closely linked. The common thread is the protection offered by the powerful to the downtrodden in exchange for their political support. Although political patronage goes back to the early stages of modern Greek history, it cannot be claimed that it was accompanied by populism in the sense that it developed in other countries. If we accept the definition that populism is a political stance emphasizing

[4] See newspaper article based on a report of the agency of Public Administration Inspectors and Controllers (E. Karamanoli, Municipalities as sources of corruption: Irregularities in projects, commissions, construction, and spatial planning. Newspaper Kathimerini, 21 September 2018).

the idea of "the people" in opposition to "the elite" or "the establishment",[5] then to claim that populism existed in Greek politics during the periods when the first priority was to build a new nation state would be a rather bold assertion. It does of course surface in the twentieth century, especially under dictatorial regimes. But even then, the similarities with North American agrarian populism or other classical populist ideologies like the Argentinian Peronism are scant. There were no such movements in Greece attempting to attract a "shirtless", industrial working-class population as in Argentina (Niedergang 1969, 141–144), for the simple reason that the structure of society was totally different and the interests of disadvantaged social groups were represented by political parties of a different persuasion. Analyzing the distinctive marks of populism, Worsley (1967, 164–167) singles out the claim that there are no divisions in society, that indigenous society is a community, a "natural" Gemeinschaft, and that co-operative forms of organization should be given priority. The populist, Worsley tells us, is "some kind of radical, hostile to Big-ness in general". A populist political party must represent "everyone, or at least, the threatened and disinherited mass … The party then becomes co-extensive with 'the nation' or 'the society'; it is a *mass* organization both in terms of numbers, and in so far as it maximizes direct contact between the rank-and-file and the leadership, and minimizes bureaucracy. The rank-and-file constantly participate in policy-making. Finally, a mystical top dressing of quasi-religious appeal to the unity of the people, land, and society is not unusual". The emphasis on "the people" permeates populist analysis. Worsley's comments will serve us later when we return to populism as a new vehicle of clientelism (see also Ionescu and Gellner 1970).

It is in response to academic studies on clientelism by Mouzelis (1977) and Tsoucalas (1977), that Kotzias (1995), whose views were quoted earlier, spoke of the emergence within the Greek system of political institutions of relatively advanced forms of social organization. Kotzias disagrees with the view that the state is exceptionally large and with the interpretation that considers the clientelist system as a key parameter in the study of the Greek state. If there is a problem, in his view, it concerns the structure of the state. Traditional functions like policing and repression or economic surveillance are indeed especially pronounced. Functions which characterize the more advanced phases of capitalism and globalization are under-represented (op.cit., 132–133). He clearly implies here the substitution of modern welfare policy with traditional tactics of favouritism and dubious legality; he gives the example of the abandonment of the social policy of workers' housing and its replacement with the legalization of unauthorized buildings. The law is abused via the promotion of a "society of complicity". For Kontogiorgis (2005), clientelism and corruption reinforce each other through a variety of mechanisms. The individual citizen perceives the client connection as a net of protection and is prepared to compromise (Malindretou and Malindretos 2005, 413). Kotzias refers to the case of land which is illegally occupied, then used for cultivation, and finally turned into building plots. Illegal structures are then erected and in due course the plot is integrated in the official statutory town plan. This is a process which takes the

[5] Wikipedia.

place of an organized welfare approach, but, nevertheless, receives official approval, which amounts to a blessing of illegality. The citizen is thus enmeshed in a network of immoral, if not openly unlawful, behaviour of the state and becomes an accomplice in a series of consequences not only of lawlessness but also, e.g., of ex post unplanned public spending on infrastructures. This sequence of events will occupy again our attention in the chapter on the planning system and on unlicensed building. In fact, Kotzias accepts the existence of the practices of patronage but places them in a wider framework of analysis. Strangely, his theoretical scheme is akin to that of public choice theorists although he would certainly refute this similarity. An extract from Hindmoor (2006, 86–87) explains the public choice approach: "State interventions in the economy – whether they come in the form of tariffs, quotas, subsidies, price supports, import licences, export credits, health and safety directives, planning requirements, regulatory pricing agreements or any one of a hundred other forms – create both economic winners and losers … The … public choice answer is that the state intervenes to create special economic privileges, or rents, benefiting its political supporters … Rent-seeking, the investment of resources by firms and pressure groups in the expectation of securing economic privileges, is, public choice theorists maintain, economically crippling".

The aforementioned approach which uses the large and clientelist state as an explanatory variable is rejected not only by Kotzias, but also by Kostis (2018, 16–17), in an equally ambivalent and two-sided analysis. If the state was really so big, he argues, the logic of clientelism would not be justified even if Greece was to be described, once again, as an exception. His argument seems rather unsound, but he argues that the clientelism parameter, which he claims has been dropped in the international political literature, has nothing to add to an understanding of the evolution of the modern Greek state. He accepts, however, the existence of a "parallel economy", particularly in public administration. He refers not only to scandals at the higher political echelons, but also to everyday corruption and to mechanisms which enable the civil servant to increase his mediocre legal income, by offering more efficient services than those of normal circumstances. These services, among many other categories, include building licencing. A "parallel" income thus accrues to the recipient, which facilitates the daily life of the citizen, and this explains the acceptance of the rules of the game even by those who shoulder the unfair cost. The parallel economy naturally operates also beyond the daily routine of public administration at the higher level of taxation policy. Kostis maintains that the parallel economy seems to have reached its limits in the conditions of the economic crisis of 2010 (op.cit., 845–847 and 854). It is difficult to side with this analysis and disassociate the circuit of the parallel or shadow economy (*paraoikonomia*) from the existence of a client-supportive state. Illicit income generation in the hidden economy goes hand-in-hand with patronage-enabled activity, tolerance of dubious business and professional transactions, and tax evasion. Rent seeking is made possible through a broad range of tacit exemptions and planning derogations. A large array of activities revolving around the value of real estate and land development generate undeclared income and wealth, which turn the wheels of the hidden economy. The land rent accruing to owners of illegally developed properties is an excellent example of this process. In his analysis of town planning in the 1950s, Economou (2000) concentrates on a range of social factors which had an impact on it, like clientelism in the political system, fragmentation of real estate, family-centred social perceptions, and mistrust of the state. The result was occupation of public land, illegal land parcelling, unauthorized building, and subsequent legalization by the state. Some of the parameters singled out by Economou are linked with the existence of a parallel and hidden economy, which accounted in that period for almost 25% of total GDP (Vavouras et al. 1990, 375). The hidden economy has not of course gone away and it is estimated currently at about 20% of national income.[6] The attitude of the citizens in relation to the power of the state, which they consider oppressive and try constantly to overcome, is a fundamental ingredient in the use of land, legally owned or illegally occupied, as well as in issues of employment and tax-evasion (Vavouras and Petrinioti 1990, 338).

Clientelism and the underground circuit of the economy are therefore seen as supplementary processes. T. Pappas (2015, 91–99) adds a third pole, that of populist politics, in order to explain the particularities of the Greek case. Classical international theory, he explains, views client relations between politicians and voters on the basis of material concessions to individuals or small citizen groups.[7] This is a mode of transaction in democratic practice – we might add in the context of welfare policy –, which involves the targeted distribution of material private goods (e.g., jobs or pensions), since of course public goods, such as economic development or full employment, cannot be distributed through the client network. According to Pappas, in full contrast to this classical approach, Greek populist democracy represents a much more complex client system, involving not just specific social groups and not only material concessions. Client concessions and privileges in this case can even include immaterial rewards, like law immunity. A second particularity, is the alternation of groups of beneficiaries depending on the political party in power. Greek society was the recipient of three types of concessions: real incomes (mainly salaries and pensions) granted through clientelist processes; privileged protection from the vagaries of the market; and legal impunity. Immunity and impunity were, for instance, enjoyed by those violating urban planning regulations. This, in the view of Pappas, is a typical symptom. The practice of unlicenced building, especially on illegally occupied land, is a way of deriving private benefits at the expense of public resources. As with tax evasion, the approach of Greek governments to illegal construction was a form of legalization, in exchange for a payment to public coffers. This is a typical case where populism, clientelism and the parallel economy converge. In its old form, political patronage had already contributed substantially to the warped development of the country's socio-spatial organization. Mistrust between state and citizenry, clientelism, the diffusion of informal economic and building activity, ad hoc corporatist interest coalitions, and the centralized and inefficient structure of the administration, all played their role (Leontidou 2011, 290).

[6] See relevant reports in newspaper Kathimerini, 10 October 2018 and 17 December 2019.

[7] Pappas quotes the work of H. Kitschelt and S.I. Wilkinson.

As indicated already, the transformation of the traditional client system and its perpetuation in postwar Greece had been studied by Nicos Mouzelis (1977, 145–146; also Mouzelis 1989, 2002; Mouzelis et al. 1989), who observed that although the profile of the intermediaries between rural society and the centre of power had begun to change, the "go-between" figure always secured dependence on the centre. This was the beginning of a combination of bureaucratic and political party patronage. In essence, what Mouzelis was saying is that the political party clientelism which becomes dominant in the 1980s was already evident in the 1960s and 70s. There was a continuity with the past which, even then, transformed the character of political parties (Demetrakos 1977, 235). We saw in Chap. 6 on history that this past goes back to the days of local primates of the nascent Greek state and includes the long years when it relied on the mediation of local bosses with whom it had to bargain (Kostis 2018, 907–908). Under a still small and weak state, a "give and take" practice grows at the interface of power and society (Sotiropoulos 2019, 28). The growth of clientelism under the Ottomans and then in the early nineteenth century is an interpretation shared by the majority of students of Greek history, but there are exceptions. Kontogiorgis (2020, 47) detects an idiosyncratic element of refusal to be a member of a herd which obeys blindly a central authority and seeks ways to subvert it. In the years of the pandemic of Covid-19 and the anti-vaccination campaign his comments gain in interest.

In those bygone years, local customs of exchange, consumption, and mutual support in rural society always revolved around the nuclear or extended family. In due course they were transported to urban environments (Terkenli 1998, 189). This return to the past of the nineteenth century serves as a reminder that even today the old traditions are used as an explanation of the culture of public administration. This culture is interpreted as the product of two conflicting traditions, that of the Weberian model which was imported after the war of independence and that of the Ottoman system (Sotiropoulos and Christopoulos 2017, 168–169). In the late twentieth century, conditions have naturally changed, but it is still impossible to comprehend the practice of spatial planning without a reference to these parameters. Dependence on interpersonal relations, based on family bonds or political party affiliation, remains strong and the capacity of the planning system to bypass it is limited (Wassenhoven 2008). What has changed is that the network of relations is now far more complex and that the nexus of clientelism and the parallel economy has received an injection of populism supported by the efficient apparatus of large political parties.

T. Pappas (2015, 26 and 237) defines populism as "democratic anti-liberalism" which is rooted in decades-old prejudices. This definition probably embellishes an ugly reality, but whatever the cause of populism, the fact remains that invocation of the wisdom of "the people" and of the instinct of the common man or woman, typical populist arguments, has found a privileged place in Greek political life (Veremis and Koliopoulos 2013, 374). Some people claim that populism has affected the Greek DNA. Populism can be blamed for the country's economic crises and offers a tool for the understanding of the state functions (Pappas 2015, 90–91). It is nevertheless necessary to point out that the workings of populism have undergone a deep change since the 1980s. D.P. Sotiropoulos (2019, 193) claims that it reached new heights in the period of the Memoranda of the 2010 crisis. The propagation of populism highlighted by Veremis and Koliopoulos does not mean that it appears for the first time, but rather that it has undergone a significant mutation. Populism is directed openly against the members of the elite class, referred to as "the *elites*", in a hellenicized plural form. Traditional client relations, vividly described by D.P. Sotiropoulos (2019, 136–137), are giving their place to the recruitment of client-followers through the machine of political parties. This is a process which had gradually started earlier: "In Greece, the power of the parties was based partly on a spoils system which seems to have been still more deep-rooted than in the other southern European countries. For most rank-and-file members, the main motive to join was the desire for official appointments or favours" (Close 2002, 155–156). However, increasingly, the individual political boss, usually a parliamentary deputy, takes a back seat and political cadres – party apparatchiks – become the nodes of client networks. This transformation enables the survival of clientelism in a new form (Kalyvas 2015, 53). Parties penetrate their electorate base through the state apparatus they have managed to control (Makrydimitris 1999, 119). Mouzelis (1989, 25–32) had drawn attention to this mutation in the late 1980s and Veremis and Koliopoulos (op.cit., 378) describe it as vertical mass recruitment and absorption in the party and as a new configuration of political actors (see also Sotiropoulos 2019, 144, and Kontogiorgis 2020, 271),). The old system was a slow process of recruitment and conservative in its outlook because it tended to preserve the established order. The bond between political patron and client produced enduring links between members of parliament and voters. Supporters of the new form of networking can therefore claim that the party which introduces it is no longer dependent on old-fashioned forms of clientelism (Gallant 2017, 430).

The end-result is a populist, conflict-prone political system, of which clientelism is a basic ingredient, a view which naturally is not universally accepted. E.g., Kotzias (1995, 141) disputes the validity of the argument that client relations declined in the West but remained dominant in Greece. The implication of the populism argument is that clientelism is a purely Greek phenomenon, while the country's democratic system is imported and alien. He rightly reminds us of the presence of clientelism in other countries, especially Mediterranean. It is a condition that all capitalist countries, especially in Western Europe, experience in the postwar period in their effort to legitimize their political system and pave the way to a welfare state (Hay and Lister 2006, 6). Let us not forget that all liberal democratic states tend "to extend their reach into more areas of social life", a frequently controversial trend (Dryzek and Dunleavy 2009, 30). The point is that in Greece the enhanced presence of clientelism is explained, according to Kotzias, by the low level of welfare policy and by internal social conflicts. It can be argued that the opinions expressed by Kotzias and those which he criticizes are perfectly compatible. It is possible to accept both his views on the causes of clientelism and the latter's role in shaping the political system and the functions of the administration.

7.5 Societal Attitudes and the State-Society Nexus

The complex relation of Greek society with the state is a constant concern for a researcher of the planning system. Is it necessary to explore the character of the Greeks as historically shaped in order to understand this relation? It may be, but the present author is not up to the task, although he tried to approach it in Chap. 6 on history. Beaton (2020) has attempted it as he shed light on many of the folds of the Greek collective identity and on the role of the nation state. State and nation have been frequently interpreted as facets of the same phenomenon. Beaton finds it surprising that notwithstanding this Janus relationship the state is still a frequent object of contempt. Nationalist propaganda of past dictatorial regimes was treated by the people with irony (Beaton 2020, 263). Be that as it may, state and populism cannot be isolated from the citizen-power nexus. From very early all students of Greek society placed clientelism at the centre of their attention. In the 1960s and 70s the same set of concepts was present in the work of sociologists, anthropologists and political scientists in their anatomies of Greek social and political life (Wassenhoven 1980, 269–287). In what the author labelled the "village-centred" approach, they focused on rural Greece, the village, its relations with Athens, the role of the state, and the protectors of local communities in bygone ages. In their essays one finds repeated references to customs and concepts like patronage and client relations, family, *philotimo* (love for honour), *rousfetia* (favours),[8] *koumbaros* (best man), and electoral bargaining. In many cases the level of analysis was folkloristic (Carey and Carey 1968; Sanders 1962), in others better documented in case studies of particular settlements (Friedl 1965), and in some cases well founded theoretically (Legg 1969; Vatikiotis 1974; Campbell 1964; Campbell and Sherrard 1968; Burgel 1965). Familism, patronage, poverty, insecurity, and state are the usual points of departure of these studies. Urbanization ensured that family networks were transferred from the villages to the cities (Sotiropoulos 2019, 131). In the following decade the level of analysis gained in sophistication in the work of Greek social scientists and historians (Filias 1974; Mouzelis 1977, 1978; Dertilis 1977).

A critical point worthy of investigation was always the manipulation by governing politicians and the state of client relations, both for forging an informal people-friendly policy and for maintaining control of the administrative apparatus. Successive governments have been accused of trying to "conquer" or "colonize" public administration through client networks and party machines, with disastrous results for administrative efficiency (Sotiropoulos and Christopoulos 2017, 164–165). The influence of party machines on the administration has never ceased to be a reality (Sotiropoulos 2019, 60). This is a vicious circle because with the fall of efficiency, ministers rely only on their inner circle of advisors. This may not be a Greek originality, but in Greece it tends to become a pathological situation. Yet, the institutions of democracy after the end of the military dictatorship in 1974 exhibit impressive resilience, without however putting an end to clientelism and state-party connections (Voulgaris 2013, 181–183 and 218).

---

[8] A word of Turkish origin.

The attitude of society, as manifested through client relations, does not only express consent to the practice of concessions. It also takes a form of rejection of, or indifference to, actions which over a long-time horizon might benefit the country. Social pressures for actions which benefit economic activity are almost absent or, at any rate, not comparable to demands for the satisfaction of consumption (Kyrtsis 1997, 241–242). Such short-sightedness can be more damaging when future developments are beyond the perception of the average individual, as in the case of climate change, and his/her readiness to exert pressure for action, particularly when even the scientific community is not ready to fix specific time horizons, although in Greece, as of course elsewhere, there are voices calling for urgent attention (Zerefos et al. 2019, 16).

To these realities of short-sighted, to some extent understandable, attitudes, we must add the logic of short-term choices which favour benefits in the immediate future. We are back to the co-existence of quick benefits for the citizen-client and electoral calculations by politicians and party cadres, with an eye on the next national election contest, which may be around the corner. This is an explosive mixture of interests favouring constant political struggle and a permanent climate of social divisions. This struggle takes a form of naïve adversarialism, which Chantal Mouffe could have described as "antagonistic" (see Chap. 4), between the faithful of "statism" and supporters of a liberal "anti-statism". Popular demands are channeled through party-political adversarial struggles and create, as Spourdalakis (1989, 70–72) remarks, a condition of "over-politicization", which ultimately descends to an a-political mentality, the chief feature of Greek populism. Lipovats (1989, 56) claims that populism is the breeding ground for the formation of collectivities in which an individual finds a niche and feels justified in his belief that his natural rights are set aside by hostile third parties. This, Lipovats asserts, is a process that helps explain illegal behaviour, as in the case of land occupation and unauthorized building. The aggrieved individual feels, in Spourdalakis' expression, that "the state owes him" preferential treatment, within "a political modus operandi" of which the ingredients are an extreme aggressive individualism and an aversion towards any form of planning and systematic organization. The inevitable results include a demand for solutions "here and now", improvisation on the side of the administration, centralization of decision-making as an expedient for quick answers, and scandals.

Hostility to long-term planning can easily cause violent conflicts, when, e.g., the creation of necessary environmental infrastructures is at stake. Ephemeral protest movements are then mobilized which are auto-designated as social, but do not have the faintest resemblance with movements of genuine long-term environmental goals, which started growing in the West in the 1960s (Dryzek and Dunleavy 2009, 136), but soon appeared in Greece too. The confusion may be deliberate, although in all cases the real target is the ultimate enemy: the state. Political parties intervene to manipulate protest movements, the members of which are potential clients. As pointed out by this author in the past (Wassenhoven 2010, 118–121), this vicious circle of movement mobilization, conflict with the central or local state, and political party interests cannot be explained without reference to the role of the state. The strategy of civil disobedience and conflict rests on the refusal to obey the strictures of power, and is thus closely related to state regulations, concerning the economy, as in the case of market controls, or society, as in the case of environmental protection and hygiene. Regulations, as we know, are made up of legal constraints, land use limitations, incentives etc. Violent resistance of genuine social movements to formal and strictly legal state regulations is no doubt evidence of law-breaking, but underneath this lawlessness there is a sense of injustice, at least with regard to inflexible and legalistic state attitudes and deficient participation. Sadly, this resistance is highjacked by groups which defend a clientelist political *acquis*, hidden behind a screen of social conscience. Mobilization to defend established, sacrosanct self-interests is widespread in matters of land development and urban sprawl. Its social and environmental effect is great, as this author had the opportunity to emphasize repeatedly and will do so again. The acceptance of illegality as a regular practice provides the framework of the way in which the citizen evaluates his/her action. He/she may even end up considering illegality as moral vindication. He/she is the person that will be described later as *homo individualis* (Chap. 10), engaged in continuous process of bargaining and compromise with the state.

Mobilization to defend a variety of causes is not of course the preserve of interest-seeking groups. Environmental concerns have played a great role in the creation of all sorts of volunteer organizations defending the protection of the natural and cultural heritage. The broad range of associations created in the recent past and dedicated to protect nature is impressive. They aim at the rescue of all kinds of species, from bears and sea-turtles to rare flowers and wildfowl. This a most welcome characteristic of modernization, conducive of course to conflicts with government policies to support economic development at all cost or to create infrastructures with an environmental content, e.g., renewable sources of energy. This is, in a sense, a healthy conflict, that bears no connection with rent seeking motives. Urban and regional planning is often caught in this tension between economic development and environmental protection priorities.

Suspicion for the regulatory activities of the state and disobedience are part-and-parcel of the perennial love-hate relationship of the citizen with the state, a relationship of dependence, grudging toleration, and enmity. The state is there to give a helping hand to every individual need, but in times of crisis is seen as failing miserably and blamed accordingly. When the 2010 economic crisis reached Greece, the revolt of the non-privileged classes, wrote T. Pappas (2015, 157–163), was an expression of deep hostility for the state which ceased to satisfy their expectations. In a crisis situation, traditional bonds with political parties are broken, alternative initiatives multiply, and new fields of conflict appear. Feelings submerged under the surface of a compromise equilibrium suddenly surface, as the father-figure of the state is shown unable to cope. All the deep fissures and rifts, long hidden under a blanket of compromise, complicity, and exchange, suddenly widen. This is a "fragmented society", Veremis and Koliopoulos (2013, 427–438) argue, which hates the "state of the law" and any obligation it imposes to combat unlicensed behaviour. This fragmentation, which is suddenly exposed to the public eye in conditions of crisis, is, in their opinion, the main obstacle to modernization which accompanies the Greek state from the day it was born. We have lived since then, they continue, in a condition of schizophrenic dualism of a state built with materials of a modernist

European Enlightenment, which co-exists with a pre-modern society fragmented in groups, with their own hierarchy, client connections, and supporters. These groups antagonize each other to manhandle the state machinery. Fragmentation survived in spite of the triumph of a centralized state and is in fact rewarded by a rising populism, which undermines the state of the law and encourages a split society. Citizens turn their back to legal institutions and side with extra-institutional guarantors of their personal and family welfare. In this context, even the meaning of corruption has diverse interpretations in public conscience. It may signify breaking the rules of the law or the respect of unwritten laws of kinship and friendship solidarity.

Whether or not this account resembles an exaggerated indictment by a public prosecutor, the problem is that in these conditions the innocent and the guilty are given the same treatment, because the state of the law is not prepared to make the necessary distinctions and to take into consideration that it must share the responsibility of illegality, because it has over the years tolerated it or given its indirect blessing, thus providing an excuse of a quasi-custom law (see also Chap. 9). When the citizen is convinced that the actions of the administration are patently unjust, then we have to think carefully if the notion of the "state of the law" should be replaced by that of the "state of justice".[9] The citizen, even if he/she derives a benefit from the tacit toleration and complicity of the state and from his/her clientelist connections, is frequently justified to consider administrative decisions as "patently unjust". When entire settlements, with properly licensed houses and with regular public supply of energy, water and transport facilities, are suddenly declared illegal and designated as forest land, the insubordination of the citizens cannot come as a surprise. The state then has to follow a different argumentation and a different conception of "justice as fairness" – to recall John Rawls – and not hide behind a façade of rigid legality. The question is whether the state, imprisoned as it is in adversarial political battles, sterile legalism, and populist rhetoric, can pursue a course of action to ensure a smooth transition to the kind of modernization it is supposed to serve. The perpetuation of an environment in which any innovative initiative is condemned a priori may lead to an inevitable dead-end (Kostis 2018, 844). The question posed by Nicos Marantzidis, who borrows a concept from the work of Michel Crozier, whether Greece is a *société bloquée*, trapped in its own contradictions, seems fully justified.[10] Crozier had referred to the problem of state bureaucratic machineries which fail to learn from their own mistakes (Makrydimitris 1999, 419). Planning, which is after all our concern in this book, is deeply embedded in these contradictions. The reasons are made clear in a quotation from Shepherd et al. (2020, 13), who, naturally, did not have Greece in mind: "[P]lanning is 'an institutionalised set of ideas and practices' which is 'concerned with mediating the relations between social, economic, political and environmental pressures relating to land and property' … This means planning has deep connections with fundamental political questions about the proper relationships between arrangements of property ownership rights, individual and collective economic freedoms, the rights of the wider

[9] These observations are inspired by the writings of the eminent jurist Petros Pararas (2014, 11–15), but this author by no means claims that they fully reflect his position.

[10] N. Marantzidis, Is Greece a "blocked" country? Newspaper Kathimerini, 23 September 2018.

community or society and the role of the state; all long-standing concerns of political ideologies".

Epilogue

In this chapter we looked at the state as a basic factor in the construction of a spatial planning theory suited to the conditions of Greece. Building a new nation state was a long and painful process marked by conflicts between, on one hand, local groups and traditional power holders, and, on the other, westernized Greeks, usually resident on foreign soil, and modernizers. The process of constructing a state was constantly under the supervision of foreign protecting powers. It is a miracle that despite constant tensions a democratic state finally emerged. However, its central government and administration were arguably oversized, ineffective, riddled with favouritism and patronage. The implications for law-making, legislative processes, proliferation of laws and never-ending law amendments were disastrous, a default that had a particular impact on spatial planning. Efforts to consolidate and codify legislation failed, but they have been resumed lately. The clientelist functions of the state gradually changed from an interpersonal bond between families and protectors to a party-dominated interaction, accompanied by a rise of populism, of a distinctive Greek variety. Attitudes to the state are always ambivalent and complex. The state is simultaneously a hated figure and an organism from which everything is expected. When something goes wrong, as in the aftermath of natural disasters or financial crises, or even after an accident, the standard question is "where is the state". Grievances do not always emanate from a disobedient attitude and a blind opposition to authority. This is a facile explanation, although correct in many respects. They also result from a feeling of injustice, given the tradition of the state to tolerate illegality for clientelist reasons. Whatever the shape of clientelism, it remains decisive in the exercise of spatial planning, especially at the local level, and of central value for the theoretical interpretation of the planning system. It is to this system thar the author draws the attention of the reader in the following chapter.

References[11]

Alivizatos N (2011) The constitution and its enemies in modern Greek history, 1800–2010. Polis, Athens (in Greek)

Avgoustos I (2003) The civil servant's anarchy: communication in public administration. Propobos, Athens (in Greek)

[11] **Note**: The titles of publications in Greek have been translated by the author. The indication "in Greek" appears at the end of the title. For acronyms used in the references, apart from journal titles, see at the end of the table of diagrams, figures and text-boxes at the beginning of the book.

Abbreviations of journal titles: AT Architektonika Themata (Gr); DEP Diethnis kai Evropaiki Politiki (Gr); EpKE Epitheorisi Koinonikon Erevnon (Gr); EPS European Planning Studies; PeD Perivallon kai Dikaio (Gr); PoPer Poli kai Perifereia (Gr); PPR Planning Practice and Research; PT Planning Theory; PTP Planning Theory and Practice; SyT Synchrona Themata (Gr); TeC Technika Chronika (Gr); UG Urban Geography; US Urban Studies; USc Urban Science. The letters Gr indicate a Greek journal. The titles of other journals are mentioned in full.

Babalioutas L (2018) The contemporary institutional framework of Greek public administration. Vol. A. Sakkoulas, Athens (in Greek)
Babalioutas L (2019) The contemporary institutional framework of Greek public administration. Vol. B. Sakkoulas, Athens (in Greek)
Beaton R (2020) Greece: biography of a modern nation. Penguin Random House
Burgel G (1965) Pobia: Étude géographique d'un village crétois. Centre des Sciences Sociales d'Athènes, Athènes
Campbell J (1964) Honour, family and patronage. Clarendon Press, Oxford
Campbell J, Sherrard P (1968) Modern Greece. Ernest Benn, London
Carey JPC, Carey AG (1968) The web of modern Greek politics. Columbia University Press, New York
Chlepas N-K (1999) Local administration in Greece. Ant. N. Sakkoulas, Athens (in Greek)
Chlepas N-K (2003) Constitutional framework and institutional architecture of metropolitan administration. In: Getimis P, Kafkalas G (eds) Metropolitan governance: international experience and Greek reality. Panteion University, Athens, pp 83–107. (in Greek)
Close D (2002) Greece since 1945: politics, economy and society. Longman/Pearson Education, London
Demetrakos D (1977) The ideology of political parties in modern Greece. In: Kontogiorgis GD (ed) Social and political forces in Greece. Hellenic Society of Political Science and Exantas, Athens, pp 219–241. (in Greek)
Dertilis G (1977) The autonomy of politics from social antitheses in 19th century Greece. In: Kontogiorgis GD (ed) Social and political forces in Greece. Hellenic Society of Political Science and Exantas, Athens, pp 39–71. (in Greek)
Dryzek JS, Dunleavy P (2009) Theories of the democratic state. Macmillan International/Red Globe Press, London
Economou D (2000) Town planning policy in the 1950s. In: Deffner A et al (eds) Town planning in Greece from 1949 to 1974. Thessaly University Press, Volos, pp 39–48. (in Greek)
Filias V (1974) Society and power in Greece: I. The false urbanization 1800–1864. V. Makryonitis, Athens (in Greek)
Fioravantes V (1990) Elements of a contemporary look at the state. In: Psycharis Y The functions of the state in a period of crisis: theory and Greek experience. Conference proceedings. Sakis Karagiorgas Foundation, Athens, pp 39–48 (in Greek)
Flogaitis SI (1989) Local government legislation. Ellinikes Panepistimiakes Ekdoseis, Athens (in Greek)
Friedl E (1965) Vasilika: A village in modern Greece. Holt, Rinehart and Winston, New York
Gallant TW (2017) Modern Greece: from independence to the present. Bloomsbury Academic, London. (first published 2001)
Giannakourou G (1994) The new spatial policy framework in the 1990s: institutional rearrangements and uncertainties. Topos 1994(8):15–40
Hay C, Lister M (2006) Introduction: theories of the state. In: Hay C, Lister M, Marsh D (eds) The state: theories and issues. Macmillan International/Red Globe Press, London, pp 1–20
Hindmoor A (2006) Public choice. In: Hay C, Lister M, Marsh D (eds) The state: theories and issues. Macmillan International/Red Globe Press, London, pp 79–97
Holgersen S (2015) Spatial planning as condensation of social relations: a dialectical approach. PT 14(1):5–22
Inch A (2018) "Cultural" work and the remaking of planning's "apparatus of truth". In: Gunder M, Madanipour A, Watson V (eds) The Routledge handbook of planning theory. Routledge, London, pp 194–206
Ionescu G, Gellner E (eds) (1970) Populism. Weidenfeld and Nicolson, London
Jessop B (2002) The future of the capitalist state. Polity Press, Cambridge
Kalyvas S (2015) Modern Greece. Oxford University Press, New York
Kerameus KD, Kozyris PJ (1993) Introduction to Greek law, Revised edn. Kluwer, Deventer

Kontiadis X (2009) Deficient democracy: state and political parties in modern Greece. I. Sideris, Athens (in Greek)
Kontogiorgis G (2005) Corruption and political system. In: Koutsoukis C, Sklias P (eds) Corruption and scandals in public administration and politics. I. Sideris, Athens (in Greek), pp 131–143
Kontogiorgis G (2020) Hellenism and Helladic state: two centuries of quarrel, 1821–2021. Poiotita, Athens (in Greek)
Kostis C (2018) "The spoiled children of history": the formation of the modern Greek state, 18th–21st centuries. Patakis, Athens (in Greek)
Kotzias N (1995) A society of complicity: state legitimacy and legitimation in postwar Greece. In: Sakis Karagiorgas Foundation (1994–95) Greek society in the first postwar period 1945–1967. Conference proceedings. Vol. B. Athens, pp 127–152 (in Greek)
Kyrtsis A (1997) Political legitimation and economic modernization. In: Lampiri-Dimaki I (ed) Sociology in today's Greece 1988–1996. Vol. B. Papazisis, Athens, pp 237–246. (in Greek)
Ladi S, Katsikas C (2017) Is it true that the public sector in Greece is oversized? In: Katsikas C, Filinis K, Anastasatou M (eds) Understanding the Greek crisis. Papazisis, Athens, pp 79–101. (in Greek)
Legg KR (1969) Politics in modern Greece. Stanford University Press, Stanford
Leontidou L (2011) *Ageographitos Chora* [Geographically illiterate land]: Hellenic idols in the epistemological reflections of European Geography. Propobos, Athens (in Greek)
Lipovats T (1989) Populism: an analysis from the standpoint of political psychology. In: Mouzelis N, Lipovats T, Spourdalakis M (eds) Populism and politics. Gnosi, Athens, pp 47–61. (in Greek)
Makrydimitris A (1994) Government and administration: the administrative machinery of the state during the reconstruction period. In: Sakis Karagiorgas Foundation (1994–95) Greek society in the first postwar period 1945–1967. Conference proceedings. Vol. A. Athens, pp 471–479 (in Greek)
Makrydimitris A (1999) Administration and society: public administration in Greece. Themelio, Athens (in Greek)
Malindretou V, Malindretos P (2005) Integrated economic approach to the phenomenon of corruption. In: Koutsoukis C, Sklias P (eds) Corruption and scandals in public administration and politics. I. Sideris, Athens (in Greek), pp 371–419
Mouzelis N (1977) Class structure and political clientelism system: the case of Greece. In: Kontogiorgis GD (ed) Social and political forces in Greece. Hellenic Society of Political Science and Exantas, Athens, pp 113–150. (in Greek)
Mouzelis N (1978) Modern Greece: facets of underdevelopment. Macmillan, London
Mouzelis N (1989) Populism: a novel way of integrating the masses in political processes? In: Mouzelis N, Lipovats T, Spourdalakis M (eds) Populism and politics. Gnosi, Athens, pp 19–45. (in Greek)
Mouzelis N (2002) From change to modernization.[12] A collection of articles. Themelio, Athens (in Greek)
Mouzelis N, Lipovats T, Spourdalakis M (1989) Populism and politics. Gnosi, Athens (in Greek)
Niedergang M (1969) The twenty Latin Americas, vol 1. Penguin Books, Harmondsworth. (first published in French 1962)
Papademetriou G (1993) The principle of the social state in the political regime transition period of Greece (1974–1991). In: Sakis Karagiorgas Foundation (ed) Dimensions of present-day social policy. Conference proceedings, Athens, pp 131–140. (in Greek)
Pappas E (1995) Decentralization and prefectoral self-government. In: SEP (ed) Regional development, spatial planning and environment in the framework of united Europe. Proceedings of a conference. Vol. II. Panteion University. SEP and Topos, pp 121–130. (in Greek)
Pappas T (2015) Populism and crisis in Greece. Icarus, Athens (in Greek)
Pararas P (2014) Res Publica I: the law state. Sakkoulas, Athens (in Greek)

[12] A less literal but more accurate translation might be "From the political slogan of change to that of modernization".

Samaras Y (1985) State and capital in Greece. Synchroni Epochi, Athens (in Greek)
Samaras Y (1990) The objective necessity of the economic functions of the state. In: Psycharis Y (ed) The functions of the state in a period of crisis: theory and Greek experience. Conference proceedings. Sakis Karagiorgas Foundation, Athens, pp 59–65. (in Greek)
Sanders I (1962) Rainbow in the rock: the people of rural Greece. Harvard University Press, Cambridge, MA
Schmidt V (2006) Institutionalism. In: Hay C, Lister M, Marsh D (eds) The state: theories and issues. Macmillan International/Red Globe Press, London, pp 98–117
Serafetinidou M (2005) The social roots of state corruption. In: Koutsoukis C, Sklias P (eds) Corruption and scandals in public administration and politics. I. Sideris, Athens (in Greek), pp 79–93
Shepherd E, Inch A, Marshall (2020) Narratives of power: bringing ideology to the fore of planning analysis. Plan Theory 19(1):3–16
Sotiropoulos DP (2019) Phases and contradictions of the Greek state in the 20th century, 1910–2001. Estia, Athens (in Greek)
Sotiropoulos DA, Christopoulos L (2017) Plethoric law production and bad legislation in Greece: a plan for a better and more effective state. diaNEOsis, Athens (in Greek)
Spourdalakis M (1989) Greek populism in the conditions of an autocratic state. In: Mouzelis N, Lipovats T, Spourdalakis M (eds) Populism and politics. Gnosi, Athens, pp 63–76. (in Greek)
Terkenli T (1998) The question of cultural identity and the sense of belonging in modern Greece. In: Aegean, Seminars of the (ed) Space, inequality and difference: from 'radical' to 'cultural' formulations? 1996 Milos seminar, Athens, pp 185–193
Tsoucalas C (1977) The problem of political clientelism in Greece of the 19th century. In: Kontogiorgis GD (ed) Social and political forces in Greece. Hellenic Society of Political Science and Exantas, Athens, pp 73–112. (in Greek)
Tsoulfidis L (2015) Economic history of Greece. University of Macedonia Press, Thessaloniki (in Greek)
Vatikiotis PJ (1974) Greece: a political essay, Washington papers no 22. Sage, London
Vavouras I, Manolas G (2005) Corruption and its relations-impacts on the official economy and the Para-economy. In: Koutsoukis C, Sklias P (eds) Corruption and scandals in public administration and politics. I. Sideris, Athens (in Greek), pp 349–369
Vavouras I, Petrinioti X (1990) An attempt to trace an invisible activity: undeclared employment in Greece. In: Vavouras G (ed) Parallel economy. Kritiki, Athens, pp 315–348. (in Greek)
Vavouras I, Karavitis N, Tsouchlou A (1990) An indirect method for the estimation of the size of the parallel economy and application in the case of Greece. In: Vavouras G (ed) Parallel economy. Kritiki, Athens, pp 367–379. (in Greek)
Veremis T, Koliopoulos I (2013) Modern Greece: a history since 1821. Patakis, Athens (in Greek)
Voulgaris G (2013) Post-regime transition Greece 1974–2009. Polis, Athens (in Greek)
Wassenhoven L (1980) The settlement system and socio-economic formation: the case of Greece. PhD thesis. London School of Economics and Political Science. University of London, London
Wassenhoven L (2008) Territorial governance, participation, cooperation and partnership: a matter of national culture? Boletin de la AGE 46:53–76
Wassenhoven L (2010) Violence and civil disobedience: the case of land use regulation and locational decisions. In: Demetriadis EP, Kafkalas G, Tsoukala K (eds) The *Logos* of the *Polis*. Volume in honour of Prof. A.Ph. Lagopoulos. University Studio Press, Thessaloniki, pp 113–123. (in Greek)
Worsley P (1967) The third world. Weidenfeld and Nicolson, London
Zerefos C, Kapsomenakis I, Antonakaki T (2019) Man-made destabilization of Greek climate. In: Serraos C, Melissas D (eds) Natural disasters and spatial policies. Sakkoulas, Athens, pp 1–18. (in Greek)

Chapter 8
The Greek Planning System: A Case Study at the Tip of the Balkan Peninsula

Abstract This chapter is an overview of the Greek spatial planning system, which is the "empirical reality" which the author will try to explain theoretically in Part III of the book. It is there that he will strive to develop a "compromise planning" theory. A progressive urban development law was promulgated in Greece in 1923 but the troubled period until the 1960s, even mid-1970s, prevented any worthwhile spatial planning effort, apart from the gigantic task of refugee settlement and agrarian reform. It was after the mid-1970s that a statutory planning system of some consistency was introduced. The Constitution of 1975, with its emphasis on spatial planning as an obligation of the state, was a critical turning point. Modern urban and regional planning legislation was enacted, local planning activity expanded in the mid-1980s, but for a variety of domestic reasons it was not until 2000 that hectic planning activity took place across all geographical scales. Apart from constitutional requirements, some other parameters were instrumental, i.e., the policies of the European Union, that Greece had joined in the meantime, the rise of environmentalism, spurred by international conventions and the EU, and the active role of administrative courts, especially of the Council of the State, which made planning a prerequisite of any project approval. Soon an arsenal of planning instruments and policies was in place, mostly imported from foreign countries and often ineffective in local conditions. By 2009, on the eve of the economic crisis, and even more so in 2020, it could be claimed that a rational-comprehensive planning system had been installed. But, was it? A parallel circuit always exists, based on a political organization dependent on patronage and clientelism that unites the state and its citizens, in a mutual compromise arrangement. Given its importance for Greece, a section is dedicated at the end of the chapter to the experience of planning after natural disasters.

Keywords Greece · Planning system · Legislation · Planning instruments · Plan production · European Union · Administrative courts · Environmental concerns · Foreign influence · State and citizens

L. C. Wassenhoven, *Compromise Planning : A Theoretical Approach from a Distant Corner of Europe*, https://doi.org/10.1007/978-3-030-94331-8_8

Prologue

The aim of this chapter is to present a panorama of the Greek urban and regional planning system, as a basis for the development of a suitable "indigenous" planning theory. This is not the first time that the author attempted to present this system to an English-speaking public. In the mid-1990s, a European Commission project was undertaken for the production of a compendium of spatial planning systems and policies of the then European Union member states. The main body of conclusions was published in 1997 and 13 individual volumes, one for each country, were published in 2000 (European Commission 1997a, 2000b). The present author wrote the volume on Greece (Wassenhoven 2000a) and opened the preface with the following paragraphs:

1. The Greek system of spatial planning is dominated by an emphasis on purely physical aspects. Its main concerns are, on one hand, the interface between private and public land ownership, and, on the other, the development rights of land owners. The key concepts are the statutory *schedio poleos* (town plan),[1] its never-ending *epektaseis* (extensions), the *oroi domisis* (building conditions) attached to it, the *oikodomiki adeia* (building permit), and the existence of extensive *ektos schediou* (out-of-plan) areas, where building can take place under a variety of conditions. Widespead *afthaireti domisi* (unauthorized building) restricts the scope of official town planning.
2. In the mind of the average citizen owning a piece of real estate, the key questions associated with spatial planning are:
   - how to build an "*afthaireto*" (unauthorized building), on his/her *ektos schediou* land parcel;
   - how to get into the *schedio poleos*, if his/her land is still *ektos schediou*;
   - what the building conditions will be when this happens and how the street line (*rymotomiki grammi*) will be fixed;
   - what the conditions are for his/her *entos schediou* (within-the-plan) plot, especially building height, plot ratio and floor-area ratio (*syntelestis domisis*), and how they can be improved.

These paragraphs had been written before important spatial planning legislation was voted in the late 1990s or a series of regional and urban plans were approved a few years later. In fact, the report was based on information collected in 1994. Nevertheless, reality at the level of the individual land property and of the average owner has not changed in practice. What is impressive is the way in which, given a reality which is inimical to integrated spatial planning, the latter came to be seen by one government after the other as a "key instrument for development and effective governance", as Stead and Nadin (2008) describe it, somewhat optimistically. A large part of this chapter is dedicated to explaining this contradiction.

The chief purpose of this chapter is to provide an outline of the Greek planning system, which is used in this book as a case study. This outline is based on prior work of the author and on a large number of publications, the great majority of which was written in Greek. To avoid saturation of the text with endless references

[1] In effect, a street-layout plan.

and to make reading more comfortable the key works of either the author[2] or other writers[3] are mentioned in the list of references.

8.1 The Legacy of the Past and Its Impact on Planning

This chapter would be hard to follow if it had not been preceded by Chap. 6 on the history of modern Greece. Planning can only be understood in a historical perspective. "History is important to urban planning practitioners in several distinct ways … More than is generally recognized, a professional is a historian, making almost daily use of past cultural knowledge. The fact that this may be done tacitly, carelessly, or offhandedly makes it no less important. History is also vital cultural knowledge in the large sense of grasping the social forces and movements which

[2]The main works of the author in English are: Wassenhoven L 1984, 1997a, 2000a, b, 2008a, 2014a, b, 2018a, and 2019. His most comprehensive works in Greek are: Wassenhoven L 1998b, 2007, 2009, 2010a, 2017a, 2018b, and 2022. For other works, without or with co-authors, see the list of references.

[3]The main Greek writers and organizations and agencies whose publications and/or reports the author consulted for this chapter are (in alphabetical order): Andrikopoulou, Angelidis, Aravantinos, Avgerinou-Kolonias, Beriatos, Christofilopoulos, Coccossis, CPER (Centre of Planning and Economic Research), Delladetsima, Demathas, Economou, European Commission, Gerardi, Getimis, Giannakourou, Giaoutzi, Hadjimichalis, D. Katochianos, N. Katochianos, Komilis, Komnenos, Kourliouros, Kyvelou, Lagopoulos, Leontidou, Loukakis, Melissas, Menoudakos, NTUA (National Technical University of Athens). Panayotatos, Petrakos, Psycharis, Sapountzaki, Sarigiannis, Serraos, TEE (Technical Chamber of Greece), Theodora, Thoidou, Tsartas, Voivonda et al, WWF Hellas, and YPECHODE (Ministry of Spatial Planning, the Environment, and Public Works). Andrikopoulou (1992, 1993, 1994, 1995, 2000, 2003 2004), Andrikopoulou and Kafkalas (1985, 2004), Angelidis (1985, 1991, 1997, 2002, 2004a, 2004b), Angelidis and Karka (2001), Aravantinos (1968, 1970b, 1973), Aravantinos (2001), Aravantinos and Gerardi (2007), Avgerinou-Kolonias (1998, 2001, 2011), Avgerinou-Kolonias and Melissas (2011), Bainbridge (1998), Beriatos (1985, 1993, 1994a, 1995a, 1995b, 1998, 2000, 2001, 2004, 2010), Beriatos and Papageorgiou (2010), CEC (1995), Christofilopoulos (1979, 1988, 2000, 2002), Coccossis (1994a, 1994b, 2001, 2009, 2017), Coccossis (2002), Coccossis and Parpairis (1993, 1995), Coccossis and Psycharis (2008), Coccossis and Tsartas (2001), CPER (1965, 1967, 1972a, 1972b, 1972c, 1976a, 1980b, 1980c, 1989b, 1990, 1991), CPER (1970), Cullingworth and Nadin (2006), Davoudi et al. (2020), Deffner et al (2000), Delladetsima (1991, 1994a, 1994b, 2000), Delladetsima and Loukakis (2013), Demathas (1995a, 2011), Demitriadis and Yerolympos-Karadimos (1989), Dunford and Kafkalas (1992), Economou (1995, 2000d, 2002a, 2002b, 2004b, 2009b, 2010, 2017), Economou and Papamichos (2003), Economou and Petrakos (2012a, 2012b), Economou et al. (2001), EKPAA (National Centre for the Environment and Sustainable Development) (2020), European Commission (1995, 2000a, 2001a, 2001b, 2004, 2005, 2006, 2008, 2010a, 2010b, 2015, 2017, 2022), European Community (1993), European Union (1997), Fabietti (1997), Georgoulis (1995), Gerardi (2002), Gerardi and Gialyri (1998), Getimis (1985, 1989, 1990, 1993, 1994), Getimis and Economou (1994), Getimis and Gravaris (1993), Getimis, et al. (1994), Getimis and Kafkalas (2001), Getimis and Kafkalas (2003a, 2003b), Giannakourou (1994a, 2003, 2004, 2008c, 2009), Giannakourou et al. (2008), Giaoutzi (1984, 1988), Giaoutzi and Nijkamp (1993), Giaoutzi and

influence, shape, and guide knowledge and ideas in general" (Bolan 2017, 109). If historical knowledge is vital for the practitioner, it is a hundred times more essential for the planning theorist! Besides, it is a central position of Southern theory and of the view from the Global South (see Chap. 4) that in order to formulate an alternative theory a historical perspective is essential. It is not only information about the physical and human-made environment that history provides. It is also knowledge on the evolution of institutions that is critical. Planning institutions, processes, and practices are introduced or set aside in the course of history and, thus, the form of the planning system at any given moment in time can only be appreciated or theorized about on the basis of past developments. Sorensen (2018, 260) is not writing about Greece, but his comment is quite relevant: "In perspective, urban planning systems can be understood as multifaceted sets of institutions that regulate the production, use, and change of urban space. These sets of institutions developed over extended periods, both during major 'critical junctures' of institutional innovation such as the development of modern planning in response to the nineteenth century

---

Nijkamp (2008, 2017), Giaoutzi and Sapio (2013), Giaoutzi et al. (1988), Governa et al. (2009), Hadjimichalis (1987, 2001, 2018), Hadjimichalis (1992, 2010), Haselsberger (2017), Kafkalas (1981, 1983, 1984, 2001, 2004a), Kafkalas (2004a), Kafkalas (2004b), Kafkalas and Andrikopoulou (2000), Kafkalas and Komnenos (1993), Kafkalas and Pitsiava (2013), Kafkalas et al. (1983), Kalokardou (1988, 1995, 2011), Katochianos D (1969, 1970, 1992, 1994, 1995), Katochianos N (1970, 1973, 1985), Katochianos D and Katochianos N (1967), Knieling and Othengrafen (2016), Komilis (1986, 1987, 1992, 1995, 2001), Komnenos (1993b, 2007), Kourliouros (1989), Koutsopoulos (1995), Koutsopoulos (n.d.), Koutsopoulos and Nijkamp (1984), Kyvelou (2001, 2003, 2016), Kyvelou and Gourgiotis (2019), Kyvelou and Papadaki (2016), Lagopoulos (1981a, 1981b, 1992, 1994, 2011), Leontidou (1990, 1996, 2011, 2020), Leontidou-Emmanuel (1982), Leontidou-Gerardi (1977), Loukakis (1994, 2004, 2010, 2017), Melissas (2010), Melissas (2010a, 2010b, 2015, 2018, 2019, 2021), Melissas and Serraos (2017), Menoudakos (2017a, 2017b, 2018), NTUA (1973, 1986, 1990, 2004, 2011), NTUA/SPE (1969, 1971a), OECD (1997, 2001b), Panayotatos (1982, 1983, 1984, 1988, 1989), Panteion University (2011), Petrakos (2000), Petrakos (2000), Petrakos and Artelaris (2008), Petrakos and Psycharis (2003), Psomopoulos (1991), Psycharis (1990, 2000), Psycharis (1990), Psycharis et al. (2000), SADAS (2002), Sakellaropoulou (2018a, 2018b), Sakellaropoulou et al. (2018), Sapountzaki (2001), Sarigiannis (1988, 2000, 2009, 2017a), SEP (1995a), Serraos (2018), Serraos and Greve (2016), Serraos and Melissas (2018, 2019), Serraos et al. (2016), Sofoulis (1967, 1979), Spilanis (1995, 2000, 2012), Stamatelatos and Vamva-Stamatelatou (2006), TEE (1978, 1980, 1985, 1995), Theodora (2008, 2011, 2018a, b), Theodora and Loukakis (2011), Thoidou (2004, 2010, 2011), Topalov et al. (2010), Tsartas (2000, 2001), Tsartas (2000), University of Thessaly (2009), Vaiou and Mantouvalou (2000), Vliamos (1988), Vliamos et al. (1991), Voivonda et al. (1977a, b), Wassenhoven (1994c, 1995a, 1995e, 1996b, 1997b, 1997c, 1998a, 2002a, 2002b, 2003a, 2004d, 2004e, 2008c, 2010b, 2016, 2018d, 2019, 2021a), Wassenhoven and Georgoulis (1997), Wassenhoven and Kourliouros (2007), Wassenhoven and Sapountzaki (2009), Wassenhoven et al. (2004), WWF Hellas (2019b), Wynn (1984), YCHOP (1984b), YPECHODE (1985, 1992, 2001a, 2001b, 2009a, 2009b), YPEN (2017-2020), YPETHO (1983, 1985), Zampelis (2001a, 2001b) and Zeikou (2011).

[4] The ninety-page article by Demitrakopoulos is a most valuable document on the history of town planning in Greece, from independence up to the eve of the Second World War. He mentions all the town plans produced in this period and lists all the failures of effective implementation and the inability of the government to capture the surplus value created by its own planning initiatives.

urban crisis, and through longer periods of incremental and evolutionary change. Because institutions are unequal in their distributive impacts, providing greater benefits to some actors than others, institutional change is contested, and those who benefit from particular institutions can be expected to try to prevent changes that weaken their position, while those who benefit less are more likely to press for change .... Urban planning institutions are the product and legacy of recurring political conflict. The exercise of power at contingent points of institutional change is important, but so are ongoing contests over the implementation and enforcement of rules on the ground". Greek planning is a reflection of a troubled history, which, nevertheless, did not prevent it to absorb influences coming from the West, from neoclassical city design in the 1830s to the acceptance of "spatial" planning as a catchword in the late twentieth century. As Davoudi (2018, 16) wrote, "by the end of the 20th century spatial planning became *à la mode* ... However, the extent to which its ideals and progressive intents managed to change planning practices is questionable. Indeed, in many parts of the world (including some countries in Europe) master planning remained the dominant mode".

Whether Greece is one of the countries that Davoudi has in mind is an open question. The Greek planning system has been repeatedly classified in a Mediterranean category of "urbanism" (Newman and Thornley 1996; European Commission 1997a). But of course, so has the Italian planning system, a classification to which Palermo and Ponzini (2010, 25) reacted when they made it their aim "to provide a less distorted image of a field of theory and practice that has been widely misunderstood or underestimated by the neo-orthodox planning schools". They consider the current theoretical literature, in particular the Anglo-Saxon, "superficial" with regard to alternative models like the Italian one. They add further that "according to widespread opinion in international planning, the Italian case was a variant – and not a particularly significant one – of a presumed Mediterranean model that was still too conditioned by traditional problems of physical planning generally handled with over-simplified techniques and excessively rigid solutions. This classification was portrayed in the European Union's comparative studies on spatial planning systems ... In truth, it was a superficial representation that avoided the problems of rigorous definition and testing. We might doubt whether the planning systems in force in Mediterranean countries fully belong to a common ideal type, but what appears clear is the influential role of a set of issues and features that seem less important in Northern Europe or US systems: the traditional pre-eminence of physical planning over functions of spatial management; the regulatory rather than strategic nature of the main planning tools; command and control action by public authorities rather than coordination and partnership" (op.cit., 111). A number of these characteristics can be found in the Greek planning system, and they have historical roots.

Ideologies are also impregnated by historical events and become a background against which planning materializes as a state practice and is often resisted by society. A typical reaction is caused by what is seen as infringement of individual liberties. As in the case of Sorensen, authors like Davoudi, Galland and Stead (2020, 21) do not write about Greece, but their comments on ideology remain useful: "A salient example of how ideologies work is the creation of urban and regional planning. The formation of planning was legitimized by ideological decontestation of some core concepts. Among them is the repositioning of the concept of liberty in relation to concepts of private property and the state. While today we may take planning intervention in property markets for granted, for the classical liberals of the late nineteenth and early twentieth centuries it was seen as an ideologically unacceptable erosion of individual liberties". It was equally unacceptable for the small property holder of traditional nineteenth, even twentieth, century Greece.

8.2 First Steps of Town Planning

For a romantic town planner, the story of urban planning in Greece starts with the plan of Athens after independence in 1832. Two architects, the Greek Kleanthis and the Prussian Schaubert were commissioned to prepare a plan. This was planning "from above and from nought", as Monioudi-Gavala (2015, 105–109) put it. Athens was not the only case, since plans were drawn for other towns as well, and a town planning policy was officially legislated. It is symbolical that a plan was prepared, also by a foreign architect, the Bavarian Staufert, for the town of Sparta, the ancient arch-rival of Athens. For centuries, Athens had grown without a plan, in fact enclosed in a wall in the last phase of Ottoman rule. The new plan was approved by royal decree in 1833 and extended the city area in at least three directions. A second plan for Athens was commissioned a year later to Von Klenze, yet another German architect, but the implementation of these plans was hampered by private interests of property owners, often speculators, Greek and foreign, who hastened to acquire land. In essence, this was largely a new town: "The new city envisaged in the … plan partly overlapped with the existing town" (Karidis 2014, 104). Tsiomis (1986, 20) places it in the same class as Washington or Brasilia, a view which is a rather daring time-shift (see also: Demitrakopoulos 1934–1937, 417–429; Biris 1995, 19–39; Great Greek Encyclopaedia 1928, v.II, 218–221; Wassenhoven 1984, 8–10; Markezinis 1966, v.1, 126–128). The activity of the Bavarian administration embraced legislation, first with a decree in 1833 regulating building activity and, two years later, with a second decree on the hygiene of towns and townships, which laid down rules of town planning (Melissas 2012). This was pioneering work, which, however, was challenged by sharp criticism that the Bavarian plan of Athens disfigured the traditional town pattern and that plan implementation was authoritarian and violent. Judged by popular reactions it is obvious that a typical social dualism dominated. This did not prevent Athens and its buildings to become a prototype to be emulated (Monioudi-Gavala 2015, 109 and 123). Since then, the Athenian model retained its attraction. These initial experiments should not give the impression that Greek, in fact mostly foreign, town planning was on a modernizing path in the course of the nineteenth century (Monioudi-Gavala 2012). In rural Greece there was a sporadic and small-scale population movement from the mountains to the plains, where, occasionally, new settlements were formed under the supervision of architects of the Ministry of the Interior. It was centrally planned, but not systematic, action (Kafkoula 1990, 79–81). As mentioned in Chap. 6, Greece remained territorially crippled until the Balkan Wars of 1912–1913. Towns, particularly Athens, continued to grow by small accretions, either illegally or by gradual and uncoordinated extensions of their official town plans. Demitrakopoulos (1934–1937, 423) mentions 150 piecemeal extensions.[4] Eventually, it would take violent events to trigger off major planning reforms. These events included international and civil wars, social schisms, territorial expansion, natural disasters, migration waves, and destruction.

When Greeks, Serbs, Montenegrins, and Bulgars defeated the Ottomans in the first Balkan War of 1912 and then Greeks, Serbs, Montenegrins, even Ottomans and Romanians, defeated the Bulgars in the second Balkan War of 1913, conditions in the newly annexed territories changed rapidly, and so did the prospects of planning. The inevitable movements of populations, which in fact had started even before 1910, made necessary the creation of new settlements for which two planning laws were enacted in 1914 and land expropriation followed, when the minister of transport Alexandros Papanastasiou, with a portfolio which included planning, pressed ahead with a series of groundbreaking initiatives. He was influenced by German socialists and English Fabians and was aware of the Garden Cities movement. It was during his period in office that dozens of plans were designed for Macedonia. Apart from Greek architects and engineers (e.g., Anargyros Demitrakopoulos and Marinos Delladetsimas), Papanastasiou hired the services of several foreign ones, English (e.g., Thomas and John Mawson), French (e.g., Ernest Hébrard, who later worked in Indochina), and possibly of Russian origin (Kafkoula 1990, 83–126; Demitrakopoulos 1934–1937).[5] A large number of "specialist intellectuals" were recruited in those days, including architects, engineers, town planners, surveyors, and agriculturalists. It would not be an exaggeration to claim that it was then that the planning profession emerged in Greece. Foreign-trained Greek town planners worked not only in Northern Greece but also in Athens, like Kostas Sgoutas, for the settlement of refugees (Sotiropoulos 2019, 72–88). In addition, new planning instruments were imported, to be used in agrarian reform and settlement planning, e.g., land adjustment plans and land-property co-ownerships, and were incorporated in national planning legislation. Ironically, in the interwar years, the English prototype of the Garden City was of great influence also in the design of garden suburbs for the affluent classes.

---

[5] Symeon Dolgopolov who was a member of the Papanastasiou team of architects, may have been the same person who around 1950 designed an allotment plan for a large estate in the vicinity of Athens to create parcels for sale. It was in the 1950s that the author had the opportunity to make his acquaintance.

[6] A short review of planning initiatives is provided by Melissas (2002, 37–46) and Sarigiannis (1980 and 1995).

[7] The decree was signed on July 17th, 1923 and published in the Government Gazette on August 16th of the same year. An interesting question is the identity of those who actually drafted the text of the 1923 decree. A reasonable guess is that two persons played a significant role, both active in

Text-Box 8.1: Garden Suburbs

The garden suburbs of Athens, a distant relative of the English equivalent which was certainly their source of inspiration, are the flip side of refugee settlement planning in the interwar period. In sharp contrast to the refugee settlements, their residents were the affluent society of the Greek capital. Even the town design of the suburbs betrays sophistication and a desire for social distinction. The suburbs are found both in the north of Athens (e..g., Psychiko, Filothei, or Ekali) and in the south (e.g., Voula). The most interesting example is the adjacent suburbs of Psychiko and Filothei, the history of which is to some extent common. They were the outcome of a combination of initiatives by a group of Greeks of the diaspora in Egypt, an association of bank employees, private banks, and a private real estate firm which also owned nearby stone quarries. Psychiko started with a transitional town plan in 1907, but its final structure followed a more elaborate plan of 1923. The real estate firm which acquired a very large chunk of the area sold plots to individual purchasers who were obliged to build a house within a prescribed period and respect the firm's specifications. Even today the suburb is an exclusively residential area with a mixture of villas and blocks of flats, but with no town centre facilities with the exception of two very small markets. Fairly recently, embassies were allowed and this has turned the suburb into an ambassadorial precinct. Meticulous transportation planning with one-way streets has effectively blocked through-traffic. Recently, Psychiko and Filothei were amalgamated into a single municipality (see Fig. 8.1).

The Balkan Wars caused the first phase of an exchange of populations which was to acquire dramatic dimensions after the 1922 Asia Minor disaster which was presented extensively in Chap. 6. By tragic coincidence, it was in 1917 that a great fire ravaged the city of Thessaloniki and a major planning and rebuilding effort was implemented, based on a plan by Hébrard (Yerolympos 1996, 99–105 and Yerolympos 2008). Coming as it did only a few years after the city's liberation the symbolism of the task was evident, from a planning, architectural, and political viewpoint (Lagopoulos 2017, 154–182).

8.3 Ground-Breaking Town Planning Legislation in the Interwar Period

Apart from town planning activity to settle the refugees, already dealt with in Chap. 6, the interwar years are those of major legislative initiatives, obviously related to the refugee problem. These issues aside, it could be argued that in this period urban development consists of some abortive attempts to produce a master plan of Athens, a continued incremental and haphazard process of statutory plan extensions, and illegal building activity, noticeable even long before the arrival of refugees. Regional

On the left Voula; in the middle the first plan of Psychiko; on the right an aerial photo of Psychiko in 1932

On the left the present plan of Psyshiko. On the right the plan of Filothei.

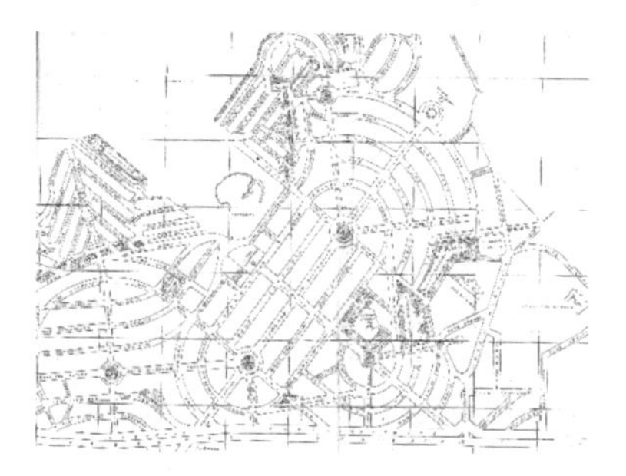

Fig. 8.1 Plans and aerial photographs of garden suburbs of Athens (see Text-Box 8.1)

spatial planning was not yet on the agenda.[6] Town planning, however, was endowed with a pathbreaking act, of which several articles still remain in force. As Leontidou remarked in 1989, the statutory town planning and urban development in Greece took their form in the interwar period (Leontidou 2001, 203).

The cutting point was the Legislative Decree of 17 July 1923 concerning the plans of towns, townships and housing estates and their building regulations (Kampourakis 1967, 3–130; Stingas 2004, 295–330; Raptis 1940).[7] The decree was based on similar French, German and Swiss legislation (Melissas 2012, 1109; Karadimou-Yerolympos 2012). The 1923 decree has been rightly described as the act with which Greek planning law came of age (Giannakourou 2019, 61). Although it did not address the regional planning dimension and focused on the internal structure of settlements, it is of importance to emphasize some elements which in due

the Papanastasiou planning team. The first was Hébrard who certainly brought with him the influence of French legislation. The second was Demitrakopoulos, a civil engineer and later government minister and professor of town planning at the National Technical University of Athens (NTUA), who had contacts with Germany, and was no doubt aware of German legislation. Apparently, he actually wrote the proposal for the new decree. Hébrard too taught for a period at NTUA (Karadimou-Yerolympos 2012).

[8] Website of the Ministry of the Environment and Energy (accessed on 1 Aug. 2021) https://ypen.gov.gr/chorikos-schediasmos/poleodomia/

[9] Krispis used the term *choronomic* instead of *chorotaxikos*, preferred by Doxiadis, who was literally translating the German term Raumordnung.

[10] N. Vatopoulos, The decade of 1960. Newspaper Kathimerini, 2 November 2019.

course were destined to become significant for wider planning, like the building of new settlements, second home zones, industrial areas, the peri-urban zone which surrounds towns with an approved plan, and the coastal areas. The decree's stipulations for the regulation of town extensions is also worth stressing, because it introduced a concern for the peri-urban zone (Papageorgiou 1993, 146). Sadly, in practice, town planning continued to be exercised with an ad hoc production of uncoordinated street layout plans (Voivonda et al. 1977, 137). It was for this reason that the 1923 decree was blamed for adopting an old-fashioned, narrow and static approach (Lagopoulos 1984, 138). Of interest is the official view of the 1923 decree, as stated in the site of the Ministry of the Environment: "A landmark year for urban planning was 1923, when the well-known legislative decree of 16.8.1923 'On City Plans' was adopted: Article 1 of the above decree includes the essence of urban planning, i.e., that the city must be developed on the basis of a plan produced under the responsibility of the State. The principle of urban planning has as its main complement the establishment of rules for the rational preparation of urban planning plans, for categories of land use, for the creation of public and communal spaces, etc. All the above elements constitute the 'normal' urban law of the country, which is proportional to European standards. However, the above 'system' of urban planning does not work effectively, as a parallel system of individual layouts, tools and regulations has gradually been created, on the basis of which the content of the design is adulterated or even overturned. This second part of urban law is held responsible for the labyrinthine, ineffective and opaque nature of urban planning legislation with very adverse effects on the area of cities and settlements".[8] The reference to a "parallel system" is absolutely justified, but this was the standard official approach to urban problems. Furthermore, the "labyrinthine, ineffective and opaque nature of urban planning legislation" was also the product of the approach of successive governments, who cared more for the satisfaction of their voters in a typical compromise and clientelist mentality.

The decree contained important provisions. The most important was that the approval of a plan for a new settlement must be preceded by a reconnaissance survey to determine the need for such an action. If the new settlement is planned at the initiative of a group of real estate owners or developers, then the official approval will be granted on conditions imposed on properties and buildings. In peripheral zones, development is permitted under special conditions, although land cannot be parcelled below a specified limit. Special provisions apply in coastal zones. As a result of the decree, land was effectively divided in three categories, as explained by Giannakourou (2012, 460–461). The first was land within-the-plan, regulated by a statutory, street layout plan, i.e., the *schedio poleos* which was mentioned at the beginning of the chapter. The second included settlements existing before 1923, which enjoyed certain privileges of ad hoc planning. The third category included all land not belonging to the first and second categories. These were the infamous

[11] The recent interest in the prospects of a provincial centre, in this case Alexandroupolis in Thrace, is an instructive lesson. Its growth seems dependent not on domestic spatial planning, but rather on

out-of-plan areas (*ektos schediou*) where building was not entirely forbidden, but limitations were imposed which aimed especially at preventing the construction of private settlements. However, building was permitted along approved roads and railway lines. It is worth pointing out that the problem of pre-1923 settlements is still bothering the public administration almost a century later, as it becomes evident in its handling of forest maps. The reason is that the land occupied by these settlements is regularly considered forest land by the forestry authorities. A new attempt to solve the problem of these enclaves, when their borders have not been properly fixed, was made in 2020 and met by fierce reactions.

As explained in Text-Box 8.2, much more serious than the case of pre-1923 settlements, is a problem which also has its roots in the 1923 decree, that of out-of-plan areas. The decree stipulated that ad hoc building conditions can be used in such areas. This was the birth certificate, as written a long time ago (Wassenhoven 1995b, 21) of a monster which has destroyed the Greek countryside. The loophole introduced in 1923 was widened by a 1928 decree, which allowed building on plots facing national, provincial, or municipal roads in out-of-plan zones. Subsequent amendments of the law made these provisions a real scourge. As if this was not enough, the 1928 decree allowed a variety of special purpose buildings to be erected in the peri-urban zones, to which residential homes were added in 1952, initially for rural dwellings, and later for a generalized class of residences. Gradually, the notorious out-of-plan construction and the associated illegal activity became a rule, in spite of its unconstitutionality (Rozos 2017), and remains a cause of vicious disagreement. It can be claimed that it is the reflection of patronage relations between the state and citizens. We shall have to return to this problem in a subsequent section.

Text-Box 8.2: Out-of-Plan Areas

Out-of-plan areas, i.e., land without a statutory local plan, are the hallmark of Greek space, as explained in the main text of the book. Out-of-plan land is everywhere: in suburban zones ripe for development, in agricultural areas, in forest zones, or in bushland. The absence of a local plan does not mean that development and construction are totally unregulated, although this is where unauthorized and unlicensed construction is most conspicuous. Regulation is provided in special legislation, mainly two presidential decrees of 1978 and 1985, repeatedly amended. The decrees, applicable in the entire national territory, allow the full range of land uses, with planning and building conditions attached to each use. Derogations are frequently in force making possible several ad hoc exceptions and turning the whole edifice of out-of-plan activity into an embarrassing politically sensitive issue. An attempt to place a ceiling on derogations even in environmentally protected zones cost the portfolio of an environment minister in the early 2010s.

(continued)

Text-Box 8.2 (continued)

In out-of-plan areas the critical considerations for a private piece of real estate narrow down to the holy words "whole and buildable". If a plot has these two attributes its value and the happiness of the owner rise accordingly. Out-of-plan construction tends to turn vast suburban zones around Athens and elsewhere into informal towns of which the construction density does nor substantially differ from legal suburbs. Suburbanization and sprawl are not of course an exclusive feature of the Greek capital. As they were once a regular occurrence in Western Europe, they still are in Eastern Europe (Slaev et al. 2018). The main use is residential, permanent or for vacation. Second homes end up regularly as permanent ones. Haphazard and speculative building activity are the order of the day and accessibility is problematic. Access to individual plots is usually provided either through municipal roads of dubious legality, as ruled by the Council of State, or through rights of way agreed between contracting parties when a plot is purchased. The implications for servicing such residential areas with essential infrastructures are serious. Escaping in case of a fire emergency or the access of a fire brigade engine can be a problem, as proved in recent lethal fires near Athens. Yet, out-of-plan land use and building activity remains a sacrosanct talisman (see Fig. 8.2).

Fig. 8.2 Eastern Attica: An out-of-plan area (see Text-Box 8.2)
Source: Google Earth

8.4 The Unfinished Decade of the 1960s and the Emergence of Regional Planning

The impressive economic development of the late 1950s and early 1960s and out-of-plan urbanization account for the intensive house-building construction. In spite of warnings (Papagiannis 1968), the environment was still out of the agenda. It is well known that the construction sector has acted as an instrument of development and mobilization of family savings, as indeed it has happened in many developing countries (Wassenhoven 1980, 166–171). As Filias (1980, 134) has remarked, the savings of internal migrants and emigrants were invested in urban land and apartments and contributed to the heating of the economy, depriving the creation of new productive and employment opportunities of precious resources which would keep in the country young people, who had no alternative but to emigrate. According to an official report on housing (CPER 1976c, 44–45, 53, 60), the housing sector absorbed almost one third of gross fixed capital investment and was used systematically by the government to re-heat the economy, regardless of long-term negative repercussions for the economy and spatial planning. More than 40% of building permits, about the same percentage of the number of housing loans, and 60% of the value of loans were concentrated in the greater areas of Athens and Thessaloniki. Investment in land and buildings dominated the economy, as emphasized in a variety of studies with a radical perspective (Voivonda et al. 1977; Bouzenberg 1967; Tritsis 1967; Kloutsinioti 1981), but the social aspects of housing, with the exception of refugee settlements, were on the whole ignored (Wassenhoven 1976), or limited to Marxist critiques based on the work of Engels!

Various town planning studies were produced in the mid-1960s, as a result of the priority attached by the then centrist government to planning in general (Voivonda et al. 1977; Vagianos 1968; Wassenhoven 1984: 21–24; Kontaratos 1981; AT Journal 1967, 1968, 1977, 1981). Worth mentioning, in the 1960s, is the increasing activity of the planning department of the then ministry of public works and the production of a master plan for Athens (Katochianos et al. 1967), which was resumed at the end of the 1970s, not so much in the form of a plan, but rather as a policy document with declared priorities of "common consensus" (Leontidou-Emmanuel 1981, 71; see also Mitaraki-Bazou and Basoukea 2001, Michail 1994, and Sapounaki-Drakaki and Stamatiou 2000). Academic research activity led to the production of spatial plans, particularly referring to the regional scale, and to related publications, e.g., by Antonis Kriesis (1963), who created the first university research centre with a focus on town planning.

The urban plans of large and middle-sized centres (Herakleion, Kavala, Chania, Patras, Ioannina, Preveza etc) were a notable achievement of the period, in spite of their doubtful influence in practice. They represented an attempt to relate the towns with their catchment areas, the importance of which had been stressed in a 1969 article (Wassenhoven 1969a; Lefantzis 2000). At the regional level, spatial planning activity was negligible and limited either to studies which were in fact about economic development or university research projects (see Text-Box 9.3). Planning methodology at this level was a matter of academic interest (Wassenhoven 1969b), but it is significant that a professional and pioneering spatial regional plan was produced in 1965 for the prefecture of Elia in the Peloponnese (Aravantinos et al. 1967a, b). At about the same time a series of plans were produced for tourism development in Crete, following an earlier economic development plan for the island (AT Journal 1968). It is worth mentioning that a document outlining a quasi-spatial national plan, described as *choronomic* plan, had appeared in 1945 (Krispis 1944–1945).[9] It was an ancestor of regional spatial planning which had to wait for another thirty years. The emerging planning activity of the 1960s came to an abrupt end with the military coup of 1967. The decade remained unfinished. Problems of regional inequalities, growth poles, planning for Athens and Thessaloniki, spatial distribution of productive activity, environmental protection, and conflicting national and regional priorities remained open (MPW 1973). To a large extent, they were in conflict with each other and still beyond the active planning capacity of the state (Andrikopoulou 1990).

8.5 Athens

There are two reasons why this chapter contains a section on Athens. The first is that the capital was the main stage of a large number of urban planning initiatives in the course of the history of modern Greece. The second is that its dominant position in the country was the key cause of regional planning activity and of endless professional and academic arguments and studies (Maloutas and Economou 1992). The debate about Athens concentrated both on its "formal" structure and on its "informal" sector. Already in the days of the refugee settlement, out-of-plan construction was evident in Athens too, thus, according to J. Papaioannou (1971, 263), violating elementary building rules, which, as he put it, were so complex as to end up simplistic. The "official", within-the-plan, Athens was changing rapidly. Legislation allowing separate horizontal properties opened the way for the construction of condominium apartment buildings, which in later years became the capital's hallmark and a model imitated in other towns (Marmaras 1989, 1991, 24). In the interwar period, under the guidance of minister Papanastasiou, a great number of foreign and Greek architects and engineers was mobilized to prepare a master plan of Athens (Biris 1995; Papageorgiou-Venetas 2010; Marmaras 2002, 2012; Karidis 2008; Ipsen et al. 2007; Burgel 1976; Sarigiannis 2017b). They included the German Ludwig Hoffmann, the British Thomas Mawson, and, above all the French Ernest Hébrard. These first attempts were criticized for being out of touch with Greek realities (Yerolympos 2008), a criticism generally familiar in the Global South, as outlined in Chap. 4. Hébrard's plan was a relatively serious attempt, but it too

international factors, i.e., a US military base and the development of a node on an oil transportation route.

encountered reactions which ruined the prospects of implementation, particularly in the conditions of the mounting economic crisis of the 1930s. As Yerolympos explains, public opinion played a key role. Enlightened views clashed with the interests of land owners who considered speculation as an inalienable right. In these circumstances, only minor improvements were feasible and important innovations remained a dream. Biris, who had been a critic of previous efforts, praised Hébrard's plan for its solid values and its brave aims to control land development for the benefit of the community. The fact that some of Hébrard's minor proposals were approved was ultimately of no consequence.

Much later, as mentioned already, a new master plan for Athens was produced in the mid-1960s, just before the military dictatorship (Mitaraki-Bazou and Basoukea 2001). A plan was also prepared for Thessaloniki, the second largest centre of the country (Papamichos and Hastaoglou 2000; Tsoulouvis 1981). As emphasized earlier, the goal of restraining the growth of Athens was a constant adage (Chiotis 1970, 85). This author had recommended the removal of some administrative functions from Athens to other urban centres (Wassenhoven 1978). In the digital era, such possibilities should be reconsidered. The effort for the planning of Athens, was resumed when the military junta commissioned Doxiadis Associates to prepare a master plan (Psomopoulos 1977, 2019), which inevitably carried with it a political stigma, and then after the 1975 Constitution, when new extensive planning studies were first produced for all the sub-sections of the capital. The new master plan that followed these studies elicited a barrage of critical comments. In official statements, notable for their fear-mongering rhetoric, it was stressed that the capital is growing in size at the rate of a new village per day and a major city, like Patras, per year. Officials and citizens were of course in a state of shock when a pollution cloud appeared over Athens at the end of the 1970s (Citizens' Movement 1991). These "apocalyptic" forecasts were criticized by Leontidou-Emmanuel (1981, 75) as indicative of the obsession of politicians, planners, and journalists with the parasitic role of Athens. This author's comments on the "gigantism" of Athens were mentioned in Chap. 6, but it is true that the spectre of Athens haunts the Greek imagination even today. Finally, master plans for both Athens and Thessaloniki were finalized in 1985, and, in what was a rather unusual exaggeration, were approved by law voted in parliament. Special agencies were set up in both cities to oversee the implementation of the plans. The Athens plan soon became irrelevant. Even the new Athens international airport "Eleftherios Venizelos" at Spata, which replaced the old Hellinikon airport in 2001, had not been foreseen in the master plan. Drastic ad hoc amendments were made in view of the 2004 Olympic Games. However, the master plan guidelines had still to be respected by lower-level local plans of municipalities. It took almost 30 years to revise the master plan and approve a new one (for Athens and the region of Attica) in 2014, again by law. Given that under recent legislation master plans are no longer included in the planning system (see below), the role of the Athens master plan has been taken over by the Regional Spatial Plan of Attica. The master plan of Thessaloniki was incorporated in the Regional Spatial Plan of Central Macedonia. Naturally, Athens never ceased to attract the interest of planning academics and government agencies (Chorianopoulos and Pagonis 2020; NTUA and University of Thessaly 2006; ORSA 1996).

8.6 Regional Development

National and regional spatial plans were introduced in the ill-fated legislation of the late-1970s, but took their final form in 1999. From 1960, five-year economic development plans, at the national level or with a regional focus, which were earlier mentioned in Chap. 6, were the forerunners of spatial planning instruments (Kanellopoulos 1962, 21–22). Their spatial content was initially insignificant, but gradually the spatial dimension became apparent (see, for instance, MinCoord 1960, 1979). It took some time before the environmental dimension was appreciated (Liodakis 1994; Modinos and Efthymiopoulos 1998, 2000). When national and regional spatial plans finally found their place in the corpus of legislation the interaction between the "economic" and the "spatial" elements increased, although it remains, in the view of the present author, inadequate.

The distance between the economic and spatial planning activities became even more obvious when national five-year plans were effectively replaced by the Community Support Frameworks and later the Regional Economic Partnership Agreements of the European Regional Development Fund (MinDevComp 2014; MinEcDev 2019; MinEcDevT 2016; MinEcFin 2007). This was not because the spatial dimension was ignored in economic planning, but rather because the synchronization of economic and spatial planning was always problematic, with spatial planning lagging behind. The same point can be made with regard to transportation planning which was always a separate field of policy (Tselios et al. 2017; Wassenhoven 1999). On the other hand, it is noteworthy that large-scale spatial planning did actually emerge, in spite of the initial difficulties of the reconstruction period of the 1950s and 1960s. The seeds of spatial planning were sown in those years. The crops harvested later became obvious in legislation, in public administration, and academic teaching and research. The distance between what was happening in Western Europe and Greek realities was closing. In the 1960s it was still wide open, to the point that a French geographer, who had done a lot of research work in Greece, made the ambivalent comment that Greece is more the appendix of Europe, than a bridge to Asia (Kayser 1965, 93; see also Wassenhoven 1995c, 97). This is a version of the somewhat haughty question whether Greece is the East of the West or the West of the East (Sotiropoulos 2019, 37). The work done in the 1960s was inspired by a sense of optimism, not limited to planning, a sense acknowledged in retrospect by contemporary commentators.[10]

[12] For the regional policy of the period 1960–1980 see Kalokardou 1980, Carter 1989 and Konsolas 1985. For a more comprehensive review over a longer period, see Giaoutzi and Stratigea

Text-Box 8.3: Development Programmes and Spatial Planning Frameworks

National economic planning, usually involving five-year development plans, was an activity of the Greek state which started in the 1950s, long before the emergence of spatial national and regional planning after the 1975 Constitution. The spatial content of the five-year plans was poor, although regional development was a steady objective. Five-year plans continued until the late 1980s, when their place was taken by the European Community Support Frameworks and later the National Strategic Reference Frameworks of the European Union. The spatial dimension in EU support instruments was important and gradually made them competitive to the purely spatial frameworks which the Greek government started producing after 2000. Spatial frameworks are supposed to be based on national economic policy documents, but, on the other hand, are binding for all public activity. An awkward relationship developed, but the problem with spatial planning was that spatial frameworks always appeared with considerable delay in comparison with EU-funded programmes which followed strict EU programme periods. Naturally, the EU programmes also had the advantage of relying on secure funding sources, which were agreed in advance. Within the government structure, this division of tasks was reflected in the allocation of responsibilities among ministries, i.e., the ministry of the national economy, the ministry of transport, and the ministry of the environment, though their names frequently changed (Arvanitaki 1995).

The balance between these approaches, "economic" and "spatial", has a wider significance, since it underlies the difficulty of choosing priorities. Spatial planning frameworks have to strike a balance between, e.g., environmental protection and the provision of a spatial platform which favours economic growth and investment attraction. This is compromise of a different order of magnitude than the usual run-of-the-mill land use compromises in urban and regional planning. At a time of economic crisis, as the one which started in 2010, this compromise is all the more tantalizing. The challenge of environmental threats, so ubiquitous under climate change, inevitably makes this equilibrium more fragile, but, unlike other planning challenges, it is much less affected by historical and societal features. How planning will adjust in a new, more global, context and how its future form will be theoretically explained are open and provocative questions. The state, caught in the environment-development pincers, will remain a key variable, but it will probably be a different one.

Between 1961 and 1967 new regional development services and a central regional development department were created, but the centralist mentality of the governmental system limited their task (Mavrakis 1970, 1971; CPER 1976b, 1980a; Voivonda et al. 1977). The first regional development plan, for the period 1960–1964, was that of the region of Epirus (Ward 1963, 150–154). An important contribution in the first steps of regional policy was the development plan 1965–1975 of the island of Crete, produced by the Israeli firm of consultants Agridev (AGRIDEV 1965; Mavrakis 1971, 127). Insular space had always been a most problematic zone and the depopulation of several islands had attracted international attention and encouraged research activity (Kolodny 1974; Wassenhoven 2004c). An important initiative of the period was the creation in 1961 of the Centre of Planning and Economic Research (CPER, known as KEPE in Greece) which played a very significant role as the research arm of successive governments (Liodakis 1994; Angelidis 2000; Sarigiannis 1995; Lefeber 1966).

Apart from the insular regions, an equally sensitive zone was always the northern frontier zone which was adjacent to the Balkan countries of Bulgaria, the then Yugoslavia and Albania. Opening up opportunities for Greek entrepreneurs in these countries or attracting tourists to Greek Macedonia and Thrace were goals of great importance. The accession of Bulgaria to the European Union was an additional opportunity (Lamprianidis 2000; Petrakos 2000; Petrakos 1996; Wassenhoven 1995c; SEP 1995b). To a large extent the development of the frontier was seen as dependent on the growth of local urban centres, hence on industry.[11] The instruments which were used repeatedly to check the dominance of the large cities and encourage development in the provinces, especially in the northern frontier zone, was industrial development incentives and the associated tool of industrial estates. Incentives of the early years were not integrated in a cohesive plan (Argyris 1986, 28).[12] Balanced regional development was a leitmotiv of government policies (Andrikopoulou 1990, 311). Development from below was also being advocated (Sidiropoulos and Papadaskalopoulos 1990). It should be remembered that in 1961, the Athens region had about 25% of the country's population, but produced almost 57% of manufacturing products and about 41% of total GDP. Percentages were much higher in the service sector. In the 1970s, the country entered a phase of economic crisis and conditions changed radically (Lamprinidis and Pakos 1990, 175). The regional problem changed character and regional policy faced the need of adjustment (Kafkalas 1990, 305). The policy of industrial estates had started in the early 1960s (Konsolas 1973, 1994; Konsolas and Kyriazopoulos 2011; Mourtsiadis 2012; Kartakis 1970; Meynaud 1966, 294–295; Alexakis and Katrivesis 1992; Klabatsea 2001, 2003) and the first estates were made operational near the cities of Thessaloniki and Volos. A large number of estates followed, often for purely political reasons, but their success remained doubtful at best. One reason was that the individual investor found it more convenient and cheaper to build a factory outside a designated estate, on out-of-plan land. The objective to direct new industrial

2011, ch. 13.

[13] See http://www.oecd.org/fr/regional/oecd-regional-outlook-2016-9789264260245-en.htm

[14] Greek Statistical Authority, Census results http://www.statistics.gr/documents

[15] The author still remembers the complaint lodged by the municipality of Kalamata in the Peloponnese, for the failure of the central ministry to include Kalamata in the list of rival cities.

investments into planned industrial zones and to discourage the scatter of factories in out-of-plan areas is still being pursued, notably in 2020 legislation. The absence of clear technological and locational criteria in industrial development policy was evident from the beginning (Koutsogiannis 1984; see also HBID 1993, HBID – Industrial Estates sa 2004, Panayotatos et al. 1993). As to the policy of incentives in broadly designated zones, it was not in fact a policy of conscious and well-argued locational choice, but rather one of discouragement to locate in central regions (Economou 1983). It was a difficult task given that industry in Greece was still undergoing a crisis which had started in the 1970s (Giannitsis 1983). The designated incentive zones lacked cohesion and the potential of agglomeration economies (NTUA/EChOA 1989). A similar situation obtained in industrial estates. Locational decisions were not supported by an analysis of external economies and diffusion potential. Their justification was weak, with the possible exception of the existence of location-fixed potential for extraction of minerals or for processing local agricultural products (Wassenhoven 1994a). There was no evidence of a clear policy to support polycentric or corridor development. Locational policies suffered from poor preparation or from purely political priorities. In a later assessment it was concluded that the lack of attention to the mutual interaction of land uses led to squandering of resources and environmental degradation (CPER 1977, 53–54). The interest in industrial estates was followed by attempts to develop technopoles, which met with limited success, with the possible exception of technology- and innovation-centred research nuclei in universities (Konsolas 1990; Komnenos 1993a; Kyrgiafini and Sefertzi 2003).

As mentioned in Chap. 6, regional development policy, towards the end of the twentieth century, was under pressure to change because of international realities and European Union priorities (see, e.g., European Union 2008; Plaskovitis 2008). Greece, like other countries, was seeking, and still does, a "creative" and productive role for lagging regions (Clifton et al. 2015). The development of rural areas was another urgent priority (Eliopoulou 2001; Maravegias 2011; Maravegias and Mermingas 2001). The role of local agencies was now considered of prime importance. But, at the beginning of the new millennium, spatial and regional inequalities were as acute as ever and concentration of industrial activities along the Athens-Thessaloniki axis was always a source of concern (Petrakos and Psycharis 2004, 25; OECD 2001a, 66; Maloutas 2000, 21–23; Tsoulouvis 1998, 67–68). The persistence of regional inequalities, regardless of how the regions were defined, was recognized in official documents (MinEcFin 2007, 212). Regional cooperation and complementarity were a necessity (Monastiriotis 2008). Anxiety was particularly acute in relation to the future of sensitive border regions like Thrace (Academy of Athens 1993; Papagiannakis 1994a, b; Wassenhoven 1994b). Regional inequality and population over-concentration were identical concepts in official pronouncements and public opinion. A large number of studies relate inequalities to national demographic decline. The demographic problem has been present in the literature for a very long time (Drettakis 2012; Kotzamanis 2009; Kotzamanis and Androulaki 2000; Tziafetas 1990), but in a recent study it was predicted that the population of the country will fall from 10.9 m. in 2015 to 9.3 m. in 2050 (diaNEOsis 2016, 87).

This fall seems to affect particular regions, while others remain unscathed. The trouble is that official statistical figures often conceal reality. An example is the size of GDP p.c. in 2003 in selected small or larger areal units, as compared to the national figure of € 14,100. In approximate figures, Attica (essentially Athens) stands at € 15,000, the island of Corfu at € 39,000, the South Aegean region at € 16,000, and the former prefectures of Corinthia and Boeotia at € 20,000 and 37,000 respectively (MinEcFin 2007, 230). The problem is that Corinthia and Boeotia are effectively part of the Athens metropolitan region, which means that official figures obscure the real image (Petrakos and Psycharis 2004, 38–39). Towns like Corinth or Chalkis are in reality part of Greater Athens, benefiting from what is referred to as "borrowed" size (Meijers and Burger 2017). In 2013, according to OECD data, 48% of GDP was produced in urban areas.[13] After the economic crisis, between 2009 and 2013, declared income fell dramatically, but the share of Attica and Thessaloniki remained remarkably stable (Psycharis and Pantazis 2016). According to the chairman of the Regions' Union, rich regions grow richer, while intra-regional inequalities never cease to be disturbing (MOD 2018).

As to the last 2011 census results of total population, out of a total of about 10.8 m. inhabitants, Attica had 35.4% and Central Macedonia (including Thessaloniki) 17,4%.[14] In other words, two regions concentrate 52.8% of the total, but to this must be added part of the population of adjoining regional units (former prefectures) which belong to the Athens catchment area. At the same time, the population of regions like Epirus, the Ionian Islands and Thrace is now below prewar levels (Polyzos 2011, 155–156). In explaining the concentration of population, Petrakos (2009, 380–381) refers to geographical, historical, and political factors, of which the latter include the concentration of public administration, the main reason of regional atrophy. He later placed emphasis on the ability of Attica to exploit the advantage of scale economies and on the systematic failure of regional policies to address the problem (Petrakos 2010, 57). Successive population studies have more or less confirmed these conclusions. From very early, the Athens phenomenon had also attracted attention in international comparative studies (Bethemont 2000; Blanc 1965).

8.7 Network of Urban Centres

The "settlement system" or "network of urban centres" (see Fig. 6.1) has been from the start a constant preoccupation of Greek planners and academics. This preoccupation was influenced not only by the model of other European countries (see Text-Box 8.4), but also by the interest, old and recent, in urbanization policies in the

[16] St. Manos, One step ahead and 40 years back. Newspaper Kathimerini, 25 October 2020.

[17] In 1978, the author had the opportunity to work for a year in that company, on a leave of absence from University College London.

developing world (e.g., Renaud 1981; United Nations Centre for Human Settlements 2001; UN Habitat III 2016).

Text-Box 8.4: The Network of Urban Centres
The network of urban centres has been consistently a central variable in Greek national and regional spatial planning and research since the 1960s. It occupies the core of the spatial model as proposed in all supra-urban plans and is given priority in the official guidelines for the preparation of plans. The logic and theory of growth poles, in the François Perroux tradition, is the foundation of this approach. The influence of French literature, starting from the 1950s and 1960s until today, on the *réseau urbain*, the *armature urbaine*, and national spatial planning in general, has been enormous (Coppolani 1959; Rochefort 1960; Hautreux and Rochefort 1964; Mercadal 1965; Guichard 1965; Subra 2014). This is not to underestimate the effect of the work of American scholars (e.g., Berry and Horton 1970) or the enduring interest in the prolific work of geographers like Peter Hall, recorded in countless successive versions (see Hall and Tewdwr-Jones 2011). The underlying belief in Greek planning was of course that the urban network and vigorous regional centres would accelerate regional development and counterbalance the weight of the metropolis of Athens. Policies were announced favouring the so-called rival cities, sometimes bi-polar, i.e., rival to Athens and Thessaloniki (see Fig. 8.4). The expected, or hoped, role of a balanced urban hierarchy was expressed in all economic development programmes and spatial plans. Both contained visions of a national urban hierarchy.

Athens was seen habitually as an organism sucking the life fluids of the national territory. The influence of another seminal French geographical study, namely Gravier's *Paris et le désert français* (Gravier 1947) was, once again, great. Here, however, we must pay attention to a Greek particular trait. Within the national urban network, the antagonistic role of Athens and Thessaloniki is not simply a matter of rival cities. The history of this antagonism goes back to the period of the First World War (see Chap. 6), when the country found itself with two governments, the liberal government of Thessaloniki and the royalist government of Athens. The feelings of resentment to the "State of Athens" are deep-rooted in the North of the country. The two cities and the spatial corridor which links them are the locus of most economic activity, but it is in Athens in particular that investment and productive services are concentrated. It is this historical background that makes the Greek example of concentration in the capital very different from western countries like the United Kingdom or France (see Fig. 8.3).

Many of the Greek urban centres, in addition to Athens and Thessaloniki, were the object of master plans produced by consultants in the 1960s (Argyropoulos 1967; Provelengios 1967; Skiadaresis 1967). For reasons that will be apparent later

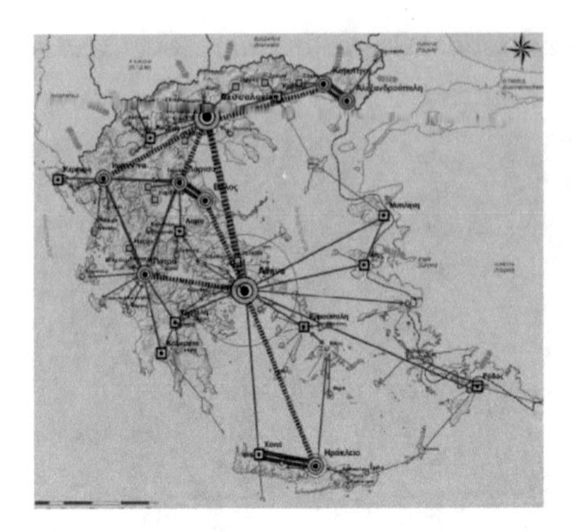

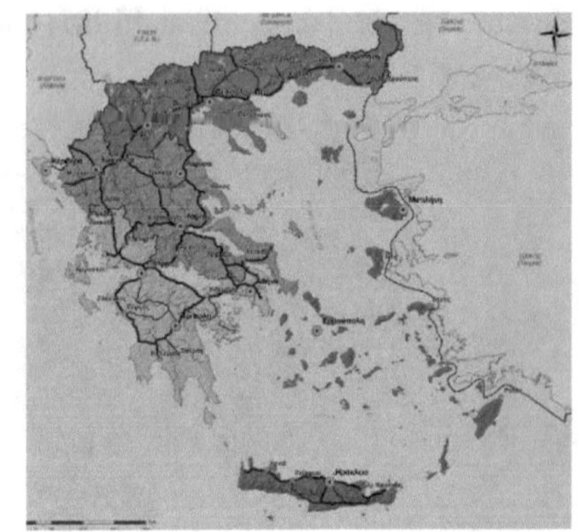

Fig. 8.3 Settlement system (see Text-Box 8.4)
(the network of urban centres in the national spatial planning framework and in the national strategic reference framework)
Sources: Ministries of the Environment and National Economy

I single out the city of Chania in Crete the plan of which was the work of Thalis Argyropoulos and Antonis Tritsis. In these plans, the planners tried to address the problems not only of the city but also of its area of influence. They were not of course regional studies in the sense of that for the Elia sub-region, which was the first professional example in this category (Aravantinos 1967, 1970a; Aravantinos et al. 1967a, b).

Typical of the same decade are the spatial and urban studies for the development of tourism zones in Crete (AT Journal 1968). Of interest is a comment, reminiscent of a point by Brian McLoughlin mentioned in Chap. 2, about the departure of consultants after the submission of their work, who then have no contact with the implementation of their plans (Vagianos 1968, 43). It should be re-emphasized that all these studies were intended to have an impact on regional space, but this impact was not yet appreciated or measured sufficiently by the planning authorities.

In the 1960s, the settlement system was elevated to the status of an independent variable in all discussion of regional spatial planning (Andreadis 1966). The debate centred mainly on the cities of Volos in Thessaly, Patras in the Peloponnese, Chania and Herakleion in Crete, Kavala in Eastern Macedonia, and Preveza in Epirus (on the Ambracian Gulf, or Gulf of Actium), and on the growth axis linking the cities of Patras, Athens, Larisa, Thessaloniki and Kavala (Hoffman 1972, 204–205; Kayser et al. 1971, 222; Ministry of Public Works 1973, 27). The writings of N. and D. Katochianos, researchers of CPER, played an important role. N. Katochianos (1966) proposed a model of "polarized regions" with five levels, of which the poles were Athens (level of the country), 3 metropolitan centres, 8 regional poles, 50 local urban centres, 500 constellations of townships, and 500 head-villages (see also Katochianos 1967, 1970; Tomprogiannis 1967). The model of polarized regions was obviously based on the assumption that urbanization and industrialization proceeded in tandem and on the analysis of growth poles of François Perroux (Economou 1994b, 68–80). The logic of growth poles permeated a good deal of work emanating from the "think tank" of CPER, especially the seminal reports of D. Katochianos (1969, 1970, 1992), concerning the network of urban centres, and E. Kalliga on rural ones (Kalliga 1969; CPER 1968). These reports opened a fruitful dialogue which continued for decades (Wassenhoven 1979b; Lagopoulos 1986; Sarigiannis 1995, 134; Papadaskalopoulos and Christofakis 2011; Anastassiadis and Burgel 2001; Asprogerakas 2003; Katsikas and Lamprianidis 1994; Pantazis 2004).

The importance of city size, which had a long history in international literature (Parkinson et al. 2015), was the background of this dialogue. Invariably, the objective was that towns should grow, because this was a sign of good health. Stable size, let alone, "shrinkage" was never considered as desirable (Sousa and Pinho 2015), although it did happen in practice. An understanding, which was always present in the Greek case was that the absence of networking and the fragmentation of space had historical roots, going back to a time when towns were oriented sometimes to the West, sometimes to the East, a precedent which delayed a genuinely national hierarchy (Kostis 2018, 13). The dialogue about the settlement system also revealed, once again, the gap between a narrowly conceived policy for regional economic development and a spatial perspective. Evidence of this divorce in conceptualizing the two dimensions of development comes from the fact that while regional economic development was discussed, building activity was producing its own geographical landscape. Mobilization of private savings in the building sector, as in many developing countries, was an independent propulsive instrument (Wassenhoven 1980, 166–171; Filias 1980, 134). From an official report (CPER 1976c, 44–60) we learn that about one third of fixed capital formation was accounted for by housing construction and that more than 60% of the value of housing construction loans and over 60% of legally constructed dwellings were concentrated in the two largest urban centres.

The official response, in the late 1970s, to the warped structure of the settlement system was the policy of intensive regional development centres (known as KEPA), described as *antipales poleis* or "rival cities" (see Fig. 8.4), clearly influenced by the French policy of the equilibrium metropolises. The rival cities were to be Kavala, Ioannina, Herakleion, and the twin-poles of Patras-Aigion, Larisa-Volos, and Kozani-Ptolemaida. The real economy was to prove much more influential for the future of these cities than official policy, but the rival cities caused a lot of excitement in academic literature (Wassenhoven 1979a; Leontidou-Gerardi 1981; Panayotatos 1983; Argyris 1986, Sarigiannis 1995; Angelidis 2000; Andrikopoulou 2010; Hadjivasileiou 2015; Papagiannakis 1996).[15] The official policy also included lower-level centres, urban or rural. The model was clearly of foreign inspiration and its implementation in practice was inadequately supported. Its empirical foundation was clearly shaky. Regardless of the rather quixotic policy of the rival cities, the

[18] The author is fully aware that the repeated use of the adjectives "special" and "spatial" in the nomenclature of planning instruments may confuse the reader.

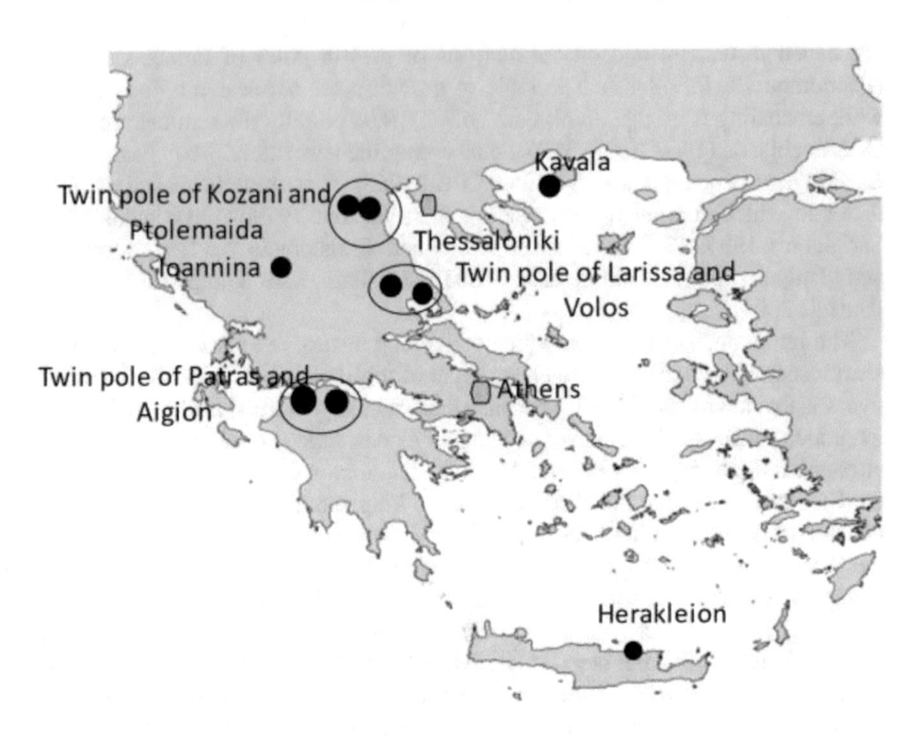

Fig. 8.4 The policy of "rival" cities

urban network remained a central parameter in official policy, spatial planning and academic analysis (Nicolacopoulos and Tsouyopoulos 1976; Angelidis 2000, 25, 125–131; S. Polyzos 2011, 159; Gerardi 1995, 1996 and 2002; Manola 1996).

More recent statistical studies underline that while a reversal of downward population trends is observed in certain secondary centres, these centres are mostly satellites of Athens and Thessaloniki and are located in their catchment areas (Petrakos and Mardakis 2012). In a study of GDP spatial distribution, it was found that when estimates for the informal, "parallel" or "hidden" economy are included the dominance of Attica is shown to "explode" (Michaelidis 2009). It must be remembered that the para-economy and administrative corruption are tightly related phenomena (Vavouras and Manolas 2005). It is clear that this is a factor that spatial planning cannot touch. Policies to fight tax evasion would have a far greater effect on the distribution of income. Nevertheless, the settlement system is still an issue of great interest for spatial planning policy, a reflection of the international interest in the role of "second rank cities" (Camagni and Capello 2015; Camagni et al. 2015). It is significant that the 2020 reform of planning legislation requires the production of a Special Spatial Framework to address the structure of the settlement system on a national scale. A possible approach is to focus attention on the fate of rural townships and their more effective networking (Gousios 2012).

Of exceptional weight within the national urban network and the country's spatial structure is the S-shaped corridor linking Patras, Athens, Larisa, Thessaloniki and Kavala, the development of which is supported by the country's most important motorway (Getimis and Economou 1992; Angelidis 2000 123–124; Skagiannis 1994, 2009; Polyzos 2011, 171). Transportation infrastructures may well have a decentralizing effect, mainly because of the Egnatia motorway linking Igoumenitsa in the northwest with the Turkish border in the northeast and a diagonal motorway which will link the Athens-Thessaloniki line with the Egnatia Motorway and Epirus. According to Skagiannis (2009) the north-south S-shaped corridor may well take the form of a loop resembling the Greek letter σ (sigma). Combined with the improvement of the north-south western corridor linking Epirus and the Peloponnese, this will be a tremendous structural change, due, of course, not to spatial planning in its narrow institutional sense, but rather to transportation policies, closely related to those of the European Union.

8.8 Planning Legislation in the 1970s and 1980s and the Operation of Town Planning Reorganization

Urban planning until the first years immediately following the military junta had remained static (Aneroussi et al. 1977; Greek National Committee for the U.N. Conference "HABITAT" 1975). The government policy for the rival cities, initiated by minister Stefanos Manos, coincided with a major breakthrough in urban and regional planning legislation, with laws enacted in 1976 and 1979. The military junta had issued in 1972 a legislative decree about urban master plans, but it was Law 360/1976, after the dictatorship, that introduced regional spatial planning and Law 947/1979 that dealt with the planning of "ekistic" (settlement) areas, i.e., town planning (Angelidis 2000; Skouris 1991; Stingas 2004; Aneroussi 2004). For the first time, Greece had a complete planning system encompassing both the regional and urban planning components. According to the new law of 1979, owners of land properties brought into the boundaries of a statutory plan had to contribute either a share of the land for social amenities or pay a levy in cash. Maybe this cost the minister the accusation of "socialmania" and ultimately his portfolio.[16] His own party had practically disowned him (Gartzos 1981). The fact remains that his reform remained in limbo (Economou 2000a, b; Nenou and Samarina 1995), and a new one was to emerge under a socialist government, which voted Law 1337/1983. In the meantime, in 1980, urban and regional planning and environmental policy had been endowed with their own separate ministry. The 1976 regional planning law introduced for the first time a hierarchy of national, special, and regional spatial plans. Special plans were also national but were to address particular sectors of the economy or infrastructure networks. Land uses were to be fixed for both land-based and

[19] G. Lialios, Greece as a model of town planning anarchy. Newspaper Kathimerini, 4 November 2018.

maritime activities. At about that time a new public agency (known as DEPOS) was created to undertake town planning, settlement and housing projects (Marmaras 2002, 239–279).[17]

Legislation in the late 1970s introduced two instruments which are a typical example of imported practices, similar to so many others in the Global South. The first was the Active Planning Zones (known as ZEP), a copy of the French ZUP (priority urban development zones) (Sofoulis 1993); the other was the Transfer of Building Coefficient (MSD), an imitation of the American TDR (transfer of development rights). The use of ZEP faced the intractable problem of excessive fragmentation of real estate property and was quickly abandoned, while MSD's use in practice was blocked repeatedly by resolutions of the Council of State. Its result was an accumulation of unrealized MSD legal titles (Papademetriou 1996; Menoudakos 1996; Angelidou 2014).

Governmental procrastination in the implementation of the new legal framework was perhaps due to inter-ministerial antagonisms. In 1980 the responsibility for the national spatial plan returned to the ministry of national economy (Skouris 1991), while environmental and economic development policies were often in opposition (Rozos 1994, 127–130). A local government reform in 1986, bearing the name of modern Greece's first governor Capodistrias, transferred powers to local and regional authorities and established the institution of elected prefects. The problem was that all these reforms were promoted while the national economy had not yet recovered from the 1973 oil crisis (Samaras 1985, 23).

The decade of 1980 bears the stamp of the minister Antonis Tritsis, a US trained planner. It was the first time that at the head of the ministry of planning and the environment was a professional planner, who later served for a while as Mayor of Athens, until his premature death. He considered spatial planning as the "crystallization of a way of life, which reflects social life and social expectations" (TEE 1976, 45). He was a visionary, who once said that a statue of the illegal settler should be erected. Tritsis launched the so-called operation of town planning reorganization (or reconstruction), known as EPA (TEE 1993; Economou 2004a and 2009a; Getimis and Economou 1992; Giannakourou 2012; Aravantinos 1997; Melissas 2007; Petmezidou and Tsoulouvis 1990). In accordance with Law 1337/1983, General (much later renamed Local) Town Plans were to be produced for all cities and townships, followed by local lay-out and implementation plans for all within-the-plan areas. It was these layout plans that constituted the *schedio poleos*, which was mentioned earlier, when the point was made that being in the *schedio poleos* was the dream of every land owner. The general plans could also designate settlement control zones (known as ZOE), around the planned areas, which were differentiated from out-of-plan land, in the sense that certain developments could be allowed under specified conditions. The ZOE were thus a *sui generis* tool lying between the urban and the regional levels and for this reason were also

[20]The triptych of the plethora of laws, their poor quality and the associated lawlessness was recently the object of a special report in the newspaper *Kathimerini* (24 Oct. 2021).

used in the first small-scale, regional spatial studies to protect environmentally sensitive zones, mainly in islands (Angelidis 2000, 161; Melissas 2010a; Voulgaris 1994, 218; Nikolaou 1994; Rozos 1994, 109). In later years they were gradually absorbed in the mainstream town planning activity. They had been clearly a transitional arrangement. The whole town planning operation proved over-ambitious, because of the shortage of sufficient qualified consultants, resistance from private real estate interests, the deplorable state of cadastral surveys, unrealistic land-use standards, and the consequent failure to implement the plans on the ground. Local pressures resulted in the inclusion within the planned area of excessive amounts of land for plan extension, adequate to house many times the population of the country. Several plans produced in the 1980s remain unimplemented even today. Others have become obsolete and are in need of urgent updating (Aravantinos 1985; Getimis and Economou 1992; Getimis 1992; CPER 1989a; Karadimou-Yerolympos and Kafkoula-Vlachou 1983). Yet, the whole operation mobilized a large number of civil servants and consultants and fuelled a feeling of optimism that town planning was coming of age. The new plans became an element of rational urban organization (Melissas 2007, 100). Years later, Economou (2009a) attempted an assessment of the operation and drew attention to the massive extension of urbanized land, the abandonment of the logic of more intensive development within urban areas, and the shift to a policy of expansion, for which should be blamed not only the planners, but also the briefs they received from the planning authorities. It is true to say that there were no plans at a higher, regional, scale, which might provide a framework within which town plans would be formulated. The ministry in charge hastened to produce in 1984 a series of outline spatial plans for all prefectures of the country, which had no statutory character, but emphasized the need for a polycentric organization of space and the servicing of rural areas (Angelidis 2000, 160–161 and 191; YCHOP 1983, 1984a, b). They offered a good "schooling" of the cadres of the planning ministry in the practice of spatial planning. Much more important and professionally ambitious were the Special Spatial Studies[18] of the late 1980s and early 1990s, most of which were funded by the EU Initiative ENVIREG (Zampelis 2003; Economou 1994a; Voulgaris 1994; Avgerinou-Kolonias 1997; Kloutsinioti 1989). Although they covered areas of fairly limited size, they proved to be a further step towards the regional spatial planning activity of the 2000 decade. Some of them had to deal with problems which provoked enormous controversy. The most famous example of bitter opposition was the area of Chalcidiki in Macedonia, the site of copper and gold extraction and of processing plants, constructed and managed by international mineral extraction firms. This activity was fiercely resisted by the local prosperous tourism industry and was accompanied by violent clashes between workers employed in the mines and those working in hotels and tourism resorts. Environmental organizations threw their weight in the conflict and appealed

[21] See, e.g.: Camhis 1994, 2004, 2007, and 2008; Giannakourou 1995, 2000, and 2008a; Kafkalas 1992; Getimis and Kafkalas 1993; Kyvelou 2010; Kritikos 1994; Wassenhoven 1996a, 1997a,

repeatedly to the Council of State. Successive governments legislated to overcome objections against the project.

8.9 Realities on the Ground and Out-of-Plan Building Activity

The reality of planning, not on the drawing board or in legal texts, but rather as reflected in the transformation of geographical space, was totally different from the impression conveyed by innovative legislation. The abuse of plans to enable the expansion or urbanized land to the detriment of the countryside has been mentioned already. But even more scandalous was the possibility offered by a 1985 Presidential Decree for settlements of less than 2,000 inhabitants, known as the "decree of the villages". The aim in theory was to secure a sort of orderly planning for small settlements which were not to acquire a General Town Plan. The problem was that the delineation of boundaries, rudimentary planning and land-use designation was entrusted to elected prefects. The result was over-generous boundaries which made possible to designate agricultural land as suitable for building (Wassenhoven 2022). The powers exercised by the prefects were later found to be unconstitutional by the Council of State, but the damage had been done and litigation followed when private properties in the zone surrounding the "villages" had to be registered in the official cadastre.

Out-of-plan construction activity is a much more serious factor in the deterioration of the countryside. The marginalization of the latter has been a cause of concern in developed countries, at various stages in their history. In Britain, Selman (1995) points out, "the distinction between town and country has become more blurred, especially in the more scenic and accessible countryside". In Greece, this "blurring" has been violent and anarchic, due to the haphazard expansion of uncontrolled construction. A lot of it is unlicensed, i.e., plainly illegal, but there is a great deal which takes place with a building permit, in accordance with presidential decrees of 1978 and 1985, later amended several times. The decrees were a good example of piecemeal legislation (Rozos 1994. 109). Such building was clearly beyond the ambit of General (now Local) Town Plans, and therefore far more exposed to favouritism and clientelism. In most cases, a minimum requirement is simply the ownership of a land parcel of at least 0.4 hectares, but even that is bypassed by derogations for "exceptional" cases. Exceptions can also be made in relation to land use or building coefficients. The problem, it must be stated, does not affect Greece only (Preteceille 1982). Out-of-plan construction has been repeatedly condemned, as one of the most important and diachronically persistent problems that degrade Greek nature in a brutal manner, while also imposing enormous costs on the Greek economy, as a result of the necessary technical infrastructures to service mushrooming settlements without any prior planning (WWF Hellas 2020). Given that, in addition, out-of-plan areas are littered with *afthaireta* (illegal buildings) which are constructed in forested land, and therefore constitute an ominous fire hazard, the cost of unauthorized development multiplies (Sapountzaki 2007, 2019a, b; NTUA/EChOA 1994). The tragic events of the forest fires of the summer 2021 provided ample evidence and not for the first time. As Getimis and Economou (1992, 25–28) remarked, the unchecked spreading of construction was the outcome of the almost uncontrolled right of land owners to choose the land use they fancied, with industrial and tourist activities enjoying special privileges, with catastrophic results for the environment. Illegal building was almost the rule after the Second World War, with successive governments turning a blind eye (Romanos 1975). Land use in out-of-plan areas was determined by special decrees for these areas of nationwide application or by ad hoc decisions at the time of granting a building permit in accordance with the categories of a decree classifying uses (Koudouni 2017).

Legal construction in out-of-plan areas was in itself a mechanism which induced and multiplied unauthorized development, with multiple environmental consequences at all levels, from local to national (Economou 2009a, 251; Economou 2000c). Demolition of illegal structures was invariably postponed, because of political pressures (Grammaticaki-Alexiou 1993, 141). In a planning law textbook, exclusively dedicated to illegal building, Christofilopoulos (1999) underlines in his foreword the confidence that illegal activity remains unpunished, lawlessness is identified with cunning, and illegality is perpetuated because of patron-client relations.

Particularly in the early postwar years, illegal activity flourished in peri-urban zones where old land estates were subdivided by speculators into tiny parcels and then sold to low-income internal migrants who built an illegal house, almost a hovel (Gartzos 1981). These settlers then waited patiently until their land became integrated in the within-the-plan area. Its value would then rise and the prospect was opened, with a favourable building coefficient, for the erection of a block of flats. In the late 1960s almost half of the urban land of large centres was still out of the limits of the official plan (Skiadaresis 1967; NTUA/SPE 1971b). Peri-urban space was always the first candidate for illegal construction and, in general, activities detrimental for the environment. It must be recognized that this is not an entirely Greek phenomenon (Allen 2017; Leaf 2017). There have been also cases of out-of-plan areas in the heart of urban agglomerations. The Elaionas district in Athens is a famous example (see Text-Box 8.5).

Text-Box 8.5: Urban Renewal and the Elaionas District in Athens

Elaionas means olive grove and the district under that name is the site of the old sacred olive grove of Ancient Athens, where the felling of trees was prohibited by Peisistratus in sixth century BC. At the beginning of the twentieth century there still were 50.000 trees, mostly used as firewood or felled to make room for vegetable gardens and later wholesale sheds or industrial plants. The 870 Ha area is intersected by the Sacred Way (*Iera Odos*), followed in ancient times by the ceremonial procession marching from Athens to

(continued)

Text-Box 8.5 (continued)

Eleusis for the celebration of the Eleusinian mysteries. The district is adjacent to the cemetery of ancient Athens and bordered on the west by the river Kifissos and on the east by the ancient Long Wall linking Athens and Piraeus. Elaionas is located only 2 kms away from the centre of Athens, but until 1991 it remained an out-of-plan area surrounded by within-a-plan zones. A statutory street layout plan was approved in that year, but was opposed by the five municipalities of which the district was part.

The reason the Elaionas case is interesting and instructive is because it offered a splendid opportunity for a major urban renewal project which was sadly missed. Such a project was planned by a research team of the National Technical University of Athens (NTUA), commissioned by the Athens Master Plan Organization in the early 1990s. Instead, the familiar well-tried method of extending the existing town plan and of designating land uses was chosen and implemented. This was a compromise imposed by practicalities and by the inability of the public sector to proceed to a more ambitious planning gesture. Curiously, the word "compromise" was in fact used by the head of the research team – and author of the present book – in a seminar presentation at the time.

The district in those days was a heavily polluted (and polluting) industrial land area with a mixture of uses: Manufacturing industry (40% of the land), wholesale trade, scrap yards, vegetable gardens, limited housing, the central Athens vegetable market, bus and track depots, road freight terminals, military facilities, power stations, the Agricultural University campus, sports grounds, derelict land, camps of Roma communities etc. Appoximately 45,000 people were employed in the area. There were few main roads on the edges of the area or intersecting from east to west and a rudimentary secondary network of service roads. One of the roads on the western side was the covered up old river Kifissos. The so-called Prophet Daniel dry stream crossed the area, used as a waste dump. Social facilities were primitive and environmental conditions appalling.

Protection of healthy manufacturing activities (within industrial parks), environmental improvement, creation of a large linear urban park, provision for educational and leisure facilities, removal of all polluting industry, and relocation of polluting plants to industrial estates outside Athens were the main objectives of the plan produced by NTUA. The creation of a public management agency was envisaged, to work closely with the private sector especially in commercial development zones, which would be developed as Active Town Planning Zones (ZEP), in accordance with existing legislation. This instrument had been modelled on the French Priority Urban Development Zones. Obstacles quickly appeared: Poor cooperation of government agencies, lengthy procedures for plant relocation, pressures by local authorities

(continued)

Text-Box 8.5 (continued)

and business firms which favoured a speedy legalization of existing activity, and resulting inertia of the Elaionas management agency. The successive research projects of NTUA, which included, apart from the land use plan, a thorough economic analysis, a study of the management agency, and the specific problem of the relocation of haulage firms, were thwarted by lack of coordination and inactivity problems. The end result was the adoption by the government authorities of a conventional land use and street layout plan, as an extension of the town plan of the surrounding areas. The building of a football ground in Elaionas had its own story (see Fig. 8.5 with plans for the Elaionas district and Text-Box 8.6 on the Council of State).

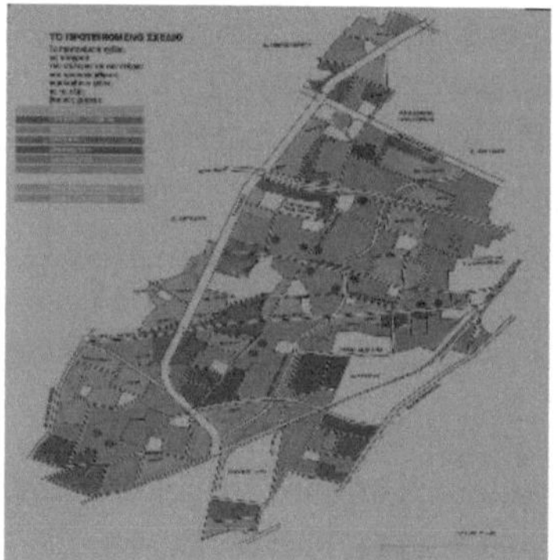

Fig. 8.5 The Elaionas district (see Text-Box 8.5)
(In the middle, the situation in the early 1990s, left, the plan of the National Technical University of Athens with its green park spine, and, right, the finally approved land use and street layout plan, with its scattered uses)
Sources: National Technical University of Athens and Organization for the Master Plan of Athens

Anarchy and illegality are still a sad reality (Serraos and Melissas 2017; Melissas 2017; Sakellaropoulou 2017),[19] in spite of the clear violation of the Constitution. Governments have repeatedly tried not to "legalize" illegal buildings, an action which would be blocked by the Council of State, but rather to "regularize" their existence for the foreseeable future, with the imposition of a special financial levy.

A key aim of a planning law voted in 2020 was to put an end to out-of-plan development and to cover practically all national land with a new generation of Local Town Plans. Sadly, the clauses of the law aiming at the abolition of out-of-plan construction were obstinately resisted in parliament and by outside interests, with

2008b, 2009, 2017a, 2022; Mitsos 1995; Laskaris 1994; Economou 2000e; Asprogerakas 2020; Asprogerakas and Zachari 2020; EKDDA 2017; Eliopoulou 1992; Fysekidou 2015.

the result that they were watered down. The game was in progress in 2021, as new law amendments were made almost daily. The new local plans should make a difference, but, although the first batch was commissioned to consultants in the summer of 2021, it will take years before they are all produced and approved, and even longer before they are implemented on the ground. When that happens, it is hoped that the extraordinary status of out-of-plan areas will disappear and that planning parameters like density, land-use, plot coefficient, building height, and floor ratio will be fixed for all buildable land. These parameters are the “measurements” that planners work with, as engineers work with “stresses and strains” and doctors with “blood pressure and heartbeats”, in Needham’s (2017) very pragmatic expression. There will still be the process of issuing building permits, which is largely dictated by the Greek national Building Regulation. This process is largely the responsibility of local authorities, at least in large municipalities, which have only an advisory role in the approval of local plans. The present haziness of planning legislation frequently makes the issuing of building permits subject to arbitrary judgements and a headache for practicing architects (Karadimou-Yerolympos 2000). Here too, it is hoped that a recent avalanche of amendments of the building regulations will contribute to more clarity.

8.10 Legal Quandaries

Planning legislation is a major puzzle for planners, administrators, judges and land owners alike. When the author analysed the Greek planning system in the 1990s (Wassenhoven 2000a), he quoted a phrase from a legal study by Choromidis (1994, 9), which is still topical: "It is not simply difficult, but almost impossible to acquire a complete overview of, but also to ‘tame’, the legal chaos of town planning law, so as to process it systematically and with lasting value. Laws, regulatory statutes (presidential decrees and ministerial decisions), circulars etc., the documents concerning issues of town planning legislation, are produced in a torrential, I would say industrial line, process, especially in view, every time, of an electoral period...". Electoral priorities are frequently the cause of urgent legislation (Chlepas 1994, 319–320). The reputation of the Greek planning system had reached the European international literature, in which reference was made to a labyrinth of amendments, exceptions and ad hoc statutes (Newman and Thornley 1996, 57; Balchin et al. 1999, 71; see also, on planning in Europe, Reimer et al. 2014, and Shaw et al. 2000).

Choromidis’ shout of despair was about planning legislation, but the problem was equally true for all legislative activity, as observed by many authors (Pangakis 1991, 109–110; Kontiadis 2009, 244; Sotiropoulos and Christopoulos 2017, 23–27, 46–47 and 76). They all draw attention to the problems of plethoric law production, the negative effect on the performance of public administration, the vicious circle of inefficiency, the extra bureaucratic load, the inevitable consequences for the average citizen and for economic activity, the urgent need for law codification, the delays in the dispensation of justice, and the legal insecurity suffered by individuals and business firms.[20] D.A. Sotiropoulos and Christopoulos counted over 4000 new laws (not of course about planning) between 1974 and 2012. It is indicative that the 1976 law about regional spatial planning carried the number 360, but the numbers of the planning laws of 2016 and 2020 were 4447 and 4759. Laws are riddled with endless amendments difficult to decipher even for experts. The author had personal experience of the difficulty of spotting an amendment that even experienced administrators could not find. The laws are made up of hundreds of articles and pages, most of which have no relation whatsoever to the law's main object, e.g., the ratification of an international convention. There have been cases when a law was amended by another law within the same day. This sorry situation is of immediate consequence for planning. Giannakourou (2019, 33) comments on the difficulties facing urban and regional planning as a result of the plethora of statutes, the overlaps between them, and the difficulties of interpretation even for legal experts. The infamous "multi-bills" or "broomstick acts" remain a regular practice. So-called "transitional" clauses are a source of confusion, as they are typically complex, perhaps intentionally, full of references to prior statutes, and prone to continuous revisions (Sotiropoulos and Christopoulos 2017, 109–112). A code of urban and regional planning legislation was produced in 1999, but became outdated within a few years. A new committee of experts was formed in 2020 with a mandate to prepare a new code for 2022. The committee has to amalgamate provisions contained in the new planning law of 2020, but, in the meantime, the law was being constantly amended in 2021, incidentally without a change of government. The author, as a member of the committee, made up of distinguished legal experts – which he is not! – had the impression that he was following a training course. One is reminded of the case of an internationally known planning and planning law expert – Rachel Alterman – who after reaching the top of her profession and academic career as a planner decided to study planning law: "Rather than viewing the law as an outsider, I decided to harness the knowledge of law to enlighten my research and empower my students and the planning profession" (Alterman 2017, 271).

Spatial planning was no longer the same after the 1975 Constitution, but, as was to be expected, new difficulties of interpretation of the legislation emerged, which increased the role of the administrative courts, particularly of the Council of State (*Symvoulio tis Epikrateias* or *StE*), i.e., of the Supreme Administrative Court, modelled on a French prototype (see Text-Box 8.6).

[22] The role of the European Union in the year of the publication of the ESDP, which coincided with new Greek planning legislation, was debated in 1999 in an international conference held in Athens. The proceedings of this most interesting conference have not been published so far.

Text-Box 8.6: The Council of State

The role of the Council of State (*Symvoulio tis Epikrateias*, known as *StE*) is of paramount importance in the Greek planning system, but difficult to explain. StE is the apex of the administrative justice and the court of last resort appeals against legislative and administrative acts. Appeals are very frequently lodged by individuals, firms, or NGOs. Even a simple planning permit can be overturned, if the law or presidential decree on the basis of which it was granted is found unconstitutional by StE. Equally, a spatial regional or local plan approved as prescribed by law can be declared null and void by StE if its legal basis or approval procedure is found faulty. This was the case of the national Special Spatial Framework for tourism of 2013, which was nullified by the court two years later for reasons not of substance but procedure. The previous framework for tourism of 2009, which returned to force when its 2013 successor was annulled, had the same fate somewhat later.

The opportunity for an appeal, which affects a particular offending article of a planning law, may arise years after the publication of the law, when an aggrieved person objects to an administrative act, e.g., a building permission, of which the legal basis is this very article. The permission may be declared null by the court precisely because the offending article was found unconstitutional. This was the case of an article concerning coastal zones contained in the planning law of 1983, i.e., the law on the basis of which hundreds of town plans were produced all over the country. It is conceivable that a building may have been erected in the meantime on the assumption that the permission was valid. A similar after-the-event scrapping of a building permission occurred in the case of a large multi-floor building block in central Athens which is still standing proudly for everyone to admire. Delays in the proceedings of administrative courts, not necessarily at the highest level of StE, affect other important planning decisions. If a private property is compulsorily acquired for the purpose of creating a park or erecting a school, the municipality concerned must compensate the owner within a fixed time limit. When the deadline lapses and no compensation has been paid, the owner can apply to the administrative courts and demand that the compulsory acquisition be lifted. The court may take years before a verdict is reached. The construction of a complex of shopping malls, offices, and a major football ground in the Elaionas district of Athens (see Text-Box 8.5) has been another example which attracted enormous publicity. The reason was that the football ground was to be used by a famous football club, of which the old ground is located in central Athens and would then be converted to a public park. Very complex negotiations for this "double renewal" continued for years. They involved the football club, the Municipality of Athens, the Ministry of the Environment, the Region of Attica, two banks, and a large construction firm. A plan for the project was approved by law in 2006, the construction firm proceeded to the building of a

(continued)

Text-Box 8.6 (continued)

large commercial centre, but in 2009 the project was declared illegal by StE, following a number of appeals. The construction firm went bankrupt and its commercial centre remained unfinished. A new presidential decree was signed in 2013 putting the project on a new base. In 2020, the interested parties reached an agreement and the prospect of a project including football and basket-ball grounds came nearer, with funding from the EU Recovery Fund. Meanwhile, the old historic football ground of Alexandras Avenue is still awaiting its conversion to a park with underground parking facilities.

Leaving aside such unfortunate planning accidents, the importance of the Council of State, from a spatial planning perspective, lies in the body of its case law and its long-standing contribution to the disentanglement of chaotic and often contradictory planning legislation. StE case law is not an independent source of law, but it can be argued that it is a sui generis parallel body of legislation, produced under the pressure to re-examine legislation. This could be considered a heretical view, but is not meant to degrade the great value of the court's work. StE is a "corrective" factor of undoubted significance.

The administrative courts are implicated when appeals are lodged against acts of the administration, e.g., the granting of a building permit or licensing of a business firm. Initially, StE showed patience or reluctance to annul such acts until all urban and regional plans demanded by legislation were in place (Menoudakos 2008). Gradually, the rulings of StE showed their teeth. Innovatory case law began to open new directions (Giannakourou 1994c, 23–27; see also Zygoura 1994, Siouti 1994 and Skouris and Tachos 1988). For years legal rules in the domain of town planning were in essence adapted to already existing situations. Reality, remarks Giannakourou, acted as a regulatory force. When this "reality" broadened its scope, as more activities entered the demand and supply markets, the use of legal instruments rapidly expanded. This development was reflected in the activity of StE and the production of case law, which, arguably, amounts to a body of parallel law. In the view of Giannakourou (1994c, 36–37), expressed in a tactful and discreet manner, it appeared that the constitutional principle of spatial organization had two sources of definition and further refinement. To the legislative function of translating the constitution into law was added the control of constitutionality in the form of case law, which, effectively, "concocted" an additional legal corpus.

In the opinion of a senior judge of StE, spatial planning aims at organizing, in a rational way and within definite geographical boundaries all human activities, to secure a balance between human and societal needs, while respecting natural resources and the requirements of economic activity (Pikramenos 2018, 105). These are the issues that the case law of StE consistently addressed, with the result that the rules that regulate planning practice are defined by both government and administrative justice. It has to be added that, for a period, the court sometimes invoked documents derived from international declarations, e.g., the Agenda 21 of the 1992 UN Rio de Janeiro conference, which apparently was not legally binding for the signatories (United Nations 1992). This period has been described as StE's phase of activism.

The importance of the role of the courts is not limited to Greece. An interesting example is Britain, as presented by Booth, although it must be made clear that the concept of case law of the Greek Council of State is by no means identical to that of the English common law: "I have attempted to show in this paper the ways in which spatial planning as practiced in Britain is a creature of the English common law. The insistence on justice that started from ascertainable facts to which the law could be applied and the understanding that rules did not always deliver justice and needed to be tempered by equity are at the heart of this inheritance. So too, is the way in which judges began to generalize on the basis of specific cases and categorize actions as a guide for future decisions. And then we have seen how in attempts to give structure to an otherwise apparently unconstrained process of judgment, judges were at pains to derive principles from the accumulation of experience. It explains why discretionary freedom has been a valued attribute of planning in Britain and the nature and place of rule-making within British planning. However, the argument can be taken further. There seem to be a number of clear lessons to be learned from law that we planners should take to heart. The first is that simply applying rules is not enough to ensure justice, and that sometimes the application of a rule may lead to a serious injustice. Two things flow from this idea. The first is that we need to have explored what justice means in terms of the allocation of space and the distribution of activities. We need, in other words, to make explicit what the end goal of spatial planning should be in order to know whether the rules we have devised match the particular circumstances we are called to deal with. The second lesson is the need for transparency and accountability. In a general way, planning activity is 'accountable before the law' and the National Planning Policy Framework specifically enjoins transparency, particularly in the sections dealing with local plan preparation. But what does this mean in practice? The great worry that members of the judiciary had, historically, about the exercise of equity was that it escaped judicial oversight, and this worry has been frequently expressed in more recent times in relation to public policy" (Booth 2016).

In Greece, as explained, the courts become a supplementary policy maker (Wassenhoven 1995d, 16). The main reason, as pointed out, is the legislation itself, but also the social importance of land property and the psychology of citizens, who seem to derive an obsessive pleasure from appeals to the courts. A side effect is that planners end up by paying attention, above all else, to making their proposals immune to appeals (Wassenhoven 2021b). An appeal to the courts may condemn the citizens concerned to years of delay and uncertainty, which is a wider problem in the dispensation of justice (Sotiropoulos 2019, 167). In the case of private urban land within the limits of a statutory plan (*entos schediou*) compulsorily acquired for reasons of public benefit, e.g., to build a school or create a park, compensation is due to the owners. When the municipal authorities fail to pay it within a specified period, the owner can lodge an appeal to the courts and demand that the compulsory acquisition be revoked (see Text-Box 8.6). There have been cases of delays of more than ten years until the property is freed from this burden. The case of unrealized legal titles from the transfer of development rights has been already mentioned. Here too, the owners of titles were trapped in endless periods of waiting, when the relevant laws were annulled as unconstitutional.

It has been made clear that responsibility for the policy making role that the Council of State is forced to assume rests on the government and the legislature and not on the judiciary. The position that the end-result is a parallel law-making system may be disputed or considered as an exaggeration. The fault for this development can be attributed to the chaotic nature of legislation rather to judicial failures, since the court often has to steer in uncharted waters left "muddy" by the legislators. Further complications arose when the court had to take into account not only national legislation but also European Union law and international conventions. Planning cases were the responsibility of a particular Chamber of StE and although many cases were submitted to the plenary session of the court, others were judged at the level of this chamber (Decleris 2000). The criteria used by the court, especially those based on international declarations of a non-binding nature, were often a cause of dispute. For some commentators they were evidence of judicial "activism", as in the interpretation of the meaning of sustainable development (Giannakourou 2019, 40; Melissas 2002, 166–167). The term "judicial activism" is not a Greek invention, as G. Voulgaris (2013, 201–202) remarked, but the case law of Chamber V of StE remains a typical phenomenon.

Regardless of the "activism" interval, the case law of StE has made an enormous contribution to the establishment of planning both in practice and in the public conscience. One further reason is the standard practice of the administration to approve plans at the urban level not by a simple ministerial decision but by Presidential Decree. This is not only in order to endow them with a higher status, but also because a PD is subjected to prior control of constitutionality by a committee of StE. This acts as a sort of protective shield in case there is an appeal against the specific plan. As urban plans are divided into higher level General (now Local) Town Plans and lower-level implementation plans, the practice until recently was to approve the latter by PD, because they were considered as the plans which affect private real estate interests. This has been changed in 2016 and it is now the higher-level Local Town Plan which is approved by PD, because its guidelines concerning land use, densities or development ratios are imperative for the lower-level implementation plan. The advance control of the legality of PDs is one more example of the crucial responsibilities of StE. It is not surprising that the examination of any contentious planning decision cannot be limited to legal statutes affecting the particular case but has to embrace all the relevant case law of StE. Planning law thus becomes a vast field of legal expertise. Concentration of planning powers in an over-centralized state was a further factor. The administration itself was often blamed for inefficiency and for impeding modernization of the planning system (Evangelidou 2002).

8.11 European Union

The impact of the European Union on the planning system has been multiple, although the EU does not impose a uniform urban and regional planning system on its member-states. The successive change of priorities endorsed at the highest level (see, e.g., Presidency Conclusions 2000, 2001, 2005) had trickle-down effects of serious ultimate impact on domestic policy. The unification of the internal market and European integration posed a threat to relatively backward regions (GREMI 1992; Holmes 1999; Luukkonen 2015, 2017). It is not possible to cover all space-related activities of the EU in a chapter on Greek spatial planning. Still, their influence has been very large in a number of direct or indirect ways, especially regarding regional planning, as it will be shown in a relevant section later. The effect of the "rescaling" of planning guidance is sufficient evidence (see Brenner 1998, 1999, 2004, and Brenner et al. 2003). As well known, since the 1990s, the EU has initiated its own spatial planning activity, albeit mostly advisory. This activity has had a serious impact on planning thinking in Greece, both in government and in publishing in Greek language. The process of successive European Commission reports (CEC 1991a, b and 1995) culminating in the European Spatial Development Perspective (ESDP) (European Commission 1997b, 1999) and later in the inclusion of territorial cohesion in EU Cohesion Policy (Davoudi 2005; Faludi 2002, 2004a, b, 2007, 2016), has been well documented by Greek authors.[21] A similar policy document by the Council of Europe (2000b) also had an effect. The influence of the EU extends to a number of policy areas, which can only be touched upon in this summary. The effect on regional development policy has been hinted at earlier.

The ESDP (European Commission 1999) was not a binding document but its importance for Greek planning is manyfold. It introduced goals, concepts and terminology which entered the objectives and parlance of Greek planning (Giannakourou 2008a 46, 48, 84, 90–91 and 175; Andrikopoulou and Kafkalas 2008, 227; Kyvelou 2010, 237). The influence had become obvious even at the stage of its preparation (Komnenidis 1995; Demathas 1995b). Concepts like "polycentricity", which had been underlying the decades-old discussion on the network of urban centres, became of central interest in spatial planning (Andrikopoulou 2010; Zampelis 2004; Demathas 2004; Brezzi and Veneri 2015). Of equal importance were the follow-up activities of the European Spatial Planning Observation Network (ESPON). European policies for sustainable development, cohesion and territorial governance (Giannakourou 2008b; Camhis 2008; Andrikopoulou and Kafkalas 2008; Wassenhoven 2008a, b, 2010a, 2017b) had a direct impact on the national and regional spatial plans after the year 2000. Problem-setting and policy-making at EU level seeped into national planning thinking through a variety of channels,

[23] The author was at that time chairman of the National Council of Spatial Development.

[24] Greece: Memorandum of Understanding on Specific Economic Policy Conditionality. Draft of 9 February 2012 (Q3-Q4 / 2012).

ostensibly unrelated to urban and regional planning. Regional development policy underwent successive mutations. Already by the end of the 1980s, the EU instruments had replaced national development plans. It was in many respects a change affecting the European South in its totality, but Greece had to undergo a fundamental reshuffling of established practices, especially those with a strategic character (Kyvelou 2010, 18). Greece, with its strictly regulatory tradition of urban planning, could easily lapse into reliance on EU guidelines and project-focused support. It could be argued that this created an illusion of EU top-down transfer of wisdom, which, however, was highly selective. Komnenos (1994, 97) had drawn attention to the fact that European integration was producing a spatial restructuring of Greek development.

EU influence was not solely the result of official policies and documents. The full gamut of formal and informal cooperation events, at ministerial or civil servant level, generated new ways of thinking or acting and a "copying" mentality. The entire edifice of spatial planning was thus affected (Giannakourou 2008a, 109–110; Rozos 1994, 137–146; Georgiadou 2002; Niessler 2004). But so was, perhaps more spectacularly, environmental policy, at least in its legal form. The introduction by an EU Directive in 1985 of the obligation to carry out an assessment of the effects of public and private projects on the environment and to impose environmental terms and conditions is an excellent example (Melissas 2002, 179; Council of the European Communities 1985), regardless of its questionable application in practice. EU transportation policy too had a direct impact, especially that of the Trans-European Networks (European Commisssion 2018; Dionelis and Giaoutzi 2008, 121–122; Kritikos 1994), although in the Greek government structure transport had remained a responsibility outside the confines of spatial planning. Other EU policies also touched sensitive issues of Greek planning. Coastal management was one, either through direct EU activities, or through its participation in United Nations programmes, like the Mediterranean Action Plan (Wassenhoven 2017a; Beriatos and Papageorgiou 2010). Another one was cross-border cooperation, through the EU Initiative INTERREG (Kostara 1995), or, later, the European Groupings of Territorial Cooperation. The EU impact on Greek environmental legislation is of direct importance for spatial planning (Melissourgos and Chasiotis 2019; Nantsou et al. 2018; WWF Hellas 2005, 2018, 2019a, b). Finally, of great significance for Greece were the directives on the maritime environment and maritime spatial planning (Wassenhoven 2017a), now incorporated in Greek national legislation.

8.12 Resuscitation of Regional Spatial Planning in the 1990s and New Planning Instruments

In the 1990s, while new ideas on spatial planning were emerging in the European Union and with regional spatial planning in a state of neglect, it was time to take stock of the situation (Habitat II Greek Committee 1996; Wassenhoven and Papagiannis 1996) and to prepare the next steps. The laws on urban and regional planning of 1997 and 1999 introduced a full-scale system of planning and an array of plans and instruments, the proliferation of which caused concern. It was on the basis of this legislation that throughout the 2000 decade, before the 2010 economic crisis, a large number of plans were produced at national, regional and urban levels. Expectations were high, because of the obvious need to secure a hierarchical structure of spatial policy choices and a functional relationship between planning levels (OKE 2007). The then minister for the environment declared that the 2000s would be the decade of the "operation of regional planning reorganization", as the 1980s had been the decade of the "operation of town planning reorganization". The dire reality at the level of urban planning is described by Economou (2000b, 29–52): Urban planning functioned in a vacuum without any input from above, i.e., from the national and regional level. One good example was the uncoordinated designation in urban plans of industrial parks and their inevitable failure. There are no mechanisms to relate local choices with a national policy and ensure a minimum of cohesion. According to the new legislative framework of 1997, the urban planning level included Master Plans for a number of cities – not only Athens and Thessaloniki –, General Town Plans (GPSs, for urban municipalities), and Plans for the Spatial and Urban Organization of Open Towns (SCHOOAPs). The SCHOOAPs were reserved for essentially rural areas with several settlements, of which the main centre did not qualify for a GPS. Still, they had the same legal weight as a GPS. Below these plans, there were local (layout) plans – approved by Presidential Decree – and their implementation administrative acts, urban renewal plans, plans of industrial parks and other special purpose plans, for instance plans for Areas of Specially Regulated Urbanization (PERPOs), which were essentially areas for private development, e.g., second-home zones.

At the level above cities and towns, on the national and regional scale, the new legislation of 1999 provided for so-called Spatial Planning Frameworks (SPFs), considered as "spatial planning means". At the top of the hierarchy was the National SPF, the Special SPFs again at the national level which addressed sectors (industry, tourism etc) or geographical zones (islands, coasts etc) and regional SPFs (Economou 2000b; Wassenhoven 2010a; Angelidis 2000; Gartzos 2004; Hatzichristos et al. 2017). The Special SPFs (e.g., on renewable energy sources or tourism) generated a great deal of discussion in the planning community (Housianakou 2010; Kloutsinioti 2010; Maraka-Romanou 2011). Regional spatial planning also included "mechanisms", like planning studies for the organized development of productive activities, spatial studies for areas of special interventions, and plans for integrated urban interventions. Economou (2000b) listed this confusing range of instruments – even managed to present a flow diagram – in which he included rightly the instruments left as a legacy of previous legislation of the 1980s and 1990s, like the settlement control zones around urban areas, the special spatial plans which had been funded from the EU Initiative ENVIREG, the special environmental plans introduced in environmental legislation, or the plans for areas of organized tourism development. He reached the conclusion that this elaborate system was in full contrast with real achievements. In 1999 a code of planning legislation had been produced but was soon out of date (Stingas ed 2004; Kousidonis 2009, 208).

Text-Box 8.7: Regional Plans

The influence of the planning systems of France, Germany, Britain, even the United States has been discussed elsewhere (see also Text-Boxes 9.1 and 9.2). Regional planning in Greece owes a great deal to the French model of the *aménagement du territoire*. The British model, with its emphasis on the urban and local level of planning and the much lower interest in the regional spatial level, was not in this case an influential factor. The United States and the Federal Republic of Germany are federal countries, totally different from centralized republics. In Greece, there has been an interest in regional economic development since the 1960s, but in successive five-year plans the spatial interest was scant. The production of regional plans was erratic and not supported by a legal statutory framework. It was after legislation on spatial planning and sustainable development in 1999, coinciding with the publication of the European Spatial Development Perspective (ESDP), that spatial planning at the regional level took off.[22] Spatial plans for 12 regions were approved by the central government in 2003. The region of Attica was the object of the master plan of Greater Athens. When these plans were approved there were no national spatial plans for individual economic sectors or types or areas, e.g., coastal or mountain zones, e.g., the Pindos mountain range (Wassenhoven 1993), as required by law. This was in itself a *non sequitur* and an inconsistency, which made the already problematic compatibility of regional plans even more precarious.

A problem with the regional plans, which has not been entirely solved even today, is their character as either regulatory binding plans or guidance documents. Their success is difficult to assess because their implementation is a long-term process, extremely sensitive to national or international conditions. As it became evident, when new regional spatial plans were produced in the late years of the 2010 decade and an ex-post assessment of the 2003 plans was carried out, the performance of plans was extremely sensitive to external factors, such as the changing geopolitical circumstances, the effect of the European Union and international conventions, and the vagaries of the economy, e.g., in the tourism sector. Domestic factors were also decisive, often related to the status of spatial planning. Decisions regarding the transport sector, e.g., the planning of motorways, had far more weight in spatial development than the content of mainstream spatial plans, which often amounted to mere wishful thinking. Still on the domestic front, the implementation of regional plans relied on the active use of instruments, such as economic activity estates and "containers", which were very seldom activated. The fatal blow

(continued)

[25] Memorandum of understanding between the European Commission acting on behalf of the European Stability Mechanism and the Hellenic Republic and the Bank of Greece. Done in Athens

Text-Box 8.7 (continued)

was the 2010 crisis, partly imported, partly due to chronic domestic management blunders, which quickly turned the first generation of regional plans to an obsolete set of documents. The fate of the second generation of plans (see Fig. 8.6), some of which were still at the approval stage in 2020, is difficult to determine as economic circumstances are a terrain of shifting sands in the days of the coronavirus epidemic. Yet, the production of regional spatial plans can be a fruitful lesson as it brings to the surface weaknesses of coordination and empowerment of regional authorities. These weaknesses are part and parcel of the structure of the Greek state and its effect on spatial planning, but also of the international dependence of the country at the edge of the European Union in a geopolitically sensitive location.

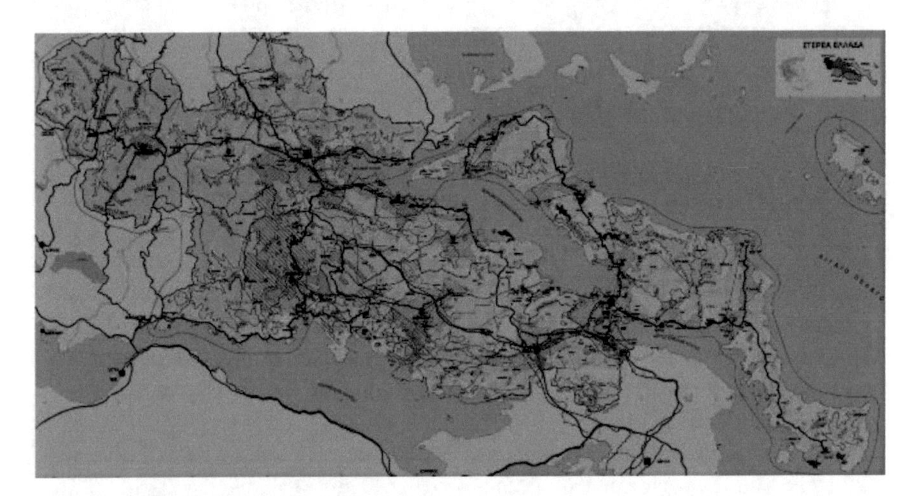

Fig. 8.6 Regional spatial plan for the region of Central Greece – 2015 (see Text-Box 8.7)
Source: Consultancy firm EDP s.a.

A few years after the classification by Economou, a special committee of the Technical Chamber of Greece was in real pains to provide a full presentation of planning instruments (Evangelidou et al. 2003). It was in 2015 that a university planning professor with a great deal of experience in government wrote that in our country we have an abundance of plans but deficient planning (Beriatos 2015, 119).

In the 2000s, national and regional frameworks were prepared and approved. Although the discussion for a national framework started early (YPECHODE 2000), it was curiously the regional frameworks that were first compiled and signed, followed by special spatial frameworks for industry, tourism, fish farming and renewable sources of energy, although a framework for the coastal zone was unfortunately not completed. The management of coastal zones remains a problem, one for which social consensus is a prerequisite (Wassenhoven and Sapountzaki 2001), but is usually open to clientelist deals. A national spatial framework was finalized in 2007. Long before that, the military junta had commissioned the firm of Doxiadis Associates to produce a national spatial plan (Psomopoulos 2019; Tortopidis 2009; Wassenhoven 1973), but as expected, the plan was quickly shelved after the fall of the dictatorship. It should be added that the framework for tourism produced in the decade of 2000 was soon replaced by a new one, but both were declared unconstitutional by the Council of State. The frameworks for tourism had aroused fierce opposition because of provisions to facilitate tourist resorts combining hotel accommodation with private villas for sale. It is sad that the debate on organized resorts diverted the attention of the wider public away from the much-needed emphasis on cultural and environmental tourism (Konsola 1993; Vliamos 2017).

The process of national and regional plan approval required the prior expression of opinion by the newly created National Council of Spatial Development (see Text-Box 8.8), an important consultation innovation. It was during its deliberation for the fish farming special framework that private interests interrupted violently its proceedings and made impossible a final vote. Instead of an official opinion, it was only the minutes of the debate that were at the disposal of the Ministry of the Environment, which, however, proceeded to the plan's final approval. An appeal to the Council of State alleged illegality but was turned down because the council's work had been subjected to external *force majeure*.[23]

During that period, the regional planning frameworks were approved by decision of the Minister for the Environment and the nationwide special frameworks by a joint ministerial decision of the ministers involved. It must be noted that during the same period there have been important administrative reforms. The number of municipalities was reduced drastically, the old prefectoral system was abolished, 13 self-governed regions were created, but, at the same time 5 larger deconcentrated regions were set up, covering the entire national territory and run by coordinators appointed by the central government. As far as urban and regional planning is concerned the exclusive power of the central state remained intact, because the Council of State had ruled that spatial planning is not a "local affair" in the spirit of the Constitution.

Text-Box 8.8: Consultation and the National Spatial Planning Council
Consultation is not merely a matter of inviting written comments on a plan document uploaded on the web. Important though they can be, these comments are not a substitute for discussion and deliberation. Equally, consultation is not simply the expression of views on a finalized plan or the scheduling of private discussions of the decision-taker, e.g., a minister, with important stakeholders, e.g., a powerful business association. Discussion behind closed

(continued)

on 19 August 2015 and in Brussels on 19 August 2015. Section 4.2 "Product markets and business

Text-Box 8.8 (continued)

doors is in fact the essence of high-profile compromise. Participation is not just an exchange of views of parliament members with their local electorate in informal encounters, no doubt important in a democratic society, but easily debased to low-profile compromise. Consultation and participation should be a continuous process of civil society engagement and building of awareness extending beyond narrow private interests. These plain truths demonstrate the distance separating, on one hand, bureaucratic expediency, narrow application of the law, and petty-politics, and, on the other, public mobilization in the face of social challenges. This is the reality of consultation and participation in a country like Greece. There is room for optimism considering the multiplication of non-governmental organizations campaigning for environmental and cultural protection and for other humanitarian causes. The independence of these organizations from party-political manipulation is of course a precondition.

The creation of the National Spatial Planning Council, which was convened for the first time in 2001, was a major innovation in planning consultation. The council included representatives of all the important employers' associations, labour unions, trade chambers, professional associations, and environmental organizations, as well as academics. Its chairman and two experts were appointed by the government. The council's task was to provide expert opinion and advice on national and regional spatial plans, at the request of the government, but also to submit proposals on its own initiative. The opinion of the council was not binding for the government, but it was mandatory before a final decision was reached for a particular plan. The planning minister was empowered to ask for its opinion on ad hoc policy issues. It was created by law in 1999, but its status was reconfirmed in the planning acts of 2014, 2016 and 2020. Delegates from the main political parties, also attended at the initial stages of its operation, albeit without voting rights. The council's chairman had the power to invite all those who were in a position to offer useful opinion to the council; they regularly included officials from other ministries, expert advisors, or delegates of business organizations not represented ex officio.

In spite of the hopes that surrounded the council's creation, its operation over the years was not as rosy as originally hoped. It was only once that the council managed to debate on key planning issues, on its own initiative. In some cases, a new minister showed little enthusiasm for the council's potential contribution. On other occasions, the political leadership of the planning ministry made clear its displeasure with the council's independence and interfered angrily in its proceedings. Soon, the government amended the law and ruled that the political secretary general of the planning ministry would act as chairman. As explained in the main text, in a notorious case, external elements interrupted violently its debates and systematically prevented the chairman's efforts to reconvene it.

A theory based on the Greek experience of planning cannot ignore the repeated revisions of planning legislation and the endless changes in terminology. An equally amusing phenomenon is the dispute between political parties and their expert followers about the paternity of reforms. What is obvious is that behind legislative reforms, particularly their sudden amendments there is a constant substratum of patronage and compromise. Plans fail not only because the planners are unable to predict the consequences of their proposals, but also because of unforeseen circumstances and geopolitical developments, e.g., the 2010 economic crisis or changes in the country's international position, not to mention natural disasters or pandemics like those of the years 2020 and 2021. They also fail because their implementation is undermined by a continuous bargaining between the state and the citizens.

8.13 The Apogee of External Pressure After the Economic Crisis and the Parallel System of Planning

In the 2010s the violent adaptations that the Greek state had to endure under the pressures of the lenders and the Eurozone leadership were unprecedented, but concerned mainly fiscal management and insurance policies and much less the political system, the administration, and the state structures (Sotiropoulos 2019, 192–193). However, they did concern spatial planning and geographical space (Gialis and Gourzis 2010). The Memoranda of Understanding signed with the Troika imposed measures which were the precondition for the granting of loans. These measures included spatial planning, e.g., the revision of regional plans and the simplification of legislation, not only of land use planning as such, but also of land registration. Time schedules for town plan production and approval had to be shortened, the number of levels of planning had to be reduced, and the attraction of investment had to be facilitated (Giannakourou 2019, 77–78).[24] Regardless of the Troika's pressures, the need for reform had also matured in the literature on planning (Gourgiotis and Tsilimingas 2016; Papageorgiou 2017).

The ills of the planning system that had to be cured were summarized in internal documents of the Ministry of the Environment and guidelines were issued to a law drafting committee, which soon reached its own conclusions, in spite of internal disagreements. A draft bill was prepared which focused on the simplification of the system. The bill attracted criticism the thrust of which was that the government was surrendering to private pressures. The reduction of levels of planning from seven to four and the introduction of the instrument of Special Town Plans for ad hoc – mainly investment – purposes, legally on a par with the municipal Local Town Plans, were the main target. Some planning academics, including a professor who

environment".

Text-Box 8.9: Natural Disasters and Land Use Planning

Natural disasters in Greece have been a regular cause of planning efforts. Planning for disasters has been an interest in a vast international research and literature, often related to the theoretical dimension of planning (Lewis et al. 1976) and to United Nations guidance (United Nations 1976). Seismic events in Greece have been a frequent phenomenon which left behind ruins and loss of life. Floods and fires in the recent past have been equally destructive. Urban areas have been badly hit by earthquakes in cities ranging from Athens, Thessaloniki, Volos and Corinth to small townships. Places with precious traditional architecture have been repeatedly damaged. It is not possible to list here the locations that suffered damage caused by natural disasters, nor to describe the planning efforts to restore and rebuild the affected settlements. The important point to make is that the story of reconstruction after natural disasters is an important chapter in Greek urban planning, but also regional spatial planning, since an earthquake or a forest mega-fire may pose questions of restructuring a settlement system in its entirety and open "windows of opportunity" for mitigation planning. The ties and necessary interactions between civil protection and spatial planning are made even more urgent in the conditions of rapid climate change.

The examples used here to make clear the effect of natural disasters on planning thinking are those of the city of Corinth, the island of Santorini, the township of Andritsaina, the mountain range of Tzoumerca (South Pindos), and Mati, a suburb of Athens (see Fig. 8.12). Corinth and Santorini have repeatedly attracted the interest of planning and architectural historians and an extensive literature exists. Corinth was ruined in 1858 and again in 1928. The Santorini earthquake of 1956 was the worst in Europe in the whole of the twentieth century. The Andritsaina and Tzoumerca planning projects have been a great learning opportunity for the author of the present book, as a member of the town planning research centre of the National Technical University of Athens in the 1960s, when the Peloponnese and Epirus suffered the impact of earthquakes. Andritsaina was a typical case of a small but historic township of great architectural value. The Tzoumerca case showed how the survival of a whole network of mountain villages, damaged by an earthquake and landslides, can be made doubtful. Concentration of the population in head villages was considered as an appropriate strategy. Mati is a metropolitan suburb of Athens, where the fire of 2018 left more than 100 victims. It is now replanned with a new type of town plan, the so-called Special Town Plan, an innovation of very recent legislation.

was later to assume a high-ranking role in the left-wing government of 2015, spoke of a reform which was anachronistic, anti-environmental, non-scientific, and served private interests. Revisions were attempted after 2015, but the new government finally signed a 3rd Memorandum.[25] It also retained the Special Town Plans the use of which extended to a variety of purposes, including the planning of the islands of Myconos and Santorini, the prime foreign tourism destinations of the country.

In the conditions of the economic crisis of the 2010 decade, urban and regional planning could not remain unaffected, especially in the European South (Knieling and Othengrafen eds 2016; Kramer 2011; Rivolin 2017). Urban centres in Greece were faced with new challenges (Papaioannou and Nikolakopoulou 2016). Melissas (2012) spoke of the "end of town planning". On the eve of the crisis, various organizations had already put their proposals for the planning system on the table of public debate (see, e.g., ISTAME 2009a, b or ITA 2006). A well-documented report of an expert committee of the Local Government Institute contained a clear proposal for the introduction of a system bypassing routine planning procedures and facilitating investment initiatives of both the public and private sectors (ITA 2006). When the crisis arrived in Greece, the country's creditors (see Chap. 6) increased their pressure to produce a lean and efficient planning system. The state was bankrupt and the margins of resistance to pressure were very narrow. The country was under foreign supervision. Soon the conditions under which planning could and would be exercised were totally transformed. An attempt started to rationalize the planning system, through what was labelled a "regional and urban planning reform", both to respond to the new economic landscape of crisis and to satisfy external pressure, particularly after the 2nd Memorandum of 2012. The outcome was Law 4269/2014, which critics considered unnecessary, since, in their judgement, the country already possessed a satisfactory legislative framework since the end of the 1990s (Beriatos 2015). Nevertheless, when a left-wing government came to power, this law was replaced by yet another one, i.e., Law 4447/2016 (Giannakourou 2019; Karatsolis 2020). The differences between the two successive laws were insignificant. This 2016 law is still in force, although amended by Law 4759/2020, which was itself being amended in 2021. With the amended 2016 law made their appearance the Local Town Plans and the Special Town Plans, to which we will return. The fundamental structure of the system was unaltered.

The importance of these years, from our present point view, lies mainly in the pressure under which the planning system was being redrafted, and, even more so, in the emergence of a parallel system of planning as a response to the need to attract investment.

---

[26] WHO (World Health Organization) (1999) Disaster prevention and mitigation. WHO/EHA/EHTP https://apps.who.int/disasters/repo/5514.pdf

[27] **Note**: The titles of publications in Greek have been translated by the author. The indication "in Greek" appears at the end of the title. For acronyms used in the references, apart from journal titles,

Text-Box 8.10: A Parallel System of Planning for Development Zones

In 2010, the Greek society awoke to the reality of a severe economic crisis. The attraction of private large-scale investment became an urgent necessity. The implication for spatial planning was the introduction of instruments which would facilitate fast-track investment projects outside the slow customary planning process. Within two years a number of laws were enacted to speed up the process. In 2010 the company Invest in Greece s.a. was set up, by a law the purpose of which was the acceleration and transparency of strategic investments. The same law introduced special plans of spatial development for such investments. A year later, three more special-purpose zones and plans were introduced by law for organized zones of industrial and business activity, for public real estate properties and mixed-use tourism resorts. The Hellenic Republic Asset Development Fund was created to manage public assets. Special legislation was enacted for the redevelopment of the old Hellinikon airport (see Text-Box 8.11). Further legislation for the encouragement of strategic investments was being debated in 2021.

Development of tourism facilities was of obvious priority both in the case of developments on privately-owned land and in that of public properties, which would be sold or leased to private firms (Kalantzi and Tsiotas 2010). Proposed developments, not surprisingly, were challenged in the Council of State which had to judge the constitutionality of the relevant laws and presidential decrees. Islands were the most attractive location for developments of strategic priority (see Fig. 8.7). Afantou in Rhodes is one of the best-known public properties destined for tourism development. Appeals against it were rejected by the Council of State in 2016. A section was exempted from development because of an archaeological site. Three years later an agreement was signed for the development of a strategic tourism project in Elounda, Crete, in this case on private land. Appeals were lodged in this case too.

Fig. 8.7 Examples of tourism projects under "fast-track" legislation (see Text-Box 8.10) (Afantou in Rhodes and Elounda Hills in Crete)
Source: Afantou and Elounda Hills websites

Ad hoc planning regulations for public land properties to enable tourism development have a long history. In 1993 even detailed building conditions and plot ratios were fixed by law for such properties. This law sanctioned a bypass planning procedure and was the ancestor of 2010 and 2011 laws which established special plans for the integrated development of areas of strategic investments and for public real estate properties (known in Greece as ESCHASE and ESCHADA respectively), as well as of the 2014 and 2016 planning laws which introduced the Special Town Plans. All special purpose zones, enjoying a preferential planning treatment, known as "organized activity containers" for tourism, manufacturing, or business services, as well as the ESCHASE and ESCHADA, were placed under the umbrella of Special Town Plans. All these cases enjoyed a status of "fast track" procedures and formed a parallel system of planning, outside the rules of the slow and cumbersome mainstream system. In fact, this parallel system, by its sheer integration in the category of Special Town Plans, was quickly made part of the main body of planning legislation. Soon, under the pressure of circumstances, the use of Special Town Plans was extended to other ad hoc cases like a park in Athens or the planning of the resort of Mati in Attica, where a tragic fire ravaged a large housing area and cost the lives of more than one hundred persons. The decision was later made to employ the instrument of Special Town Plans (rather than the regular Local Town Plans, with which they have the same legal status) in the case of over-burdened tourist destinations, e.g., the islands of Santorini and Mykonos, or in other municipal entities, selected for unknown reasons.

The "fast track" planning alternative had additional precedents besides those mentioned above. One was the organization of the Athens 2004 Olympic Games. Faced with tight time schedules and the impossibility of relying on normal planning processes, the government legislated in 1999 and introduced a new breed of plans, the special integrated development plans, in order to avoid the trodden path of normal planning.

The other spectacular precedent was the redevelopment of "Hellinikon", the old Athens international airport, which closed after a 1995 law sanctioned the contract for the construction of a new airport. After a long period of endless studies and competitions, the critical moment came, in the middle of the economic crisis, to vote a law which opened the way to a major private development and used the "special integrated development plan" as a medium (see Text-Box 8.11). By 2021, the Hellinikon project was under way, as an upper-class housing and business precinct, in fact a new town. Its skyscrapers, luxury apartments, and casino will soon enjoy the view of the Saronic Gulf and dispute the landmark status of the Acropolis and the Parthenon.

The end of the 2010 decade found the country badly wounded by the economic crisis and faced with its long-term consequences. The public debt remains high and real investment was lower by 60% than in the pre-crisis period (OECD 2018, 4). The pandemic of Covid-19 which started in 2020 and still claims thousands of new cases and the effects of the 2021 wildfires (e.g., more than 50,000 hectares burnt in Euboea) will render the recovery much more difficult. The spatial effects of the continuing crisis are difficult to predict, but an overview of the most recent developments in planning activity and law is in order.

Text-Box 8.11: The Redevelopment of Hellinikon Airport

The redevelopment of the former international airport of Athens at Hellinikon is a classical case of a multi-purpose urban project of a neoliberal character. The project includes the old airport itself and a coastal strip amounting to a total of over 600 hectares. The area includes a leisure boat marina and is located on the coast of Saronikos Gulf, next to a golf course. The airport ceased to function when a new one was built on the east of Athens, before the 2004 Olympics Games of which certain facilities were in fact located in the old airport itself. It was in 1995 that a law had been passed for the construction of the new airport and the conversion of Hellinikon to a park and various ancillary uses.

A plan proposal for the redevelopment of the area was submitted in 1997 by a research laboratory of the National Technical University of Athens (NTUA). The key concept of the proposal was the creation of a large green park and of leisure, art, and education facilities. Commercial development areas were to be included, only to the extent of covering the cost of the project. In the future the area was to be managed by a public agency, with the participation of the private sector. The project had been commissioned by the Master Plan Organization of Athens, but was never completed because of the then government's decision to launch an international architectural competition. Several proposals were in the meantime submitted, including one by another NTUA laboratory on behalf of the municipalities of the Hellinikon precinct. A new plan was produced in 2006 by the government based on the first prize of the competition.

The whole process came to a halt when the economic crisis of 2008 hit the Greek economy. The responsibility of the project was assigned to the new organization of public real estate property and an ad hoc agency was set up to handle the planning process of the site. To support the task, a special land use planning law was enacted in 2012 bypassing normal legislation. The ad hoc planning agency drafted its own plan, but once again economic priorities and the urgent need to attract foreign investment imposed a new approach. Private investment bids were invited and an investment consortium (Hellinikon Global) under the company Lamda Development was selected. The privatization process was bitterly criticized by the then left-wing opposition, which, however, finalized the deal when it came to power in 2015 with yet another special law. Repeated rounds of appeals to the supreme administrative court (Council of State) were rejected and the project entered a phase of implementation. Apart from a park, the new plan provides for a casino, several 200 metres-tall skyscrapers, private housing for 15.000 people, commercial centres, office accommodation, hotels etc. The creation of a luxury suburb is beginning to take shape more than a quarter of a century after the initial decision to close the airport (see Fig. 8.8).

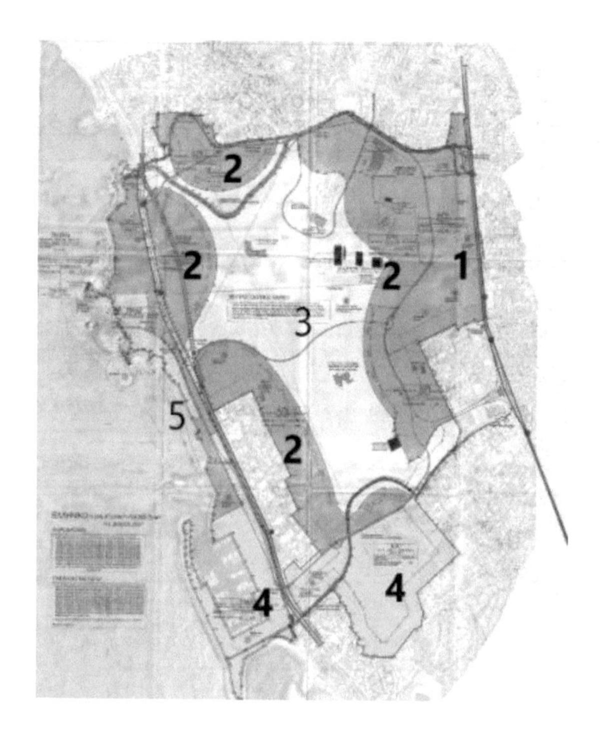

Fig. 8.8 The Hellinikon Global land use plan for the redevelopment of Hellinikon airport (see Text-Box 8.11)
(1: Urban centre. 2: Mixed housing. 3: Metropolitan park. 4: Mixed development zones. 5: Exclusive housing)
Source: Ministry of the Environment and Energy

8.14 The Rational and Hierarchical Edifice of Greek Planning

Starting from 2017, updated regional spatial frameworks were finished and approved. Conditions in the economy were far from ideal and it remains to be seen how they will affect the implementation of the new plans. Their legal framework, that of the law of 2016, did not change in the 2020 new law, but their initial assumptions can prove doubtful. At the same time, the revision of special spatial frameworks on a national scale (e.g., for industry, tourism, renewable sources of energy, fish farming) was under way, which means that the pattern of the 2000s is being repeated, because regional plans will have to be adjusted to national guidelines. A new special plan has been commissioned for the sector of mineral extraction, an economic activity of great importance and international implications given the potential of oil drilling in maritime zones and the agreements with neighbouring countries for the delimitation of Exclusive Economic Zones (Maniatis 2012; Wassenhoven 2017a). The overall national plan has been in the meantime replaced by an indicative guiding strategy. A much-expected novelty in the field of regional spatial planning is maritime planning. Originally, this was to be the object of a national special plan, but in the planning Law 4759/2020 it is stipulated that maritime space will be dealt with in several plans classified as regional. Difficulties

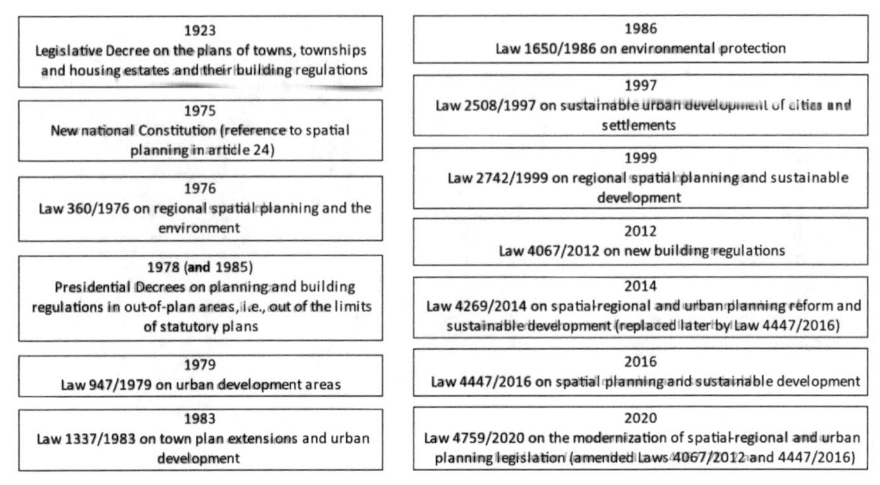

Fig. 8.9 Key acts and decrees in the twentieth and twenty-first centuries
Note: Selection by the author

related to exclusive economic zones may have been the reason. The importance of maritime planning is more than evident in the case of Greece (Wassenhoven 2017a, 2018c; Kyvelou 2016; Papageorgiou 2016a, b; Beriatos 2016; Kanellopoulou 2018; Kyvelou and Pothitaki 2016; Banousis 2016; Bolanou and Kiousopoulos 2014; Konstantinidis 2012; Parri 2015). An issue which provoked a debate was the interaction between land planning, coastal zone management, and maritime planning (Economou 2018; Piperis 2018; Pournara 2018; Coccossis 2018). The solution chosen in Law 4759/2020 was to limit maritime planning to sea waters and avoid potential conflicts. Maritime planning will put to the test inter-ministerial cooperation because of the large number of ministries with a say over sea space (Fig. 8.9).

The intention of the new government which took over in 2019, to revise planning legislation was clear from the beginning. Although Law 4759/2020 did change the law in a number of areas, the main structure of the planning system was not really modified. It retained a hierarchical, rational and systemic character, which had first appeared years earlier. The political leadership was attacked first of all for the absence of advance consultation. It must be remembered that the discussion leading to the final vote in parliament took place while extraordinary measures were in force to check the spreading of the Covid-19 pandemic. The bill was balancing on a tightrope between the aim of making a contribution to the effort of attracting investment, the pressure to protect the environment, and the urgent need to make social concessions after a prolonged economic crisis. It was not the easiest of circumstances. The environmental goal can be summarized in the effort to combat out-of-plan construction which damages the countryside, particularly the plague of a variety of derogations which facilitated unchecked building (Fig. 8.10).

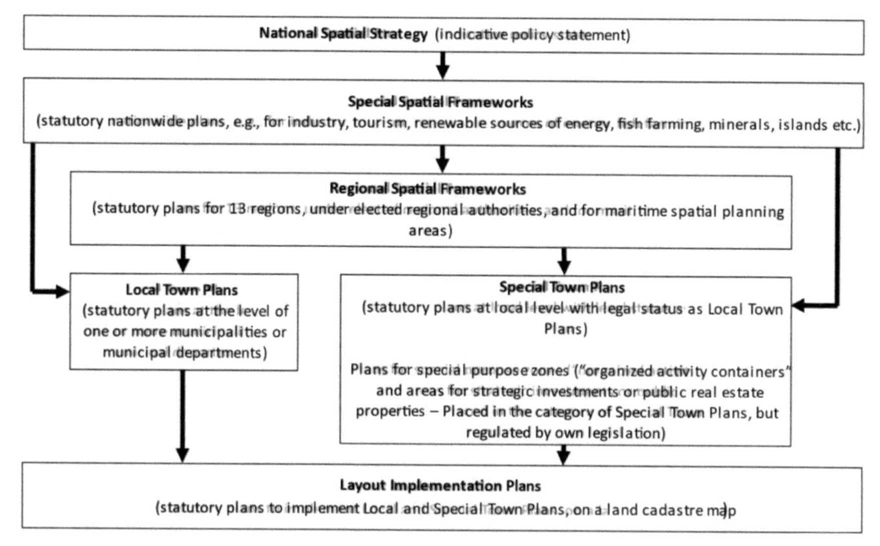

Fig. 8.10 The present hierarchy of spatial plans
Note: Diagram drawn by the author

Out-of-plan building is a real curse for the countryside which has caused enormous environmental damage (Wassenhoven 1995b, 1997d, 2014b; Michail 2008; Karra 2008). During the debate for the 2020 law, there were supporting voices in favour of the abolition of out-of-plan construction, both in the press and the social media, as well as reports of some environmental organizations, but the reactions were violent and frequently originated not only in the parliamentary opposition but also in the ruling party itself and in professional associations, like the Technical Chamber of Greece and the Association of Greek Urban and Regional Planners. The limitations imposed were described by some as blatant expropriation or as a *coup d'état*. The end result was that not only some of the law provisions were withdrawn, but also that mistakes were made in the final act of parliament, which caused problems later when amendments were voted to correct them. The deputy minister who had shouldered the task of drafting the bill was quickly replaced.

A fundamental aim of the new law was to simplify and accelerate all processes of plan production and revision and to clarify the hierarchical relations of planning documents. As mentioned, the new law did not depart substantially from the logic of its predecessor of 2016, but rather reorganized its structure. A National Spatial Strategy was to replace the old national plan, as a simple non-binding policy document. Changes were also introduced regarding consultation and adjudication organs and procedures.

Text-Box 8.12: Town Plans – A Slow Process of Disillusionment

As explained in the main text of the book, a major effort was undertaken in the 1980s, on the basis of the 1983 planning law, to produce "general town plans" for all Greek towns. The effort continued in the 2000 decade, this time founded on the 1997 law. Most of these plans never reached the stage of full implementation on the ground. What's even worse is that plans of the first batch were never updated and tend to become detached from reality. The government launched in 2021 a new programme of commissioning "local town plans", under a new name, to cover the total of the national territory. The first such plans were being commissioned to private consultancy firms in the autumn of 2021, but the procedure adopted soon encountered the fierce resistance of the Greek Association of Urban and Regional Planners.

An example of a "ageing" plan and of the so far abortive efforts to produce a new one is that of the town of Nafplion, the seat of the first Greek government after the independence war of 1821. A "general town plan", drawn by private consultants, was approved in 1987, which the municipality decided to update in 2003, when a contract was signed with another consultancy firm. The consultants' proposal was submitted in 2006 and resubmitted three years later, but it was met with objections and never reached the stage of approval. In 2012, the municipality decided to pursue the effort with its own technical service, with the assistance of external advisors. Following an extensive participation effort, the new plan was completed in 2017 and received the agreement of the municipal council in 2019. However, the final approval of the central government is still being awaited. The nearby town of Argos had a better luck. Its plan was approved in 2006. The example of Nafplion is instructive. For a variety of reasons, plan making can be an exhaustive and demoralizing experience. Ambitious goals are faced with hard administrative realities (see Fig. 8.11).

The disagreements reached their peak when the bill was introduced in parliament. For those following the debate it was a most illuminating seminar on the pros and cons of the planning system. From the author's point of view the most disappointing outcome was the failure to put an end to the infamous out-of-plan building activity. Otherwise, the existing legislation concerning the Special and Regional Spatial Frameworks and the Local and Special Town Plans was not amended. Soon afterwards, a long-term programme to prepare the new Local Town Plans was launched and the first ones were commissioned in the summer of 2021. Unfinished plans under prior legislation will have to be completed before the end of 2022. There are towns where the statutory plan is totally out of date. The whole process shows the extreme slowness of the planning process, with the inevitable uncertainty which prevails in the use of land. The new Local Town Plans will take years before

The 2006 plan of Argos and the 1987 plan of Nafplion

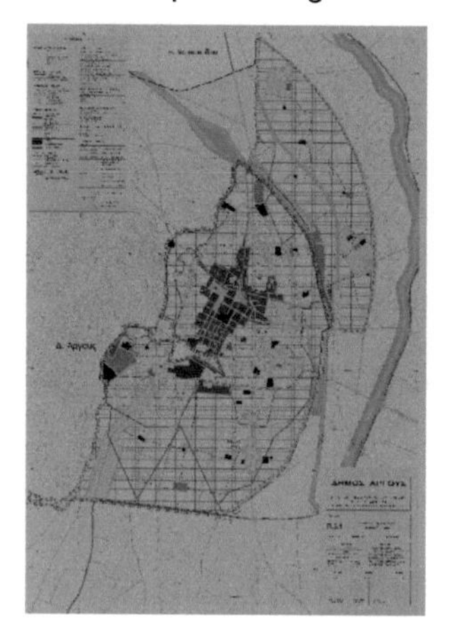

The Nafplion plans of 2006 and 2017

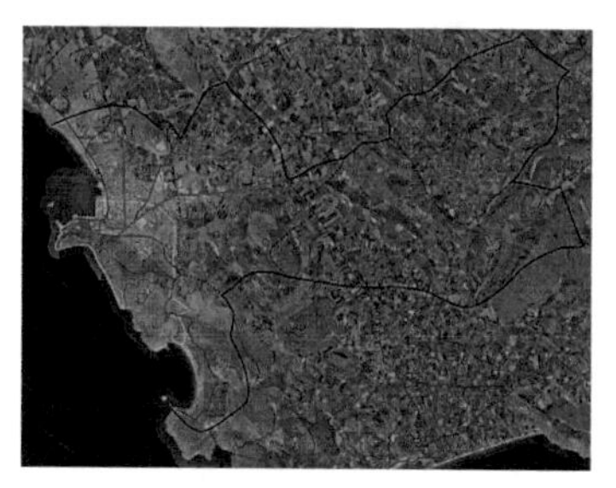

Fig. 8.11 Examples of town plans (see Text-Box 8.12)

they are completed and in the meantime the country will have towns with old-type plans and towns with the new form of plans.

The new local plans will adhere to the boundaries of municipalities or municipal departments and will all be approved by Presidential Decree, in other words by the central government. This means that there will be no chance of producing plans independently of official administrative boundaries, e.g., based on interpretations with cultural or historical foundations. But then this is not a problem in Greece only. The rigidity of the administrative structure and the limits it imposes on planning is not an exclusive Greek drawback. Davoudi (2018, 19 and 24), referring to Britain, mentions "the difficulties of translating relational space into planning practices whose spatial imagery is still dominated by fixities and certainties of absolute space and bounded place". She uses the conclusions of a study, which shows "that 'planners' conceptual interpretations of the socio-spatial processes have remained surprisingly similar to the ones formed in the mid-twentieth century by a positivist view of the world". Davoudi adds that "planners find it difficult to translate the relational space into the administrative, legal confines of planning practice". An obvious case of planning where adherence to administrative boundaries is questionable is the areas affected by natural disasters. Planning for such areas has always been a special chapter in Greek town and regional planning, which deserves a place in the presentation of the country's planning system. This, however, has to be done in the context of environmental protection. The following section, in a sense, is a follow-up of the last section of Chap. 5 on the "climate current".

8.15 Spatial Planning, Environment and Natural Disasters

Environmental protection and disaster risk prevention and mitigation are vast policy and research areas in themselves, which cannot be possibly addressed adequately in a short section. We can only touch on the interactions with spatial planning and the contribution of the latter to support them. In Greece, the years after the 1967–1974 dictatorship are notable for an increased legislative activity for the protection of the environment (see OECD 1983, 2000, 2009), prompted by EU policies and

1967 plan of the town of Andritsaina (NTUA/SPE 1969)

Fig. 8.12 Plans for disaster-stricken areas (see Text-Box 8.9) (The 1858 plan of Corinth, local town plan in Santorini after the 1956 earthquake, the Mati plan of 2020, and a plan for the Tzoumerca region (1968). NTUA/SPE 1971a)

international conventions, e.g., the landscape and heritage conventions (Angelidis 2000, 339–340; Beriatos 1994b; Papageorgiou et al. 2010; Council of Europe 2000a; Tsilimingas and Gourgiotis 2014; UNESCO 1972). Environmental law retained a role independent from mainstream planning policy (Siouti 1995, 2011), although it is often claimed that planning law reserves for planning the main privilege to protect the environment (Karatsolis 2020, 39). It is noteworthy that the codification of urban and regional planning legislation which was in process in 2020–2022 did not include environmental legislation. As Giannakourou (1994b) has remarked. this produces a "sectoralization", evident in government structures and policies. This is particularly true in two areas of protection of the natural and cultural environment, i.e., in the protection of forests and archaeological sites (Getimis and Economou 1992, 30–31). Parenthetically, it is worth remembering that the first law on archaeological protection dates from the days of Bavarian administration in 1834 (Wassenhoven ME 2019). There is no doubt that the environmental factor, with the exception of the protection of forests and the sea shore, with very poor results, entered late in the domain of statutory planning. International conventions and the case law of the Council of State accelerated this process. Environmental protection soon acquired a legislative framework (WWF Hellas and Greek Ombudsman 2009; Siouti 2011). The law of 1986 on the protection of the environment introduced important innovations directly affecting spatial planning. Gradually, the emphasis on environmental concepts was bound to force planning to incorporate them into its own conceptual framework (Angelidis 1994). The directives of the European Union regarding the maritime environment and maritime spatial planning, both now part of national legislation, added a new field which spatial planning will have to deal with (Wassenhoven 2017a, 2018c; Papageorgiou 2015, 2016a, b; Stefani and Tsilimingas 2015; Theodora 2018c).

The overlaps between environmental protection and urban and regional planning are obvious in three areas: protected zones, planning tools, and planning process. Although protected zones, e.g., in forest land, had existed for a long time, the Natura zones were a major breakthrough (Douros 2007; Efthymiopoulos and Modinos 2007; Kollias and Klabatsea 1995; Maniatis 2009; Papagiannis 2009). Land use policy in protected zones was acquiring priority in the European Union (Maksin et al. 2018). The so-called Habitats Directive of the EU (1992) and the consequent Natura 2000 Network created a new landscape for Greece, frequently condemned by the European Court of Justice for her failure to implement the directive (Wassenhoven 2000b). The Habitats Directive extended the provisions of the Birds Directive of 1979, which had already imposed the creation of Special Protection Areas, by creating Special Areas of Conservation, all of them now part of the Natura Network, which was included in the Greek law on biodiversity of 2011. The total area of the Natura zones amounts to about 5.5 m. hectares, out of a total area of the country of 13.2 m. hectares. The biodiversity law requires the elaboration of special plans, which are in fact a version of spatial plans and are now called Special Environmental Studies after an amendment of 2020. These special plans will be in the future embedded in local plans which are to cover the whole of the national territory in the years to come. Special plans of environmental upgrading and revival for declining or abandoned settlements were introduced in 2014 – but never used – thus adding to the confusing multiplicity of instruments.

The influence of environmental policy, all emanating from the European Union, on new planning tools is not limited to special environmental plans. The process itself of all spatial planning is fundamentally altered, as all plans have to be accompanied by strategic environmental assessment studies, studies for the delineation and control of water currents, and special geological studies. In the latter cases, we have an additional interpenetration of environmental and spatial planning, which is closely related with another major interest, viz. planning for disaster risk prevention. For some time already, municipalities were supposed to have civil protection plans and urban plans had the obligation to designate areas of escape in the case of earthquakes, but all spatial plans have now to give increased attention to civil protection. The expanded environmental and risk prevention content was evident in the new generation of regional spatial plans of the 2010 decade, but is now explicitly demanded in the guidelines for the new local town plans. These guidelines, issued in the summer of 2021, specify that the planning study should include plans for climate change adaptation (see also the last section of Chap. 5), for emergency situations, flood and fire risk, and soil erosion.

Greece has had a long experience of planning for the reconstruction of settlements destroyed by seismic events, already before the Second World War (Corinth) or soon after (Ionian Islands, Volos, Santorini), but in the last few years was severely struck by lethal floods and forest fires in Attica, the Peloponnese, Euboea and elsewhere. In the summer of 2021, the large fires which devastated more than 100.000 hectares have opened a debate on the future of disaster risk mitigation and prevention, which will not leave spatial planning unaffected. Assuming that mitigation is limited to the reduction of the severity of human and material damage, it is wiser to give emphasis to prevention in order to ensure that human action or natural phenomena do not result in disaster.[26] The key, however, as Sapountzaki et al. (2022) explain, lies in the realization that risk is ultimately a function of the factors of hazard, vulnerability and exposure. "Hazard is related to the size, duration, location and time of the extreme phenomenon. Vulnerability is an endogenous feature of people, communities, territories, infrastructure etc.; while exposure indicates the geographical location of the entity at risk compared to the location and range of the hazard" (Sapountzaki 2019b). Reducing vulnerability and increasing resilience are both physical and social parameters. In Greece, as elsewhere, the systemic nature of risk, disasters, and climate change is painfully coming home.

---

see at the end of the table of diagrams, figures and text-boxes at the beginning of the book.

Epilogue: New Challenges and the Danger of déjà vu

This chapter was a real challenge for the author, because it was here that he had to outline the "empirical reality" on which he will try to base his theoretical contribution in Part III of the book. As in the Chap. 6 on history, emphasis was placed on the influence of the past on the Greek planning system and on the importance of the town planning achievements of the period between the two World Wars, either for the settlement of refugees in rural and urban areas, or for the planning of suburbs of western inspiration. These early experiments were based on path-breaking legislation enacted in 1923 by a modernizing, liberal administration. In spite of its modern character, this legislation, especially its subsequent amendments, opened the way for the creation of a most grievous phenomenon, the environmentally disastrous, dispersed construction in out-of-plan, unregulated areas. The efforts in the same period to produce a master plan for Athens, capital of the country, had poor results and were resumed in the 1960s, after the intervening reconstruction period, which followed the Second World War and the Civil War of the 1940s. The 1960s were a most interesting experimental period for town planning, which came to an end with the military coup of 1967. It was before the dictatorship that a most important ferment of new ideas appeared, focused on regional development and the networking of cities and growth poles. Genuine innovations in legislation and planning practice had to wait until the late 1970s and 1980s, but plan implementation continued to suffer and disorderly urbanization was the rule. Problems were caused by problematic complex legislation and difficulties of legal interpretation, which the supreme administrative court (Council of State) had to disentangle.

Under the influence of European Union policies new legislation appeared in the late 1990s which introduced new planning instruments. Sustainable development was now a priority, with environmental concerns and natural hazards receiving increasing attention. The streamlining and consolidation of the planning system became an urgent demand in the post-2010 economic crisis and tremendous pressures in this direction were exerted on the Greek government by international lending institutions. After a number of legislative initiatives until, and including, 2020, the country seemed to be equipped with a rational and comprehensive planning system, but the underlying social and political processes of bargaining and favouritism were still present. A deplorable indication is that a new flood of law amendments surfaced in 2021, supposedly justified by the financial bankruptcy and the economic impact of the Covid-19 pandemic. Nevertheless, both these developments and the rising threat of climate change (see Chap. 5), had two beneficial effects. One was the awakening of central government in the face of the climate crisis and a renewed determination to adapt the planning system to new challenges. The other was not exclusive to planning, but it is bound to modernize planning practice too. Technological innovations are gradually being introduced, making use of advanced digital information technology, a development which had been on the cards many years earlier (Sellis and Georgoulis 1997). The central state, and, even more so, regional governments, face an urgent need of adaptation (Rodakinias et al. 2008; Skordili 2008). Apart from the already established use of Geographical Information

Systems (Maniatis and Myridis 1990; Pappas 1995), a digital information platform has now been announced in which all plans will be recorded, while the so-called "electronic identity" of buildings is now obligatory. Digitalization has received an enormous impetus in the conditions of the pandemic of the coronavirus and the impressive activity of a new ministry of digital policy is spreading in all areas of government policy.

References[27]

Academy of Athens (1993) Conclusion of the Academy's scientific committee at the end of the project on the development of Thrace. Statement of A. Angelopoulos, J. Pespazoglou and X. Zolotas, members of the Academy, 10 June 1993

AGRIDEV (1965) Crete: development plan 1965–67, Volume 1/Summary / Draft, Agricultural development Company (International) Ltd., Tel Aviv

Alexakis P, Katrivesis V (1992) Development incentives for industrial areas in Greece. Topos 1992(4):77–111. (in Greek)

Allen A (2017) Peri-urbanization and the political ecology of differential sustainability. In: Parnell S, Oldfield S (eds) The Routledge Handbook on cities of the Global South. Routledge, London. (first published 2014), pp 522–538

Alterman R (2017) Planners' beacon, compass and scale: linking planning theory, implementation analysis and planning law. In: Haselsberger B (ed) Encounters in planning thought: 16 autobiographical essays from key thinkers in spatial planning. Routledge, New York/London, pp 260–279

Anastassiadis A, Burgel G (2001) Dynamiques urbaines et dynamiques régionales 1961–1991. In: Burgel G, Demathas Z (eds) Greece in the face of the 3rd millennium. Panteion University, Athens, pp 31–47

Andreadis S (1966) Regional development and the structure of the urban network in Greece. Koinoniologiki Skepsi 1966(2):153–175. (in Greek)

Andrikopoulou E (1990) Regional policy transformations and the prospects of local development policies. In: Psycharis Y (ed) The functions of the state in a period of crisis: theory and Greek experience. Conference proceedings. Sakis Karagiorgas Foundation, Athens, pp 310–322. (in Greek)

Andrikopoulou E (1992) Whither regional policy? Local development and the state in Greece. In: Dunford M, Kafkalas G (eds) Cities and regions in the new Europe: the global-local interplay and spatial development strategies. Belhaven Press, London, pp 195–212

Andrikopoulou E (1993) New local development institutions and social policy problems. In: Sakis Karagiorgas Foundation (ed) Dimensions of present-day social policy. Conference proceedings. Athens, pp 527–542 (in Greek)

Abbreviations of journal titles: AT Architektonika Themata (Gr); DEP Diethnis kai Evropaiki Politiki (Gr); EpKE Epitheorisi Koinonikon Erevnon (Gr); EPS European Planning Studies; PeD Perivallon kai Dikaio (Gr); PoPer Poli kai Perifereia (Gr); PPR Planning Practice and Research; PT Planning Theory; PTP Planning Theory and Practice; SyT Synchrona Themata (Gr); TeC Technika Chronika (Gr); UG Urban Geography; US Urban Studies; USc Urban Science. The letters Gr indicate a Greek journal. The titles of other journals are mentioned in full.

[28] The title of the article in the table of contents was different.

Andrikopoulou E (1994) Regional policy in the 1980s and prospects. In: Getimis P, Kafkalas G, Maravegias N (eds) Urban and regional development: theory, analysis and policy. Themelio, Athens, pp 347–383. (in Greek)
Andrikopoulou E (1995) The regions in the European Union. Themelio, Athens. (in Greek)
Andrikopoulou E (2000) Cohesion policy and its prospects in New European Space. In: Andrikopoulou E, Kafkalas G (eds) The new European space: enlargement and the geography of European development. Themelio, Athens, pp 339–380. (in Greek)
Andrikopoulou E (2003) The meaning of cohesion. In: Getimis P, Kafkalas G (eds) Space and environment: globalization, governance, sustainability. Athens, Topos and Institute of the Urban Environment and Human Potential, Panteion University, pp 67–77. (in Greek)
Andrikopoulou E (2004) Territorial cohesion and spatial development in the European Union. In: Kafkalas G (ed) Issues of spatial development. Kritiki, Athens, pp 173–212. (in Greek)
Andrikopoulou E (2010) Polycentricity of the settlement network: the contribution of the founders of spatial science and its application in Greece. In: Demetriadis EP, Kafkalas G, Tsoukala K (eds) The *Logos* of the *Polis*, Volume in honour of Prof. A.Ph. Lagopoulos. University Studio Press, Thessaloniki, pp 87–98. (in Greek)
Andrikopoulou E, Kafkalas G (1985) The regulation of regional space: theory and practice. Paratiritis, Thessaloniki. (in Greek)
Andrikopoulou E, Kafkalas G (2004) Organization of the settlement network and urban-rural relations: the impact of European policies on Greek space. In: Angelidis M (ed) New polycentricity and spatial development. Conference proceedings (2001). YPECHODE, Ministry of Economy, NTUA and Université de Paris X / Nanterre. Athens, pp 75–93 (in Greek)
Andrikopoulou E, Kafkalas G (2008) "Constructing" the conceptual content of territorial cohesion. Diethnis kai Evropaiki Politiki 10(April–June):218–234
Aneroussi F (2004) The march of town planning and the impasses. In: NTUA (ed) City and space from the 20th to the 21st century. Volume in honour of Prof. A.I. Aravantinos. NTUA, University of Thessaly, and SEPOCH, Athens, pp 25–39 (in Greek)
Aneroussi F et al (1977) Greek report: analysis of planning and implementation procedures in Greece, with special reference to the Athens region. Cooperative action programme: Joint Activity on Urban Management, Symposium on implementation of urban plans (Athens, 10–14 October 1977). Organization for Economic Cooperation and Development
Angelidis M (1985) Spatial programming and planning, and decentralization in Greece: problems, prospects and proposals. In: TEE (ed) Regional programming and planning: the role of the administration and local self-government. One-day conference proceedings. TeC April-June 1985: 43–48. (in Greek)
Angelidis M (1991) Spatial regional planning. Symmetria, Athens. (in Greek)
Angelidis M (1994) The interdisciplinary approach to peak issues of spatial development in Greece. In: Rokos D (ed) Sciences and the environment at the end of the century: problems and prospects. NTUA and Enallaktikes Ekdoseis, Athens, pp 251–260. (in Greek)
Angelidis M (1997) Transformation of the urban system and regional policy in Greece. In: Sellis T, Georgoulis D (eds) Urban, regional and environmental planning and informatics to planning in an era of transition, Proceedings of the Athens International Conference, 22-24.10.1997. NTUA, Athens, pp 78–98
Angelidis M (2000) Spatial regional planning and sustainable development. Symmetria, Athens. (in Greek)
Angelidis M (2002) Sustainable development of cities, European Union, and architecture. In: SADAS (ed) Architecture and the Greek city in the 21st century. Proceedings of the 10th Panhellenic Architectural Conference (1999). SADAS and TEE, Athens, pp 85–102. (in Greek)
Angelidis M (2004a) New centralities in Greece: Urban areas, small and middle-sized towns, and countryside areas. In: Angelidis M (ed) New polycentricity and spatial development. Conference proceedings (2001). YPECHODE, Ministry of Economy, NTUA and Université de Paris X / Nanterre. Athens, pp 20–36. (in Greek)
Angelidis M (2004b) European Union policies for spatial development. University Press NTUA, Athens. (in Greek)

Angelidis M, Karka G (2001) The settlement system and spatial management. In: Burgel G, Demathas Z (eds) Greece in the face of the 3rd millennium: space, economy, society in the last 40 years. Panteion University, Athens, pp 309–347. (in Greek)
Angelidou V (2014) Town planning intervention mechanisms and modern legislation, with emphasis on the transfer of development rights and the transfer of the building coefficient. Doctoral thesis. NTUA, Athens (in Greek)
Aravantinos A (1967) Problems of regional planning, town planning and housing in Greece. TeC (general edition) 5–6, May–June 1967 (in Greek)
Aravantinos A (1968) Coordination of settlement and industrial developments. Labour Housing Organization, Athens. (in Greek)
Aravantinos A (1970a) Regional development and town planning. In: CPER (ed) The framework of regional development. Lecture series. CPER and Gutenberg, Athens, pp 175–200 ((in Greek))
Aravantinos A (1970b) Planning objectives in modern Greece. Town Plan Rev 41(1):41–62
Aravantinos A (1973) Mutual impacts of industrial and urban development. In: NTUA (ed) Industrial development in space. Lecture series. NTUA, Athens, pp 125-147. (in Greek)
Aravantinos A (1985) Town and land use planning in Greece: actual state, problems and trends. In: Malusardi F (ed) Urbanistica, Territorio e Crisi dei Processi di Sviluppo. Officina Edizioni, Roma
Aravantinos A (1997) Town planning for sustainable development of urban space. Symmetria, Athens. (in Greek)
Aravantinos A (ed) (2001) European and Greek cities in the 21st century: the anticipated role of planning. Continued education programme. NTUA, Athens (in Greek)
Aravantinos A, Gerardi K (2007) Town planning (*poleodomia*). Lemma in the Encyclopaedia Papyrus – Larousse – Britannica (new edition 2007). Vol. 55. Annex on Greece. Papurus, Athens, pp 513–518 (in Greek)
Aravantinos A et al (1967a) Regionalplanung und Vorschlag für eine Siedlungsumordnung im Raum Ilia (Elis) im West-Peloponnes. Die Erde 1967(4):281–291
Aravantinos A, Desyllas N, Kontargyris D, Lampakis A, Loukakis P (1967b) Spatial regional plan and settlement restructuring of the Elia prefecture. AT 1967(10):32–38. (in Greek)
Argyris Th (1986) Regional economic policy in Greece: 1950–81. Adelfoi Kyriakidi, Thessaloniki (in Greek)
Argyropoulos T (1967) Town planning and the Greek city. AT 1967(1):40–43. (in Greek)
Arvanitaki A (1995) Regional policy and spatial planning in Europe and the inadequacies of the Greek spatial planning system. In: SEP (1995a) Regional development, spatial planning and environment in the framework of united Europe. Vol. III. Proceedings of a conference, Panteion University, 1995. SEP and *Topos,* pp 134–146. (in Greek)
Asprogerakas E (2003) Characteristics and development potential of Greek middle-sized cities. Doctoral thesis. NTUA, Athens (in Greek)
Asprogerakas E (2020) Strategies of integrated interventions in Greece: tools and governance schemes. PPR 35(5):575–588
Asprogerakas E, Zachari V (2020) The EU territorial cohesion discourse and the spatial planning system in Greece. EPS 28(3):583–603
AT Journal (1967) Issue 1. Sections on regional planning, regional development and town planning (in Greek)
AT Journal (1968) Issue 2. Sections on regional planning, regional development and town planning (in Greek)
AT Journal (1977) Issue 11. Special issue on the regulation of space in Greece (in Greek)
AT Journal (1981) Issue 15. Special issue on the regulation of space in Greece – II (in Greek)
Avgerinou-Kolonias S (1997) Questions of spatial development and spatial regional planning. NTUA, Athens. (in Greek)
Avgerinou-Kolonias S (1998) Les villes moyennes en Grèce: Changements démographiques et emploi urbain. In: Petites et grandes villes du bassin Méditerranéen. École Française de Rome, Rome

Avgerinou-Kolonias S (2001) Spatial dynamics of tourism 1961–1991. In: Burgel G, Demathas Z (eds) Greece in the face of the 3rd millennium: Space, economy, society in the last 40 years. Panteion University, Athens, pp 243–269. (in Greek)

Avgerinou-Kolonias S (2011) Preconditions and proposals for a new special spatial framework for tourism. In: Avgerinou-Kolonias S, Melissas D (eds) Economic conjuncture and spatial organization of tourism. Propobos, Athens, pp 81–95. (in Greek)

Avgerinou-Kolonias S, Melissas D (eds) (2011) Economic conjuncture and spatial organization of tourism. Propobos, Athens. (in Greek)

Bainbridge T (1998) The Penguin companion to European Union. Penguin, Harmondsworth

Balchin P, Sýkora L, Bull G (1999) Regional policy and planning in Europe. Routledge, London

Banousis D (2016) Nomothetic rules for maritime spatial planning in the European Union: grouping the initiatives. Nomos+Physis, 24-3-2016 (available in the web) (in Greek)

Beriatos E (1985) Political dimensions of spatial regional planning. In: TEE (ed) Regional programming and planning: the role of the administration and local self-government. One day conference proceedings. TeC April-June 1985, pp 16–19 (in Greek)

Beriatos E (1993) Spatial planning, administration, self-government: past and future of territorial organization in Greek space. Paper presented at the international conference "Greece in Europe: Spatial planning and regional policy towards the year 2000". December 1993. Athens (in Greek)

Beriatos E (1994a) Administrative restructuring at the local level: re-organization and replenishing of local self-government cells. Topos (special issue on spatial planning and the environment) 1994(8):41–70. (in Greek)

Beriatos E (1994b) Towards a comprehensive view of spatial and environmental planning: new approaches at a theoretical and practical level. In: Rokos D (ed) Sciences and the environment at the end of the century: Problems and prospects. NTUA and Enallaktikes Ekdoseis, Athens, pp 261–271. (in Greek)

Beriatos E (1995a) The role of local and regional self-government in the promotion of transboundary and transregional cooperation in the Balkans. In: SEP (ed) Frontier regions. Proceedings of a conference, 1995. SEP, Thessaloniki, pp 197–205. (in Greek)

Beriatos E (1995b) Spatial planning and its management: The problem of administrative and organizational structures. In: SEP (ed) Regional development, spatial planning and environment in the framework of united Europe. Proceedings of a conference, Panteion University, 1995. SEP and *Topos*. Vol. II, pp 187–198 (in Greek)

Beriatos E (1998) Organization and functions of agencies for the administration and management of metropolitan areas. Research paper series. University of Thessaly, Volos (in Greek)

Beriatos E (2000) Administrative geography and spatial planning in Greece: the evolution of spatial-administrative structure in the modern Greek state. In: Psycharis Y, Gospodini A, Christopoulou O (eds) Seventeen essays on planning, cities, and development. Thessaly University Press, Volos, pp 23–61. (in Greek)

Beriatos E (2001) The Greek insular space and the problems of small islands. In: YPECHODE (ed) Coastal and island space. Proceedings of a one-day conference. 29-4-1999. Athens, pp 13–20 (in Greek)

Beriatos E (2004) Effects of polycentric development policies in Greece to this day: Towards a new policy. In: Angelidis M (ed) New polycentricity and spatial development. Conference proceedings (2001). YPECHODE, Ministry of Economy, NTUA and Université de Paris X / Nanterre. Athens, pp 205–212. (in Greek)

Beriatos E (2010) Protected areas and management agencies: developments and prospects. In: Demetriadis EP, Kafkalas G, Tsoukala K (eds) The *Logos* of the *Polis*. Volume in honour of Prof. A.Ph. Lagopoulos. University Studio Press, Thessaloniki, pp 147–158. (in Greek)

Beriatos E (2015) The ekistic AIDS and planning culture (collection of articles in the daily press). Andy's publishers, Athens (in Greek)

Beriatos E (2016) International and European framework, Greece and spatial planning of coastal and maritime areas – between land and sea: a critical approach of coastal and maritime space.

In: Kyvelou S (ed) (2016) Maritime spatial issues: the maritime dimension of territorial cohesion – maritime spatial planning – sustainable blue development. Kritiki, Athens, pp 64–76. (in Greek)

Beriatos E, Papageorgiou M (2010) Spatial planning of coastal and maritime space. the case of Greece in the Mediterranean. In: Beriatos E, Papageorgiou M (eds) Spatial planning – town planning – environment in the 21st century. Thessaly University Press, Volos, pp 189–204. (in Greek)

Berry BJL, Horton FE (1970) Geographic perspectives on urban systems. Prentice-Hall, Englewood Cliffs

Bethemont J (2000) Géographie de la Méditerranée. Armand Colin, Paris

Biris C (1995) Athens: from the 19th to the 20th century. Melissa, Athens (first published 1966)

Blanc A (1965) Géographie des Balkans. Presses Universitaires de France, Paris

Bolan RS (2017) Urban planning's philosophical entanglements: the rugged, dialectical path from knowledge to action. Routledge, London

Bolanou C, Kiousopoulos J (2014) Marine spatial planning in Hellas; recent facts and perspectives. Paper presented at the FIG Congress 2014 "Engaging the Challenges – Enhancing the Relevance". Kuala Lumpur, Malaysia, 16–21 June 2014 (available on the web)

Booth P (2016) Planning and the rule of law. PTP 17(3):344–360

Bouzenberg C (1967) State, land and urban development. AT 1967(1):78–81

Brenner N (1998) Global cities, glocal states: global city formation and state territorial restructuring in contemporary Europe. Rev Int Polit Econ 5(1):1–37

Brenner N (1999) Globalisation as reterritorialisation: the re-scaling of urban governance in the European Union. US 36(3):431–451

Brenner N (2004) New state spaces: Urban governance and the rescaling of statehood. Oxford University Press, Oxford

Brenner N et al (eds) (2003) State/space: a reader. Blackwell, Oxford

Brezzi M, Veneri P (2015) Assessing polycentric urban systems in the OECD: country, regional and metropolitan perspectives. EPS 23(6):1128–1145

Burgel G (1976) Athens: the development of a Mediterranean capital. Exantas, Athens (Greek translation)

Camagni R, Capello R (2015) Second-rank city dynamics: theoretical interpretations behind their growth potentials. EPS 23(6):1041–1053

Camagni R, Capello R, Caragliu A (2015) The rise of second-rank cities: what role for agglomeration economies? EPS 23(6):1069–1089

Camhis M (1994) Towards a European spatial policy. Topos 1994(8):207–214. (in Greek)

Camhis M (2004) European Union and cities: the urban dimension of Community policies. Paper presented at a conference on Sustainable Development Policy of Cities in Greece, NTUA and YPECHODE, Athens, 10.9.2004 (in Greek)

Camhis M (2007) The unification of European space 1986–2006: a designed undertaking on a grand scale. Kritiki, Athens. (in Greek)

Camhis M (2008) European space and territorial cohesion: from the Single European Act to the European Reform Treaty. Diethnis kai Evropaiki Politiki 10(April–June):201–217. (in Greek)

Carter FW (1989) Post-war regional economic development: a comparison between Bulgaria and Greece. In: Dimitriadis EP, Yerolympos – Karadimos A (eds) Space and history: Urban, architectural and regional space. Prodeedings of the Skopelos symposium. Aristotle University of Thessaloniki, Thessaloniki (papers in English and Greek), pp 203–210

CEC (1991a) The regions in the 1990s. Luxembourg, DG Regional Policy

CEC (1991b) Europe 2000: development prospects of the European territory. Luxembourg (Greek edition)

CEC (1995) Europe 2000+: cooperation for regional development in Europe. Luxembourg (Greek edition)

Chiotis GP (1970) Regional development policy and the organizational – institutional framework of regional programming. In: CPER (ed) The framework of regional development. Lecture series. CPER and Gutenberg, Athens, pp 67–107. (in Greek)

Chlepas N-K (1994) Multi-level self-government. Ant. N. Sakkoulas, Athens (in Greek)
Chorianopoulos G, Pagonis Th (2020) In the footprints of the Mediterranean city: urbanity, planning, and governance in the Athenian metropolis. Kritiki, Athens (in Greek)
Choromidis KG (1994) The law of town layout plans and urban planning. Thessaloniki (in Greek)
Christofilopoulos D (1979) La planification urbaine en Grèce. Revue Héllénique de Droit International. 32ème année: 196–208
Christofilopoulos DG (1988) The new legal framework of town planning. P. Sakkoulas, Athens
Christofilopoulos DG (1999) Building law – Vol. B: Unlicenced construction. P.N. Sakkoulas, Athens. (in Greek)
Christofilopoulos DG (2000) Building law – Vol. C: General Building Regulation. P.N. Sakkoulas, Athens (in Greek)
Christofilopoulos DG (2002) Cultural environment – Spatial planning and sustainable development. P.N. Sakkoulas, Athens (in Greek)
Citizens' Movement (1991) Cloud [of pollution] and active citizen. Conference proceedings 11-12/11/1991. Αθήνα (in Greek)
Clifton N, Comunian R, Chapain C (2015) Creative regions in Europe: challenges and opportunities for policy. EPS 23(12):2331–2335
Coccossis H (1994a) Environmental protection in urban and regional development policies. In: Getimis P, Kafkalas G, Maravegias N (eds) Urban and regional development: theory, analysis and policy. Themelio, Athens, pp 385–408. (in Greek)
Coccossis H (1994b) The environment in spatial planning: Coastal management policy. Topos 1994(8):191–205. (in Greek)
Coccossis H (2001) Tourism and the environment: impacts, planning, and policies of planning and management. In: Coccossis H, Tsartas P (eds) Sustainable tourism development and the environment. Kritiki, Athens, pp 131–281. (in Greek)
Coccossis H (ed) (2002) Man and the environment in Greece. Ministry for the Environment, Spatial Planning and Public Works, Athens
Coccossis H (2009) Spatial plan of tourism: priorities and directions. In: University of Thessaly (ed) 25 texts on spatial planning and development. Thessaly University Press, Volos. pp 57-68. (in Greek)
Coccossis H (2017) Tourism, technological change and regional development in islands. In Giaoutzi M, Nijkamp P (eds) Tourism and regional development: New pathways. Routledge, London (first published 2006), pp 271–279
Coccossis H (2018) Maritime spatial planning and integrated coastal zone management: prospects and challenges. In: Serraos C, Melissas D (eds) Maritime spatial planning. Sakkoulas, Athens, pp 67–73. (in Greek)
Coccossis H, Parpairis A (1993) Environment and tourism issues: preservation of local identity and growth management – Case study of Mykonos. In: Konsola D (ed) Culture, environment and regional development. Regional Development Institute, Athens, pp 79–100
Coccossis H, Parpairis A (1995) Carrying capacity: catalyst for shaping harmonious relations between tourism and the environment. In: TEE (ed) Tourism and environment: Choices for sustainable development. Proceedings of a one-day conference. TeC 1995(5): 68–79 (in Greek)
Coccossis H, Psycharis Y (eds) (2008) Regional analysis and policy: the Greek experience. Physica-Verlag, Heidelberg
Coccossis H, Tsartas P (2001) Sustainable tourism development and the environment. Kritiki, Athens (in Greek)
Coppolani J (1959) Le réseau urbain de la France: Sa structure et son aménagement. Les Éditions Ouvrières, Paris
Council of Europe (2000a) European Landscape Convention
Council of Europe (2000b) Guiding principles for sustainable spatial development of the European continent. Conférence du Conseil de l'Europe des Ministres responsables de l'aménagement du territoire (CEMAT). Hannover
Council of the European Communities (1985) Council Directive of 27 June 1985 on the assessment of the effects of certain public and private projects on the environment (85/337/EEC) (subsequently amended)

CPER (1965) Draft of Economic Development Plan for Greece 1966–70, Athens (in Greek)
CPER (1967) Settlement (*Oikismos*). 5-year plan studies. Athens (in Greek)
CPER (1968) Study of small settlement network / Part A: Network structure (by Efi Kalliga). National spatial plan studies. Mimeograph. Athens (in Greek)
CPER (ed) (1970) The framework of regional development. Lecture series. CPER and Gutenberg, Athens (in Greek)
CPER (1972a) Long-term Development Plan for Greece 1973–1987. 2 volumes. Athens (in Greek)
CPER (1972b) Urban development organization. Studies for Long-term Development Plan for Greece 1973–1987. Athens (in Greek)
CPER (1972c) Spatial and cultural aspects of the environment. Studies for Long-term Development Plan for Greece 1973–1987. Athens (in Greek)
CPER (1976a) The environment. Development Plan 1976–1980. Athens (in Greek)
CPER (1976b) Regional development. Development Plan 1976–1980. Athens (in Greek)
CPER (1976c) Housing. Development Plan 1976–1980. Athens (in Greek)
CPER (1977) Economic and Social Development Plan 1976–1980: 1977–1980 period – General statement (in Greek)
CPER (1980a) Regional Development Plan 1981–1985. CPER and Ministry of Coordination. Athens (in Greek)
CPER (1980b) Regional Development Plan: land use policy. Athens (in Greek)
CPER (1980c) Regional Development Plan: industrial location 1963, 1973, 1978. Athens (in Greek)
CPER (1989a) Report No 2 on the 1988–1992 Plan: Town planning. Athens (in Greek)
CPER (1989b) Report No 6 on the 1988–1992 Plan: Transport. Athens (in Greek)
CPER (1990) The development of Greece: past, present, and policy proposals. Athens (in Greek)
CPER (1991) Report No 24 on the 1988–1992 Plan: Regional policy. Athens (in Greek)
Cullingworth B, Nadin V (2006) Town and country planning in the UK. Routledge, London
Davoudi S (2005) Understanding territorial cohesion. Plan Prac Res 20(4):433–441
Davoudi S (2018) Spatial planning: the promised land or rolled-out neoliberalism? In: Gunder M, Madanipour A, Watson V (eds) The Routledge handbook of planning theory. Routledge, London, pp 15–27
Davoudi S, Galland D, Stead D (2020) Reinventing planning and planners: ideological decontestations and rhetorical appeals. PT 19(1):17–37
Decleris M (2000) The law of sustainable development. Ant.N. Sakkoulas, Athens (in Greek)
Deffner A et al (eds) (2000) Town planning in Greece from 1949 to 1974. Thessaly University Press, Volos (in Greek)
Delladetsima PM (1991) Forms of state intervention for the removal of obstacles due to land ownership: the case of Greece in the postwar period 1944-1952. Doctoral thesis. NTUA, Athens (in Greek)
Delladetsima PM (1994a) Views, theories and practices of spatial planning during the reconstruction period of the 1950s. In: Sakis Karagiorgas Foundation (1994–95) Greek society in the first postwar period 1945–1967. Vol. A. Conference proceedings. Athens pp 561–572 (in Greek)
Delladetsima PM (1994b) The role of extraordinary interventions in the evolution of planning and environmental policy. In: Rokos D (ed) Sciences and the environment at the end of the century: problems and prospects. NTUA and Enallaktikes Ekdoseis, Athens, pp 243–250. (in Greek)
Delladetsima PM (2000) The Ministry of Reconstruction: state intervention intentions in reconstruction period 1945-1951. In: Deffner A et al (eds) Town planning in Greece from 1949 to 1974. Thessaly University Press, Volos, pp 127–138. (in Greek)
Delladetsima PM, Loukakis GD (2013) The policy for commercial centres in Greece. Institute of Small Business Firms. Hellenic Confederation of Professionals, Craftsmen & Merchants. Athens (in Greek)
Demathas Z (1995a) The new territorial organization of Europe and the frontier areas of the Southern Balkans. In: SEP (ed) Frontier regions. Proceedings of a conference, 1995. Thessaloniki, pp 87–96 (in Greek)
Demathas Z (1995b) Greece in the European territorial organization: problems and prospects. In: SEP (ed) Regional development, spatial planning and environment in the framework of

united Europe. Vol. II. Proceedings of a conference, Panteion University, 1995. SEP and *Topos*, pp 28-39. (in Greek)
Demathas Z (2004) Greek urban centres: Development, competitiveness, polycentricity, potential, and planning. In Angelidis M (ed) New polycentricity and spatial development. Conference proceedings (2001). YPECHODE, Ministry of Economy, NTUA and Université de Paris X / Nanterre. Athens, pp 113–134 (in Greek)
Demathas Z (2011) Regional development, spatial or territorial development and urban questions. In: Panteion University (ed). Volume in honour of Prof. P. Loukakis. Gutenberg, Athens, pp 153–177 (in Greek)
Demitrakopoulos A (1934–1937) Town plans: town planning in Greece. Techniki Epetiris tis Ellados, 1934–1937 Vol. A (II), 359–449. TEE, Athens (in Greek)
Demitriadis EP, Yerolympos-Karadimos A (eds) (1989) Space and history: Urban, architectural and regional space. Proceedings of the Skopelos symposium. Aristotle University of Thessaloniki, Thessaloniki (papers in English and Greek)
diaNEOsis (2016) Population evolution in Greece (2015-2050) – Summary presentation of results. Laboratory of Demographic and Social Analysis, University of Thessaly (in Greek)
Dionelis C, Giaoutzi M (2008) The enlargement of the European Union and the emerging new TEN transport patterns. In: Giaoutzi M, Nijkamp P (eds) Network strategies in Europe: developing the future for transport and ICT. Ashgate, Aldershot, pp 119–132
Douros G (2007) The protection of forest areas: conflicts, contradictions and antinomies. In: Efthymiopoulos E, Modinos M (eds) Mountain space and forests. Interdisciplinary Institute of Environmental Research and Ellinika Grammata, Athens, pp 129–148. (in Greek)
Drettakis M (2012) Crisis hour for Greece. En Plo, Athens (in Greek)
Dunford M, Kafkalas G (eds) (1992) Cities and regions in the new Europe: the global-local interplay and spatial development strategies. Belhaven Press, London
Economou D (1983) The spatial organization of industry and incentive zones. Poli kai Perefereia 1983(6) (January–April):31–53 (in Greek)
Economou D (1994a, 1994) The system of spatial regional planning and special spatial planning studies. PeD (1):41–86. (in Greek)
Economou D (1994b) Urban development and spatial structure of the settlement network. In: Getimis P, Kafkalas G, Maravegias N (eds) Urban and regional development: theory, analysis and policy. Themelio, Athens, pp 43–93. (in Greek)
Economou D (1995) Land use and out-of-plan construction: The Greek version of sustainability. In: SEP (ed) Regional development, spatial planning and environment in the framework of united Europe. Proceedings of a conference. Vol. II. Panteion University, 1995. SEP and *Topos*, pp 63–73 (in Greek)
Economou D (2000a) Town planning policy in the 1950s. In: Deffner A et al (eds) Town planning in Greece from 1949 to 1974. Thessaly University Press, Volos, pp 39–48. (in Greek)
Economou D (2000b) Spatial planning system: Greek reality and international experience. Epitheorisi Koinonikon Erevnon 2000(101–102):3–57. (in Greek)
Economou D (2000c) The environmental dimension of urban development policy in post-war Greece. In: Modinos M, Efthymiopoulos E (eds) The sustainable city. Stochastis and Interdisciplinary Institute of Environmental Research, Athens, pp 47–70. (in Greek)
Economou D (2000d) The Master Plan of Athens: experiences and prospects. In: Psycharis Y, Gospodini A, Christopoulou O (eds) Seventeen essays on planning, cities, and development. Thessaly University Press, Volos, pp 161–193. (in Greek)
Economou D (2000e) The regional impact of the Community's agricultural policy: the case of Greece. In: Shaw D, Roberts P, Walsh J (eds) Regional planning and development in Europe. Ashgate, Aldershot, pp 15–34
Economou D (2002a) The statutory framework and the tribulations of spatial regional planning. Aeichoros 2001(1):116–127. (in Greek)
Economou D (2002b) The structural features of the spatial policy and development model.[28] In: SADAS (ed) Architecture and the Greek city in the 21st century. Proceedings of the 10th Panhellenic Architectural Conference (1999). SADAS and TEE, Athens, pp 73–84 (in Greek)[29]

Economou D (2004a) Urban planning policy in Greece: Structural characteristics and present trends. In: NTUA (ed) City and space from the 20th to the 21st century. Volume in honour of Prof. A.I. Aravantinos. NTUA, University of Thessaly, and SEPOCH, Athens, pp 371–382 (in Greek)
Economou D (2004b) The international role of Athens: problems, prospects and impacts. In: Angelidis M (ed) New polycentricity and spatial development. Conference proceedings (2001). YPECHODE, Ministry of Economy, NTUA and Université de Paris X / Nanterre. Athens, pp 151–158 (in Greek)
Economou D (2009a) Re-evaluation of Law 1337/1983, 25 years after. In: University of Thessaly (2009) 25 texts on spatial planning and development. Thessaly University Press, Volos, pp 241–262 (in Greek)
Economou D (2009b) Spatial planning and settlement network: the early views of C.A. Doxiadis. In: Kazazi G (ed) Constantinos Doxiadis and his work, vol B. TEE, Athens, pp 46–55. (in Greek)
Economou D (2010) New forms of spatial regulation in the period prior to the 2nd World War. In: Demetriadis EP, Kafkalas G, Tsoukala K (eds) The *Logos* of the *Polis*, Volume in honour of Prof. A.Ph. Lagopoulos. University Studio Press, Thessaloniki, pp 44–54. (in Greek)
Economou D (2017) Interactions between the Presidential Decree draft for land use and the Law 4447/2016. In: Melissas D, Serraos (eds) Organization of land uses and activities in the spatial planning process. Sakkoulas, Athens, pp 19–26 (in Greek)
Economou D (2018) Maritime spatial planning and traditional planning instruments. In: Serraos C, Melissas D (eds) Maritime spatial planning. Sakkoulas, Athens, pp 95–102. (in Greek)
Economou D, Papamichos N (2003) Spatial planning of metropolitan areas. In: Getimis P, Kafkalas G (eds) Metropolitan governance: International experience and Greek reality. Panteion University, Athens, pp 169–191. (in Greek)
Economou D, Petrakos G (2012a) Urban development policies and urban organization in Greece. In: Economou D, Petrakos G (eds) The development of Greek cities: interdisciplinary approaches of urban analysis and policy. Thessaly University Press, Volos (first published 1999), pp 413–446 (in Greek)
Economou D, Petrakos G (eds) (2012b) The development of Greek cities: interdisciplinary approaches of urban analysis and policy. Thessaly University Press, Volos (first published 1999) (in Greek)
Economou D et al (2001) The international role of Athens: present situation and the strategy for its development. YPECHODE and Thessaly University Press, Volos (in Greek)
Efthymiopoulos E, Modinos M (eds) (2007) Mountain space and forests. Interdisciplinary Institute of Environmental Research and Ellinika Grammata, Athens (in Greek)
EKDDA (2017) Strategies of Integrated Territorial Development: report from an expert meeting (in Greek)
EKPAA (National Centre for the Environment and Sustainable Development) (2020) Environment and health 2019. Athens (in Greek)
Eliopoulou P (1992) Spatial policies of the European Economic Community and development projects. In: Koutsopoulos C (ed) Development and planning: an interdisciplinary approach. Papazisis, Athens, pp 454–466. (in Greek)
Eliopoulou P (2001) Rural space in Greece (1961–1991). In: Burgel G, Demathas Z (eds) Greece in the face of the 3rd millennium: space, economy, society in the last 40 years. Panteion University, Athens, pp 187–203. (in Greek)
European Commission (1995) Development prospects of the Central Mediterranean Regions (Mezzogiorno – Greece). Office for Official Publications of the European Communities, Luxembourg

[29] No pages are mentioned in the table of contents.

[30] The year is indirectly deduced from the text.

European Commission (1997a) The European Union compendium of spatial planning systems and policies. Office for Official Publications of the European Communities, Luxembourg
European Commission (1997b) European Spatial development perspective, First Official Draft, Brussels
European Commission (1999) European spatial development perspective: towards balanced and sustainable development of the territory of the European Union, Prepared by the Committee on Spatial Development. Office for Official Publications of the European Communities, Luxembourg
European Commission (2000a) Communication on the Integrated Coastal Zone Management. EU 27.09.2000 COM (2000) 547
European Commission (2000b) The European Union Compendium of Spatial Planning Systems and Policies. 13 volumes on Austria, Belgium, Denmark, Finland, France, Germany, Greece, Italy, Netherlands, Portugal, Spain, Sweden, and United Kingdom. Office for Official Publications of the European Communities, Luxembourg
European Commission (2001a) White Paper on European governance. COM (2001) 428 Final / 25.7.2001
European Commission (2001b) Communication on European governance and the community method. COM (2001) 727 Final, 5.12.2001
European Commission (2004) Communication on a thematic strategy for the urban environment. 11.2.2004 – COM (2004)60/final
European Commission (2005) Communication on cohesion policy in support of growth and jobs: Community strategic guidelines, 2007–2013. COM(2005) 0299/5.7.05
European Commission (2006) Governance of Territorial and Urban Policies from EU to Local Level, Final Report. ESPON Project 2.3.2
European Commission (2008) Green book on territorial cohesion. COM (2008) 616 final/6-10-08
European Commission (2010a) Communication: Europe 2020 – A strategy for smart, sustainable and inclusive growth. COM (2010) 2020, Brussels, 3.3.2010
European Commission (2010b) Investing in Europe's future: Fifth report on economic, social and territorial cohesion. Publications Office of the European Union, Luxembourg
European Commission (2015) Territorial Agenda 2020 put in practice: Enhancing the efficiency and effectiveness of Cohesion Policy by a place-based approach. Volume I – Synthesis Report. European Commission, Brussels
European Commission (2017) My region, my Europe, our future: seventh report on economic, social and territorial cohesion. Publications Office of the European Union, Luxembourg
European Commission (2018) Transport in the European Union: current trends and issues. Brussels
European Commission (2022) Cohesion in Europe towards 2050: Eighth report on economic, social and territorial cohesion. Publications Office of the European Union, Luxembourg
European Community (1993) Towards Sustainability. A European Community programme of policy and action in relation to the environment and sustainable development. Official Journal of the European Communities. No C138/5/17.5.1993
European Union (1997) Consolidated Treaties. Luxembourg
European Union (2008) Working for the Regions: EU Regional Policy 2007–2013. Publications Office
Evangelidou M (2002) The administrative system of urban planning in Greece: a fundamental adverse factor for the promotion of modern concepts. In: SADAS (ed) Architecture and the Greek city in the 21st century. Proceedings of the 10th Panhellenic Architectural Conference (1999). SADAS and TEE, Athens, pp 427–433 (in Greek)
Evangelidou M, Kosta K, Basoukea E, Moysiadou T, Chlepas K (2003) Study of the levels of spatial planning and of related nomothetic processes in Greece. Polis programme. KEDKE and TEE, Athens (in Greek)
Fabietti W (ed) (1997) Urban challenge in Europe. Vol. II. Istituto Nazionale di Urbanistica
Faludi A (2002) Positioning European spatial planning. EPS 10(7):897–909
Faludi A (2004a) Territorial cohesion: old (French) wine in new bottles? Urban Stud 41(7)

Faludi A (2004b) The impact of a planning philosophy. PT 3(3):225–236
Faludi A (ed) (2007) Territorial cohesion and the European model of society. Lincoln Institute of Land Policy, Cambridge, MA
Faludi A (2016) EU territorial cohesion, a contradiction in terms. PTP 17(2):302–313
Filias V (1980) Aspects of maintenance and change of the social system. Vol. B. Nea Synora – A. Livanis, Athens (in Greek)
Fysekidou M (2015) European territorial cooperation as a form of territorial governance in Europe and Greece. Doctoral thesis. National and Kapodistrian University of Athens (in Greek)
Gartzos C (1981) Recent developments of the statutory framework of urban and regional spatial planning in Greece. AT 1981(15):58–63. (in Greek)
Gartzos C (2004) The evolution of regional spatial planning in Greece and the case of Chalkidiki. In: NTUA (ed) City and space from the 20th to the 21st century. Volume in honour of Prof. A.I. Aravantinos. NTUA, University of Thessaly, and SEPOCH, Athens, pp 167–172 (in Greek)
Georgiadou Th (2002) The funding of the Community Support Framework and its impact on spatial policy and urban centres. In: SADAS (2002) Architecture and the Greek city in the 21st century. Proceedings of the 10th Panhellenic Architectural Conference (1999). SADAS and TEE, Athens, pp 103–116 (in Greek)
Georgoulis D (ed) (1995) Texts on the theory and implementation of urban and regional spatial planning. Papazisis, Athens (in Greek)
Gerardi K (1995) Planning of the metropolitan area of Athens. In: SEP (1995a) Regional development, spatial planning and environment in the framework of united Europe. Proceedings of a conference, Panteion University, 1995. Vol. II. SEP and *Topos*, pp 261–270 (in Greek)
Gerardi K (1996) Strategic planning of the metropolitan area of Athens for sustainable development. In: ORSA (ed) Athens – Attica: Strategic planning for sustainable development. Proceedings of an international conference. ORSA and YPECHODE, Athens, pp 36–42 (in Greek)
Gerardi K (2002) Prospects for the planning of the Capital. In: SADAS (ed) Architecture and the Greek city in the 21st century. Proceedings of the 10th Panhellenic Architectural Conference (1999). SADAS and TEE, Athens, pp 401-406 (in Greek)
Gerardi K, Gialyri Th (1998) Planning for spatial development as an essential instrument of a strategy for the sustainable development of national space. Interdepartmental graduate course on Urban and Regional Planning. NTUA, Athens
Getimis P (1985) Spatial planning in the new conditions of distribution of administrative and political powers in Greece. In: TEE (ed) Regional programming and planning: the role of the administration and local self-government. One day conference proceedings. TeC, April–June 1985 (in Greek): 37–39
Getimis P (1989) Settlement policy in Greece. Odysseas, Athens (in Greek)
Getimis P (1990) Settlement policy in Greece: preconditions of formulation, reform efforts and crisis. In: Psycharis Y (ed) The functions of the state in a period of crisis: theory and Greek experience. Conference proceedings. Sakis Karagiorgas Foundation, Athens, pp 323–330 (in Greek)
Getimis P (1992) Social conflicts and the limits of urban policies in Greece. In: Dunford M, Kafkalas G (eds) Cities and regions in the new Europe: the global-local interplay and spatial development strategies. Belhaven Press, London, pp 239–254
Getimis P (1993) Social policies and local state. In: Getimis P, Gravaris D (eds) Social state and social policy. Themelio, Athens, pp 91–121. (in Greek)
Getimis P (1994) Urban development and policy. In: Getimis P, Kafkalas G, Maravegias N (eds) Urban and regional development: theory, analysis and policy. Themelio, Athens, pp 307–333. (in Greek)
Getimis P, Economou D (1992) New geographical inequalities and spatial policies in Greece. Topos 4:3–44. (in Greek)
Getimis P, Economou D (1994) Economic dimensions of building activity in the 1950s. In: Sakis Karagiorgas Foundation (1994–95) Greek society in the first postwar period 1945–1967. Conference proceedings. Vol. A. Athens, pp 593–598 (in Greek)

Getimis P, Giannakourou G (2014) The evolution of spatial planning in Greece after the 1990s. In: Reimer M, Getimis P, Blotevogel HH (eds) Spatial planning systems and practices in Europe. Routledge, London, pp 149–168
Getimis P, Gravaris D (eds) (1993) Social state and social policy. Themelio, Athens. (in Greek)
Getimis P, Kafkalas G (eds) (1993) Urban and regional development in the new Europe. Topos – Special series, Athens (in Greek)
Getimis P, Kafkalas G, Maravegias N (eds) (1994) Urban and regional development: theory, analysis and policy. Themelio, Athens. (in Greek)
Getimis P, Kafkalas G (2001) Scientific thinking on Greek space 1974-2000: an evaluation attempt and thoughts for the future. Topos 2001(16) (in Greek)
Getimis P, Kafkalas G (eds) (2003a) Metropolitan governance: international experience and Greek reality. Panteion University, Athens (in Greek)
Getimis P, Kafkalas G (eds) (2003b) Space and environment: globalization, governance, sustainability. Topos and Institute of the Urban Environment and Human Potential, Panteion University, Athens (in Greek)
Getimis P, Kafkalas G, Economou D (1994) Regional spatial planning and environment: new institutions and future symbiosis. Topos 1994(8):5–13. (in Greek)
Gialis S, Gourzis K (2010) Cities, regions and informal work in contemporary Greece: towards a regional geography of crisis. In: Beriatos E, Papageorgiou M (eds) Spatial planning – town planning – environment in the 21st century. Thessaly University Press, Volos, pp 125–138. (in Greek)
Giannakourou G (1994a) The new spatial policy framework in the 1990s: institutional rearrangements and uncertainties. Topos 1994(8):15–40
Giannakourou G (1994b) Sectoral and spatial logics in Greek environmental policy. In: Rokos D (ed) Sciences and the environment at the end of the century: problems and prospects. NTUA and Enallaktikes Ekdoseis, Athens, pp 272–280. (in Greek)
Giannakourou G (1994c) Regional spatial planning and administration court rulings: from legality control to reshaping of spatial policy? PeD 1994(1):23–40. (in Greek)
Giannakourou G (1995) European integration, competitiveness and spatial law: institutional dilemmas and challenges for European spatial policy. In: SEP (ed) Regional development, spatial planning and environment in the framework of united Europe. Proceedings of a conference. Vol. II. Panteion University, 1995. SEP and Topos, pp 74–85 (in Greek)
Giannakourou G (2000) Spatial policy in an enlarged Europe. In: Andrikopoulou E, Kafkalas G (eds) The new European space: enlargement and the geography of European development. Themelio, Athens, pp 401–424. (in Greek)
Giannakourou G (2003) Planning of metropolitan areas: institutions and policies. In: Getimis P, Kafkalas G (eds) Metropolitan governance: International experience and Greek reality. Panteion University, Athens, pp 63–82. Also published in Kafkalas G (ed) (2004). (in Greek)
Giannakourou G (2004) Planning of metropolitan areas: institutions and policies. In: Kafkalas G (ed) Issues of spatial development. Kritiki, Athens, pp 351–372. First published in Getimis P, Kafkalas G (eds) (2003a). (in Greek)
Giannakourou G (2008a) Regional spatial planning in the European Union. Papazisis, Athens (in Greek)
Giannakourou G (2008b) The institutional dimension of territorial cohesion: looking for a new form of territorial governance in Europe. Diethnis kai Evropaiki Politiki (2008)10, April-June: 235–246 (in Greek)
Giannakourou G (2008c) The institututional framework of regional spatial planning in Greece: actual dilemmas and challenges for the future. In: Giannakourou G, Menoudakos C, Wassenhoven L (eds) Regional spatial planning in Greece: legal framework and implementation in practice. Nomos+Physis and Ant.N. Sakkoulas, Athens, pp 13–29 (in Greek)
Giannakourou G (2009) Territorial cohesion of European space: a new policy field for the European Union. In: University of Thessaly (ed) 25 texts on spatial planning and development. Thessaly University Press, Volos, pp 35–55. (in Greek)

Giannakourou G (2012) The statutory framework of city planning in Greece: historic transformations and modern demands. In: Economou D, Petrakos G (eds) The development of Greek cities: interdisciplinary approaches of urban analysis and policy. Thessaly University Press, Volos (first published 1999), pp 457–480 (in Greek)
Giannakourou G (2019) Regional-spatial and urban planning law. Nomiki Vivliothiki, Athens (in Greek)
Giannakourou G, Menoudakos C, Wassenhoven L (2008) Regional spatial planning in Greece: legal framework and implementation in practice. Nomos+Physis and Ant.N. Sakkoulas, Athens (in Greek)
Giannitsis T (1983) Greek industry: development and crisis. Gutenberg, Athens (in Greek)
Giaoutzi M (1984) Evolution of regional structures: the case of Greece. In: Koutsopoulos KC, Nijkamp P (eds) Regional development in the Mediterranean. Phebus Editions, Athens, pp 153–167
Giaoutzi M (1988) Regional dimensions of small and medium-sized enterprises in Greece. In: Giaoutzi M, Nijkamp P, Storey DJ (eds) Small and medium size enterprises and regional development. Routledge, London, pp 264–281
Giaoutzi M, Nijkamp P (1993) Decision support models for regional sustainable development: an application of geographic information systems and evaluation models to the Greek Sporades Islands. Avebury, Aldershot
Giaoutzi M, Nijkamp P (eds) (2008) Network strategies in Europe: developing the future for transport and ICT. Ashgate, Aldershot (UK)
Giaoutzi M, Nijkamp P (eds) (2017) Tourism and regional development: new pathways. Routledge, London (first published 2006)
Giaoutzi M, Sapio B (2013) In search of foresight methodologies: riddle or necessity. In: Giaoutzi M, Sapio B (eds) Recent developments in foresight methodologies. Springer, New York, pp 3–9
Giaoutzi M, Stratigea A (2011) Regional spatial planning: theory and practice. Kritiki, Athens. (in Greek)
Giaoutzi M, Nijkamp P, Storey DJ (eds) (1988) Small and medium size enterprises and regional development. Routledge, London
Gourgiotis A, Tsilimingas G (2016) A new approach to spatial regional planning in Greece. Aeichoros 2016(26):103–122
Gousios D (2012) Countryside, rural space and small towns: From "agriculturation" to local development. In: Economou D, Petrakos G (eds) The development of Greek cities: Interdisciplinary approaches of urban analysis and policy. Thessaly University Press, Volos (first published 1999), pp 157–207 (in Greek)
Governa F, Rivolin UJ, Santangelo M (eds) (2009) La costruzione del territorio Europeo: sviluppo, coesione, governance. Carocci, Roma
Grammaticaki-Alexiou A (1993) Regional and urban planning and zoning. In Kerameus KD, Kozyris PJ (eds) Introduction to Greek law. Rev edn. Kluwer, Deventer, pp 135–142
Gravier J-F (1947) Paris et le désert Français. Flammarion, Paris
Great Greek Encyclopaedia (1928). 2nd edition. "O Foinix" Publishing, Athens (in Greek)
Greek National Committee for the U.N. Conference "HABITAT" (1975) Greece: National report. Athens
GREMI (1992) Development prospects of the Community's lagging regions and the socio-economic consequences of the completion of the internal market. Research project directed by R.P. Camagni. Final report, Groupe de Recherche Européen sur les Milieux Innovateurs, Milan
Guichard O (1965) Aménager la France. Laffont – Gonthier, Paris
Habitat II Greek Committee (1996) Greek national report. UN World Conference on Human Settlements "Habitat II". Athens (in Greek)
Hadjimichalis C (1987) Uneven development and regionalism. Croom Helm, London
Hadjimichalis C (ed) (1992) Regional development and politics. Exantas, Athens (in Greek)
Hadjimichalis C (2001) Geography, development and politics. O Politis, Athens (in Greek)

Hadjimichalis C (ed) (2010) Greekscapes: aerial photography atlas of modern Greek landscapes. Charokopeion University and I.S. Latsis Foundation (available on the web) (in Greek)
Hadjimichalis C (2018) Landscapes of crisis in South Europe. Alexandreia, Athens (in Greek)
Hadjivasileiou E (2015) Greek liberalism: the radical current, 1932–1979. Patakis, Athens (in Greek)
Hall P, Tewdwer-Jones M (2011) Urban and regional planning, 5th edn. Routledge, London
Haselsberger B (ed) (2017) Encounters in planning thought: 16 autobiographical essays from key thinkers in spatial planning. Routledge, New York/London
Hatzichristos T, Giaoutzi M, Mourmouris JC (2017) Delineating ecoregions for tourism development. In: Giaoutzi M, Nijkamp P (eds) Tourism and regional development: new pathways. Routledge, London, pp 153–175. (first published 2006)
Hautreux J, Rochefort M (1964) Les métropoles et la fonction régionale dans l'armature urbaine française. Ministère de la Construction, Paris
HBID (1993) Investment incentives in Greece: law 1892/1990. ETVA, Αθήνα (in Greek)
HBID – Industrial Estates sa (2004) Orientation study of the project of the Operational Programme "Competiveness" in the sector of new Industrial and Business Parks of national significance. Project team of ETVA Industrial Estates sa (participation of the author) (in Greek)
Hoffman GW (1972) Regional development strategy in southeast Europe. Praeger, New York
Holmes P (1999) The political economy of the European integration process. In: Dyker DA (ed) The European economy. Addison Wesley Longman, Harlow, pp 43–63
Housianakou M (2010) The critical approach of SEPOCH to the statutory Special Spatial Framework for the renewable energy sources. In: Melissas D (ed) Special spatial plan for the renewable energy sources: a year after, Proceedings of a one-day conference. Papasotiriou, Athens, pp 85–90. (in Greek)
Ipsen D et al (2007) Athens: the social creation of a Mediterranean metropolis. Kritiki, Athens. (in Greek)
ISTAME (2009a) For a different spatial organization model of the country. Proceedings of a one-day conference. Athens (in Greek)
ISTAME (2009b) Green development: Consultation documents. Athens (coordinator: J. Maniatis) (in Greek)
ITA (2006) Local self-government proposals for urban planning in Greece. Report of an expert committee. ITA and KEDKE, Athens (in Greek)
Kafkalas G (1981) The regional organization of the Greek economy, 1948–1974. PoPer 1981(2): 7-38. (in Greek)
Kafkalas G (1983) Basic concepts and levels of spatial programming. In: Kafkalas G, Komnenos N, Lagopoulos AP (eds) Urban programming: theory, institutions, methodology. Paratiritis, Thessaloniki, pp 19–33. (in Greek)
Kafkalas G (1984) Regional organization of industry. Doctoral thesis. Aristotle University of Thessaloniki (in Greek)
Kafkalas G (1990) Regional development and regional policy in the 1980s: crisis or transition to a new era? In: Psycharis Y (ed) The functions of the state in a period of crisis: theory and Greek experience. Conference proceedings. Sakis Karagiorgas Foundation, Athens, pp 302–309 (in Greek)
Kafkalas G (1992) Regional development and spatial integration. Thessaloniki (in Greek)
Kafkalas G (2001) Strategic spatial planning of metropolitan areas. In: Demetriadis EP, Kafkalas G, Tsoukala K (eds) The *Logos* of the *Polis*. Volume in honour of Prof. A.Ph. Lagopoulos. University Studio Press, Thessaloniki, pp 67–75. (in Greek)
Kafkalas G (2004a) The scientific field of spatial development: evolution and main components. In: Kafkalas G (ed) Issues of spatial development. Kritiki, Athens, pp 15–37. (in Greek)
Kafkalas G (ed) (2004b) Issues of spatial development. Kritiki, Athens. (in Greek)
Kafkalas G, Andrikopoulou E (2000) Spatial impacts of European policies: the Greek experience 1989-1999. YPECHODE and Aristotle University of Thessaloniki (in Greek)

Kafkalas G, Komnenos N (1993) Local development strategies and social policy problems. In: Sakis Karagiorgas Foundation (ed) Dimensions of present-day social policy. Conference proceedings, Athens, pp 515–526 (in Greek)
Kafkalas G, Pitsiava M (2013) Spatial impacts of transport and policies for the promotion of sustainable spatial development. Aeichoros 2013(18):94–115. (in Greek)
Kafkalas G, Komnenos N, Lagopoulos APh (eds) (1983) Urban programming: theory, institutions, methodology. Paratiritis, Thessaloniki (in Greek)
Kafkoula K (1990) The garden city idea in interwar Greek town planning. Doctoral thesis, Aristotle University of Thessaloniki. National Archive of Doctoral Theses (didaktorika.gr)
Kalantzi O, Tsiotas D (2010) Critical review of the institutional framework of the Integrated Tourism Development Areas (POTAs): the case of Messenia in SW Greece. In: Beriatos E, Papageorgiou M (eds) Spatial planning – town planning – environment in the 21st century. Thessaly University Press, Volos, pp 257–272. (in Greek)
Kalliga E (1969) Thoughts on restructuring the network of small settlements. AT 1969(3):112–113. (in Greek)
Kalokardou R (1980) Government policy for regional development and its implementation. In: TEE (1980) Spatial planning and development. Conference on development – 2nd preconference one-day meeting. Athens, pp 41–53. (in Greek)
Kalokardou R (1988) Tourism: policy and implementation; SyT 11(34) (May): 15–20
Kalokardou R (1995) The role of the environment for tourism, in tourism policy, in location planning, and in regional development: The Greek experience 1975–1995. In: SEP (ed) Regional development, spatial planning and environment in the framework of united Europe. Proceedings of a conference (vol. III). Panteion University, 1995. SEP and Topos, Athens, pp 238–246
Kalokardou R (2011) Features of tourism activity in European and Greek space: a spatial planning view. In: Avgerinou-Kolonias S, Melissas D (eds) Economic conjuncture and spatial organization of tourism. Propobos, Athens, pp 25–34. (in Greek)
Kampourakis VN (1967) Legislation of town plans, 2nd edn, Athens. (in Greek)
Kanellopoulos A (1962)[30] Views on the development of the Greek economy. Athens. (in Greek)
Kanellopoulou C (2018) Incorporation of the European directive on maritime spatial planning in Greek law. In: Serraos C, Melissas D (eds) Maritime spatial planning. Sakkoulas, Athens, pp 59–66. (in Greek)
Karadimou-Yerolympos A (2000) The General Building Regulation (GOK) and the Greek city: from comprehensive planning conception to regulation of private profiteering. In: Deffner A et al (eds) Town planning in Greece from 1949 to 1974. Thessaly University Press, Volos, pp 151–165. (in Greek)
Karadimou-Yerolympos A (2012) Anargyros Demitrakopoulos: town planning as ideology. In: NTUA (ed) 170 years of Polytechnic, engineers and technology in Greece. Vol. B. National Technical University of Athens, pp 149–169 (in Greek)
Karadimou-Yerolympos A, Kafkoula-Vlachou C (1983) Historical evolution of planning legislation and modernization attempts: the new town planning law 1337/1983. In: Kafkalas G, Komnenos N, Lagopoulos AP (eds) Urban programming: theory, institutions, methodology. Paratiritis, Thessaloniki, pp 279–299. (in Greek)
Karatsolis C (2020) Introduction to town planning law in Greece and Cyprus: planning the city, constructing the civitas. Nomiki Vivliothiki, Athens. (in Greek)
Karidis DN (2008) The seven books of town planning. Papasotiriou, Athens. (in Greek)
Karidis DN (2014) Athens from 1456 to 1920. Archaeopress, Oxford
Karra L (ed) (2008) Vulnerable land. ELLET, Athens. (in Greek)
Kartakis EA (1970) Le développement industriel de la Grèce. Centre de Recherches Européennes, Lausanne

[31] Pages are not mentioned in the table of contents.

Katochianos N (1966) The relationship of general economic and spatial programming of the country with city master plans. Paper presented at the 2nd Panhellenic Conference of Architects. Athens, 16–23 January 1966 (in Greek)

Katochianos D (1967) Spatial study of the national network of urban centres. CPER, Athens. (in Greek)

Katochianos D (1969) The urban network of urban centres: research methodology. AT 1969(3):102–111. (in Greek)

Katochianos N (1970) National spatial planning. In: CPER (ed) The framework of regional development. Lecture series. CPER and Gutenberg, Athens, pp 139–173. (in Greek)

Katochianos N (1973) Spatial planning of industrial development. In: NTUA (ed) Industrial development in space. Lecture series. NTUA, Athens, pp 83–123. (in Greek)

Katochianos N (1985) The role of KEPE in regional and spatial planning. In: TEE (ed) Regional programming and planning: the role of the administration and local self-government. One-day conference proceedings. TeC, April–June 1985: 20–22. (in Greek)

Katochianos D (1992) The Greek system of cities. Ekistics 59(352/353):56–60

Katochianos D (1994) Development in space and environment in Greece. Topos 1994(8):285–301. (in Greek)

Katochianos D (1995) Economic and spatial development of tourism in Greece: a primary approach. SyT 8(55) (April–June): 62–71. (in Greek)

Katochianos D, Katochianos N (1967) Planning the development of metropolises in the framework of the national system of urban centers: a case study of Greece. Centennial Study and Training Programme on Metropolitan Problems. Section: Comments on planning and urban design. Bureau of Municipal Research, Toronto

Katochianos D, Katochianos N, Marcopoulou N (1967) Planning for metropolitan Athens. Centennial Study and Training Programme on Metropolitan Problems. Section: Comments on planning and urban design. Bureau of Municipal Research, Toronto

Katsikas E, Lamprianidis L (1994) Population movements in the interior of the countryside and trends for the creation of middle-sized urban centres. In: Sakis Karagiorgas Foundation (1994–95) Greek society in the first postwar period 1945-1967. Conference proceedings. Vol. A. Athens, pp 490–501 (in Greek)

Kayser B (1965) La Grèce en voie de développement. L'Information Géographique, No 3:93–103

Kayser B, Pechoux P-Y, Sivignon M (1971) Exode rural et attraction urbaine en Grèce. Centre National de Recherches Sociales (EKKE), Athènes

Klabatsea E (2001) Geographical evolution of investment and employment in industry during the period 1961–1998. In: Burgel G, Demathas Z (eds) Greece in the face of the 3rd millennium: space, economy, society in the last 40 years. Panteion University, Athens, pp 213–233. (in Greek)

Klabatsea E (2003) Declining industrial areas: Sustainability of illusion? In: Getimis P, Kafkalas G (eds) Space and environment: globalization, governance, sustainability. Topos and Institute of the Urban Environment and Human Potential, Panteion University, Athens, pp 145-155. (in Greek)

Kloutsinioti R (1981) Constitutional impositions and political reality: from mismatch to retreat. AT 1981(15):87–89. (in Greek)

Kloutsinioti R (1989) Special spatial planning study (EChM) for the wider zone of Oinofyta in the prefecture of Boeotia. Phase A. YPECHODE, Athens (in Greek)

Kloutsinioti R (2010) The special spatial planning framework (EPChS) for the renewable energy sources. In: Melissas D (ed) Special spatial plan for the renewable energy sources: A year after. Proceedings of a one-day conference. Papasotiriou, Athens, pp 79-84. (in Greek)

Knieling J, Othengrafen F (eds) (2016) Cities in crisis: socio-spatial impacts of the economic crisis in southern European cities. Routledge, London

Kollias L, Klabatsea E (1995) Spatial planning in Greece and forest development. In: Greek Forestry Society (ed) Forest development: land property and spatial planning. Athens, pp 105–115. (in Greek)

Kolodny ÉY (1974) La population des îles de la Grèce. 3 volumes. EDISUD, Aix-en-Provence
Komilis P (1986) Spatial analysis of tourism. Scientific studies. CPER, Athens (in Greek)
Komilis P (1987) The spatial structure and growth of tourism in relation to the physical planning process: the case of Greece. PhD thesis. University of Strathclyde, Glasgow
Komilis P (1992) Tourism and regional development: theoretical, methodological and programmatic approaches. In: Koutsopoulos C (ed) Development and planning: an interdisciplinary approach. Papazisis, Athens, pp 247–272. (in Greek)
Komilis P (1995) Tourism policy and Integrated Tourism Development Areas (POTA). SyT 18(55):77–80. (in Greek)
Komilis P (2001) Eco-tourism: the alternative perspective of sustainable tourism development. Propobos, Athens. (in Greek)
Komnenidis N (1995) Progress of the European Spatial Development Perspective. In: SEP (ed) Regional development, spatial planning and environment in the framework of united Europe. Proceedings of a conference. Vol. II. Panteion University, 1995. SEP and *Topos*, pp 40–62 (in Greek)
Komnenos N (1993a) Technopoles and development strategies in Europe. Gutenberg, Αθήνα (in Greek)
Komnenos N (1993b) Innovative growth in the peripheral regions: some implications for Greece. In: Getimis P, Kafkalas G (eds) Urban and regional development in the new Europe. Topos – Special series, Athens, pp 193–206. (in Greek)
Komnenos N (1994) Single European Market and spatial restructuring of Greek development. Topos 1994(8):97–116. (in Greek)
Komnenos N (2007) Regional innovation poles in Greece 2001–2009: Planning for focussed innovation systems. Aeichoros 2007(2):10–33. (in Greek)
Konsola D (1993) Cultural tourism and regional development: some proposals for cultural itineraries. In: Konsola D (ed) Culture, environment and regional development. Regional Development Institute, Athens, pp 19–43
Konsolas N (1973) Industrial zones and estates. In: NTUA (ed) Industrial development in space. Lecture series. NTUA, Athens, pp 149-178. (in Greek)
Konsolas N (1985) Regional economic policy: a general view. Papazisis, Athens (in Greek)
Konsolas N (1990) The technopoles as an instrument of ETVA development policy. VIPETVA, Athens (in Greek)
Konsolas N (1994) The role of industrial estates in the development process: The Greek experience. In: Getimis P, Kafkalas G, Maravegias N (eds) Urban and regional development: theory, analysis and policy. Themelio, Athens, pp 335–346. (in Greek)
Konsolas N, Kyriazopoulos E (2011) The standards of industrial spatial concentration and their application in Greece. In: Panteion University (2011). Volume in honour of Prof. P. Loukakis. Gutenberg, Athens, pp 416–443 (in Greek)
Konstantinidis Ch (2012) Maritime policy (available on the web) (in Greek)
Kontaratos S (1981) The regulation of space in Greece: preface. AT 1981(15):56–57. (in Greek)
Kontiadis X (2009) Deficient democracy: state and political parties in modern Greece. I. Sideris, Athens (in Greek)
Kostara M (1995) INTERREG Initiative. In: SEP (ed) Frontier regions. Proceedings of a conference, 1995. Thessaloniki, pp 109–115 (in Greek)
Kostis C (2018) "The spoiled children of history": the formation of the modern Greek state, 18th-21st centuries. Patakis, Athens (in Greek)
Kotzamanis V (2009) The population variable in spatial planning in postwar Greece: planning with virtual data? In: University of Thessaly (ed) 25 texts on spatial planning and development. Thessaly University Press, Volos, pp 131-165. (in Greek)
Kotzamanis V, Androulaki E (2000) Spatial dimensions of demographic developments in Greece 1981–1991: a first approach. In: Psycharis Y, Gospodini A, Christopoulou O (eds) Seventeen essays on planning, cities, and development. Thessaly University Press, Volos, pp 63–110. (in Greek)

Koudouni A (2017) From the presidential decree of 1987 to present day land uses for space regulation. In: Melissas D, Serraos C (eds) Organization of land uses and activities in the spatial planning process. Sakkoulas, Athens, pp 47–54. (in Greek)
Kourliouros E (1989) The development of space and spatial planning: issues of scientific method, systems of analysis, and epistemological critique of theories. Doctoral thesis. NTUA, Athens (in Greek)
Kousidonis Ch (2009) The rhetoric of space production in Greece and the charm of higher-level planning. In: University of Thessaly (ed) 25 texts on spatial planning and development. Thessaly University Press, Volos, pp 195–214. (in Greek)
Koutsogiannis P (1984) Basic guidelines of industrial policy. CPER, Athens (in Greek)
Koutsopoulos C (ed) (1995) Greece in maps and numbers. Artia, Athens (in Greek)
Koutsopoulos C (ed) (n.d.) Development and planning: An interdisciplinary approach. Papazisis, Athens (in Greek)
Koutsopoulos KC, Nijkamp P (eds) (1984) Regional development in the Mediterranean. Phebus Editions, Athens
Kramer E (2011) Economic, social and territorial situation of Greece. Directorate-general for internal policies. In: Policy Department B: structural and cohesion policies. European Parliament, Brussels
Kriesis A (1963) A regional planning scheme for a country under development. Patras
Krispis K (1944–45) A 20-year spatial (*choronomic*) plan for Greece. Athens (in Greek)
Kritikos A (1994) European spatial planning policy: the long road from Rome to Maastricht. Topos 1994(8):223–233. (in Greek)
Kyrgiafini L, Sefertzi E (2003) Changing regional systems of innovation in Greece: the impact of regional innovation strategy initiatives in peripheral areas of Europe. EPS 11(8):885–910
Kyvelou S (2001) Spatial planning and spatial re-composition: realities, prospects, and modern instruments of analysis for physical planning. Topos 2001(17) (in Greek)
Kyvelou S (2003) The concept of territorial/spatial cohesion in Europe and spatial management. In: Getimis P, Kafkalas G (eds) Space and environment: Globalization, governance, sustainability. Topos and Institute of the Urban Environment and Human Potential, Panteion University, Athens, pp 133–143. (in Greek)
Kyvelou S (2010) From regional spatial planning to space-management: the concepts of strategic spatial planning and territorial cohesion in Europe. Kritiki, Athens (in Greek)
Kyvelou S (2016) Maritime spatial planning under the lens of geo-philosophy, geography and geopolitics: thoughts on spatial planning of "mare liberum" and of maritime spatial zones. In: Kyvelou S (ed) Maritime spatial issues: the maritime dimension of territorial cohesion – maritime spatial planning – sustainable blue development. Kritiki, Athens, pp 37–62. (in Greek)
Kyvelou SS, Gourgiotis A (2019) Landscape as connecting link of nature and culture: spatial planning policy implications in Greece. USc 3(3):81
Kyvelou S, Papadaki M (2016) Maritime spatial planning in practice: experiences, practical instruments, and evaluation of eco-systemic services. In: Kyvelou S (ed) Maritime spatial issues: the maritime dimension of territorial cohesion – maritime spatial planning – sustainable blue development. Kritiki, Athens, pp 606–617. (in Greek)
Kyvelou S, Pothitaki V (2016) Incorporation of maritime spatial planning in national planning systems: convergence, divergence, and prospects. In: Kyvelou S (ed) Maritime spatial issues: the maritime dimension of territorial cohesion – maritime spatial planning – sustainable blue development. Kritiki, Athens, pp 254–296. (in Greek)
Lagopoulos APh (1981a) Town planning manual – Part B. Planning interventions: Vol. II: Space-functional planning and planning of built structures. Aristotle University of Thessaloniki (in Greek)
Lagopoulos APh (1981b) The administrative and statutory framework of physical programming in Greece. PoPer 1981(1):47–66. (in Greek)
Lagopoulos APh (1984) Greece. In: Williams RH (ed) Planning in Europe: Urban and regional planning in the EEC. Allen and Unwin, London, pp 128–143

Lagopoulos APh (1986) Social formation and settlement network in Greece. Geoforum 17(1):39–56

Lagopoulos APh (1992) The programming and planning system in Greece and urban planning theory. Teaching manual, Thessaloniki. (in Greek)

Lagopoulos APh (1994) Analysis of urban practices and uses. In: Getimis P, Kafkalas G, Maravegias (eds) Urban and regional development: Theory, analysis and policy. Themelio, Athens, pp 123–166. (in Greek)

Lagopoulos APh (2011) Development typology of geographical areas. In: Panteion University (ed). Volume in honour of Prof. P. Loukakis. Gutenberg, Athens, pp 502–535. (in Greek)

Lagopoulos APh (2017) Urban planning theory and methodology: from political economy to the semiotics of space. Patakis, Athens (in Greek)

Lamprianidis L (2000) The reconstruction of the Balkans and the role of Greece: a critical approach. In: Petrakos G (ed) The development of the Balkans. Thessaly University Press, Volos, pp 371–396. (in Greek)

Lamprinidis MI, Pakos Th (1990) The Greek fiscal crisis. In: Psycharis Y (ed) The functions of the state in a period of crisis: theory and Greek experience. Conference proceedings. Sakis Karagiorgas Foundation, Athens, pp 175–193. (in Greek)

Laskaris C (ed) (1994) National and Community policies and programmes for the development of urban centres. European Social Fund and NTUA, Athens (in Greek)

Leaf M (2017) The urban, the periurban and the urban superorganism. In: Rangan H et al (eds) Insurgencies and revolutions: Reflections on John Friedmann's contributions to planning theory and practice. Routledge, New York, and The Royal Town Planning Institute (RTPI Library Series), London, pp 119–128

Lefantzis M (2000) Town planning in Greece as seen by the Athenian press: 1944–1974. In: Deffner A et al (eds) Town planning in Greece from 1949 to 1974. Thessaly University Press, Volos, pp 111–125. (in Greek)

Lefeber L (1966) Location and regional planning. Training seminar series. CPER, Athens

Leontidou L (1990) The Mediterranean city in transition. Cambridge University Press, Cambridge

Leontidou L (1996) Alternatives to modernism in (southern) urban theory: exploring in-between spaces. Int J Urban Reg Res 20(2):178–195

Leontidou L (2001) Cities of silence: working class colonization of Athens and Piraeus, 1909–1940. ETVA Cultural Technological Foundation, Athens (first published 1989) (in Greek)

Leontidou L (2011) *Ageographitos Chora* [Geographically illiterate land]: Hellenic idols in the epistemological reflections of European Geography. Propobos, Athens (in Greek)

Leontidou L (2020) Urban planning and pandemic in the compact Mediterranean city: anthropogeographical lateral losses of Covid-19. EpKE 154:11–27. (in Greek)

Leontidou-Emmanuel L (1981) Master plans for Athens, 1950–1980: a comment on their political function and their social roots. AT 1981(15):70–78. (in Greek)

Leontidou-Emmanuel L (1982) Urban growth and socioeconomic structure of Greater Athens, 1834–1981: a reinterpretation of urban history. Conference paper. 12th International Fellows Conference, Johns Hopkins University, Center for Metropolitan Planning and Research, Athens

Leontidou-Gerardi K (1977) Functional classification of Greek settlements. Doctoral thesis. NTUA, Athens (in Greek)

Leontidou-Gerardi K (1981) The network of urban centres of the country: development and state-sponsored programmes. AT 1981(15):64–69. (in Greek)

Lewis J, O'Keefe P, Westgate KN (1976) A philosophy of planning. Occasional paper No 5. Disaster Research Unit, University of Bradford (UK)

Liodakis G (1994) Regional development and environment. Topos 1994(8):159–190

Loukakis P (1994) Implementation issues relevant to territorial levels of spatial programming and planning in Greece. Topos 1994(8):71–95. (republished in Loukakis 2017) (in Greek)

Loukakis P (2004) The "urban alliances" as a perspective of Greek settlements' competitiveness in the framework of spatial development in Europe: the case of the triangle of Drama-Kavala-Xanthi. In: Angelidis M (ed) New polycentricity and spatial development. Conference proceedings (2001). YPECHODE, Ministry of Economy, NTUA and Université de Paris X/Nanterre, Athens, pp 37–45. (in Greek)

Loukakis P (2010) Critical examination of the completion processes of spatial planning and its implementation in contemporary Greece. In: Demetriadis EP, Kafkalas G, Tsoukala K (eds) The Logos of the Polis, Volume in honour of Prof. A.Ph. Lagopoulos. University Studio Press, Thessaloniki, pp 101–111. (in Greek)
Loukakis P (2017) Urban and regional spatial developments: Greece 1952–2012 – Experiences of action. Thessaly University Press, Volos. (in Greek)
Luukkonen J (2015) Planning in Europe for 'EU'rope: spatial planning as a political technology of territory. PT 14(2):174–194
Luukkonen J (2017) A practice theoretical perspective on the Europeanization of spatial planning. EPS 25(2):259–277
Maksin M et al (2018) The role of zoning in the strategic planning of protected areas: lessons learnt from EU countries and Serbia. EPS 26(4):838–872
Maloutas T (ed) (2000) Social and economic atlas of Greece: the towns. EKKE and Thessaly University Press, Volos. (in Greek)
Maloutas T, Economou D (eds) (1992) Social structure and urban organization in Athens. Paratiritis, Thessaloniki. (in Greek)
Maniatis Y (2009) The challenge of green development. Livanis, Athens. (in Greek)
Maniatis Y (2012) Energy and mineral wealth: national development pillars. Livanis, Athens. (in Greek)
Maniatis Y, Myridis M (1990) Geographical methods and information systems: the problem of development planning. In: NTUA (ed) The interdisciplinary approach to development. Proceedings of an Interuniversity Interdisciplinary Conference. Papazisis, Athens, pp 457–469. (in Greek)
Manola K (1996) The role of two Greek metropolises in relation to the other urban centres. In: ORSA (ed) Athens – Attica: strategic planning for sustainable development. Proceedings of an international conference. Athens, pp 60–65 (in Greek)
Maraka-Romanou M (2011) The Special spatial plan of tourism under the lens of tourism policy. In: Avgerinou-Kolonias S, Melissas D (eds) Economic conjuncture and spatial organization of tourism. Propobos, Athens, pp 51–56. (in Greek)
Maravegias N (2011) search of a development strategy for the countryside. In: Panteion University (ed) Volume in honour of Prof. P. Loukakis. Gutenberg, Athens, pp 644–659. (in Greek)
Maravegias N, Mermingas G (2001) The Greek rural economy towards the 3rd millennium: developments in the last 40 years. In: Burgel G, Demathas Z (eds) Greece in the face of the 3rd millennium: space, economy, society in the last 40 years. Panteion University, Athens, pp 205–211. (in Greek)
Markezinis S (1966) Political history of modern Greece 1828–1964. 4 volumes. Papyrus, Athens (in Greek)
Marmaras E (1989) The privately-built multi-storey apartment building: the case of inter-war Athens. Plan Perspect 4(1):45–78
Marmaras M (1991) The urban apartment building of interwar Athens. ETVA Cultural and Technological Foundation, Athens. (in Greek)
Marmaras M (2002) Planning and settlement space. Ellinika Grammata, Athens. (in Greek)
Marmaras M (2012) About architecture and town planning in Athens. Papazisis, Athens. (in Greek)
Mavrakis P (1970) Regional development programming in Greece. In: CPER (ed) The framework of regional development, Lecture series. CPER and Gutenberg, Athens, pp 201–239. (in Greek)
Mavrakis P (1971) Regional programming and regional development policy. EVEA Economic Library, Athens. (in Greek)
Meijers EJ, Burger MJ (2017) Stretching the concept of 'borrowed size'. US 54(1):269–291
Melissas D (2002) Fundamental issues of regional spatial planning. Ant.N. Sakkoulas, Athens. (in Greek)
Melissas D (2007) Land uses and the General Town Plan (GPS). Sakkoulas, Athens. (in Greek)
Melissas D (ed) (2010) Special spatial plan for the renewable energy sources: a year after. Proceedings of a one-day conference. Papasotiriou, Athens (in Greek)

Melissas D (2010a) Zones of Settlement Control (ZOE) and General Town Plans (GPS): unobstructed co-existence? In: Demetriadis EP, Kafkalas G, Tsoukala K (eds) The *Logos* of the *Polis*, Volume in honour of Prof. A.Ph. Lagopoulos. University Studio Press, Thessaloniki, pp 125–133. (in Greek)
Melissas D (2010b) Questions regarding the special APE plan. In: Melissas D (ed) Special spatial plan for the renewable energy sources: a year after, Proceedings of a one-day conference. Papasotiriou, Athens, pp 115–124. (in Greek)
Melissas D (2012) The end of town planning. Dioikitiki Diki 2012(5):1105–1118. (reprint) (in Greek)
Melissas D (2015) New building regulation (L.4067/2012 as amended), 3rd edn. Sakkoulas, Athens. (in Greek)
Melissas D (2017) The legislation for building "regularization" and legalization is testing the credibility of the Greek state. In: Serraos C, Melissas D (eds) Unlicensed building. Sakkoulas, Athens, pp 29–43. (in Greek)
Melissas D (2018) A conceptual approach to maritime spatial planning. In: Theodora G (ed) Maritime space, coastal urban front, port-cities. NTUA, Athens, pp 31–41. (in Greek)
Melissas D (2019) TPS and EPS plans. Sakkoulas, Athens. (in Greek)
Melissas D (2021) Land uses. Sakkoulas, Athens (in Greek)
Melissas D, Serraos C (eds) (2017) Organization of land uses and activities in the spatial planning process. Sakkoulas, Athens. (in Greek)
Melissourgos G, Chasiotis G (2019) Spatial planning. In: WWF Hellas (ed) Environmental legislation in Greece. Annual report. Athens, pp 35–50 (in Greek)
Menoudakos C (1996) The Council of State case law on the transfer of development rights. In: Papademetriou G (ed) The mechanism of the transfer of development rights. Nomos+Physis/Ant.N. Sakkoulas, Athens, pp 37–45. (in Greek)
Menoudakos C (2008) Spatial planning in Council of State case law. In: Giannakourou G, Menoudakos C, Wassenhoven L (eds) Regional spatial planning in Greece: legal framework and implementation in practice. Nomos+Physis/Ant.N. Sakkoulas, Athens, pp 31–50. (in Greek)
Menoudakos C (2017a) Unlicensed building: legality and reality. In: Serraos C, Melissas D (eds) Unlicensed building. Sakkoulas, Athens, pp 6–9. (in Greek)
Menoudakos C (2017b) Organization of land uses and activities in the spatial planning process. In: Melissas D, Serraos (eds) Organization of land uses and activities in the spatial planning process. Sakkoulas, Athens, pp 3–5. (in Greek)
Menoudakos C (2018) Some thoughts with reference to the book "Maritime Spatial Planning – Europe and Greece" by Prof. Louis Wassenhoven. In: Serraos C, Melissas D (eds) Maritime spatial planning. Sakkoulas, Athens, pp 1–11. (in Greek)
Mercadal G (1965) Les études d'armature urbaine régionale, Consommation, No 3, 1965. Dunod, Paris, pp 3–42
Meynaud J (1966) Political forces in Greece. Byron, Athens (in Greek; translated from the French original)
Michaelidis G (2009) Crisis, space and development: questions and answers (?). In: University of Thessaly (ed) 25 texts on spatial planning and development. Thessaly University Press, Volos, pp 395–421. (in Greek)
Michail Y (1994) Urban development of Athens in the first two postwar decades. In: Sakis Karagiorgas Foundation (1994–95) Greek society in the first postwar period 1945–1967. Conference proceedings. Vol. A. Athens, pp 573–582 (in Greek)
Michail Y (2008) Towards a sustainable spatial planning policy. In: Karra L (ed) Vulnerable land. ELLET, Athens, pp 47–51. (in Greek)
MinCoord (1960) 5-year economic development plan 1960–1964. ETyp (National Press), Athens. (in Greek)
MinCoord (1979) Economic and social development programme 1978–1982 – Preamble. ETyp, Athens (in Greek)
MinDevComp (2014) ESPA 2014–2020 – Partnership agreement. Athens (in Greek)

MinEcDev (2019) Greece: National strategy for sustainable and fair development – 2030. (in Greek)
MinEcDevT (2016) ESPA 2014–2020: Factsheet. Athens (in Greek)
MinEcFin (2007) ESPA 2007–2013. Athens (in Greek)
Ministry of Public Works (Greece) (1973) Greece: National monograph (prepared by a working party of experts). Programme: Housing, building and planning problems and policies in the less developed countries of Southern Europe. United Nations Economic Commission for Europe – Housing, building and planning committee. Published by the Technical Chamber of Greece, Athens
Mitaraki-Bazou F, Basoukea E (2001) The story of the creation of the Athens Master Plan Office in 1965. Topos 2001(17) (in Greek)
Mitsos A (1995) The redistributive and developmental policy of the European Union in the new phase of European integration. In: SEP (ed) Regional development, spatial planning and environment in the framework of united Europe. Proceedings of a conference. Vol. I, Panteion University, SEP and Topos, pp 35–45 (in Greek)
MOD (2018) One-day information conference on Cohesion Policy in the period 2021–2027. Athens (in Greek)
Modinos M, Efthymiopoulos E (eds) (1998) Ecology and environmental sciences. Interdisciplinary Institute of Environmental Research, Athens. (in Greek)
Modinos M, Efthymiopoulos E (eds) (2000) The sustainable city. Stochastis and Interdisciplinary Institute of Environmental Research, Athens. (in Greek)
Monastiriotis V (2008) The geography of spatial association across the Greek regions: patterns of persistence and heterogeneity. In: Coccossis H, Psycharis Y (eds) Regional analysis and policy: the Greek experience. Physica-Verlag, Heidelberg, pp 17–39
Monioudi-Gavala T (2012) Town planning in the Greek state 1833–1890. Athens (bilingual)
Monioudi-Gavala T (2015) The Greek city from Hippodamus to Cleanthis. Hellenic Academic Ebooks (www.kallipos.gr) (in Greek)
Mourtsiadis A (2012) Industrial and business estates: Location, urban planning, environmental planning. Ath. Stamoulis, Athens. (in Greek)
MPW (1973) Greece: national monograph (prepared by a working party of experts). Programme: housing, building and planning problems and policies in the less developed countries of Southern Europe. United Nations Economic Commission for Europe – Housing, building and planning committee. TEE, Athens
Nantsou Th, Chasiotis G, Marangou P (2018) Spatial planning. In: WWF Hellas (ed) Environmental legislation in Greece. Annual report, in cooperation with Nomos+Physis, pp 50–62 (in Greek)
Needham B (2017) A renegade economist preaches good land-use planning. In: Haselsberger B (ed) Encounters in planning thought: 16 autobiographical essays from key thinkers in spatial planning. Routledge, New York/London, pp 165–183
Nenou E, Samarina A (1995) Law 360/1976 on spatial planning and the environment: Implementation results and prospects. In: SEP (ed) Regional development, spatial planning and environment in the framework of united Europe. Proceedings of a conference. Vol. II. Panteion University, SEP and Topos, pp 131–135 (in Greek)
Newman P, Thornley A (1996) Urban planning in Europe. Routledge, London
Nicolacopoulos I, Tsouyopoulos GS (1976) Structural aspects of the network of Greek cities. EpKE 1976(26–27) (reprint)
Niessler R (2004) Urban actions in the European Structural Funds: Lessons and prospects. Conference on Urban Sustainable Development Policy in Greece. NTUA and YPECHODE, Athens, 10.9.2004 (in Greek)
Nikolaou A (1994) Evolution of the environmental dimension in spatial planning: critical view of the mechanism of ZOE. In: Rokos D (ed) Sciences and the environment at the end of the century: problems and prospects. NTUA and Enallaktikes Ekdoseis, Athens, pp 231–242. (in Greek)
NTUA (1973) Industrial development in space, Lecture series. NTUA, Athens

NTUA (1986) Scientific event dedicated to the memory of Prof. Antonis Emm. Kriezis (in Greek)
NTUA (1990) The interdisciplinary approach to development. Proceedings of an Interuniversity Interdisciplinary Conference. Papazisis, Athens (in Greek)
NTUA (2004) City and space from the 20th to the 21st century. Volume in honour of Prof. A.I. Aravantinos. NTUA, University of Thessaly, and SEPOCH, Athens (in Greek)
NTUA (2011) Spatial development – Spatial policies. Proceedings of a one-day conference, Athens (in Greek)
NTUA, University of Thessaly (2006) Integrated development operational plan of the Prefecture of Athens. Research report (coordinators M. Angelidis and P. Skagiannis). Prefecture of Athens (in Greek)
NTUA/EChOA (1989) Relations of economic and physical planning: Interactions of urban renewal policies, controls and parameters with the operations of small manufacturing firms. Phase B research report. Coordinators L. Wassenhoven and E. Panayotatos. NTUA and GGET, Athens (in Greek)
NTUA/EChOA (1994) Estimation of fire hazards in forest land and spatial fire mitigation planning for prevention and preparedness: A preliminary study of the Attica region. Research report. General Secretariat of Research and Technology, 1991–1994 (researchers: K. Sapountzaki, N. Pangas, L. Wassenhoven)
NTUA/SPE (1969) Andritsaina (Elia Prefecture): Master plan, town plan and area of influence. Research report. Athens (in Greek)
NTUA/SPE (1971a) Regional analysis of Epirus and Thessaly, settlement restructuring of earthquake-stricken areas, and town plans of three townships for the relocation of population. Research report. Athens (in Greek)
NTUA/SPE (1971b) Master development plan of the city and influence zone of Kalamata in the Peloponnese. Research report. Athens (in Greek)
OECD (1983) Environmental policies in Greece. Paris
OECD (1997) Managing across levels of government. Collection of national reports, Paris
OECD (2000) Environmental performance reviews: Greece. Paris
OECD (2001a) Territorial outlook. Paris
OECD (2001b) Towards a new role for spatial planning. A collection of essays. Paris
OECD (2009) OECD environmental performance reviews. Greece, Paris
OECD (2018) Economic surveys: Greece – Overview. April 2018. Paris
OKE (2007) Opinion report on spatial regional and urban planning. Athens (in Greek)
ORSA (1996) Athens – Attica: Strategic planning for sustainable development. Proceedings of an international conference. ORSA and YPECHODE, Athens (various languages)
Palermo PC, Ponzini D (2010) Spatial planning and urban development: critical perspectives. Springer, Dordrecht
Panayotatos E (1982) Introduction to regional spatial planning. NTUA, Athens. (in Greek)
Panayotatos E (1983) Issues of spatial development. NTUA, Athens. (in Greek)
Panayotatos E (1984) A critique of regional planning in Greece. Habitat Int 8(1):35–44
Panayotatos E (1988) Contribution to a comprehensive view of space and a different planning practice. NTUA, Athens. (in Greek)
Panayotatos E (1989) Economic interdependence and space integration: Implications in theory and practice. Seminars - The Graduate School for Architecture, Planning and Preservation, Columbia University
Panayotatos E, Zacharatos G, Markou M, Sagias I, Spourdalakis M (1993) Evaluation of investment initiatives' programmes in key sector of the Greek economy: An alternative method of intervention. Research report. NTUA and GGET, Athens (in Greek)
Pangakis G (1991) Introduction to public administration. Ant.N. Sakkoulas, Athens. (in Greek)
Pantazis A (2004) Analysis and planning of the settlement system of a Greek region: The example of Western Greece. In: Angelidis M (ed) New polycentricity and spatial development. Conference proceedings (2001). YPECHODE, Ministry of Economy, NTUA and Université de Paris X/Nanterre, Athens, pp 52–62 (in Greek)

Panteion University (2011) Volume in honour of Prof. P. Loukakis. Gutenberg, Athens. (in Greek)
Papadaskalopoulos A, Christofakis M (2011) Development models and urban centres in Greek regional programming after 2000. Aeichoros 2011(15):8–41
Papademetriou G (ed) (1996) The mechanism of the transfer of development rights. Nomos+Physi/Ant.N. Sakkoulas, Athens. (in Greek)
Papageorgiou G (1993) Elements of town planning. Evgenidis Foundation, Athens. (in Greek)
Papageorgiou M (2015) Maritime spatial planning and coastal land use: applications and practices in Greece. In: Proceedings of the 4th panhellenic conference on urban and regional spatial planning and development, 24–27 September 2015. University of Thessaly, Volos (in Greek)
Papageorgiou M (2016a, 2016) Maritime spatial planning and maritime uses: conceptual and theoretical issues. Aeichoros (23):41–63
Papageorgiou M (2016b) Spatial planning and protected maritime areas in Greece. In: Kyvelou S (ed) Maritime spatial issues: the maritime dimension of territorial cohesion – maritime spatial planning – sustainable blue development. Kritiki, Athens, pp 148–158. (in Greek)
Papageorgiou M (2017) Spatial planning in transition in Greece: a critical overview. EPS 25(10):1818–1833
Papageorgiou M, Giannoula M, Telianidou (2010) The role of spatial planning in natural heritage management and protection. In: Beriatos E, Papageorgiou M (eds) Spatial planning – town planning – environment in the 21st century. Thessaly University Press, Volos, pp 797–812. (in Greek)
Papageorgiou-Venetas A (2010) Athens: a vision of neoclassicism. Kapon, Athens. (in Greek)
Papagiannakis L (1994a) People and their activities: on the road to crisis. In: Papagiannakis L (ed) The development of the region of Thrace: challenges and prospects. Research Centre of the Greek Society, Academy of Athens, Athens, pp 15–37. (in Greek)
Papagiannakis L (ed) (1994b) The development of the region of Thrace: challenges and prospects. Research Centre of the Greek Society, Academy of Athens, Athens. (in Greek)
Papagiannakis L (1996) The role of Athens in the framework of Euro-Mediterranean relations. In: ORSA (ed) Athens – Attica: Strategic planning for sustainable development. Proceedings of an international conference. Athens, pp 47–54 (in Greek)
Papagiannis T (1968) Introduction to the study of the environment. AT 1968(2):92–95. (in Greek)
Papagiannis T (2009) With respect for man and nature: the case of Prespa. Athens (in Greek)
Papaioannou JG (1971) Evaluation of recent trends in metropolitan planning in Greece. In: The mastery of urban growth. Report of the International Colloquium, Brussels, 2–4 December 1969. Meus en Ruimte, V.Z.W. (M+R International). Brussels, pp 263–293
Papaioannou A, Nikolakopoulou C (2016) Greek cities in crisis: context, evidence, response. In: Knieling J, Othengrafen F (eds) Cities in crisis: Socio-spatial impacts of the economic crisis in southern European cities. Routledge, London, pp 172–189
Papamichos N, Hastaoglou V (2000) Town and regional planning in the 1960s: the case of the plan for Thessaloniki. In: Deffner A et al (eds) Town planning in Greece from 1949 to 1974. Thessaly University Press, Volos, pp 49–57. (in Greek)
Pappas V (1995) Information data base for spatial and regional programming. In: SEP (ed) Regional development, spatial planning and environment in the framework of united Europe. Proceedings of a conference. Vol. II. Panteion University, 1995. SEP and *Topos*, pp 167–185 (in Greek)
Parkinson M, Meegan R, Karecha J (2015) City size and economic performance: is bigger better, small more beautiful or middling marvellous? EPS 23(6):1054–1068
Parri I (2015) Planning and application of a maritime cadastre in Greece: an instrument of maritime spatial planning. In: Proceedings of the 4th Panhellenic Conference on Urban and Regional Spatial Planning and Development, 24–27 September 2015. University of Thessaly, Volos (in Greek)
Petmezidou M, Tsoulouvis L (1990) Aspects of state programming in Greece: Historical continuity and crisis impacts. In: Psycharis Y (ed) The functions of the state in a period of cri-

sis: Theory and Greek experience. Conference proceedings. Sakis Karagiorgas Foundation, Athens, pp 288–301. (in Greek)

Petrakos G (1996) The new geography of the Balkans: Cross-border cooperation between Albania, Bulgaria and Greece. University of Thessaly, Volos

Petrakos G (2000) Spatial effects of the East-West integration in Europe. In: Andrikopoulou E, Kafkalas G (eds) The new European space: enlargement and the geography of European development. Themelio, Athens, pp 185–212. (in Greek)

Petrakos G (ed) (2000) The development of the Balkans. Thessaly University Press, Volos. (in Greek)

Petrakos G (2009) Regional inequalities and selective development: the economic forces of space and the preconditions of an effective regional policy. In: University of Thessaly (ed) 25 texts on spatial planning and development. Thessaly University Press, Volos, pp 359–393. (in Greek)

Petrakos G (2010) Development paths and landscapes: A critical discussion about economy and space. In: Demetriadis EP, Kafkalas G, Tsoukala K (eds) The *Logos* of the *Polis*. Volume in honour of Prof. A.Ph. Lagopoulos. University Studio Press, Thessaloniki, pp 55–65. (in Greek)

Petrakos G, Artelaris P (2008) Regional inequalities in Greece. In: Coccossis H, Psycharis Y (eds) Regional analysis and policy: the Greek experience. Physica-Verlag, Heidelberg, pp 121–139

Petrakos G, Mardakis P (2012) Recent changes in the Greek system of urban centres. In: Economou D, Petrakos G (eds) The development of Greek cities: interdisciplinary approaches of urban analysis and policy. Thessaly University Press, Volos. (first published 1999), pp 45–64 (in Greek)

Petrakos G, Psycharis Y (2003) Regional inequalities in Greece: an alternative method of quantification. TeC 2003(23/1–2):19–33. (in Greek)

Petrakos G, Psycharis Y (2004) Regional development in Greece. Kritiki, Athens. (in Greek)

Pikramenos MN (2018) Judicial spatial distribution: Nomothetic choice of major significance for the judicial system. In: Sakellaropoulou K et al (eds) Justice in Greece: proposals for a modern judicial system. diaNEOsis, Athens, pp 105–146. (in Greek)

Piperis S (2018) Greek maritime space: a geographical approach of its potential with a focus on energy. In: Theodora G (ed) Maritime space, coastal urban front, port-cities. NTUA, Athens, pp 63–75. (in Greek)

Plaskovitis I (2008) Change in regional policy priorities, objectives and instruments in Greece: a comparative analysis of regional programmes. In: Coccossis H, Psycharis Y (eds) Regional analysis and policy: the Greek experience. Physica-Verlag, Heidelberg, pp 141–162

Polyzos S (2011) Regional development. Kritiki, Athens. (in Greek)

Pournara SI (2018) Issues and problem-setting in relation to maritime spatial planning. In: Theodora G (ed) Maritime space, coastal urban front, port-cities. NTUA, Athens, pp 77–96. (in Greek)

Presidency Conclusions (2000) Lisbon European Council, 23–24 March 2000

Presidency Conclusions (2001) Göteborg European Council, 15–16 June 2001

Presidency Conclusions (2005) Brussels European Council, 22–23 March 2005

Preteceille E (1982) Urban planning: the contradictions of capitalist urbanisation. In: Paris C (ed) Critical readings in planning theory. Pergamon Press, Oxford, pp 129–146. (originally published in French 1974

Provelengios A (1967) The master plan of Herakleion, Crete. AT 1967(1):50–55. (in Greek)

Psomopoulos P (1977) Guiding the growth of metropolitan Athens. Ekistics 44(262):120–133

Psomopoulos P (ed) (1991) Urban networking in Europe – I & II: concepts, intentions and new realities. 2 double special issues. Ekistics 58(350/351) and 59(352/353)

Psomopoulos P (2019) Issues of Ekistics: views, proposals, lectures 1962–2011. Edited by A. Papageorgiou-Venetas. Eurasia, Athens (in Greek)

Psycharis Y (1990) Testing and interpretation of regional development policies in Greece 1976-87. In: Psycharis Y (ed) The functions of the state in a period of crisis: theory and Greek experience. Conference proceedings. Sakis Karagiorgas Foundation, Athens, pp 268–287. (in Greek)

Psycharis Y (ed) (1990) The functions of the state in a period of crisis: theory and Greek experience. Conference proceedings. Sakis Karagiorgas Foundation, Athens. (in Greek)
Psycharis Y (2000) Regional inequalities and state interventionism: review of developments over 25 years. In: Psycharis Y, Gospodini A, Christopoulou O (eds) Seventeen essays on planning, cities, and development. Thessaly University Press, Volos, pp 367–386. (in Greek)
Psycharis Y, Pantazis P (2016) The geography of declared income in Greece before and during the economic crisis. Regions 303(Summer):4–7
Psycharis Y, Gospodini A, Christopoulou O (eds) (2000) Seventeen essays on planning, cities, and development. Thessaly University Press, Volos. (in Greek)
Raptis MN (1940) Elements of town planning. S. Fotiadis, Athens. (in Greek)
Renaud B (1981) National urbanization policy in developing countries. A World Bank Research Publication. Oxford University Press, Oxford
Rivolin UJ (2017) Global crisis and the systems of spatial governance and planning: a European comparison. EPS 25(6):994–1012
Rochefort M (1960) L'organisation urbaine de l'Alsace. Les Belles Lettres, Paris
Rodakinias P, Skayannis P, Zygoura A (2008) Regional development and the information society: how Greek regions measure up in the information age? In: Coccossis H, Psycharis Y (eds) Regional analysis and policy: the Greek experience. Physica-Verlag, Heidelberg, pp 217–229
Romanos A (1975) Problems of Greek space. Reprint. Kathimerini, Athens (in Greek)
Rozos N (1994) Legal problems of regional spatial planning. Ant.N. Sakkoulas, Athens. (in Greek)
Rozos N (2017) Law maker, administration, judge and unlicensed building. In: Serraos C, Melissas D (eds) Unlicensed building. Sakkoulas, Athens, pp 11–26. (in Greek)
SADAS (2002) Architecture and the Greek city in the 21st century. Proceedings of the 10th Panhellenic Architectural Conference (1999). SADAS and TEE, Athens (in Greek)[31]
Sakellaropoulou K (2017) Remarks on unlicensed building. In: Serraos C, Melissas D (eds) Unlicensed building. Sakkoulas, Athens, pp 27–28. (in Greek)
Sakellaropoulou K (2018a) Case law of the Council of State on the planning and management of maritime and coastal space. In: Serraos C, Melissas D (eds) Maritime spatial planning. Sakkoulas, Athens, pp 39–58. (in Greek)
Sakellaropoulou K (2018b) Environmental constitution and the Council of State in the years of crisis. Nomos+Physis (available on the web) (in Greek)
Sakellaropoulou K et al (2018) Justice in Greece: proposals for a modern judicial system. diaNEOsis, Athens. (in Greek)
Samaras Y (1985) State and capital in Greece. Synchroni Epochi, Athens. (in Greek)
Sapounaki-Drakaki L, Stamatiou E (2000) The development of public administration for spatial regulation through the testimonies of cadres of the Ministry of Public Works in the 1950s and 1960s. In: Deffner A et al (eds) Town planning in Greece from 1949 to 1974. Thessaly University Press, Volos, pp 139–150. (in Greek)
Sapountzaki K (2001) The role of prefectoral self-government in the statutory spatial planning system in Greece: A preliminary empirical approach. Topos 2001(16)
Sapountzaki K (ed) (2007) Tomorrow in danger: natural and technological disasters in Europe and Greece. Gutenberg, Athens. (in Greek)
Sapountzaki K (2019a) Less vulnerable, more adaptive societies: Can spatial planning make a contribution? In: Serraos C, Melissas D (eds) Natural disasters and spatial policies. Sakkoulas, Athens, pp 49–62. (in Greek)
Sapountzaki K (2019b) Less vulnerable, more resilient territories – What can be a contribution of spatial planning? Lecture at Université Savoie Mont-Blanc, 3–5 November
Sapountzaki K et al (2022) Chapter 12: A risk-based approach to development planning. In: Eslamian S, Eslamian F (eds) Handbook of disaster risk reduction for resilience (HD3R). Vol.4. Springer

[32] Pagination problems in vol. III.

Sarigiannis G (1980) In TEE (Technical Chamber of Greece) (1980) Problem areas: meaning, causes and problem-solving prospects. In: TEE (ed) Spatial planning and development. Conference on development – 2nd pre-conference one-day meeting. Athens, pp 33–40 (in Greek)
Sarigiannis G (1988) The mountain settlements of Epirus. The problem of their survival and their study 20 years earlier. Reprint from Epirotiko Imerologio, Ioannina (in Greek)
Sarigiannis G (1995) Economic policy and spatial planning: from the Marshall Plan to Maastricht. In: Georgoulis D (ed) Texts on the theory and implementation of urban and regional spatial planning. Papazisis, Athens, pp 83–155. (in Greek)
Sarigiannis G (2000) Town planning in the 1960s: education and practice. In: Deffner A et al (eds) Town planning in Greece from 1949 to 1974. Thessaly University Press, Volos, pp 61–83. (in Greek)
Sarigiannis G (2009) The theoretical and design views of K.A. Doxiadis in relation to the town planning theories of his day. In: Kazazi G (ed) Constantinos Doxiadis and his work, vol A. TEE, Athens, pp 310–347. (in Greek)
Sarigiannis G (2017a) The change of perception of mountain zone development in the 1960s in the research programmes of SPE / NTUA: A critical evaluation. NTUA, School of Architecture "Thursday lecture" series. 23-2-2017 (in Greek)
Sarigiannis G (2017b) The master plans of Athens and the changes of their framework diakonima.gr
Sellis T, Georgoulis D (eds) (1997) Urban, regional and environmental planning and informatics to planning in an era of transition, Proceedings of the Athens International Conference, 22-24.10.1997. NTUA, Athens
Selman P (1995) Theories for rural-environmental planning. PPR 10(1):5–14
SEP (1995a) Regional development, spatial planning and environment in the framework of united Europe. 3 volumes. Proceedings of a conference, Panteion University, 1995. SEP and *Topos*[32] (in Greek)
SEP (1995b) Frontier regions. Proceedings of a conference, 1995. SEP, Thessaloniki (in Greek)
Serraos C (2018) Official planning and bottom-up practices: an antithetical trajectory in 21st century Greece? In: Theodora G (ed) Maritime space, coastal urban front, port-cities. NTUA, Athens, pp 43–55. (in Greek)
Serraos C, Greve T (eds) (2016) Metropolitan interventions Athens 2021: Exchange of European experiences. Proceedings of a scientific workshop. National Technical University of Athens, School of Architecture. Propobos, Athens
Serraos C, Melissas D (eds) (2017) Unlicensed building. Sakkoulas, Athens. (in Greek)
Serraos C, Melissas D (eds) (2018) Maritime spatial planning. Sakkoulas, Athens. (in Greek)
Serraos C, Melissas D (eds) (2019) Natural disasters and spatial policies. Sakkoulas, Athens. (in Greek)
Serraos C et al (2016) Athens, a capital in crisis. In: Knieling J, Othengrafen F (eds) Cities in crisis: socio-spatial impacts of the economic crisis in southern European cities. Routledge, London, pp 116–138
Shaw D, Roberts P, Walsh J (eds) (2000) Regional planning and development in Europe. Ashgate, Aldershot
Sidiropoulos E, Papadaskalopoulos A (1990) The strategy of development from below: a new perception for regional development. In: NTUA (ed) The interdisciplinary approach to development. Proceedings of an Interuniversity Interdisciplinary Conference. Papazisis, Athens, pp 527–541. (in Greek)
Siouti G (1994) Fundamental principles of spatial planning control in the case law of the Council of State. PeD 1994(1):9–22. (in Greek)
Siouti G (1995) Sustainable development and environmental protection. In: SEP (ed) Regional development, spatial planning and environment in the framework of united Europe. Proceedings of a conference. Vol. III. Panteion University, 1995. SEP and Topos, pp 43–50 (in Greek)
Siouti G (2011) Manual of environmental law, 2nd edn. Sakkoulas, Athens. (in Greek)

[33] In collaboration with K. Sapountzaki, E. Asprogerakas, E. Gianniris and Th. Pagonis.

Skagiannis P (1994–95) The role of infrastructures in the capital accumulation regimes of the first postwar periods in Greece. In: Sakis Karagiorgas Foundation (ed) Greek society in the first postwar period 1945–1967. Conference proceedings. Vol. A. Athens, pp 115–132 (in Greek)
Skagiannis P (2009) From S shape to "sigma" shape: Towards a new development of Greek space? In: University of Thessaly (ed) 25 texts on spatial planning and development. Thessaly University Press, Volos, pp 69–118. (in Greek)
Skiadaresis G (1967) Development problems of a Greek urban centre. AT 1967(1):44–49. (in Greek)
Skordili S (2008) Regional inequalities and the digital economy challenge: variations in Internet accessibility across Greek regions. In: Coccossis H, Psycharis Y (eds) Regional analysis and policy: the Greek experience. Physica-Verlag, Heidelberg, pp 231–248
Skouris V (1991) Spatial regional and town planning law. Sakkoulas, Thessaloniki. (in Greek)
Skouris V, Tachos AI (1988) Environmental protection in the case law of the Council of State. Paratiritis, Thessaloniki. (in Greek)
Slaev AD et al (2018) Suburbanization and sprawl in post-socialist Belgrade and Sofia. EPS 26(7):1389–1412
Sofoulis C (1967) Regional development or the spatial plan of national ambitions. AT 1967(1):24–25. και 318 (in Greek)
Sofoulis C (1979) Land as *produced* factor of production. Papazisis, Athens (in Greek)
Sofoulis C (1993) Active town planning as a method of housing policy. In: Sakis Karagiorgas Foundation (ed) Dimensions of present-day social policy. Conference proceedings. Athens, pp 635–654 (in Greek)
Sorensen A (2018) New institutionalism and planning theory. In: Gunder M, Madanipour A, Watson V (eds) The Routledge handbook of planning theory. Routledge, London, pp 250–263
Sotiropoulos DP (2019) Phases and contradictions of the Greek state in the 20th century, 1910–2001. Estia, Athens. (in Greek)
Sotiropoulos DA, Christopoulos L (2017) Plethoric law production and bad legislation in Greece: A plan for a better and more effective state. diaNEOsis, Athens. (in Greek)
Sousa S, Pinho P (2015) Planning for shrinkage: paradox or paradigm. EPS 23(1):12–32
Spilanis G (1995) Is the environment a parameter which refrains or limits regional development? The necessity of spatial planning. In: SEP (ed) Regional development, spatial planning and environment in the framework of united Europe. Proceedings of a conference. Vol. II. Panteion University, 1995. SEP and *Topos*, Athens, pp 144–154 (in Greek)
Spilanis G (2000) Tourism and regional development: the case of the Aegean islands. In: Tsartas P (ed) Tourism development: multi-disciplinary approaches. Exantas, Athens, pp 149–187. (in Greek)
Spilanis G (2012) Greek islands and political cohesion. Gutenberg – G. and K. Dardanos, Athens (in Greek)
Stamatelatos M, Vamva-Stamatelatou F (2006) Single-volume geographical dictionary of Greece. Ermis, Athens. (in Greek)
Stead D, Nadin V (2008) Spatial planning: key instrument for development and effective governance, with special reference to countries in transition. United Nations/UN Economic Commission for Europe, New York/Geneva
Stefani F, Tsilimingas G (2015) The importance of maritime spatial planning for the protection and development of Greek maritime space. Proceedings of the 4th panhellenic conference on urban and regional planning and regional development. 24–27 September 2015. University of Thessaly, Volos (in Greek)
Stingas TN (ed) (2004) Urban planning law: comments and case law. Nomiki Vivliothiki, Athens. (in Greek)
Subra P (2014) Géopolitique de l'aménagement du territoire. Armand Colin, Paris
TEE (1976) Public discussion on the national spatial plan: Commissioning process of the plan and programme to Doxiadis Associates. TeC 1976(2) (in Greek)

TEE (1978) Spatial regional planning process and plan implementation in Greece. Report of the TEE permanent spatial planning committee. TeC 1978(8):5–7. (in Greek)
TEE (1980) Spatial planning and development. Conference on development – 2nd pre-conference one-day meeting. Athens (in Greek)
TEE (1985) Regional programming and planning: the role of the administration and local self-government. One-day conference proceedings. TeC, April–June 1985 (in Greek)
TEE (1993) EPA: 10 years later – Reality and prospects. Proceedings of a two-day conference. TeC 1993(5) (in Greek)
TEE (1995) Tourism and environment: Choices for sustainable development. Proceedings of a one-day conference. TeC 1995(5) (in Greek)
Theodora G (2008) An approach to the effects of Greek regional universities on the development of the country regions. In: Coccossis H, Psycharis Y (eds) Regional analysis and policy: The Greek experience. Physica-Verlag, Heidelberg, pp 249–270
Theodora G (2011) Knowledge as a parameter in spatial development theories. In: Panteion University (ed) Volume in honour of Prof. P. Loukakis. Gutenberg, Athens, pp 241–285. (in Greek)
Theodora G (2018a) The question of teaching spatial planning. In: Theodora G (ed) Maritime space, coastal urban front, port-cities. NTUA, Athens, pp 13–29. (in Greek)
Theodora G (2018b) The Greek settlement network – An approach to its development potential. Postgraduate lecture. NTUA, Athens (available in the NTUA website) (in Greek)
Theodora G (ed) (2018c) Maritime space, coastal urban front, port-cities. NTUA, Athens. (in Greek)
Theodora G, Loukakis P (2011) Development trends of the Greek urban centres' network. Aeichoros 2011(15):102–129
Thoidou E (2004) Spatial integration of spatial development policies: European perspectives and Greek experience. In: Kafkalas G (ed) Issues of spatial development. Kritiki, Athens, pp 213–238. (in Greek)
Thoidou E (2010) Strategic spatial planning: references to contemporary Greek experience. In: Beriatos E, Papageorgiou M (eds) Spatial planning – town planning – environment in the 21st century. Thessaly University Press, Volos, pp 161–176. (in Greek)
Thoidou E (2011) Strategic spatial planning: modern framework and characteristics. In: Panteion University (ed) Volume in honour of Prof. P. Loukakis. Gutenberg, Athens, pp 304–330. (in Greek)
Tomprogiannis I (1967) Regional spatial programming and economic development. AT 1967(1):26–27. (in Greek)
Topalov C et al (eds) (2010) L'aventure des mots de la ville. Robert Laffont, Paris
Tortopidis A (2009) National spatial plan of Greece: today and 30 years earlier. In: Kazazi G (ed) Constantinos Doxiadis and his work, vol B. TEE, Athens, pp 16–28. (in Greek)
Tritsis A (1967) Organization and organic structure in wider space. AT 1967(1):28–31. (in Greek)
Tsartas P (2000) Critical evaluation of the constitutive parameters of the characteristics of postwar tourism development. In: Tsartas P (ed) Tourism development: multi-disciplinary approaches. Exantas, Athens, pp 189–211. (in Greek)
Tsartas P (ed) (2000) Tourism development: multi-disciplinary approaches. Exantas, Athens. (in Greek)
Tsartas P (2001) Tourism development: characteristics and models. In: Coccossis H, Tsartas P (eds) Sustainable tourism development and the environment. Kritiki, Athens, pp 19–129. (in Greek)
Tselios V, Rovolis A, Psycharis Y (2017) Regional economic development, human capital and transport infrastructure in Greece: the role of geography. In: Fonseca M, Fratesi U (eds) Regional upgrading in Southern Europe. Springer, Cham, pp 151–174
Tsilimingas G, Gourgiotis A (2014) Landscape management in the framework of spatial planning. Aeichoros 19:24–37. (in Greek)
Tsiomis I (1986) Parler d'Athènes de 1834, comme on parle de Brasilia de 1964. In: Burgel G (ed) De la Polis aux politiques urbaines: Renaissance et mutations de la ville grecque contem-

poraine, Villes en Parallèle. No 9. Laboratoire de géographie urbaine de l'Université de Paris X-Nanterre, Nanterre, p 20
Tsoulouvis L (1981) Thessaloniki: beyond the implementation inability of a master plan. AT 1981(15):79–86. (in Greek)
Tsoulouvis L (1998) Planning, the urban system and new forms of inequality in Greek cities. Reprint. Progress in Planning, 50(1)
Tziafetas GN (1990) The demographic problem of Greece. Foundation for Tackling the Demographic Problem, Athens (in Greek)
UN Habitat III (2016) Habitat III Policy Paper 3 – National Urban Policy. Unedited version. 29 February. United Nations Conference on Housing and Sustainable Urban Development. United Nations, New York
UNESCO (1972) The World Heritage Convention (available on the web)
United Nations (1976) Guidelines for disaster prevention: Vol. 1 – Pre-disaster physical planning of human settlements. Office of the United Nations, Disaster Relief Co-ordination, Geneva
United Nations (1992) Conference on Environment & Development: Agenda 21. Rio de Janeiro, Brazil, 3 to 14 June 1992
United Nations Centre for Human Settlements (Habitat) (2001) Cities in a globalizing world: Global report on human settlements. Earthscan, London
University of Thessaly (2009) 25 texts on spatial planning and development. Thessaly University Press, Volos. (in Greek)
Vagianos S (ed) (1968) Studies on town planning and tourism development in Crete. AT 1968(2):42–91
Vaiou D, Mantouvalou MM (2000) Postwar Greek town planning, between theory and conjuncture. In: Deffner A et al (eds) Town planning in Greece from 1949 to 1974. Thessaly University Press, Volos, pp 25–37. (in Greek)
Vavouras I, Manolas G (2005) Corruption and its relations-impacts on the official economy and the para-economy. In: Koutsoukis C, Sklias P (eds) Corruption and scandals in public administration and politics. I. Sideris, Athens. (in Greek), pp 349–369
Vliamos S (1988) Industrial estates and industrial regional policy in Greece. ETVA, Athens. (in Greek)
Vliamos S (2017) Regional development, environment and the tourist product. In: Giaoutzi M, Nijkamp P (eds) Tourism and regional development: New pathways. Routledge, London (first published 2006), pp 281–302
Vliamos S, Georgoulis D, Kourliouros E (1991) Industrial estates: Institutions, theory and planning methodology. ETVA/Papazisis, Athens. (in Greek)
Voivonda A, Kizilou V, Kloutsinioti R, Kontaratos S, Pyrgiotis Y (1977) Space regulation in Greece: a brief historical account. AT 1977(11):130–151. (in Greek)
Voivonda A, Kizilou V, Kloutsinioti R, Kontaratos S, Pyrgiotis Y (eds) (1977a) Studies produced from 1960 onwards and the views of the planners involved. AT (11):152–192. (in Greek)
Voivonda A, Kizilou V, Kloutsinioti R, Kontaratos S, Pyrgiotis Y (eds) (1977b) Reasons and objectives of state intervention in spatial regulation. AT 1977(11):193–213. (in Greek)
Voulgaris A (1994) Regional spatial planning: efforts, problems, prospects. Topos 1994(8):215–222. (in Greek)
Voulgaris G (2013) Post-regime transition Greece 1974–2009. Polis, Athens (in Greek)
Ward B (1963) Greek regional development. CPER, Athens. (in Greek)
Wassenhoven L (1969a) Preliminary work for the preparation of the master plan of a new settlement. TeC 1969(6/516):409–416. (in Greek)
Wassenhoven L (1969b) Spatial planning analysis: Principles and methodology according to Louis - Joseph Lebret. TeC 1969(9/519):603–612. (in Greek)
Wassenhoven L (1973) The commissioning to a private consultancy firm of the national spatial plan. Letter to the editor. Economicos Tachydromos, 15 March: 23 (in Greek)
Wassenhoven L (1976) Socio-political aspects of housing. Int J Environ Stud 9:37–48

Wassenhoven L (1978) Administrative decentralization with prior organization and planning. To Vima 29 June: 9 (in Greek)
Wassenhoven L (1979a) Growth poles and the policy of "rival cities". Ekonomia kai Koinonia, Aug.–Sept.: 32–39 (in Greek)
Wassenhoven L (1979b) Development and settlement system: a contribution to theory. SyT 1979(6) (Sept.-Nov.): 33–41. (in Greek)
Wassenhoven L (1980) The settlement system and socio-economic formation: the case of Greece. PhD thesis. London School of Economics and Political Science, University of London, London
Wassenhoven L (1984) Greece. In: Wynn M (ed) Planning and urban growth in Southern Europe. Mansell, London, pp 5–36
Wassenhoven L (1993) For mountainous Greece. Newspaper Kathimerini 24-11-1993 (in Greek)
Wassenhoven L (1994a) Location planning of industrial plants and estates in Greece. In: Laskaris C (ed) (1994) National and Community policies and programmes for the development of urban centres. European Social Fund and NTUA, Athens, pp 107–121. (in Greek)
Wassenhoven L (1994b) Economic geography of the region. In: Papagiannakis L (ed) The development of the region of Thrace: challenges and prospects. Research Centre of the Greek Society, Academy of Athens, pp 85–106. (in Greek)
Wassenhoven L (1994c) Environment and sciences at the end of the 20th century from the viewpoint of a planner. In: Rokos D (ed) Sciences and the environment at the end of the century: Problems and prospects. NTUA and Enallaktikes Ekdoseis, Athens, pp 78–87. (in Greek)
Wassenhoven L (1995a) Urban renewal in a modern megalopolis: The case of Elaionas in Athens. In: Georgoulis D (ed) Texts on the theory and implementation of urban and regional spatial planning. Papazisis, Athens, pp 375–401 (originally published 1993) (in Greek)
Wassenhoven L (1995b) Regional spatial planning and the countryside. In: SEP (ed) Regional development, spatial planning and environment in the framework of united Europe. Proceedings of a conference. Vol. II. Panteion University, 1995. SEP and *Topos*, pp 13–27 (in Greek)
Wassenhoven L (1995c) The "frontier" and internationalization of spatial planning. In: SEP (ed) Frontier regions. Proceedings of a conference, 1995. Thessaloniki, pp 97–107 (in Greek)
Wassenhoven L (1995d) Urban and regional spatial planning in Greece. Translation of a paper presented at a conference in the London School of Economics and Political Science. Pyrforos, Nov.–Dec. 1995: 11–17 (NTUA journal) (in Greek)
Wassenhoven L (1995e) Regional spatial planning in the 1960s. In: Sakis Karagiorgas Foundation (1994–95) Greek society in the first postwar period 1945-1967. Conference proceedings. Vol. B. Athens, pp 109-123 (in Greek)
Wassenhoven L (1996a) Regional and spatial policy: The European strategy of Greece. Pyrforos, May-June 1996: 42-50 (NTUA journal). Originally presented at a TEE conference (Dec. 1993) on "Greece in Europe: Spatial and regional policy towards the year 2000" (in Greek)
Wassenhoven L (1996b) Sustainable urban development and the notion of urban resources. In: Laskaris C (ed) Sustainable development: Theoretical approaches to a critical concept. EU Petra II Programme, Papasotiriou, Athens, pp 87–134. (in Greek)
Wassenhoven L (1997a) Greece. In: OECD (ed) Managing Across Levels of Government. OECD, Paris pp 191–202.
Wassenhoven L (1997b) Spatial planning: theory and practice. In: Aetoloakarnania Prefectoral Self-Government, Sustainable development with environmental education. Conference proceedings. Mesologhi, pp 180–201 (in Greek)
Wassenhoven L (1997c) Manual for spatial planning course. NTUA, Athens (in Greek)
Wassenhoven L (1997d) Spatial planning and problems of rural space and the countryside. Pyrforos 27, Jan.–Feb. 1997: 39–42 (NTUA journal). (in Greek)
Wassenhoven L (1998a) Order and disorder in space. In: Modinos M, Efthymiopoulos E (eds) Ecology and environmental sciences. Interdisciplinary Institute of Environmental Research, Athens, pp 57–78. (in Greek)
Wassenhoven L (1998b) Spatial planning (*chorotaxia*). Lemma in the Encyclopaedia Papyrus – Larousse – Britannica (1998). Annex on Greece. Papurus, Athens, pp 407–411. (in Greek)

Wassenhoven L (1999) Public works as a transformative factor of geographical space and the environment. In: PUCSA /Env (ed) Environment and public works: Rival concepts? PUCSA / YPECHODE. Conference proceedings. Athens, pp 57–67 (in Greek)
Wassenhoven L (2000a) The EU compendium of spatial planning systems and policies: Greece. European Commission Regional Development Studies, Luxembourg
Wassenhoven L (2000b) Protected Natural Areas and the Ecosystem of "Caretta – Caretta" Sea Turtles in Laganas Bay (Zakynthos). Case study: Protected sensitive area – Greece. E.U. SPECTRA Project
Wassenhoven L (2002a) Is there still a theory of spatial planning? In: SADAS (2002) Architecture and the Greek city in the 21st century. Proceedings of the 10th Panhellenic Architectural Conference (1999). SADAS and TEE, Athens, pp 393–399. (in Greek)
Wassenhoven L (2002b) The democratic quality of spatial planning and the contestation of the rational model. Aeichoros 1(1):30–49. (in Greek)
Wassenhoven L (2003a) Sustainable development and the theory of collaborative planning. Public lecture. University of Thessaly, 5 March 2003, Volos (in Greek)
Wassenhoven L (2003b) National and regional spatial plans in Greece: Can they contribute to sustainable development? In: ERP Environment (ed) Proceedings of the International Sustainable Development Research Conference 2003 (University of Nottingham, 24–25 March 2003). ERP Environment, Shipley (West Yorkshire UK)
Wassenhoven L (2004c) Sustainable insular development: United we stand – The case of the Northern Dodecanese islands. In: Angelidis M (ed) New polycentricity and spatial development. Conference proceedings (2001). YPECHODE, Ministry of Economy, NTUA and Université de Paris X / Nanterre. Athens, pp 135–146. (in Greek)
Wassenhoven L (2004d) The theory of urban planning: What future is there in the 21st century? In: NTUA (ed) City and space from the 20th to the 21st century. Volume in honour of Prof. A.I. Aravantinos. NTUA, University of Thessaly, and SEPOCH, Athens, pp 75–102 (in Greek)
Wassenhoven L (2004e) The planning of cities in the constellation of environmental responsibility. In: NTUA (ed) City and space from the 20th to the 21st century. Volume in honour of Prof. A.I. Aravantinos. NTUA, University of Thessaly, and SEPOCH, Athens, pp 103–113 (in Greek)
Wassenhoven L (2007) Spatial planning (*chorotaxia*). Lemma in the Encyclopaedia Papyrus – Larousse – Britannica (new edition 2007). vol. 55. Annex on Greece. Papurus, Athens, pp 518–532 (in Greek)
Wassenhoven L (2008a) Territorial governance, participation, cooperation and partnership: a matter of national culture? Boletin de la A.G.E. 46:53–76
Wassenhoven L (2008b) Territorial cohesion: New directions for spatial development and their importance for Greece. DEP 10 (April-June): 247–263 (in Greek)
Wassenhoven L (2008c) National spatial planning: content, process, plans and programmes. In: Giannakourou G, Menoudakos C, Wassenhoven L (eds) Regional spatial planning in Greece: legal framework and implementation in practice. Nomos+Physis and Ant.N. Sakkoulas, Athens, pp 51–78. (in Greek)
Wassenhoven L (2009) Regional spatial and urban planning. Teaching manual. EKDDA, Dept. of Regional Government. Athens (in Greek)
Wassenhoven L (2010a) Territorial governance: Theory, European experience and the case of Greece.[33] Kritiki, Athens (in Greek)
Wassenhoven L (2010b) Violence and civil disobedience: The case of land use regulation and locational decisions. In: Demetriadis EP, Kafkalas G, Tsoukala K (eds) The *Logos* of the *Polis*. Volume in honour of Prof. A.Ph. Lagopoulos. University Studio Press, Thessaloniki, pp 113–123. (in Greek)

[34] The year is indirectly deduced from the text.

Wassenhoven L (2014a) Assistance to coordinate legislative reform processes related to land development and environment protection: final report. Contract No VC/2014/0152. DG Employment, Social Affairs and Inclusion, European Commission

Wassenhoven L (2014b) The problem of periurban areas of mixed land use in Greece. Workshop on Land Management, European Commission Task Force for Greece, 11–12 June 2014

Wassenhoven L (2016) The systemic character of heat waves and long-term mitigation: the case of Athens. Geografies 27(Spring):89–102. (in Greek)

Wassenhoven L (2017a) Maritime spatial planning: Europe and Greece. Crete University Press, Herakleion (in Greek)

Wassenhoven L (2017b) The history of the concept of sustainable development. In: ICOMOS (2017) Cultural internationalism and Greek cultural policy in the 21st century. Proceedings of a conference to celebrate 50 years of ICOMOS (1965 – 2015). ICOMOS Hellenic National Committee, pp 161–169 (in Greek)

Wassenhoven L (2018a) The parallel universes of rhetoric and practice: lessons learnt from spatial and urban planning reforms in Greece. In: Farinós Dasí J, Peiró E (eds) Territorio y estados/Territory and states. Tirant Humanidades, Valencia, pp 347–372

Wassenhoven L (2018b) Hellinikon … passion: the development of the former Hellinikon airport, Athens. Sakkoulas, Athens (in Greek)

Wassenhoven L (2018c) Maritime spatial planning in the context of European spatial planning. In: Serraos C, Melissas D (eds) Maritime spatial planning. Sakkoulas, Athens, pp 13–37. (in Greek)

Wassenhoven L (2018d) Planning theory: The challenge of maritime spatial planning. PeD (4):562–569. (in Greek)

Wassenhoven L (2019) The ancestry of regional spatial planning: a planner's look at history. Springer, Cham (Switzerland)

Wassenhoven ME (2019) Facets of the archaeological landscape. Paper presented at an international seminar. Lefkas island (August 2019) (in Greek)

Wassenhoven L (2021a) Law 4759/2020, the rational model of planning, and the public interest. PeD 2021(95):3–13. (in Greek)

Wassenhoven L (2021b) A planner's comment on Council of State resolution 175/2021. PeD 2021(2):297–299

Wassenhoven L (2022) Putting our country in order: a history of spatial regional planning in Greece after the Second World War. Kritiki, Athens. (in Greek)

Wassenhoven L, Georgoulis D (1997) Spatial planning: theory and practice. Lectlure notes, NTUA, Athens. (in Greek)

Wassenhoven L, Kourliouros E (2007) Social and political dimensions of spatial planning. In: Terkenli T, Iosephidis T, Chorianopoulos G (eds) Human geography: man, society and space. Kritiki, Athens, pp 366–406. (in Greek)

Wassenhoven L, Papagiannis Th (1996) National experience from the implementation of the action plan Habitat I in the last 20 years. In: Habitat II Greek Committee (1996) National Greek report for the United Nations World Conference on Human Settlements "Habitat II". Annex II. Athens, pp 102-116 (in Greek)

Wassenhoven L, Sapountzaki P (2001) Securing social consensus in spatial planning: the case of coastal space. Paper presented at the 13th conference of the Panhellenic Network of Ecological Organizations. Rhodes, 19-22/10/2001 (in Greek)

Wassenhoven L, Sapountzaki K (2009) Il difficile percorso della governance territorial nei paesi europei. In: Governa F, Rivolin UJ, Santangelo M (eds) La costruzione del territorio Europeo: sviluppo, coesione, governance. Carocci, Roma, pp 141–171

Wassenhoven L, Gerardi K, Serraos C (2004) Metropolitan planning in Greece. Research document. NTUA, Athens (in Greek)

WWF Hellas (2005) Unimplemented undertakings: Environmental legislation in Greece, Athens. (in Greek)

WWF Hellas (2018) Environmental legislation in Greece. Annual report, in cooperation with Nomos+Physis (in Greek)

WWF Hellas (2019a) Environmental legislation in Greece. Annual report, Athens (in Greek)

WWF Hellas (2019b) Clean energy now: position statement on APE development in Greece, Athens. (in Greek)
WWF Hellas (2020) Contribution to consultation on the draft law about modernization of spatial regional and urban planning legislation. Athens (in Greek)
WWF Hellas, Greek Ombudsman (2009) Legislation guide on the environment. Athens (in Greek)
Wynn M (ed) (1984) Planning and urban growth in Southern Europe. Mansell, London
YCHOP (1983) The EPA process. December 1983, Athens (in Greek)
YCHOP (1984a) Arcadia Prefecture: Spatial organization proposals – Outline plan of structural interventions. May 1984. Athens (in Greek)
YCHOP (1984b) Argolis Prefecture: Spatial organization proposals – Outline plan of structural interventions. May 1984. Athens (in Greek)
Yerolympos A (1996) Urban transformations in the Balkans (1820 – 1920). University Studio Press, Thessaloniki
Yerolympos A (2008) Ernest Hébrard and the plan of Athens (Greek translation of original French version) https://www.academia.edu/23127489/
YPECHODE (1985) Guidelines of EChM. Athens (in Greek)
YPECHODE (1992) Guidelines of EChM of the Community Initiative ENVIREG. Athens (in Greek)
YPECHODE (2000)[34] GPChS (draft national plan). Athens (in Greek)
YPECHODE (2001a) Spatial planning of mountain zones. Proceedings of a one-day conference (21-4-1999). Athens (in Greek)
YPECHODE (2001b) Coastal and island space. Proceedings of a one-day conference (29-4-1999). Athens (in Greek)
YPECHODE (2009a) National spatial planning. Athens (in Greek)
YPECHODE (2009b) National strategy for biodiversity: Consultation document. February 2009. Athens (in Greek)
YPEN (2017-2020) Reports of the Regional Spatial Planning Frameworks (in Greek)
YPETHO (1983) Report on the National Spatial Plan of Industrial Areas. Athens (in Greek)
YPETHO (1985) Economic and social development programme 1983–1987. Preliminary report and final recommendations. Athens (in Greek)
Zampelis Ch (2001a) Coastal space – Islands: Problems and prospects. In: YPECHODE (ed) Coastal and island space. Proceedings of a one-day conference (29-4-1999). Athens, pp 21–28 (in Greek)
Zampelis Ch (2001b) Spatial planning of mountain areas. In: YPECHODE (ed) Spatial planning of mountain zones. Proceedings of a one-day conference (21-4-1999). Athens, pp 37–42 (in Greek)
Zampelis Ch (2003) Regional spatial planning and its present implementation. YPECHODE, Athens (in Greek)
Zampelis Ch (2004) Polycentricity and national spatial planning. In: Angelidis M (ed) New polycentricity and spatial development. Conference proceedings (2001). YPECHODE, Ministry of Economy, NTUA and Université de Paris X / Nanterre. Athens, pp 227–231. (in Greek)
Zeikou P (2011) The spatial model of tourism development and the views and proposals of YPEKA for its attainment. In: Avgerinou-Kolonias S, Melissas D (eds) Economic conjuncture and spatial organization of tourism. Propobos, Athens, pp 57–64. (in Greek)
Zygoura O (1994) Review of case law of the Council of State on matters of urban and regional spatial planning. PeD 1994(1):231–286. (in Greek)

Part III
A Theory of Compromise Planning

Chapter 9
Planners, Knowledge Transfer, Planning Culture: Looking for a New Theory

Abstract The ultimate aim of this chapter is to sketch an image of the "empirical reality" of the Greek planning system and prepare the ground for an explanatory theoretical model of "compromise planning". The chapter starts with a personal statement about the development of the author's interest in planning theory, which is then linked with the profile of Greek planners and their academic background, and with planning education. Such considerations are associated with the question of the transfer of knowledge and the creation of a distinct planning culture, in which modernist and traditionalist elements are intermingled. For the purpose of outlining the Greek system the conditions of the country, extensively presented in previous chapters, are summarized and the influence of foreign countries is stressed, this time with specific reference to imported types of plans and other planning instruments. The particularity of the Greek case as a combination of compromise, complicity, collaboration, pragmatism and improvisation warrants a theoretical approach which differs from the conventional body of theory and is rather akin to the ideas of Southern theory. The chapter ends with a diagrammatic presentation of the factors and issues constituting the planning system, as a prerequisite for the formulation of a novel theory.

Keywords Reforms · Imported practices and instruments · Foreign theories · Dependence · Periphery · Knowledge · Planning professionals · Planning education · Pragmatism and compromise · Empirical reality

Prologue

This chapter is dedicated to the transfer of knowledge effected through the students of planning and academics and to the native Greek planning culture. The author starts with a personal statement, appropriately written in the first person, which will hopefully clarify his motives for seeking a new theoretical approach, in this and the following last chapters of Part III of the book. Before entering the main core of the chapter, some key findings about the conditions of the country are summarized, analyzed of course in detail in Part II. Inevitably, certain details may reappear in this

L. C. Wassenhoven, *Compromise Planning : A Theoretical Approach from a Distant Corner of Europe*, https://doi.org/10.1007/978-3-030-94331-8_9

chapter. Many administrative reforms, from the early years of the Kingdom of Greece in the nineteenth century, e.g., the French Napoleonic prefectures, were imported from the Great Powers of the day, in an effort to establish a modern state. Reforms were frequently opposed by local interests accustomed to conditions of local power wielding, going back to Ottoman rule. Incorporating foreign practices is a difficult task in countries with a colonial or quasi-colonial past. Greece was not a colony in the strict sense, but experienced a semi-colonial – or "dependence" or "periphery" – status for long. Not surprisingly, the recent emergence of postcolonial planning theory, although mostly emanating from erstwhile colonial dominions and protectorates, is rightly touching this sensitive issue. By the 1970s the main western streams of planning theory were a fully-fledged theoretical edifice based on the totally different spatial and economic organization of advanced nations. This theoretical "super-structure" was bound to seep through the knowledge lattices dictating spatial policy in Greece, even if the impact of the European Union is not taken into account. Greek planners, few and far between, were trained in the West and were equipped with theoretical and practical knowledge largely irrelevant for the needs of the country. The situation has changed in the last few years, after independent planning faculties had been created in two Greek universities.

9.1 A Personal Journey

As the reader has been warned in the prologue, the author has chosen to present this personal statement in the first person, in order to emphasize the influence of his own experience.

> My exposure to what amounts to a theory of planning goes back to the 1960s, but my study and reading went through peaks and troughs. Although, as I shall explain later, it was inevitably dependent on the stages of my, almost exclusively, academic career, it was consistently an accompaniment of my work. It peaked during the years of my studies, research and teaching in London, declined relatively when, back in Greece, I taught mostly at undergraduate level, and blossomed again when my teaching expanded at postgraduate level. However, it took a sharp upward turn in the last 15 years in the context of writing books on diverse subjects. During this last period, I studied systematically the literature of planning theory. I read diligently a large number of books, I extracted quotations, took copious notes, which I codified carefully, compiled an extensive bibliography, and produced annotated lists of relevant terms. Planning theory books became my reading companions.
>
> Almost 50 years ago I wrote an article for a weekly journal in Greece. In it I made a plea for an integrated urban planning theory (Wassenhoven 1971). I was disappointed by what passed as theory in academic teaching, which, in my view, was no more than instructions on how to produce a land-use plan or to design an urban development. I was then under the spell of the systems approach to planning and I kept, so to speak, my copy of Brian McLoughlin's *Urban and regional planning: A systems approach* under my pillow. Since then, I watched the dizzying flood of books, articles and research reports on planning theory and was impressed by their never-ending borrowing of complex concepts from more respectable scientific disciplines, from mathematics to philosophy and from physics to economics, human geography and sociology. Theoretical edifices were being erected to explain

the foundations of urban and regional planning, which taxed my ability and knowledge to understand, let alone apply them to the real world of planning. Yet, in the 1970s and early 80s, I did my best to assimilate and teach them to an international audience of students of the Development Planning Unit (DPU), University College London. Underlying this effort was also my wish to apply them, in my personal research, to the conditions of my own country, Greece. I deliberately emphasize the *international* origin of my DPU students in London, in their vast majority from so-called less developed countries, because I was already feeling the strain of adapting western notions of planning to their needs. Thankfully, this adaptation and the appreciation of varying national conditions were a crucial element of the intellectual environment of my university department, which greatly enriched my thinking. I also recall the uneasiness I felt when I lectured on planning systems and styles to postgraduate Indonesian students at the Bandung Institute of Technology, in the context of a University College London overseas project (Wassenhoven 1983).

In the 1980s, back in Greece to teach Greek undergraduates, and later postgraduates, of the National Technical University of Athens, I found myself faced with the need to reconcile my theoretical perceptions with a particular social reality and a specific planning process. It was not only a *must* if I was to help my students to make the necessary mental connections. It was also an underlying necessity in a long series of applied research projects which I coordinated; projects dealing with Greek cities and regions, with planning at the European Union level, and with the practice of planning. Thankfully, these projects enabled me to retain a close contact with planning realities. It was gradually becoming apparent that my mostly British, American and French theoretical toolkit was of little relevance to Greek conditions. The local horizons of theoretical preoccupations being inevitably limited and the fact that most planning academics and professionals were in those days the product of foreign academic institutions, made my work, and of course that of my university colleagues, an arduous task. It was in the 1990s that things began to change, but the problem of a small academic research "market" still placed severe limitations on purely theoretical work and publishing.

Academic teaching, research and administrative activity demanded a heavy workload. These obligations and my personal choice of keeping purely professional work within very strict limits kept me fairly away from practical plan-making consultancy until my retirement. It was then that apart from engaging in hectic book-writing I got also involved in a limited amount of consultancy work at the local level. Both these activities opened my horizons, especially as far as my understanding of planning theory was concerned, to new directions. Parallel with these new endeavours, I concentrated on intensive reading of history, law and political science, which only had a parallel in the days of my doctoral research. I remained naturally an amateur in these fields, but my personal study helped me immensely to figure out what planning theory ought to be about. History is an inexhaustible source of knowledge, which explains the reason for plunging into the remote past of spatial planning and for writing a book in English, entitled *The Ancestry of Regional Spatial Planning: A planner's look at history*, published by Springer. One important aspect of this work was the focus on organized states of a variety of forms and on their role in shaping geographical space, regardless of the real reasons of their actions, e.g., military conquest, exploitation of natural resources, and colonization. The transition from the hegemonic states of antiquity and the medieval era to the nation state of modernity, was a key change which provided a mighty explanatory instrument.

I am still an incorrigible believer in the nature of theory as a medium for the interpretation and explanation of real-life phenomena, which, in the case of planning theory, are identified with planning in the real world. I do not restrict this notion to the mere practice of plan-making and implementation. I am fully aware that "space", which spatial planning intends to shape, is not produced solely by governments, public administrations and qualified planning practitioners. Ultimately there is a process through which geographical (not just physical) space is transformed and it is this that theory should explain. But, at this point, we have to be careful and clear. Socio-spatial change, which encompasses all human activity and its attendant inequalities, disparities and un-equalization effects, is a matter which is up to the theory of human and economic geography to address, regardless of the forces that initiated it, a point to which I return in Chap. 10. These forces (collective or private, local or global) have their own momentum and agendas of action. However, *spatial planning*, as a deliberate action of organized society, is a different field. This form of planning takes place through the choices and activities of state entities, at all their geographical scales, or of supra-state structures and agencies, of which states are part on their own volition. It is this form of planning that spatial planning theory is concerned with. It is this type of planning that it must explain and interpret. Geographical theory and planning theory, in spite of inevitable overlaps, are not identical. From the point of view of planning theory, as I see it, the role of the state (national, regional, local) remains central, even after taking into account the outside forces (local, nation-wide, global), that exercise their influence, which is then mediated by the state apparatus. Planning activity and practice as a social phenomenon is what planning theory should focus on, without losing touch with the spatial realities it strives to interpret, remedy, subvert, improve or replace.

9.2 A *Précis* of Conditions in Greece and Imported Planning Tools

State planning-related functions are not independent of the history and geopolitical position of the country concerned. During the early decades of the twentieth century, when planning in western countries was being developed as a practice, even as a discipline, regional and urban structures in Greece were totally different from those in industrial western countries. An analytical picture of the events of both the nineteenth and twentieth centuries is given in the relevant chapters of Part II, but here, as I am about to formulate a new theoretical approach, it is essential to summarize them.

Text-Box 9.1: Imported Plan Types

There is practically no planning procedure or instrument in the vocabulary of the Greek planning system which was not imported from German, French, British, or American practice. A reservation is in order when it comes to land ownership rights going back to the Ottoman period or to Church and monastic property rights. This is not surprising given that even in the days of the Bavarian administration of the 1830s building regulations were modelled on German prototypes. Even more important is the drafting of the 1923 legislative decree which remained for decades the cornerstone of town design legislation. Judging from the nationality and/or country of training of the main planning advisors of the liberal government of the day, one can safely deduce that the concepts and instruments introduced by this decree were of German and French origin. The French architect Ernest Hébrard played a major role as a government advisor and as an active planner. So did Anargyros Demitrakopoulos, an architect with contacts in Germany and later appointed

(continued)

Text-Box 9.1 (continued)

as professor of town planning in the National Technical University of Athens. Nevertheless, it is interesting to remember that in the planning of settlements that followed, especially suburbs, there was an influence of the English Garden City movement (Text-Box 8.1). When the Greek planning system began to be modernized in the 1960s and 1970s, especially after the 1975 Constitution, the introduction from abroad of new types of plans, procedures and tools became a real avalanche. The influence of foreign-trained planners acting as advisors to the competent ministry is beyond doubt, although we should not forget the examples of planners, who, in their ministerial capacity, left their personal mark on the laws they proposed. Under successive laws from 1976 to the end of the 1990s, the character, range and hierarchy of spatial plans from the national to the local level were fully determined and did not substantially change until today.

At the national level, the Greek national spatial framework (now reduced to a policy statement) was probably influenced by the French national territorial planning directive (*Directive Territoriale d'Aménagement*) and the German federal guidelines for spatial planning (*Raumordnungspolitischer Orientierungsrahmen*) and the operational framework for spatial planning (*Raumordnungspolitischer Handlungsrahmen*). At the national – sectoral level the Greek special spatial frameworks remind us of the French collective services scheme (*Schéma de services collectifs*), the British planning policy and minerals planning guidance notes (although not regulatory), or the German federal land use ordinance (*Baunutzungsverordnung*). Of course, the federal system in Germany makes a great deal of difference. The similarity with the French *Schémas de services collectifs*, which cover themes like health, higher education and research, culture, passenger transport, goods' transport, new technology, energy, rural and natural areas, and sports, is striking. In spite of these similarities, the Greek special spatial frameworks (for tourism, industry, alternative energy sources, minerals, fish farming, islands and coasts etc.) are a very distinctive element in the planning system. At the regional spatial level the influence of the French spatial planning and sustainable development schemes (*Schéma régional d'aménagement et de développement durable du territoire*) is more than obvious. At the local urban level, the Greek model of local town plans and subordinate implementation plans is more or less a replica of the French package of *projet d'agglomération* and *plan local d'urbanisme* (formerly *plan d'occupation des sols*), the British structure and local plans (later revised into local development frameworks), and the German set of a preparatory land use plan (*Flächennutzungplan*) and a binding land use plan (*Bebauungsplan*). Terminology is frequently changed in all countries. It was not only in the types of plans that the foreign influence was evident in Greece; it was also important in the case of planning tools or methodological instruments (see Text-Box 9.2 on imported planning instruments).

Greece until the early twentieth century was only half of what it is today, but a great deal of its geographical structure was already in place. The dominant position of Athens, secured by its choice as the capital of the modern state that emerged in the 1830s, became more and more strengthened. Planning however, of the kind practiced in western countries, was usually sporadic and uncoordinated. Urban development in Athens since the late nineteenth century was but a constant accretion of disjointed areas of unauthorized housing, in a haphazard manner. Urban planning became identified with this process of incorporation in the urban fabric. At that time, the rare examples of (western-inspired) plans were still too few and of little consequence.

A decree was promulgated in 1923 to regulate urban development plans, but the interwar period which followed was dominated by the exchange of populations which was triggered by the Balkan Wars, the annexation of northern Greece and the influx of refugees after the Asia Minor military disaster. Urban planning inevitably focused on the settlement of incoming refugees. It was only in this case that truly innovative town planning activity was undertaken. The events of the 1940s (World War, German occupation, Civil War) caused an inflow of migrants into Greater Athens and a wave of reconstruction, urban housing development, and urban sprawl which was totally unplanned and uncontrolled.

Text-Box 9.2: Imported Planning Instruments and Policies

The imitation of western planning systems was extended beyond the copying of types of plans (see Text-Box 9.1 on imported plans). West European planning systems were a model to emulate (Larsson 2006). Foreign influence is also evident both in planning policies and in individual instruments. The Greek policy concerning growth poles (expressed in all national and regional plans, but also in the "rival cities" programme of the late 1970s), owes a lot to the French policies on the urban hierarchy and the metropolises of equilibrium, but also on particular instruments (Magnan 1966). The recent interest in compact cities and passive design is due to foreign influence. Instruments and methods like zoning, land use classes, land occupation and building coefficients, planning standards, SWOT technique (strengths – weaknesses – opportunities – threats), environmental impact assessment, participation processes, and administrative appeal procedures, all have a foreign precedent or bear the European Union impact. Methodologies have also been loans, from the old trilogy of "survey – analysis – plan" to the systems sequence of "goals – alternatives – evaluation – strategy – monitoring – feedback", or to scenario planning, now incorporated in the methodology of Greek planning studies. Planning for special purpose zones such as industrial estates, enterprise zones or technology parks originated in the West. Urban planning special tools such as the active town planning zones were modelled on the French prototype of the priority urban development zones (*zone à urbaniser par priorité* or ZUP,

(continued)

Text-Box 9.2 (continued)

later replaced by the *zone d'aménagement concerté* or ZAC). Their practical application in Greece was virtually nil. The detailed design of settlements, already in the interwar years, in the form of garden suburbs, refugee townships, land parceling schemes, and land adjustment plans, had been developed and experimented with in Britain, Germany or the United States. A tool like the so-called transfer of building coefficient was an imitation of the American TDR (transfer of development rights). In Greece, it was repeatedly resisted by the Council of State and is still causing consternation because thousands of legal titles are still in the air. Modern technologies used in spatial planning and various digital and simulation applications, although not strictly speaking planning instruments, were also imported. Even the exercise of pre-emption rights (as in the case of the French *zones d'aménagement différé* or ZAD) was transferred to Greece with minimal effect. New forms of planning have recently made the journey from abroad, e.g., resilience planning, but their future remains unknown.

Unauthorized (in fact illegal) building became the rule in peri-urban zones, while in the central areas of Athens and to a lesser extent of other urban centres the system of *antiparochi* housing transformed the urban fabric. This system resulted from the policy of the government, which, in order to confront the urgent housing need, allowed land owners in the cities, especially Athens and Thessaloniki, to join forces with small scale developers who built condominium blocks of flats. The apartments were sold to incoming migrants, while the landowners kept a sizeable share of the total floorspace for themselves. On a wider scale, the government's attention, initially with considerable foreign aid, focused on national and regional development. A series of national economic development plans were produced, with negligible spatial content. Essentially, it is not until the 1960s that urban and regional planning begins to emerge. By then the country's urban system was already dominated by Athens and Thessaloniki. It is from the mid-1970s (after a seven-year military dictatorship) that a statutory planning system of some consistency was introduced.

The Constitution of 1975 was a critical turning point, because of its emphasis on spatial planning. Soon afterwards urban and regional planning legislation was enacted, local planning activity expanded in the mid-1980s, but for a variety of domestic reasons it was not until the end of the 1990s that hectic planning activity dealt with all geographical scales. As explained extensively in Part II of the book, apart from constitutional requirements, some other parameters were instrumental. The first was the influence of the policies of the European Union, that Greece had joined in the meantime. The second was the wave of environmental consciousness, spurred by international conventions and the EU itself. The third was the active role of administrative courts. It was naturally based on the clauses of the Constitution, because by now all government legislation and project permits (e.g., for a new airfield or an industrial plant) were declared unconstitutional by the Council of State (the Supreme Administrative Court), if there had been no prior approved urban and/or regional plan in the area concerned.

The arsenal of the planning system of the country was soon replete with a large number of planning instruments and policies, mostly imported from foreign countries. Many of them proved inapplicable or ineffective in the conditions of the country. It should be remembered, as stated already, that many administrative reforms of the past were also imported, particularly in the days of state-building. They were frequently opposed by local interests accustomed to conditions of local power control, going back to the Ottoman Empire. Incorporating foreign practices and instruments is a difficult task in countries which had undergone a purely colonial or quasi-colonial period. As explained, Greece was not a colony in the strict sense of the word, but the study of the planning system owes a lot to postcolonial planning theory.

When a consistent planning system appeared in Greece, in the late 1970s, a body of international planning theory literature was already available, founded on the socio-economic environments of the West. Greek planners, few and far between, were trained in Britain, France, Germany or the USA – rarely in Canada, Italy, or the Netherlands – and were equipped with theoretical and practical knowledge largely irrelevant for the needs of the country. The situation has changed in the last few years, after independent planning faculties were created in two Greek universities. A critical difference between Greece and the countries where most planning theories originate, is the composition of the body of planners and their functions. The planning profession does not occupy the same position as does its equivalent in the United Kingdom or the USA. The difference is particularly striking in the public sector, especially in the context of local and regional authorities. It is rare to find a qualified planner working in local government. A recurring theme found in planning theories is their relation with planning practice and the role of planners in government. In fact, planners are repeatedly accused of being a mere cog in the state political apparatus or servants of economic interests. The role of planners in Greek administration, at all levels of government, is a shadow of the role and power of their foreign counterparts in the West, a sign of cultural and ideological bias and of established professional interests. Even in private consultancy, where the presence of qualified planners is greater, the powers that planners have are comparatively insignificant.

Architects and engineers serving in the Greek public sector – before some qualified planners entered the field – had adequate knowledge of the practical problems, but the range of tools they were accustomed to use did not much extend beyond the incorporation of illegal housing areas into the patchwork of urban extensions. The division between "within-the-plan" and "out-of-plan" areas was the hallmark of reality on the ground. "Within-the-plan" areas were those covered by a statutory local plan. The designation "out-of-plan" does not imply agricultural or forest use only. In fact, intensive construction took place in these areas under a different statutory regime, full of exemptions, derogations and *ad hoc* regulations. It was, and still is, a Greek originality. It has never ceased to be a source of controversy, as shown in the debate leading to the latest planning law of 2020. The nature of economic development, insecurity and attachment to individual property further encouraged a form of investment focused on private houses, land-lots and apartments. Illegal land development and building extensions was not simply the rule, it was also a social need to protect family savings. It was usually the only conceivable private investment and the only feasible outlay of private savings of medium-income strata.

Among the consequences of the accumulation of planning instruments and of the distance separating the planning system on paper from its implementation of the ground is the chaotic character of planning legislation. This, however, is also due to the way governments and governed citizenry interrelate, i.e., to the operation of a system of political clientelism, which we touched upon mainly in Chap. 7, when the author introduced his concept of compromise planning. Clientelism, administrative inefficiency, the operation of a shadow economy, and erratic legislating are interconnected processes. The result of dark and ill-interpreted patches in planning legislation is the huge distance separating noble planning intentions and the practice of granting planning (in fact rather building) permits in the frontline of the so-called *poleodomies*. *Poleodomia* is the Greek word for town planning (literally meaning town-building), but for the public at large *poleodomia* is a government office charged with granting building permits. Chaotic legislation is made worse by *ad hoc* and sudden amendments added by the government in the text of acts voted in parliament, not only to correct errors, but also to respond to the pressures of political clients. Naturally, such legislation becomes the standard stuff of appeals to the courts and causes an immense work burden for the Council of State, the supreme administrative court. Codifying planning legislation into a meaningful body has become an urgent, but elusive for many years, task. Producing such a code is the job of a recently formed commission, made up of judges, other legal experts and advisors, and senior civil servants, on which the author also serves as the only academic with an urban and regional planning background. This naturally offers a splendid opportunity to make sense of the complex legal edifice of planning.

In the recent past, some would say in the neoliberal period, an increasing part of planning work is concerned with large projects, either public (mostly for infrastructures), or private, primarily in the field of tourism development. Private residential development remains small- scale and individual. Large-scale residential developments are not a feature of urban development in Greece. Large private construction firms, Greek or foreign, tend to concentrate their activities in public works, tourism resorts, transport facilities, and commercial malls. The exceptions are rare and related to specific conditions. An example is Eddison Hill, Pallini, Attica, a housing estate converted from the foreign journalists' village of the 2004 Olympic Games. The development of the former Hellinikon airport in Athens is a case apart and qualifies to be classified as a flagship project, which has attracted intense criticism (see Text-Box 8.11). This situation impacts, as expected, on the nature of work that planners can aspire to. For this reason, government activity in the field of urban and regional plan production is of crucial importance for employment opportunities of the average planner. This is the background of the current launching of an ambitious programme to produce local plans practically covering all the country's municipalities, funded by the EU Recovery Fund.

9.3 The Profile of Planners

The previous section was an overview of conditions in Greece which were analyzed in great detail in the chapters of Part II. The author referred to laws, policies, development programmes, and spatial plans, drawing both on personal experience and on a variety of written sources. There is however a less discussed field, that of planning education, which serves as substratum of planning activity and its changes over time. Up to the late 1970s, university education in the field of spatial planning was contained in the curricula of the then Chairs of Town Planning of Schools of Architectural Engineering.

Text-Box 9.3: The Planners: Profession and Education

As in other countries, town planning in Greece was for a long time in the hands of architects. From the moment Governor Capodistrias commissioned a French architect to produce a plan for the historic town of Corinth and a Greek one to work for other towns in the Peloponnese, or King Otto selected two architects, a Greek and a German, to design the plan of Athens, in the early 1830s, there has been a long series of architects who offered their services to public town planning initiatives. In the interwar years town planning was taught in the School of Architecture of the National Technical University of Athens (NTUA), known by everyone simply as the *Polytechneion*. Incidentally, the words "the Polytechneion events" are now associated with the student uprising of 1973 against the military junta. Professionals with a university degree in planning appear after the Second World War, but they are rare and naturally trained in foreign countries. Things began to change, but not, at first, in the confines of universities. In fact, it was in the Athens Centre of Ekistics, created by Konstantinos Doxiadis, that teaching of planning to graduates began in the early 1960s. Doxiadis is certainly the most internationally famous Greek planner, with a vast consultancy activity, who can also be credited with the coining in 1942 of the Greek term *chorotaxia* (spatial ordering), a translation of the German term *Raumordnung* (Doxiadis 1942, 1959, 1965, 1968; Kazazi 2009; Kyrtsis 2006; Pertsemlidis 2009; Tournikiotis 2000).

Students of architecture at NTUA – and later at other Greek universities – were regularly producing diploma dissertations on town planning, but it was only in 1961 that a diploma dissertation was submitted with a regional spatial planning content. In that year, the Centre of Planning and Economic Research was created, which soon developed intensive activity in the field of spatial planning. Shortly after that, a town planning research laboratory was established at NTUA, where a large number of researchers – and later academics – had their first contact with spatial planning projects (NTUA/SPE 1966, 1969, 1971a, b, 1973). The NTUA example was quickly emulated by the Aristotle University of Thessaloniki. During the 1960s and 1970s a growing number of

(continued)

Text-Box 9.3 (continued)

Greek architects and engineers turned to West European and American universities for postgraduate education in planning. A parallel activity was taking place in the field of economic regional development planning and in regional science, which culminated in the establishment, in 1975, of a Regional Development Institute at the Panteion University of Athens (PASPE 1984). This is important if we recall the contribution of public regional development plans to the growth of spatial planning and the fact that even architects and engineers turned frequently for postgraduate work to the Panteion institute.

In the 1980s, developments gained speed. Foreign-trained planners, but also civil servants and professionals without academic planning titles but with long planning experience in the public sector, created in 1981 the Association of Greek Urban and Regional Planners (SEPOCH), which, 10 years later, joined the European Council of Spatial Planners. In the meantime, in 1982, under a university reform, the chairs of town planning were abolished and new urban and regional planning departments were set up in Athens and Thessaloniki, albeit always under the umbrella of architectural faculties. An exception to this rule came in 1989, when the new University of Thessaly established in Volos a totally independent school of urban planning, regional spatial planning, and regional development. A similar school was created in 2004 in Thessaloniki. Their undergraduate courses now lead to the award of a purely planning degree, of which the holders anticipate planning employment opportunities. Back in Athens, a research centre for regional spatial planning and urban development started functioning at NTUA in 1985 and an interdisciplinary postgraduate programme in urban and regional spatial planning received its first students in 1999. Naturally, several Greek students still pursue planning studies in the West. Like their predecessors, they return with a heavy knowledge load of foreign planning theories, practices and instruments, which are alien to their "southern" perspective. They too, are expected to "compromise".

On the basis of the 1982 reform, new departments were created in Athens and Thessaloniki. Similar departments were created later in other universities too. However, students attending the classes of these departments were still receiving a degree in architecture, not in planning. Comparatively limited planning classes were offered in other engineering schools, mainly of surveying. Later on, postgraduate planning courses were established granting advanced qualifications. It was in 1989 that an independent planning school, offering a distinct planning undergraduate degree, was created in the University of Thessaly as already mentioned. The University of Thessaloniki followed suit. As a result, a large number of planners are now active in Greece with an undergraduate or graduate degree in planning from Greek universities.

As made clear, until these developments changed the landscape of planning education in Greece, there were of course planners educated in foreign countries, holding postgraduate – very rarely undergraduate – degrees. They were rare before the 1967–1974 military dictatorship. All these holders of foreign titles had been members of a huge crowd of international students studying planning in countries of the West. As Healey (2010, 13) remarks, following a point made by other authors, "students from different parts of the world who come to America or Europe to study planning often struggle to grasp what kind of practice their teachers are referring to and what contexts are assumed in the literature they are invited to read. Travellers themselves, they are forced to develop skills for comparing the different ways in which planning is thought about and practiced. Many academics and practitioners are unaware of the distinctiveness of the contextual assumptions implicit in the accounts and arguments about the planning field they provide about their own countries. Lecturers may live, as Sanyal suggests …, quite cloistered lives. Many 'home' students are also unaware of the specificities of their context, absorbing what is, in effect, a particular understanding as if it were a universal and 'natural' dimension of planning knowledge".

The 1967 dictatorship in Greece produced a flight of young scientists who sought to expand their knowledge in western democratic countries. Students aspiring to a planning career could not be left unaffected, particularly because of the political content of planning. Planning courses in the West attracted a large number of such students. They turned primarily to West European and American universities, where the academic environment allowed the contact with a broad range of students from more or less developed countries and exposed the Greek students to a hitherto unknown wealth of ideas and opinions on development, spatial and social differences and inequalities, urban affairs, and the natural environment. The Greek student learned a lot about urban and regional planning in the country where he/she studied, but also about the experience that his/her fellow students brought from African, Latin American, Asian, European and North American countries. It should be remembered that this was a period of international turbulence, war conflicts, and geopolitical disputes, which left their mark on the university world. It was the era of Vietnam, of anti-racist struggles in the USA, of the May 1968 events in Paris, and of the Cold War. The Greek student learned not only about planning recipes in the country of his/her study, but also about Portugal, Argentina, Cambodia or Iraq. This was valuable knowledge for someone who had left behind the "Colonels' Greece" and was now learning about the problems that united the planet. Apart from learning from fellow students originating in distant countries, there was in addition cross-learning between students of different disciplines. In the classroom there were architects, economists, political scientists, lawyers, environmentalists, even theologians, exchanging views about disciplines which had to cooperate in the planning effort.

This new generation of planners was initiated, on one hand, in a conception that their profession was an emerging "scientific discipline", totally different from traditional urban design, but nearer to a new interdisciplinary field, and on the other hand, in a politicized view which repudiated watertight technocratic and bureaucratic mentality. On their return to Greece, these students brought new concepts and ideologies which accompanied them in their careers as civil servants, university teachers and researchers, consultants, or trade unionists. Some of them may have been prone to extremism or high-flown idealism, but were undoubtedly carriers of a spirit of innovation. They were in a better position to appreciate that spatial planning is not a one-dimensional professional activity, but a unified task with multiple extensions. They could, in theory, escape from imprisonment in enclosed, traditional professional domains. In this, they frequently found themselves in opposition to established and powerful professional privileges.

In our days, planners must be capable to appreciate the global challenges of climate change, and, at the same time cope with the onerous compromises facing them in the planning trivialities of a clientelist society. Their communicative capacity at the coal face of planning will be heavily taxed. They have to be "political animals", but also technically competent. They must be capable to appreciate the international dimension, without pretending to be experts in geopolitics, but rather in order to come to grips with the implications of environmental crises and of climate change. All that is a tall order, but this is why planning is exciting and rewarding. Still, the first priority is to immerse themselves in the planning culture they will have to deal with, without becoming its accomplice. Accordingly, the first priority of their academic educators is to help them avoid drowning in murky waters.

9.4 The Culture of Planning and the Greek Peculiarity

The ideological and cross-disciplinary makeup of planners is of the utmost importance for a profession faced with the challenge of local petty-political priorities and, at the same time, with the internationalization of issues such as climate change or globalization of economic interests, at a time when a "peripheral" country like Greece faces dire development prospects. It is an irony that a much-needed exposure of planning students to world realities may be better served by their education in multi-cultural university environments, rather than by training enclosed in the national confines of the university system of their own country. Given that, first, education for all wishing to pursue a planning career cannot be provided by costly university studies in foreign countries, and, second, because studying in those countries has the disadvantages described earlier, of equipping the student with a knowledge-kit which is not transferrable, we have here a genuine dilemma. It is a dilemma that a Southern approach to planning theory must take into account. It is not possible to offer here a recipe for locally-offered curricula, but what is certain is that the problem concerns the culture of planning which is dominant in Greece, and, generally, in the countries of the global South. This culture bears the influence of past history (Stead et al. 2015), the patterns of differentiation of human geographical space as they resulted from a turbulent past, political realities, and of course physical geography. E.g., in the Greek case, insularity can be considered as an "unsolvable" geographical problem (Kyvelou 2010, 248), although offering development opportunities. These are all factors which were discussed in earlier chapters.

Far more complex than geography is the web of social conditions which impinge on planning in a myriad of ways. Getimis (1994, 316) had drawn attention to the difference of spatial planning in Greece from that of other European countries. Planning in Greece is a reflection of the integration of socio-economic processes which differ radically from countries of the West. Social and geographical conditions are more fluid, population movements more intense, property relations are constantly in doubt, the land cadastre remains a dream, and court cases take years to be finalized. This is a state of uncertainty which leaves room for under the table settlements, or, to put it differently, for a culture of endless compromise between the state administration and society. Can the introduction of planning processes from foreign contexts remedy this situation? To some extent this is happening already, frequently creating a deceptive impression that everything works smoothly. Importing foreign models has a long tradition in Greece and has received added impetus when she became a full member of the European Union. Imported models may have an Enlightenment ancestry which never reached Greece with the same intensity as in Western Europe. But even when they were more recently introduced from the West, they were still imbued with a western understanding of the functions of the modern state. When they are transferred to a place like Greece, an unstable equilibrium results between obedience to rules and traditional, unwritten loyalties. The activity of spatial planning is a good example of this underlying conflict of values. While it is modernized, against considerable resistance of course, the traditional system of clientelist and populist values is fighting a rear-guard, underground war with remarkable success. Modernization is then undermined, all too often with the complicity of the supposed "modernizer" component of the state apparatus, which turns a blind eye to reality. Imported practices, e.g., through the European Union, may be leaving a welcome and positive trace, but they are still modelled on the conditions of countries like Germany or France (Getimis et al. 1994, 13). In addition, the administrations of such countries take the culture of planning for granted, which in Greece is often resisted in conditions of perpetual state-citizen antagonism. For most citizens in Greece planning should be limited to fixing building regulations, or, at best, land use. Demathas has observed the existence of dualism in the perception of spatial problems. All too often, problems may appear, at first sight, to be identical or similar, but they can be the result of different "genetic mechanisms". When this happens, they have to be addressed in a different manner which is more sensitive to real causes (Demathas 1995, 31). In a way, Demathas, Economou and others are touching on the delicate need to overcome dualism by reaching for mutually acceptable solutions. Arguably this is what consensus should be about, which, however, in long-rooted antagonistic situations, may be very close to compromise.

9.5 Greece: A Sui Generis Case?

The ingredients of what was earlier called the "Greek peculiarity" are found most clearly in several examples of spatial organization and planning. It is not the author's intention to propose here a solution in the form of a planning manual, but rather to underline the parameters which a theory should take into account. It was implied elsewhere (Chap. 7), that the refusal to abide by a formal legal rule is often a manifestation of a sense of injustice. In such case, the citizen confronts a statutory regulation as unjust, particularly when his/her actions were systematically, and for long periods of time, tolerated, even praised, by the authorities. The minister Antonis Tritsis spoke in the 1980s of the statue of the "illegal settler" (Chap. 8), that should be erected in recognition of his contribution to the relief of housing shortages. The blanket condemnation by the authorities of unlawful land use and illegal buildings gives rise to feelings of unfairness. This is what happens when the administration has been an accomplice by tolerating illegalities. We must remember here the reference by Kotzias to a society of complicity (Chap. 7). A typical example is the designation by the authorities of private properties as public forest land and the consequent banning of all new construction, extension of existing buildings, and housing use. The problem here is not whether the protection of forest land is justified, which it is. The problem lies in the definition of what constitutes forest land and even more so in how the citizen interprets the sudden turn-around of official attitude, when for decades the administration has followed a practice of servicing this land with all the water, energy and communication infrastructures and of granting perfectly legal building permits (see Text-Box 9.4). This is a clear case of complicity, because of which the citizen feels betrayed and considers that his/her land is practically expropriated. Entire densely populated settlements are described by the forest authorities as forest land. The answer of course should be found in their designation as "mixed-use areas" which would be planned with very strict regulations ensuring the coexistence of modest housing and forested land and the protection of the environment, as indeed happens in other countries (Wassenhoven 2014). Yet the authorities procrastinate or fail to find workable solutions, which would overcome constitutional and court obstacles. The so-called "settlement condensations" (*oikistikes pyknoseis*) legislated in 2016 were soon declared unconstitutional by the Council of State.

In Chap. 7 on the Greek state, reference was made to the notion of the "state of justice", inspired from a comment by Pararas, who emphasizes rightly that the duty of respecting the rules of law obliges the administration and the courts not to act whimsically and lawlessly, i.e., not to act arbitrarily. This is what he calls the deontological element of their operation. The "state of law", however, is not simply the state which functions in accordance with positive law, hence a better term has been proposed to replace it, i.e., "state of justice". As a result, the supreme duty of the state of law is the realization of justice, which means that the mission of state organs is to transubstantiate law into justice. Pararas concludes that state actions which in the view of the average individual are patently unjust, exceed the limits of the space of the state of law (Pararas 2014, 11, 15).[1] The question is if the average individual, even that who benefits from the state tolerance and complicity, is entitled to consider the belated reactions of the state as patently unjust. This is not to dispute the obligation of the state, even with considerable delay, to protect natural resources, but rather to make clear the impasse and dilemma generated by a long-lasting, petty-political relationship between state and citizenry, which urban and regional planning has to disentangle.

Text-Box 9.4: Status and Boundaries of Settlements and Zones
A plague for the planning system is the perpetuation of unresolved problems which are left in suspense for years or decades, causing undue suffering to land owners. In 1923, a pioneering legislative decree created the framework of town planning, but placed already existing settlements in a transitional status. Many, though not all, of these settlements were gradually absorbed into the area covered by statutory plans, a process of "regularization" which continued for years. About 60 years after 1923, a new presidential decree set the framework for the normalization of the status of still pending cases of pre-1923 settlements, part of the area of which might be forested land. Areas claiming this status ought to be recognized first, and then delimited by acts of the then nomarchs (prefects). There were cases of settlements recognized by such an act, e.g., in the late 1990s, but the boundaries were not appropriately fixed, a decision which was left to the local municipality to take. Naturally, the owners of properties in these settlements mobilized, individually or through ad hoc formed associations, to press for this process to be concluded. They were asking for the boundaries to be fixed and for some form of land use planning to follow. They hired the services of surveyors, planners and lawyers, searched for historical evidence to support their arguments, and, eventually, convinced the local municipality to endorse their claim and approve the settlement boundaries. However, the final decision was in the hands of the central government which procrastinated endlessly, fearing the reaction of forest services, although any future plan could incorporate protected green spaces (see Fig. 9.1). It goes without saying, that in the meantime new houses were constructed, often with a legal building permit. In 2021, almost a century after the 1923 decree, the government eventually came up with the proposal that a series of boundary-setting plans of a new type should be drawn, so as to enable these areas to be exempted from forest designation until a formal town plan is prepared. It sounds like an irony that the programme of which boundary-setting plans will be part bears the name of the famous Greek planner Konstantinos Doxiadis.

(continued)

[1] As already done in Chap. 7, it is repeated here that the author hopes that he is not misinterpreting the views of Pararas.

Text-Box 9.4 (continued)

Boundary-setting plans are also to be the solution for another embarrassing relic of the past, although more recent, this time small settlements of less than 2,000 inhabitants, which, in the 1980s, were not to be planned with the procedures of town planning adopted under the 1983 law for towns and townships. As explained in the main text, a 1985 decree authorized the prefects to set a boundary around small settlements and approve loose land use and building conditions. These village zones acquired a status midway between fully planned areas with a statutory plan and out-of-plan areas (see Text-Box 8.2). The whole 1985 procedure was subsequently found unconstitutional by the Council of State thus leaving one more hole in the operation of town planning. These areas of semi-planned extension of villages (some of which may be pre-1923 settlements) are left in a legal vacuum. In this case too, the government intends to proceed first with a boundary-setting plan. These examples demonstrate how history haunts the practice of planning. They also prove that national particularities create conditions which no imported tools (see Text-Boxes 9.1 and 9.2) can remedy.

Fig. 9.1 The area of a pre-1923 settlement near Athens (see Text-Box 9.4). (Source: Google Earth)

If there is a phenomenon which represents the shady transactions between citizens and state then it is the illegality surrounding land use building construction, a key issue in planning. This planning lawlessness casts doubt on the credibility of public administration and the functions of the state. As put by Melissas (2017, 29), political credibility in this case means the trust that politicians must inspire in the implementation of each planning action. Credibility exists when the visible on the ground is identical to the legal, and ultimately corresponds to the truth. Illegal development is perpetual lawlessness, subject to control at any moment in time. This means that illegal land use and construction is in itself a factor differentiating the Greek situation from that of countries, where most of planning theory has been formulated. Illegal development, remarks a former president of the Council of State, is a typical example of the pathology of the political system (Menoudakos 2017, 5–6). It could be explained in periods of war, foreign occupation, abnormal political conditions, and internal strife. Yet, it continued, in fact was intensified, after the 1970s, at a time when social and economic conditions had been radically transformed. This is illegality, Menoudakos continues, which flourished in a climate of unbridled consumerism and is no longer resorted to in order to satisfy urgent housing needs. Nevertheless, as stressed several times in this book, the state was looking the other way.

A glaring example of chaotic and failed planning which causes frequent conflicts is the delimitation of the sea shoreline. Although the definition of the shoreline had been clarified in 1940 legislation, its boundaries on the ground had been fixed for only 11% of its total length 50 years later (Stamatiou 2003, 516). As late as 2014 the official estimates were even lower. Legislation in 2001, 2014, and 2016 attempted to speed up the process. Amendments were continuing in 2019, which is one more example of failed legislation, already discussed in Chap. 7. The consequences of incompetence, back-pedalling, and compromise with private interests were extensive, causing illegality, uncertainty, and legal insecurity. The responsibility of the administration has been meticulously analyzed in a report of the Greek Ombudsman (2013).

In the examples quoted here, of out-of-plan areas or of coastal zones, the failures cannot be simply blamed on the land use planning system, although its rigidities and delays played a part. Kyvelou rightly emphasized the limits of the system and quoted comments from the international literature, e.g., from Patsy Healey, which focus on the lack of creativity, excessive technical character, inability to enter into public dialogue, underestimation of local socio-economic conditions, narrow-mindedness, and absence of important actors (Kyvelou 2010, 122–123). She only accepts the relative advantage of a stable institutional framework. It is true that the regime of land use planning was, up to a point, a stabilizing factor, when of course it was implemented without exceptions. The trouble is that it was frequently bypassed and that its general rules, as fixed in the official guidelines for all planning studies, were rigid and uniform for the entire national territory without paying adequate attention to local variations. Getimis and Economou (1992, 28–29) had drawn attention long ago to the role of planning as a land rent-generating mechanism, to the pitfalls of a generalized control system, and the opportunities that it leaves open for the real estate market to discover loopholes. The undifferentiated framework of land use control and the existence of ad hoc exceptions, e.g., to locate infrastructures or legalize unauthorized developments, allowed room for clientelist political compromises and for the populist reproduction of a system of consent among small land owners.

Illegal development in the countryside and peri-urban zones, even on forest land or coastal strips, is not the only example of the erratic behaviour of the state regarding illegal or uncertain land rights. The endless saga of the production of a land cadastre and of forest maps is another, and it is pleasant to notice a major effort to remedy the situation in the last few years. All these examples demonstrate the importance of the parameter of the state for the formulation of a credible planning theory. The democratic nature and the efficiency of the state are of great significance in this context. Pluralism and efficiency are hampered by "extractive institutions", as Karadimitriou and Pagonis (2019, 13–14) point out. Such institutions include, in their view, the planning statutes which favour specific rent-seeking interests. It remains to be seen to what extent recent legislation, a more automated control system, and the use of digital technology (e.g., for a planning data base or for issuing building permits), will transform the traditional interplay between control mechanisms and private activity. The challenge is always there and it is no other than the adaptation of innovations in the particular national conditions. As Theodora (2018) warned, experience shows that we have to be wary of imported planning practices and seek logics which are compatible with the history of places. For a long time, the ideological basis of planning was the fight against unequal spatial development, which justified an increased state role. Supporters of a liberal tendency are not always in favour of this priority. The importance of spatial organization has not, however, diminished, although it now imposes a more place-adapted approach, without surrendering to the perpetuation of a traditional personalized and clientelist compromise logic, which is still the main feature of planning in Greece. However, compromise can be embedded in a pragmatist approach of action, problem-solving and learning by doing. Increased and more flexible participation will have to be a new direction, as, e.g., Serraos (2018) supports, provided it does not reproduce clientelism and populism. Meaningful and transparent participation is necessary for a changed mentality of governance, but occasional, bureaucratized participation on the occasion of a specific plan or project is not sufficient. In the age of climate change, participation must be encouraged over a broader spectrum of issues and embrace civil society regardless of narrow interests. Sadly, and in spite of encouraging signs, civil society in Greece, as Voulgaris (2013, 387–390) observed, exhibits an atrophy. Seen together with the hypertrophic state, he claims, it is a key element in any attempt to interpret the Greek peculiarity. In the absence of a radical turn in this direction, we can add, compromise will remain the crucial parameter.

9.6 Compromise, Collaboration, Pragmatism and Improvisation

It can be said that resorting to compromise is a sign of failure in the pursuit of consensus. However, when consensus-seeking leads to an impasse, compromise can be a decent and pragmatist way out and can still remain faithful to the values of a pluralist and democratic society. This is how Sanyal is putting the case: "One could …

argue that in democratic societies, compromise when reached honorably and in a spirit of honesty to all is the only fair and rational way of reaching an agreement between different points of view. In that sense, compromise is the essence of democracy. It is through compromise we arrive at what is called the public interest. Also, the act of compromise can be a learning process as one learns about the opposite and multiple sides of a problem" (Sanyal 2002, 118–119). Sanyal's argument is akin to the reasoning behind dispute settlement as observed in American communities by Ellickson (1991), also in a pragmatist line of argument. Sanyal refers to compromise as an ethical issue for policy makers and deals with a range of compromise situations and public policies generated by globalization, but it is worth noticing his arguments: "What do planners rely upon when engaged in compromising? What theory of action can they look to for guidance? The current literature on so called planning theory is rather thin and somewhat useless for this purpose … Some may disagree with this pessimistic assessment of the current state of planning theory. After all, there is a body of literature on theories of negotiation. Also, as some economists would remind us, there is 'the game theory' to guide or, at least, predict action. Also, on the left – not far left, but somewhat left of center – is the Habermasian theory of deliberative democracy according to which open and free deliberation is the key to moral choices. It was revealing to us that practicing planners troubled with moral dilemmas and looking for ethical compromises do not use any such theories" (op.cit., 120). Sanyal's argument continues: "For public policy, one has to transform principles to interests (who is paying for what, and so on) to start a discussion. In other words, the planner has to define the issues at hand in such a way that trade-offs are indeed possible. The planners must also be aware that a compromise, if it has to have legitimacy, must be transparent to the constituencies whose interests are being compromised. Good compromises do not destroy political coalitions which force the powerful to compromise. Such compromises do not give away the autonomy for future disagreements. These are merely a few ad hoc observations. It requires much more systematic, institutional research to create a good body of knowledge planners can rely upon as they are faced with new moral dilemmas created by globalization" (op.cit., 121–122).

The necessary body of knowledge may well exist in the philosophy of pragmatism to which we referred in Chap. 3. "Adopting a pragmatist line of thinking, we argue that rather than a moral problem, compromise offers an important moral resource for making future-oriented plans that seek to anticipate and cope with complex development issues such as those involving environmental and social change" (Vidyarthi and Hoch 2018, 629). Perhaps, there is here a prior need for internal compromise within the planner's mind and conscience before he/she proceeds to external compromise in his/her planning action. "Pragmatism rejects the quest for moral perfection whether tied to moral integrity, optimal utility or heroic virtue. In this line of thinking, the modern promise of self-development both as a resource and product of purposeful inquiry and practical experimentation adopts different meanings in disparate circumstances … Thus, instead of relying upon moral ideals to guide planning, we argue that practitioners can and should focus upon constructing forms of compromise that improve the provision and allocation of complex goods" (op.cit., 630). Vidyarthi and Hoch add that "our analysis is, however, predicated on the belief that participants learned from prior efforts and that the meaning of these compromises shifted in response to new moral conceptions of the public interest or common good" (op.cit., 640).

This line of thinking continues in Hoch's major textbook on pragmatic spatial planning: "Adopting a pragmatist approach, I argue that rather than a moral problem, compromise offers an important moral resource for integrating moral sensibilities and values making plans for complex practical problems. The ethical significance of compromise includes the integration of rules, goals and conduct in plans that anticipate and cope with complex environmental and social change. Compromise for planning does not abstract moral norms as guides independent of historical context and practical possibility. The shifting compromises fit within a developmental approach; testing the record for better outcomes tied to innovations in democratic conduct responsive to context over time. On this account compromise is not inherently bad even if it may yield bad outcomes. Pragmatic thought embraces enough give and take to explore the emergence of an uneven and incomplete common good sustained despite mutual antagonism. This robust conception of compromise can reconcile a diverse range of social differences and disparate moral claims" (Hoch 2019, 175). No doubt, Hoch admits, there can be "rotten" compromises and "in many cases, the idea of compromise also denotes a kind of moral failure" (ibid.). Yet, "pragmatist planning weaves compromise into the very fabric of any practical proposal composing, judging and selecting a joint project or action" (op.cit., 177). Compromising ceases to burden the planner's conscience. "Thus, compromise plays a key role in planning practice because professional practitioners who sincerely plan for common goods face the challenges of complexity and the threat of external constraints. However, compromise often generates ambivalence because those who do it combine a good act (cooperating on one goal) and a bad act (subverting another goal)" (op.cit., 182). At this point, Hoch quotes Avishai Margalit who "embraces compromise as a crucial dimension of political democracy". According to Margalit, "on the whole political compromises are a good thing. Political compromises for the sake of peace are a very good thing. Shabby, shady and shoddy compromises are bad but not sufficiently bad to be always avoided at all costs, especially when they are concluded for the sake of peace. Only rotten compromises are bad enough to be avoided at all costs" (ibid.; see also Hoch 2018).

The pragmatist viewpoint and this conception of compromise is inspiring Nicholas Low, when he writes that "the job of planners is not just to describe and understand human variety but to act on it and with it … Let us suppose, then, that effective planning of our living environments occurs through processes of negotiation and mutual persuasion in which people try to resolve their differences. How we, as planners, resolve differences between ourselves and others can be described in terms of the construct: 'compromise-collaboration'. *Compromise* is the result of an interaction in which each person's world model remains unchanged, but sacrifices of some parts of each person's model – its ends and means – are made in order to arrive at a decision. Compromise is an outcome of negotiation. *Collaboration* is an approach to the process of negotiation which may result in compromise" (Low 2020, 16). Low skilfully blends pragmatism and communicative theory. This blending is all the more interesting when undertaken by one of the champions of communicative theory, John Forester. In the 1970s, he writes in an autobiographical essay, "ambitious social theories omitted social actors; attractive political theories ignored political action". This reminded him of the frequently heard excuse "I was just doing my job" – an excuse used even by those charged in the 1940s for crimes against humanity. Years after the decade of the 70s, he continues, "still perplexed that planning theorists so often avoided this challenge of assessing what planners might really do in diverse planning contexts, I wrote polemically about 'planning theory's dirty little secret'. That secret: *No matter how much we theorize* planning systems, promising ideas, or threatening ideologies, *somebody still has to do the work*. So, theories of state power or neoliberalism, for example, without accounts of astute practical resistance, risk being not 'critical' but theoretically self-indulgent and evasive" (Forester 2017, 280–281). In his conclusions he discusses the challenges of planning: "Put most simply, without (i) dialogic understanding, we risk solving the wrong problems. Without (ii) expert debate, we risk tilting at windmills and attempting the technically impossible. Without (iii) pragmatically probing creatively crafted, jointly acceptable moves, we might settle for insight but fail to act. These are the challenges of exploring practically value, fact, and action, each historical, social, and political. Without attention to these challenges, public deliberation will fail; mediated negotiations will fail; many planning projects, I suspect, will fail. But the inescapability of these challenges – of diagnosis and dialogue, expertise and debate, negotiation and action – does not so much 'explain' creatively improvised planning as much as, I believe, it provides a framework or infrastructure for practically situated and responsive action … Improvising well in planning presumes a wise ability to listen beyond words, an ability to decipher or learn deliberatively about value. Improvising well requires assessing power and possibility, the move to free ourselves from self-limiting cynicism and the presumption that needless suffering cannot give way to better lives for many. Improvising well to respond justly as multiple, plural, and conflicting stakeholders find themselves interconnected and interdependent will require planners to integrate dialogue, expertise, and negotiation" (op.cit., 294). One can be justified to claim that Forester's concept of "improvising" has a clear pragmatist ring and an affinity with the notion of compromise. Let us return to Hoch, who reminds us that "a pragmatist theory helps people comprehend the limits of what plans do. It inspires people to study what and how plans contribute to commitment, choice, and action for places. But the pragmatist approach recognizes that plans do not make things happen. Plans offer advice and counsel, but only people and their institutions act" (Hoch 2017, 298) (see also Chap. 3).

9.7 Basis for a Theoretical Model

After a thorough study of the development of spatial planning in Greece since the 1940s, the author started sketching a first approach to a pertinent planning theory in an area which he believes is attracting interest in the international literature (see Chap. 4). It is a literature produced by theoretical communities in academic environments for which the dominant paradigms of developed countries tended for long to monopolize interest. It is in this very community that several theorists now open up new windows to embrace non-Western countries. Such theorists, e.g., Palermo, Ponzini, Yiftachel, Miraftab, Huxley, Hillier, Sandercock and others, bring into the debate a view from the South or from the South-East. It is in this direction that the author hopes to add a new perspective, when speaking of a theory of *compromise planning*. In this chapter, all the issues touched upon in the introductory sections of the book will be elaborated.

The author would like to repeat here a phrase used earlier in this chapter: "I am still an incorrigible believer in the nature of theory as a medium for the interpretation and explanation of real-life phenomena, which, in the case of planning theory, are identified with planning in the real world". The basis of this theoretical approach is what, for lack of a better adjective, can be called the "empirical" reality of spatial planning in Greece, which a purely communicative approach cannot explain. This reality is approached not as a snapshot photograph of today's planning system but in an evolutionary perspective, i.e., as urban and regional planning developed in modern Greece. In this perspective, the task of theory is to explain and interpret empirical reality, with the proviso that this reality will be examined not in a way which describes the present in a narrow fashion, but in the context of multi-sided and complex conditions, carefully *situated* in a historical, social, political and economic nexus. As indicated in Fig. 9.2 on empirical reality, the author is trying to observe, interpret and explain the Greek planning system with the intention to formulate a planning theory. In this task he had to answer some fundamental questions, which obliged him to study carefully, in Part II of the book, the history of the country, the structure and character of the Greek nation-state, and the form of its planning system, as it evolved over time. One key interest was to isolate the impact of the presence or absence of characteristics which shaped – or would have shaped differently if they were present – the space of the country and, consequently, the priorities of planning. The persistent influence of "foreign powers" throughout the country's history and the importation of alien administrative arrangements and practices is one example of omnipresent characteristics. The trust and confidence of citizens in the state, notable for their absence, is an example of a condition which would have made a difference if it could be secured consistently in the long history of the new nation-state. The creation of a state which would remain malleable, friendly to the voter-clients of the ruling political parties, populist when required, and always prepared to adjust its course to accommodate large or small private interests, was a permanent, although officially unacknowledged, objective. This may not be a Greek exclusive characteristic, but had a special flavour in this

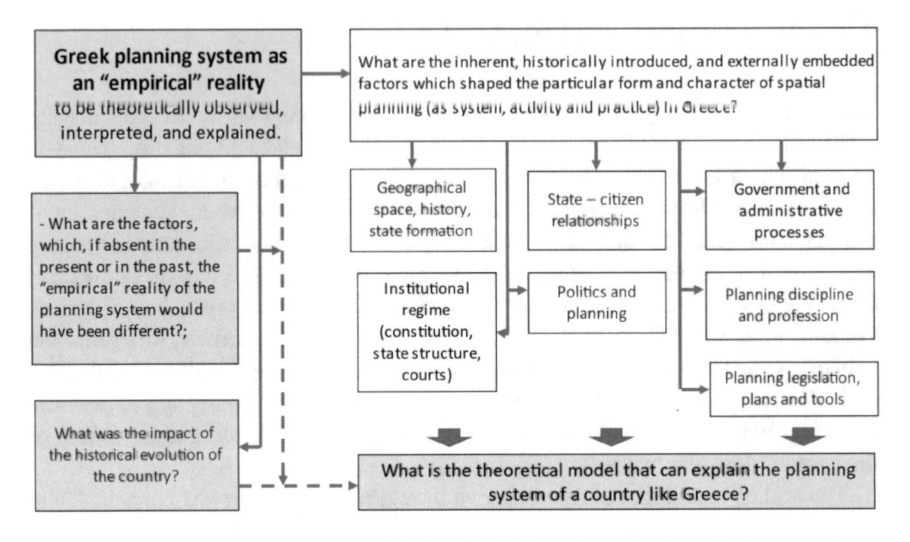

Fig. 9.2 The planning system as an empirical reality. (Figure drawn by the author)

"appendix" of Europe, because of an ambiguity which runs in the veins of every Greek. That's why, this ambiguity regarding the sense of belonging to the West or East is one more excellent example of a characteristic, of which the presence or absence is critical. Is Greece the East of the West, the West of the East, or both? It would not have mattered, if it had not become, time and time again, a contentious issue in internal politics. Sometimes it takes the form of a rationality v. irrationality division. These features, positive or negative according to the ideology of each individual, impinge on the planning system and on its evolution over the years. Half-hearted acceptance by society of the choices made by the political system, became the object of bargaining and was only obtained in exchange for preferential treatment in crucial planning situations. It is hoped that Chap. 6 of this book conveyed an adequate account of historical developments.

To answer these questions, the author felt the need to take a broad sweep over the history of the modern Greek nation-state (Chap. 6), in order to paint a backcloth of the subsequent analysis. The gradual expansion of the new state, the efforts to build its national identity, the constant interference of the so-called "protecting powers", the struggle between the local traditionalists and the westernized Greeks of the diaspora, the fervent desire of the Great Idea of a state extending on Asia Minor soils, the victories and defeats in the course of irredentist aspirations, the trauma of accommodating huge inflows of refugees, the painful internal divisions, schisms and civil conflicts, the agony of reconstruction and economic development after the Second World War, the shameful dictatorial intervals, the eventual victory of democratic liberalism – of various political colours no doubt –, the accession to the European Union, all were chapters of a history the understanding of which is essential to comprehend the setting of an appropriate theoretical approach. The role and function of the state, under its various transfigurations of the last two centuries, are of obvious importance. The state is the actor that dominated what was happening on the historical stage. As a puppet of foreign influence or an institution that crowned the road to emancipation of the Greek people it was the king of players. As adjusted to foreign influence and pressures and to the needs of domestic political forces, it demonstrated ultimately impressive versatility and capability to adapt. However, it still maintained its "eastern" mode of thinking and acting.

From the view point of planning, the state excelled in the skill of compromise, either as a sign of political intelligence, or as a ruse to survive. It is not for nothing, that the Greeks adore the figure of Karagöz, the shrewd puppet of the Turkish shadow theatre, a character with a deep sense of the act of adaptability, survival and manipulation.

This leads the author to a set of specific issues and factors (Fig. 9.2), which explain how and why particular spatial configurations and problems emerged, which the planning system had to address, from the 1920s onwards, but particularly after the 1967–1974 military dictatorship (see Fig. 8.1 in Chap. 8). The diagrammatic representation of crucial issues and factors, leads to the question of what the theoretical model will be which best explains the Greek planning system, not just as it is today but as it has gradually developed.

Epilogue

In this chapter, which started with an account of personal experience, an attempt was made to throw light on the elements which constitute the empirical reality of planning in Greece, of which the explanation will be the aim of a proposed theoretical model (see Chaps. 10 and 11). Most of these elements have been examined in previous chapters. The structure of the national territory, its gradual emergence, the succession of historical events, the formation and ideological justification of the nation-state, and the state's often controversial relations with the citizenry are the main ones. The characteristics of government and administration, the administrative processes, the importance of the national Constitution, the legislature and the judiciary, the role of politics and politicians as mediators, always with an eye on urban and regional planning, are additional, but very important, parameters. The final set of parameters, which the author had to build upon, is planning legislation itself and the hierarchy of plans and tools of the planning system, most of which are imported (see also Chap. 8). The effort of the author has been not to present this system as described in official documents, i.e., as a detached observer, but to embed it in the framework of the history of the country and its state structure. In the middle of this complex web, the planners have to play the usual game of intermediaries, experts, facilitators, and objective policy makers, either as public servants and consultants, or as academics. Hence, separate sections were devoted to the planning community and its academic background and on the Greek-style planning culture. The nature of the mixture of compromise, pragmatism and improvisation which compose the scenery of planning was then discussed. This may not be unusual in any country, but in the Greek context planners have to adapt to the "particularities" of their sociopolitical environment, as sketched earlier. They have been trained in the West, although this is changing rapidly, but wherever they studied their discipline they absorbed knowledge, methods and techniques produced in altogether different contexts from the one in which they would follow their career. This is a typical characteristic so often underlined in Southern theory (see Chap. 1). All this forms the basis of the effort to build a theory which corresponds to the Greek profile, and, hopefully, that of other countries of which the planning systems are squeezed between similar conflicting parameters.

References[2]

Demathas Z (1995) Greece in the European territorial organization: problems and prospects. In: SEP (1995a) Regional development, spatial planning and environment in the framework of united Europe. Vol. II. Proceedings of a conference, Panteion University, 1995. SEP and *Topos*, pp 28–39 (in Greek)

Doxiadis C (1942) About *chorotaxia* (spatial planning). TeC, No 241–244 (in Greek)

Doxiadis C (1959) The science of Ekistics. Architecture 1959(13): 13–25 (in Greek)

Doxiadis C (1968) Ekistics: An introduction to the science of human settlements. Hutchinson, London

Doxiadis CA (1965) Order in our Cosmos. In: Doxiadis CA, Douglass TB (1965) The new world of urban man. United Church Press, Philadelphia, pp 13–34

Ellickson RC (1991) Order without law: How neighbors settle disputes. Harvard University Press, Cambridge (Mass)

Forester J (2017) On the evolution of critical pragmatism. In: Haselsberger B (ed) (2017) Encounters in planning thought: 16 autobiographical essays from key thinkers in spatial planning. Routledge, New York and London, pp 280–296

Getimis P (1994) Urban development and policy. In: Getimis P, Kafkalas G, Maravegias (eds) Urban and regional development: Theory, analysis and policy. Themelio, Athens, pp 307–333 (in Greek)

Getimis P, Economou D (1992) New geographical inequalities and spatial policies in Greece. Topos 1992(4): 3–44 (in Greek)

Getimis P, Kafkalas G, Economou D (1994) Regional spatial planning and environment: New institutions and future symbiosis. Topos 1994(8):5–13 (in Greek)

Greek Ombudsman (2013). Coastal zone management. Special report. SynP Independent Authority. Athens (in Greek)

Healey P (2010) Introduction: The transnational flow of knowledge and expertise in the planning field. In: Healey P, Upton R (eds) Crossing borders: International exchange and planning practices. Routledge, London, pp 1–25

Hoch C (2017) Pragmatism and plan-making. In: Haselsberger B (ed) Encounters in planning thought: 16 autobiographical essays from key thinkers in spatial planning. Routledge, New York and London, pp 297–314

[2] **Note**: The titles of publications in Greek have been translated by the author. The indication "in Greek" appears at the end of the title. For acronyms used in the references, apart from journal titles, see at the end of the table of diagrams, figures and text-boxes at the beginning of the book.

Abbreviations of journal titles: AT Architektonika Themata (Gr); DEP Diethnis kai Evropaiki Politiki (Gr); EpKE Epitheorisi Koinonikon Erevnon (Gr); EPS European Planning Studies; PeD Perivallon kai Dikaio (Gr); PoPer Poli kai Perifereia (Gr); PPR Planning Practice and Research; PT Planning Theory; PTP Planning Theory and Practice; SyT Synchrona Themata (Gr); TeC Technika Chronika (Gr); UG Urban Geography; US Urban Studies; USc Urban Science. The letters Gr indicate a Greek journal. The titles of other journals are mentioned in full.

Hoch C (2018) Neo-pragmatist planning theory. In: Gunder M, Madanipour A, Watson V. The Routledge handbook of planning theory. Routledge, London, pp 118–129
Hoch C (2019) Pragmatic spatial planning: Practical theory for professionals. Routledge, New York
Karadimitriou N, Pagonis T (2019) Planning reform and development rights in Greece: institutional persistence and elite rule in the face of the crisis. EPS 27(6):1217–1234
Kazazi G (ed) (2009) Constantinos Doxiadis and his work. 2 volumes. TEE, Athens (in Greek)
Kyrtsis A (ed) (2006) Konstantinos A. Doxiadis: Texts, plans, settlements. Icarus, Athens (in Greek)
Kyvelou S (2010) From regional spatial planning to space-management: the concepts of strategic spatial planning and territorial cohesion in Europe. Kritiki, Athens (in Greek)
Larsson G (2006) Spatial planning systems in western Europe. IOS Press, Amsterdam
Low N (2020) Being a planner in society: For people, planet, place. Edward Elgar, Cheltenham
Magnan R (1966) L'évolution des conceptions urbanistiques. Conférence de stage. Ministère de l'Équipement, Secrétariat d'État au Logement, Paris
Melissas D (2017) The legislation for building "regularization" and legalization is testing the credibility of the Greek state. In: Serraos C, Melissas D (eds) Unlicensed building. Sakkoulas, Athens, pp 29–43 (in Greek)
Menoudakos C (2017) Unlicensed building: Legality and reality. In: Serraos C, Melissas D. Unlicensed building. Sakkoulas, Athens, pp 6–9. (in Greek)
NTUA/SPE (1966) Skala of Lakonia: Master plan, town plan and area of influence. Research report. Athens (in Greek)
NTUA/SPE (1969) Andritsaina (Elia Prefecture): Master plan, town plan and area of influence. Research report. Athens (in Greek)
NTUA/SPE (1971a) Regional analysis of Epirus and Thessaly, settlement restructuring of earthquake-stricken areas, and town plans of three townships for the relocation of population. Research report. Athens (in Greek)
NTUA/SPE (1971b) Master development plan of the city and influence zone of Kalamata in the Peloponnese. Research report. Athens (in Greek)
NTUA/SPE (1973) Peloponnese: Regional spatial plan. Research report. Athens (in Greek)
Pararas P (2014) Res Publica I: The law state. Sakkoulas, Athens (in Greek)
PASPE (1984) In memory of Prof. Ieronymos Pintos. Athens (in Greek)
Pertsemlidis K (2009) The foundations of Ekistics: A standardization plan for Ekistics theory and knowledge. In: Kazazi G (ed) Constantinos Doxiadis and his work. Vol. A. TEE, Athens, pp 348–375 (in Greek)
Sanyal B (2002) Globalization, ethical compromise and planning theory. PT 1(2):116–123
Serraos C (2018) Official planning and bottom-up practices: An antithetical trajectory in 21st century Greece? In: Theodora G (ed) Maritime space, coastal urban front, port-cities. NTUA, Athens, pp 43–55 (in Greek)
Stamatiou E (2003) Legal developments on the sea shore zone and the beach – town development: forecasts, omissions and impacts on coastal space. University of Thessaly Research essays 9(22):513–536. (in Greek)
Stead D, de Vries J, Tasan-Kok T (2015) Planning cultures and histories: Influences on the evolution of planning systems and spatial development patterns. EPS 23(11):2127–2132
Theodora G (2018) The question of teaching spatial planning. In: Theodora G (ed) Maritime space, coastal urban front, port-cities. NTUA, Athens, pp 13–29 (in Greek)
Tournikiotis P (2000) The ancient and the modern city in the work of Konstantinos Doxiadis. In: Deffner A et al (eds) Town planning in Greece from 1949 to 1974. Thessaly University Press, Volos, pp 85–98. (in Greek)
Vidyarthi S, Hoch C (2018) Learning from groundwater: Pragmatic compromise planning common goods. Environ Plan C Polit Space 36(4):629–648

Voulgaris G (2013) Post-regime transition Greece 1974–2009. Polis, Athens. (in Greek)
Wassenhoven L (1971) Town planning: need for an integrated theory; Economicos Tachydromos 916, 11 November: 14–16. (in Greek)
Wassenhoven L (1983) A review of styles of planning. University College London (Bandung Project) and Bandung Institute of Technology, Bandung, Indonesia, Teaching manual
Wassenhoven L (2014) The problem of periurban areas of mixed land use in Greece. Workshop on Land Management, European Commission Task Force for Greece, 11–12 June 2014

Chapter 10
Compromise Planning and Homo Individualis

Abstract The author's chief concern in this chapter is to pave the way towards the formulation of a compromise theory of spatial planning, which makes use of the Greek planning system as a case study. Building a planning theory is here related to the character of the person whose interests planning is supposed to serve. In this task, the author resorts to concepts developed by human geographers. In the relevant sections he presents the debate on the *homo* who is at the receiving end of all planning initiatives, but is also a key actor. The types of homo which appear in the literature – *homo economicus, socialis, sociologicus, publicus, creator, fabulans* – are presented and commented upon, before the introduction of the character of *homo individualis*. This latter type is found to be the person involved in processes of bargaining, adjustment, negotiation and compromise with the state and the planning authorities. The discussion is placed in a perspective of pragmatism and contains references to the public interest, the planner as a social "product", and the interplay between rationalism and exceptionalism. The chapter ends with a presentation of the factors affecting the model of compromise planning, which will be presented in the final chapter.

Keywords Human geography · *Homo economicus* · *Socialis* and other variants · *Homo individualis* · Compromise · Pragmatism · Planning theorist · Rationalism · Exceptionalism · Public interest

Prologue

The international literature on planning theory has been presented, hopefully adequately, in Part I of the book. It was followed by Part II, of which the purpose was to outline the empirical reality, based on Greece, that would help the author construct a theory of compromise planning. This reality was further distilled in the previous chapter. We have now reached the real crunch of this endeavour, that of synthesizing all this material into a cohesive outcome. Although we will occasionally return to international planning theory sources, we will here rely mainly on the dialogue among human geographers and planners, who have made a contribution to

L. C. Wassenhoven, *Compromise Planning : A Theoretical Approach from a Distant Corner of Europe*, https://doi.org/10.1007/978-3-030-94331-8_10

the understanding of space and spatial planning. The author began this review of Greek literature in the chapters on Greek history and on the Greek state (Chaps. 6 and 7) and will continue it here, the reason being, that domestic sources, although few, have made a valuable contribution. The local academic debate is perhaps of equal importance as factual knowledge.

The author described, in Chap. 9, his personal journey as an academic and lately as writer and small-scale practitioner in the world of planning. He wrote recently (Wassenhoven 2018a), that he was of course, all along, well aware of the distance separating the worlds of theory and practice, but the shifting of emphasis in his activity was an eye-opener, almost a form of crash-landing. One of the peculiarities of the Greek planning system he noticed in his career was the distance separating official planning rhetoric, as expressed in the statutes of planning, and actual reality. While observing this distance, he felt a need to explain it in theoretical language and to communicate with the ideas developed in human geography. New terms began entering his vocabulary which found their place in a new explanatory scheme. It is useful to indicate here what these terms were. As he did when discussing the rational, communicative, radical and climate currents, he lists here the key ideas and concepts which underlie the proposed, tentative "compromise current".

Keywords of the Compromise Current
Ideas, concepts, preoccupations, interpretations, concerns, and socio-political conditions embedded in the compromise current, which inspire pragmatist compromise planning theory and practice:

Pragmatism, incrementalism, agreement, instrumentalism, experience, synthesis, settlement, accommodation, blending values, individuality, mutuality, moderation, pluralism, practical knowledge, realism, inquiry, togetherness, testing, learning-by-doing, disruption avoidance, self--realization, pioneer spirit, opportunism, contingency, correlations, experimentation, truth-illusion, reciprocity, human as doer, coping, transparency, robustness, negotiation, clientelism, patronage, corruption, favouritism, improvising, complicity, knowledge accumulation, land property, private interests, storage of experience, precedent.

10.1 The Solace of Pragmatism

Throughout this book, a parade of theories and concepts was presented which swirl in the author's mind like whirling dervishes, as he settles down to put his own ideas in a theoretical mould. He had started his exploration of planning theories with the rational planning tradition (Chaps. 2 and 3), which was found being reincarnated in official planning systems, like that of Greece, in spite of the regular lambasting it received from critics of the communicative or radical schools. In Chap. 3, an amusing phrase by Brooks (2017, 81) was quoted: "Much like the creatures in horror movies, rationality is dead – but keeps showing up in public places. Despite its purported flaws, rationality is still the dominant paradigm in planning practice, and therefore continues to deserve careful scrutiny". Modernism in the case of planning seemed to resist the onslaught of post-modernism. Of interest here is a phrase by Beauregard (1996, 224), also quoted in the same chapter: "Theoretically, planning remains in a modernist mode. The literature on planning theory is devoid of attempts to view planning theory through the lens of the postmodern cultural critique. Rather, this theoretical investigation has been initiated by urban geographers". Faced with this dilemma of a rational model which seems both anachronistic and "too hard to die", the author found some consolation, as explained in Chaps. 2 and 3, in the pragmatist approach of Hoch and the contingency model of Ernest Alexander. The view of Charles Hoch that a way could be found to bring closer theory and practice is most interesting. Here is his own remark, as already quoted: "Practitioners do this kind of appropriation all the time by selectively combining policies, regulations, designs, innovations and other forms of planning activity that they find useful for their own work. Such practical synthesis fits the demands of professional work and of course my own pragmatist conception of the discipline. But I think it offers a more fruitful role for theory" (Hoch 2019, 46). Pragmatism, contingency, instrumentalism, correlation of situations, understanding of truth, realism, human being as doer (Bolan 2017, 156), practical judgement, practical comprehension, negotiation, bargaining, coping; all these concepts became for the author ingredients of an approach which he came to describe as compromise planning. This is both political compromise between the state and the body of citizens and compromise between the rationally-inspired official planning system and planning in practice. Whether it is a morally justified compromise or a pragmatist strategy is an open question, but it is a valid explanatory model, which sidesteps the pitfalls of pure rationality.

The slippery path to which rational planning in its systemic version can lead us is mostly related to its procedural character, as originally proposed by Faludi (see Chaps. 2 and 3). It is the rigor, but also the rigidity, of the rational process which makes it vulnerable. Yet, the intention of its supporters was to oppose conceptions of a deified planner, whose unchallenged authority was supposed to be the ultimate criterion, in a typical black box process. The alternative was to be a glass box approach, in which the planning phases and successive decisions were to be transparent and open to challenge. This may have led to a dangerous tendency to dress up all the steps of the process with scientifically impeccable techniques. The inevitable use of quantitative methods, perfectly justified in the handling of complex data, bred the illusion that the socio-political aspects of spatial planning could be tamed into a fault-proof model. The trouble is that the rejection of the "mathematization" of planning, should not necessarily guide us to equally inflexible explanations of the reality that planning has to address.

10.2 A Native Greek Theoretical Debate

The disputes over rationality in decision making gave birth to a dialogue that planning theorists engaged in, as outlined in all Part I of the book, and was transferred to Greece too after the years of the military dictatorship, admittedly among a limited number of academics, mostly human geographers. National and international conferences and seminars provide ample evidence. The so-called Seminars of the Aegean, organized by the department of geography of the University of the Aegean are a good example (see, for example, Aegean, Seminars of the 1995, 1998). Academic planners too were active in defining the framework of planning. Apart from refuting the mathematical twist of the planning methodology, Lagopoulos (1981, 1982), writing for an academic audience, sided with a socio-political approach to planning and adopted a Marxist perspective of historical materialism and class struggle. In his view, the common objectives of social harmony implied in the rational approach have to be transformed in line with the conflicts of class interests. Writing many years later, he remarked that the geography of the 1950s adhered to the view that geographical space followed certain inviolable rules of regularity and form of worldwide application. It was later accepted, under the influence of social science, that beyond absolute geometrical space there exists a relational space. It was this reorientation that led to a neo-Marxist interpretation emphasizing the role of capital accumulation in shaping spatial structure (Lagopoulos 1994, 123–125; see also Chap. 7). This author's concern is that when it comes to comprehend what planning is in reality, in the circumstances of a given national territory, relational space is difficult to pin down in abstract terms. He has argued that we need to base our interpretation on the interaction of several overlapping systems which affect the collective conscience, and hence the logic of planning (Wassenhoven 2018a). Real planning, as distinct from an abstraction, at least in the case of Greece, is located at the intersection of these systems which interact and feed into one another. The systems, as explained, include the social system, as a web of social practices, the political system, the administrative and professional system, the legal and judicial system, the land property system, and the economic system, particularly the creation of economic values and capital. The environmental system has not yet become part of this web of interactions in the mind of the citizen. Kourliouros too accepts that geographical space in capitalist societies is the result of condensation of multiple antitheses and conflicts, not simply the passive container of human interaction. He does not, however, endorse the view that there is an ecumenical rule of an optimal geographical distribution of activity, because one has to answer first the question "optimal for whom and with what criteria". The answer cannot be limited to a mechanistic approach to the spatial contradictions of capital and labour. Space is not simply a mirror image of the dominant relations of production (Kourliouros 2001, 46–47). In the present author's understanding, we have to wonder to what extent a theory based on the above contradictions can explain the practice of planning. Commenting on the effect of planning theory on practice, Lauria (2010, 157) has remarked that "empirical research on this question would need to specify the theories a priori (a deductive or retroductive approach) or infer them from the action of practicing planners (an inductive approach). It would need to discuss the context (including political/governance) of the theorizing and the practice. It would need to unravel the necessary and contingent relations involved and explicate the empirical evidence that led to the researcher's interpretation of the extent of influence, the fashion and mechanism by which planning theories affected practice, and the conditions under which the theories were apparent". Can the deductive approach of a theory based on capital accumulation "explicate the empirical evidence"? Can it uncover the "fashion and mechanism" by which it affects practice? Should one rather follow instead an inductive approach based on the practice of planning? These are all valid questions.

Although theoretical preoccupations about space were mainly the domain of geographers, mostly with an emphasis on thematic geography (Skordili 2007), in Greece as elsewhere, there have been tentative efforts by planners too, again academics, in spite of the initial distortions of the concept of "theory". No one has managed, in this author's estimation, to put forward a Greek version of planning theory. The interest was mostly limited to taking a cursory look at imported theories, as, e.g., in the case of Getimis and Kafkalas (2001), Panayotatos (1982), and Economou (1994). Economou refers specifically to cities and to the urban phenomenon, in relation to the influence to the capitalist mode of production. The latter changed radically the role of the territorially organized state, the relations between urban and rural areas, the style of urban life, and the importance of real estate ownership. Most theories, he asserts, are either focused on geographically and historically limited areas; but there are those which claim universal validity. Even then, they rely on conditions which are chronologically and spatially defined and on particular processes from which they derive their empirical material. The "unification" of a hitherto fragmented international space, Economou believes, does not increase the applicability of theories claiming universal value; on the contrary, it reduces it because the active forces are now of a different nature.

Developments in human geography have been complex and innovative and have had a great impact not only on the community interested in space, but on that which is charged with the task to change it, i.e., the regional and urban planners. Should the latter pay attention to the findings of the former? Certainly yes, not just out of deference to inter-disciplinarity, but also because the geographical community has everywhere, even in Greece, left its mark on the planning discipline and its object. One should not underestimate the fact that practicing planners are to a large extent restricted in their work by state institutions, while human geographers can continue undeterred to expound their theories as they choose, but the theory of space, sooner or later, brings to the forefront concepts, which gradually modify the institutions, and parameters which find their way into the methods that planners use. One of these parameters, a very significant one from this perspective, is the nature of the human being, whose activities planning is supposed to regulate. Who is that person? Is he/she a rational being interested exclusively in the maximization of utility, even though he/she often ignores how to secure and measure it? Are we talking about the *homo economicus* we inherited from nineteenth century economic utilitarian theories? Or, are we referring to a new type of person? A challenging approach is to imagine this person as an individual who adapts his/her thinking to the demands of the social group or collective organization to which he/she belongs, particularly the family unit, because it was there that his/her experiences were formed. If this is a promising explanation, we would accept that the individual's *collective* conscience is incorporated in his/her *individual* mode of conceptualizing and thinking and in his/her being as a human. The place where he/she was born, lives, works, maintains personal interests, or fantasizes becomes part of his/her individual identity and choices. The links of this individual with the power structures have been forged out of a nexus which is not the exclusive outcome of the capitalist mode of production, that, in the Greek case at any rate, has remained for most of the history of the Greek nation state a deformed system, a far cry from western capitalism. A friend, and distinguished socialist economist, once said jokingly about Greek capitalism: If we only had a genuine capitalist system, it would have been better!

We will return to these questions, but let us first point out that they have a clear bearing on the task of planning. Can planning escape from its institutional bondages and embrace a process in which men and women, as individuals or members of collective entities, have a priority? Can it, in so doing, take into account the never-ending fluidities and transformations of the capitalist system? Can we be as provocative as to assert that a system of repetitive compromises between authorities and people absorbs and subdues these transformations and, as a result, preserves a sort of stability, a kind of homeostasis in systems jargon? Can one reconcile this balancing act of compromise and maintenance of a historically formed equilibrium with a theoretical logos based on the role of capital accumulation? In this author's understanding, Hadjimichalis (2018) seeks this reconciliation in the context of a radical school of thought and of an emphasis on unequal development. Inequalities are due to capital accumulation and flows, division of labour, actions of social groups, and state policies. It is significant in his view that governments, especially in neoliberal conditions, blame wrong choices and policy failures for the perpetuation of inequalities in order to divert attention from structural deficiencies of capitalism. What he wishes to repudiate, in essence, is the deliberate attempt to separate government functions and policies from the real processes of capitalism, which is thus exonerated. The point is that government actions cannot be divorced from the operation of the capitalist market. As Lai and Lorne (2015, 45) put it, "state rules, which include, but are not limited to, property rights, can enlarge a market. State planning is broadly defined as the state making and enforcing the rules for transactions and investment in the resource market. As land is a key resource, town and country planning is therefore what is implicitly at stake". In the view of Hadjimichalis, the objects of planning (national territory, regions, cities) are open, not closed, systems, receptive of the influence of wider capitalist relations. A radical human geography focuses exactly on the causes of unequal spatial development and, as a result, explores the meaning of spatial justice. His argument is clearly influenced by the work of David Harvey, Doreen Massey, Edward Soja and Henri Lefebvre. The individuals and social groups, which stand to lose in this situation of inequality, do not suffer because of their own responsibility but because of the consequences of geographical inequalities under capitalism. Injustice finds its material expression in the living conditions of everyday life of individuals and groups, in particular localities. This approach, which no doubt reflects a great deal of the realities of capitalism in its pure form, nevertheless sweeps aside the role of individuals and groups themselves in their interaction with the state, which however remains highly interventionist in all planning matters, unlike the neoliberal state which exalts the dominance of the market. The networks of clientelism and patronage, which have been discussed repeatedly in this book, are left out of this explanatory framework. Bargaining and give-and-take arrangements between political actors and private interests, as in land development, are omitted from this picture. Important concepts like pragmatism, experience, incrementalism, accommodation, practicality, learning-by-doing, disruption avoidance, opportunism, contingency, negotiation, improvisation, complicity, and power of precedent, which were listed as key-words at the beginning of the chapter, are set aside as explanatory variables. I am reminded of a phrase by Barrie Needham (2017, 171): "[L]and-use planning (and by extension, any societal planning) should be modest, flexible and incremental". As in Chap. 4, we can quote again Jean Hillier (2008, 27): "I regard spatial practice as temporally engaged experimentation, improvisation or phronetic practical wisdom". In the analysis of Hadhimichalis, individuals as autonomous actors or as group members, as decision-taking politicians or plan-making planners, or as beneficiaries, take a back seat in the game between structure and agency, on which more will be said later. The discussion between human geography and planning discipline does not end here.

10.3 Homo Economicus and Homo Socialis

In the Greek literature on geographical space, the book by Leontidou (2011) is a good companion to help the author discuss the variables which enter his analysis.[1] We have extensively referred to the criticism levelled against official spatial planning for its rational obsessions and its inherent statism and we have presented the radical planning theoretical current (Chap. 4). The criticism against rational planning was paralleled by an advocacy of a more democratic, bottom-up approach, which brings to front-stage the human being, its groupings, and its place; a place which is much more than the sum-total of quantifiable characteristics. What the critics in fact attack is the assumption that there exists a human being capable of unlimited rationalism, the ghost of the "economic man" (homo economicus, or oeconomicus). This assumption is what underlies spatial analysis and planning and restricts the thinking of planners. "Homo economicus, or economic human, is the figurative human being characterized by the infinite ability to make rational

[1] This section, and the next one, of this chapter are a much-expanded version of a short sub-chapter in a book written by the author in Greek (Wassenhoven 2022).

decisions" The term was first proposed by John Stuart Mill in 1836.[2] "Mill … was from birth a member of a school of thought which held, first, that individuals were moved solely by a desire to obtain pleasure and to avoid pain; second, that pleasure and pain could be measured; and third, that a sovereign legislature could, by the use of penalties, encourage conduct leading to a maximum of pleasure and a minimum of pain" (Lancaster 1968, 112). Other versions of homo were later proposed to replace this abstraction.

Leontidou bases her arguments on a negation of the concept of homo economicus, which was the foundation of rationalism, modernism, positivism, and, ultimately, of quantitative geography. She supports the conception of the individual as a member of groups and collective entities, determined by place. Homo economicus, the ideological figure espoused by Fordism and Keynesianism, is a one-dimensional automaton of consumerism and is supposed to be endowed with the power of rational action and to possess all the available information. His only concern is to maximize the benefits derived from optimal location, rational choice of the place of his activity, and the minimization of the effects of the friction of distance. Positivism assumes a universal behaviour for all individuals, families, and households (Leontidou 2011, 93). As explained by Brennan (2010), the key characteristic of the economic man of neoclassical economics is the rational maximization of personal utility. This implies that all actors are constantly engaged in an effort to maximize their quantifiable benefit.

These perceptions were built into the inheritance of positivism and had enormous influence not only on pure social science, but also on the tradition of spatial planning, to the extent that positivism excluded anything that was not founded on experience and observation. Transcendental conceptions were left out of its realm. In the field of geography, Leontidou argues, positivism led to a fragmentation in watertight specializations, i.e., economic, social, historical, political, cultural, urban, or communications geographies. This imprisons the geographers in confined thought and research domains, each with its own theories, laws and languages. Each domain has its boundaries and scientific territorialities. It is of interest to note that in a classical book on human geography which, in translation, appeared in the Greek language, the thematic fields into which the material is arranged corresponds to these fields: geographies of population, agriculture, fisheries, industry, commerce, tourism, circulation, and cities (Derruau 1991). Indicative of the interaction between social geography and planning is that in these fields one can recognize the chapters of the stage of analysis of regional plans in Greece, as prescribed by the official specifications. There is a reason why this remark is relevant, because it shows the danger awaiting the planner who risks being trapped in a logic of fragmented interests in partial fields, deprived of an overall vision. His/her work may easily slip into a ritual of addressing a variety of issues, for the sake of multi-disciplinarity. Yet, genuine inter-disciplinarity still relies on his/her ability to synthesize and secure disciplinary consistency, a demanding job indeed! In this effort, the planner, still

[2] See James Chen, What is Homo Economicus? (Investopedia.com, 2019).

struggling as the direct or indirect arm of the state machinery, strives to overcome the difficulty by focusing on the final recipient of his spatial proposals. This is no other than the people who are ultimately affected by the plan, as individuals or members of groups. At this point, the planner finds himself at the interface between social science, especially human geography, and planning. The planner is back in the search for the representative *homo*.

The critique of the homo economicus, Leontidou (2011) argues, created a new dynamic in modern social geography, which proposed the replacement of the homo economicus with the *homo socialis* and developed a new perspective based on the familiar theme of structure and agency. In an article on the new type of homo, Gintis and Helbing (2013) contrast this conception, which they place in the context of a "social equilibrium model", with its predecessor (homo economicus) being the product of an "economic equilibrium model": "The general economic equilibrium model recognizes only one social institution: profit-maximizing firms. Families in this model are treated as 'black boxes,' as is government, if it is treated at all. For the general social equilibrium model, we must add, at a minimum, families, communities, as well as public institutions and private associations, such as governmental, religious, scientific, charitable, and cultural organizations".[3] In their highly sophisticated article, for which they borrow concepts and analytical tools from mathematics and physics, the "assumption of rational individual agents is supplemented by a group-level assumption that the agents communicate and coordinate with each other across social networks".[4] And elsewhere, Gintis and Helbing, who go back to the work of classical sociologists of the functionalist tradition, stress that "society is held together by moral values that are transmitted from generation to generation by the process of socialization. These values are instantiated through the internalization of norms". In contrast to the economic man, Brennan (2010) agrees, homo socialis originates in functionalist sociology. This new approach attempts to overcome ulterior utilitarian motives by laying emphasis on the behaviour and actions of actors who are guided by a wider set of values, preferences, social obligations and structures. Social practices are thus seen through the lens of structures and not merely human agency. This distinction can be crucial for spatial planning. For the planner, whose task is to transform geographical space, even to "invent" its future, it is easier to have as interlocutors those who claim to represent structures, than the isolated individual, in order to predict the acceptance of his planning proposals. Representing structures can be a very dubious assertion, but in practice it is convenient to open a dialogue with such representatives, on the assumption that they express the opinion of a cohesive group or local society of individuals behaving as "social" men and women. In a sense, the planner would rest on the risky assumption that this *individuum* is a different version of homo, a homo socialis.

[3] See also Homo Socialis: An Analytical Core for Sociological Theory – [PPTX Powerpoint] (vdocuments.net).

[4] See pcl.sitehost.iu.edu/papers/homosocialissapiens.pdf.

This syllogism on the type of homo socialis is founded on a study of individual behaviour which intends to move beyond the simplistic assumption of the possibility of recording and measuring individual actions, even of predicting them accurately. Behavioural geography, writes Leontidou, is in a way a critique of quantitative geography, to the extent that it rejects its atomistic and positivistic logic and places homo socialis at the centre of its analysis. Behavioural geography may still quantify individual behaviour, but asserts that human rationalism is explained in a realm of ideas and not observable facts. Perception and values become key parameters. "Behavioral geography is an approach to human geography that attempts to understand human activity in space, place, and environment by studying it at the disaggregate level of analysis – at the level of the individual person. Behavioral geographers analyze data on the behavior of individual people, recognizing that individuals vary from each other. A key tenet of behavioral geography holds that models of human activity and interaction can be improved by incorporating more realistic assumptions about human behavior".[5]

According to Gold (2019), behavioural geography has two contingent meanings. "The first views it *sensu lato* as a primarily historical movement with multidisciplinary leanings that enjoyed its greatest influence between the years 1965 and 1980. Roughly coterminous with what was also known as 'environmental perception', 'behavioral and perceptual geography', 'behavioral and cognitive geography' or 'image geography', behavioral geography emphasized the role of cognitive processes in shaping decision-making and behavior, for which reason its underlying approach was known as 'cognitive behavioralism'". Terms like landscape image, topophilia, psychology of space, environmental perception, cognitive mapping, and mental maps are frequently encountered. "The second and more contemporary meaning of behavioral geography defines it *sensu stricto* as a subdiscipline of human geography", which, in Gold's view, has an "increasingly marginal presence in geographical research". Still, it embraces important themes like perception of natural hazards, place and placeness, meaning of landscapes, niches in cities, or gated communities.

In her own analysis, Leontidou argues that human perception and values are important variables, but homo socialis remains an individual and is not a social actor with a historical past, a member of a social class, and a claimant of rights, who intervenes in geographical space in a variety of ways, resists cultural cliches, and adapts to, but also affects, ecological processes. In her opinion, behavioural geography remains an outsider with respect to these considerations, but is also opposed to positivism. It views society as an aggregate of individuals, but these individuals also hold cultural values and possess a social substance, and not just economic behaviour, as in the case of homo economicus. According to radical critics, Gold adds, "cognitive-behavioralism quarried dangerous and dehumanising theories (e.g., ideas about territoriality derived from ethology), led to psychologism (the fallacy by

[5] D.R. Montello (2021), Behavioral geography. Available on the web (Oxford bibliographies, https://doi.org/10.1093/OBO/9780199874002-0069).

which social phenomena are explained purely in terms of facts and doctrines about the mental characteristics of individuals), and obscured the objective economic and social conditions that operate independently of the individual. Other radical critics, who had equally shown initial support for cognitive-behavioralism, denounced its practices on similar lines. Notably, behavioral geography became branded as 'bourgeois thought' that had the ideological function of supporting the *status quo* in that, inadvertently or otherwise, its emphasis on the individual made it unlikely that the data collected would throw light on the true nature of social relations".

10.4 Critical Realism, Places and Fables

Our homo has so far preserved an individualized behaviour, but not exclusively economic, although he/she has not yet reached the status of a complex social subject intervening in geographical space in alternative ways, contrary to established cultural standards. He/she is not yet endowed with a radical attitude or autonomous creativity, that would turn him/her into a *homo creator*, about whom we shall have more to say. Homo socialis is of course no longer the narrowly defined utilitarian homo economicus. Some observers stress this homo's social propensity which sometimes finds expression in feelings of reciprocity urging him to place his/her own interest below the common, reciprocal one. In contrast to established opinion, Falk (2003) insists, reciprocity and sense of justice are standard motives in human behaviour. This is what characterizes *homo reciprocans*, who in a way repays his/her debt to fellow humans and society, in a mood of reciprocity. Cooperation within wider social groupings is a stable human quality, claim Bowles and Gintis (2002), which manifests itself in acts of philanthropy, participation in political movements and obedience to social rules. This quality can turn into retaliation, when certain individuals break the rules and damage the group. This is why Bowles and Gintis use the term homo reciprocans to describe also the person who retaliates, either because of his/her devotion to the group or clan, or in order to restrain the behaviour of those who harm the group, even to seek their punishment. When the social group is the family, nuclear or extended, the behaviour of a family member, especially of the family head, incorporates an element of mutual support, which is compatible with the logic of the economic man. The alleged absolute rationality of the homo economicus does not restrict the range of his/her preferences. Only naïve interpretations take it as given that he/she has full knowledge of the behaviour which maximizes personal benefit. It is perfectly possible that he/she extends this benefit to other persons, his/her kith and kin, which makes the concept of homo economicus compatible with that of homo reciprocans, that includes human cooperation.

We can claim that the profile of homo reciprocans must be associated with his/her care for the "common place", place of living or place of origin. This care can take the form of conservation of a place of birth or of ancestral home. But in practical everyday life, it usually takes the form of protection of good neighbourhood and of conditions which guarantee good quality of life, to the exclusion of activities that are conceived as alien, detrimental or debasing the value of property. Human geography, in the view of the present author, must embrace notions familiar to urban economists, such as external and neighbourhood economies, particularly negative externalities. It is at this point of encounter, that the personal and the collective frequently collide, a crucial factor in spatial planning and its legal framework. This observation harks back, and in a violent way, to the logic of homo economicus, because we refer here to a factor which influences the behaviour of a person in response to that of the "other" – individual or public agency – with regard to the use of land which, in turn, affects life quality and real estate values. A good deal of urban and regional planning practice is centred on this point. We cannot ignore this down-to-earth reality in a discussion of the values of place, which an ideal homo supposedly shares. This point should not escape the attention of humanistic geography, which, as Leontidou explains, differs from other branches of human geography, because of its emphasis on the social subject, on social conscience, and human creativity. It followed the steps of behavioural geography, but soon recognized the subjective judgement both of the observed homo socialis and of the observant geographer-researcher. Thus, in her view, humanistic geography places a new emphasis on the relation of people and place, on *topophilia*, and the sense of place. Topophilia, a Greek word literally meaning "friendship with place" or "love of place", is defined as "the love of or emotional connections with place or physical environment".[6] Relph (2015) follows the analysis of Yi-fu Tuan and accepts that topophilia "varies greatly in emotional range and intensity, including fleeting visual pleasure, the sensual delight of physical contact, the fondness for familiar places such as home, and joy because of health and vitality … But for all that, it strikes a chord. It is a familiar sentiment, a word that encapsulates the pleasantly varied relationships we have with particular bits of the world both as individuals and as participants in cultures with long histories". Relph approached the concepts of place and placeness from the viewpoint of phenomenology. "He argues that space is not a void or an isometric plane or a kind of container that holds places. Instead, he contends that, to study the relationship of space to a more experientially-based understanding of place, space too must be explored in terms of how people experience it" (Seamon and Sowers 2008).

The notion of place, to which we referred already in the context of the so-called cultural turn (see Chap. 3), is not just an idealistic image. It is the product of experience which can be expressed in individual accounts, i.e., personal narratives, as Leontidou and many others, would prefer to call them. This is where the individual inserts socially constructed forms of "knowledge", in other words alternative stories of reality, and acts as a tale-teller or fable-narrator. He is no other than *homo fabulans*, proposed by Brennan (2010), a sociological fabrication of the actor, whose "rationality" is re-constructed in retrospect, through an ex-post narration. This homo, one can add, refers to places which make up his own geography. As Derruau (1991) wrote in his introduction, everyone can have "his" geography, provided he

[6] Collins English Dictionary.

does not believe that it coincides with "the geography". The interest in local narratives is closely related to the analysis of places and landscapes. It acquires special impetus with recent efforts to read the landscapes of capitalism. This new direction, according to Leontidou, enters a new phase in the context of post-modernism and of the influence of the cultural turn. It affects cultural geography, especially the geography of human activity, material and technical culture, and institutions, in sharp contrast with abstract concepts of inherited cultural traits. The cultural turn shifts the emphasis of the analysis of landscapes, from an insistence on the production of space to their representation. In Greece, the work of Hadjimichalis and his collaborators on "Greekscapes" has been a notable contribution (Hadjimichalis 2010).

The physiognomy of places and their pervading spirit, or *genius loci*, so cherished in landscape architecture and architectural neo-rationalism, have become a distinct field of research. The terms genius loci and sense of place are used interchangeably, although, Parker and Doak (2012, 162) point out, "*genius loci* has tended towards a root in narrower, more objective elements of place". Places, under the cultural turn, possess soul and spirit. According to Leontidou, a landscape's unique physical and cultural features, historical connotations, atmosphere, symbolisms and psychology endow each place with an exclusive profile. We are reminded of the notion of territorial capital, to be distinguished from the economic concept of social capital, which includes locally identified common goods of a collective character, as a rule immobile, intricately connected with a particular place, as well as products of tradition and a definite historical past. Apart from fixed and natural elements, which are more easily identifiable, territorial capital encompasses social, cultural and institutional goods, in addition to immaterial traditions, which in a way bind together all the other elements in a distinct unitary whole (Wassenhoven 2010, 58). Leontidou adds, in a poetic mood, that people who spend their lives in a city, island, plain, promontory, or mountain range receive the temporal influence of the place, but act back and seal its uniqueness. This interaction, she claims, symbolized by place-spirits, deities and idols that haunt places, is nothing but geography itself as a mode of worldly reflexivity. She reminds us of the concept of "worldling", used by Roy (2017), albeit on a different scale (see Chap. 5).

It is true of course that a place embodies symbolisms and memories well beyond its material existence. A comment, written in a literary vein, speaks of the unshakeable bond of humans with the environment of their birth and the mythological landscapes of their past. "Place governs as a super-ego".[7] The planner is here faced with the trap of what is empirically testable, when he/she has to plan the future of a place. The challenge he/she has to come to terms with is the essence of the school of critical realism. "Critical Realism (CR) is a branch of philosophy that distinguishes between the 'real' world and the 'observable' world. The 'real' cannot be observed and exists independent from human perceptions, theories, and constructions. The world as we know and understand it is constructed from our perspectives and

[7] G. Veis, The imperishable bond between place and man. Newspaper Kathimerini, 8 September 2020.

experiences, through what is ‘observable’. Thus, according to critical realists, unobservable structures cause observable events and the social world can be understood only if people understand the structures that generate events”.[8] What are “observable facts and events”, which can be quantified and mapped, and what are “unobservable structures” that “generate events”? And how can these structures be taken into account in the planner’s calculus? To what extent and with what tools can he/she explain the behaviour of actors and their reaction to planning proposals? The planner is thus up against the theoretical challenge of critical theory, the domain of which “is inquiry into the normative dimension of social activity, in particular how actors employ their practical knowledge and normative attitudes from complex perspectives in various sorts of contexts. It also must consider social facts as problematic situations from the point of view of variously situated agents”.[9]

The problem for planners is that they are asked first to assess social reality, then outline the future, and anchor it in norms and procedures of the state on behalf of which they are acting. If their assessment cannot be reconciled with practices which dominate a particular socio-political culture, then the likelihood is that alternative modes of solving problems will assume preponderance and “official” planning will remain an illusion. Therefore, the more their work remains speculative and in the realm of multiple and vague interpretations, the more it will prove redundant. Yet, as stated by another Greek academic social geographer, one of the key assumptions of critical realism is that certain facets and dimensions of social reality remain independent of conventional interpretation, perception and theoretical explanation. The nature and character of social reality leads to an understanding of causality radically different from that of positivism. Critical realism bases the interpretation of social reality on the interaction between structure and agency, but also accepts the discreet existence of both material and ideal structures (Iosefidis 2019). We are entitled to wonder what the residues are of positivist and quantitative methods in this radical framework. The answer for the same author is a compromise which aims at justifying the continued use of quantitative techniques in geographical analysis, and, we could add, planning work. Quantitative methods, Iosefidis argues, are powerful descriptive tools, provided they are accompanied by a theoretical framework, to help locate “demi-regularities” at the empirical level, which presage the existence of deeper causal mechanisms. In his opinion, there are plenty of such innovative tools which assist us in discovering regularities, e.g., geographical information systems, satellite mapping, and thematic cartography. Critical realism, however, maintains that even the most advanced methods of analysis do not suffice to offer explanations of social reality which demand the search of causal relations beyond the observable and empirical. To overcome this impasse, Iosefidis recommends the use of intensive, qualitative research and exploratory, representative case studies. This seems a brave attempt to rescue an otherwise defunct positivist, quantitative geography, and,

[8] *See* https://warwick.ac.uk/.../maps/criticalrealism

[9] See Stanford Encyclopedia of Philosophy.

perhaps unintentionally, spatial planning which would find it cumbersome, not say impossible, to turn every plan production task into a fully-fledged research project.

Fully conscious of the difficult path which critical realism would like us to follow, Leontidou explains that critical realists are not merely anti-positivists but reach beyond transitional scientific paradigms which introduce hypothetical, abstract individuals, like homo economicus or homo socialis. For a critical realist, what we see, hear, or perceive through our senses is just a shadow, a secondary effect, the epiphenomenon of true reality. The real world is beyond the perception of our senses. Science is the in-depth quest for truth, the contemplation through our inner soul, which cannot be replaced by observation and measurement, in accordance with the recipes of positivism. Our senses cannot capture all the intricacies of this reality, the structures, the mechanisms, the eventualities. The meaning of truth is not conceived in identical ways in all epistemological approaches. Truth, on some occasions may be identical to what we observe, but on others it can be a social construction, something we mould and nurture in our tales and narratives, in alternative ways. Leontidou here reminds us of the Japanese movie classic *Rashomon*, of the director Akira Kurosawa, where four individuals present alternative “true” versions of the same event. The reliance on alternative narratives and storytelling has been discussed in Chap. 3, in the communicative theory section. We have to concede, in all honesty, that even in modest planning practice, the planner is frequently faced with alternative accounts of what a “problem” consists of, what are its causes, and what the remedies should be to cure it. The awkward problem is that the individual affected by the problem, as he/she perceives it, has other routes in mind for its solution which bypass the avenues of statutory planning. If this is what critical realists have in mind, the author is unable to tell.

It is not the author’s intention to study in depth the content of human geography or to put forward a critique of its theories. At the same time, we should not forget the close links between human geography and spatial planning and the effort to underline the support that the former can offer to the latter. The author is here trying to distinguish the variables that form the spine of a theoretical explanation of spatial planning praxis in the conditions of a particular country. A position that can be shared with geographers, like Leontidou, and planners alike is that a theory rooted in these conditions cannot be a mere replica of imported western theories. Nevertheless, the study of human geography places one in an impasse, when one imagines the planner who is asked to supplement his/her compulsory quantitative analyses with qualitative and intensive research methods. This is where the problem lies. Critical realism, as we saw earlier, denies the irrefutable validity of what is empirically observable and measurable. For the planner, this approach may appear doctrinaire because it can undermine the foundations of his/her own discipline. The planner’s job is to find evidence on which to base his/her proposals, not to propose theories. Let us remember, in passing, the critique against “evidence-based planning” (Chap. 3). The planner must be able to answer objections with well-founded arguments, not with vague speculations about undeterminable behaviour. He/she is up against laws, administrative courts, state services, parliaments, local councils, judgements regarding unconstitutionality, citizen appeals, and media attacks. In defending the planning proposals, the planner cannot fall back on the claim that what we observe with our senses is nothing but a shadow, the epiphenomenon of truth, when, e.g., the observed reality is the crude violation of land property rights and of legally enforced bylaws, and the destruction of the natural and cultural environment for the sake of private benefits. If the planner elevates his/her personal ideological assessment of the epiphenomenon of truth to the status of a convincing interpretation of reality, he/she risks to be accused of lack of objectivity and impartiality. Metaphysical explanations in lieu of observed reality cannot assist us in the effort to develop a theory of spatial planning practice.

The rejection of abstract interpretation does not of course imply a negation of the anxiety over a form of planning totally dependent on ironclad structures, in which the individual is but a pawn, nor does it entail an indifference for the fluidity of social change. The title of a recent book in Greek ("geographies in an era of fluidity") is highly topical (Afouxenidis et al. 2019). It does not come as a surprise that the theoretical viewpoint of critical realism does not accept the "evidence-based" reality of planning practice (see Chap. 3), something that Leontidou reiterates repeatedly, while stressing that critical realism does not necessarily reject all evidence. It does, nevertheless, insist that simple observation is not sufficient if detached from social structures and that one has to understand the mechanisms which produce observable facts. Consequently, the search for factual evidence, from a different epistemological standpoint, evolves into a more demanding process of successive rounds of intensive and extensive inquiry, to single out the mechanisms and then, and only then, check them in the field with empirical evidence. For instance, the violation of land property rights is undoubtedly an observable fact, but the mechanism that produces it escapes observation because it is caused by class inequalities undetected at first glance. Leontidou reminds us of the cave allegory in Plato's Republic, in which the shadows seen by prisoners on the walls of the cave are misinterpreted as reality. These are concerns by no means limited to the theory of human geography. They are of equal interest for planning theorists and will remain among the author's preoccupations to the end of this book.

10.5 More *Homines*, Structure and Human Agents

The theoretical currents striving to explain space and the role of the individual actor have not been exhausted yet. One of the important themes, which was discussed in Chap. 3, is regulation theory and its relationship with the concepts of structure and agency. In regulation theory we witness the dominant position of a neoliberal vision. We can add here that regulation often causes de-regulation, a phenomenon which became obvious in Greece at the beginning of the economic crisis, the legislative acts of 2011–2012 for strategic investments, and the appearance of the novelty of big projects like the redevelopment of Hellinikon airport in Athens. These policies were not only based on a private business logic akin to neoliberal regulation theory, but also brought about the deregulation of official planning, by installing a parallel planning system (Wassenhoven 2018b, section 5.3). This trend continues in 2021 with new legislation facilitating strategic investments. Yet, there are elements which differentiate regulation theory from hard-line structuralism, a point which does not escape the analysis by Leontidou (2011, 198–199). In her view, there has always been a disagreement between humanist and radical – or generally Marxist – geography, about the primacy of human agency or structure (see Chap. 3) in spatial articulation. In different words, this is about the relative importance, on one hand, of grassroots actions and the subjectivity of social movements, and, on the other, restrictive socio-economic formations, state superstructures, top-down directives by the state and the economy, which all determine spatial structure, material culture, and modes of life.

Mainstream Marxist structuralists speak of iron laws and inescapable evolution, which assume the existence of a *homo sociologicus*, a person, according to Dahrendorf, unable to shape his/her cultural environment and forced to adapt to external social conditons. "In his study from 1958 Ralf Dahrendorf claimed that there is a simplified notion of the human individual in the foundations of sociological thinking, in which he/she is considered as unilaterally determined by social forces. Dahrendorf called this idea 'homo sociologicus', and he linked it with the concept of social roles in the context of his period" (Subrt 2015). We find in a Wikipedia lemma that "homo sociologicus is largely a *tabula rasa* upon which societies and cultures write values and goals; unlike *economicus*, *sociologicus* acts not to pursue selfish interests but to fulfill social roles (though the fulfillment of social roles may have a selfish rationale – e.g., politicians or socialites). This 'individual' may appear to be all society and no individual" (Hirsch et al. 1990). Subrt refers to the alternative figure of homo duplex and wonders whether it is "possible to overcome the reductionism about which Dahrendorf spoke with a conception of the human individual that respects the dual nature (duplex) of the human self", and whether we can find a balance between the "I" and "We" identities. In contrast to the straight jacket of the homo sociologicus, the action of human agency, Leontidou emphasizes, embraces meanings, intentions, goals, and choices, is historically produced, and leaves room for human creativity. A variety of diverse spatial outcomes result from this action: Rural reform, migration, planning, land squatting, or decentralization of private activity. All currents of post-positivist geography have dwelled on the structure-agency duality, the local-global interactions, and the mutual influence of space and nature with society. Political-economic approaches and regulation theory tend to place the emphasis on wider socio-economic structures, especially governance, and on economic spatial restructuring, while cultural ones give priority to human action and creativity.

Although Leontidou does not mention yet one more type of homo, she is indirectly leading us to him. The creative man (*homo creator*), Giordano (2005) points out, is altogether different from homo economicus. The concept of homo creator originates in the conviction that humans do not obey blindly rules of natural law or inescapable social, cultural and psychological orders. Homo creator is not the puppet of socio-economic determinism. Like theatre actors, he/she is free to improvise and innovate. If this was not the case, creativity and cultural change would be devoid of meaning. Nevertheless, Giordano claims, there is still a dominant theory that individuals are embedded in a social body and a cultural system. The "embeddedness" view implies that voluntarism and radical individualism are illusions, utopias, or decorative elements serving political interests. The human being is not, therefore, free to act according to his/her free will, because he/she is closed in a social and cultural cage, which permits only a limited freedom of movement. The truth is, Giordano contends, that members of society construct the present and invent the future, by activating inherited knowledge; they do not simply reproduce it as some people wrongly believe. Homo creator energizes the past, reexamines it, revises it, rejects it, or adapts it to the needs of the present, so as to be ready to plan the future. An inevitable question is whether the actions of the creative man are severely restricted in circumstances which are beyond his/her control, as in conditions of an economic crisis. Hadhimichalis (2018) refers to the indebted homo, a term which recently appeared in the international literature (Lazzarato 2012). He/she is the *homo debtor*, although *homo debitor* would be a more correct term. This person is free to the extent that his/her way of life is compatible with the repayment of his/her debt. His/her freedom and creativity are accordingly restricted.

A valid question with regard to the creative man is whether the dominant structures, which in modern society, are mainly expressed and legitimized by the state, militate against his/her action, by imposing norms and restrictions. Government policy and planning, which the governing classes constantly praise, impose rules, e.g., on land use and on permitted activities. The social legitimation of restrictions, so as to make them more acceptable to society, needs another type of homo who acts as an organizer and mediator. For Brennan (2010), this is the political man or *homo publicus*, who is capable of putting together a narrative, a sort of top-down, hegemonic story of what is at stake, what is expected to happen, and what we should plan for the future. In order to write a story of the future, the power that be uses a dominant discourse, using the language of institutions and laws. The reader of these lines, as this author adapted them, surely finds analogies with planning practice and distinguishes the affinity of planner and homo publicus. Without risking to be accused of bad faith, we must confess that the organizer and mediator can act rationally, but at the same time harbours ideological biases, in which case his/her story ceases to be objective and ends up by distorting reality. He/she may use a vocabulary which keeps the real decision-making processes in the dark, which are ultimately purely political. In this way, he/she conveys a false image of rationality, to prevent or minimize the negative reactions of forces opposing the policies of the organizational edifice which the mediator represents. We have here an inverse dilemma for the planner; this time he/she is in danger not to follow his/her own abstract story of what is really true, but rather to embrace the official narrative of the structure which is enfolding him/her. To escape this dilemma, of which he/she is fully conscious, he/she may fall back on a bureaucratic narrow-minded, but potentially harmless, mentality.

The alleged integration or embeddedness of the individual in a social structure or cultural system was mentioned above. Leontidou relates embeddedness with networking and social capital, which is a bundle of common values which citizens follow in their daily exchanges. She is careful to keep a reservation regarding the negative aspects of clientelist relations and corruption. In his paper on embeddedness, "Granovetter (1985) attempts to find a more appropriate middle ground between economic theory that under-socializes behaviour, and much of the existing sociological theory that over-socializes behaviour. Granovetter believes that it's more accurate to view economic rationality as 'embedded' within social relationships. He sees both extremes in this debate as 'atomizing' the individual as blindly obedient either to 'perfect knowledge' decision making or social norms. His view is similar to Giddens's views on structuration in that one may gain better understanding by acknowledging that both extreme views are important and must be considered simultaneously".[10] For Leontidou, the existence of networks offers the opportunity to reassess neoclassical approaches and to reject the lonely figure of homo economicus, whose selfish behaviour is considered by the utilitarian tradition as unaffected by social relations. The subversive element that networking and embeddedness introduce is an explanation which focuses not on individual characteristics, but on the quality of inter-person ties. The actor is at the centre of communal microcosms, roles and values which organize social realities. Actors operate not as independent units or in accordance with a social category scenario. Their actions are embedded in specific but constantly evolving systems of social values, confidence or mistrust. This analysis is at the root of the cultural turn which proposes cultural and customary elements which are instilled in economic activity. There is here an undoubted touch of idealism, but this hermeneutic approach carries with it tendencies which we discover in communicative theory in planning, of which a fundamental starting point is that knowledge is socially constructed (Wassenhoven 2018b).

Who is after all the human agent, for whom human geography proposes theories and spatial planning policies for the future? Is this agent a mechanistic rational being, that cares only for personal pleasure and benefit, which he/she can forecast and quantify? Is this agent a homo socially defined, a reflection of the social groups where he/she belongs, like family, trade union, political party, compatriots and fellow villagers, ethnic or migrant group, protest movement, cultural club, or all these put together? What is this agent's attachment with the places of residence, vacation, business interests, or ownership of land and real estate? What are the values and preferences which are derived from this attachment? How can a geographer or planner diagnose, analyse, picture, measure, and in general take into account the complex characteristics of this place of attachment, if charged with the task of deciding its future?

[10] Granovetter, Embeddedness Theory (babson.edu).

10.6 The Person Next Door: Homo Individualis

The human agent is certainly not the fully rational homo economicus. But, even when the human agent's actions and desires are to some extent dictated by the sense of belonging to a social group, they are distilled and atomized in an exclusive, personal behaviour. To justify his/her actions the agent is prepared to enter into a conflict, to be a *homo conflictans*, if this author can be excused for suggesting one more Latin term. He/she will do so because of the internalized moral priorities imposed by the feeling of "belonging", but also because he/she estimates that this serves personal interests. When he/she revolts against a neighbouring land use or social group, which are considered as devaluing his/her property and reducing its market value, it is not only because this property is located in the place where the individual lives or was born. He/she adopts this attitude because of personal or family interests. A similar remark can be made for the individual who builds without permission or occupies public land or land which belongs to a neighbour. When he/she opposes an otherwise essential public infrastructure, the fears and motives are similar. When he/she bribes an official to get a building permit, this same individual may have in mind the family interest and the dowry of a daughter.

We therefore have a homo located between the extreme versions we presented earlier, with the assistance of human geographers. This homo is neither a pure homo economicus, nor an idealistic creature who incarnates deities and idols of each person's fatherland or imagined stories of the past. This is a next-door person, an *individuum* like all others, a model that nobody will publicly admit as his/her simulacrum, but who represents a vast majority of one's fellow-citizens. He/she can easily slide into irrationality, but functions for what are deemed to be his/her interest, while translating and filtering all the traits of his/her group, local society or collectivity, so as to adjust and reconcile them with personal choices and objectives. To create a new neologism, this homo is a *homo individualis*, a term the present author believes suits the Greek realities. The author does not apologize for adding one more category of homo, since recently the term *homo pandemicus* has also been coined! Having for so long discussed the individual, the homo with all his/her epithets, the author feels he may be blamed for ignoring the broader forces that determine the place this individual occupies in society and the socio-spatial inequalities to which he/she is subordinated, so as to obey unconsciously impositions, which cannot be overthrown. The analyses of social geographers which were quoted extensively are perfectly suitable to describe this dependence of the individual, even though they are prone to ascribe all responsibility to capitalism in general. However, they rightly connect social processes and inequalities with spatial policies.

Individualism is not of course a new concept. According to one definition, it is "the moral stance, political philosophy, ideology, and social outlook that emphasizes the moral worth of the individual. Individualists promote the exercise of one's goals and desires and so value independence and self-reliance and advocate that, interests of the individual should achieve precedence over the state or a social group while opposing external interference upon one's own interests by society or institutions such as the government".[11] It is derived from a focus on the individual both cultural and economic. There are views claiming that, in western society, the individual has become a cult. "The capitalist system of social and economic organization is based on the assumption that both human beings and companies are individual, rational, and independent actors that base their decisions on certain kinds of information like prices, wages, and profitability, *but not on what they see and hear of other people doing*. This mistaken assumption ignores the fact that we human beings are imitative creatures *par excellence*, and that the desire to imitate what we see other people doing is a strong and innate part of our nature. Economists have caused a great deal of harm and confusion by seeking to impose on all people a theoretical model of behaviour that ignores our strong tendency to imitate and conform to the behaviour of those who are in our realm of influence".[12] This author's view is that the behaviour of homo individualis, in Greece at any rate, is not simply a matter of imitation, but rather one of accumulated experience which has taught him/her what is, in his/her view, the optimal approach in his/her dealings with the authorities, always with the ultimate aim to serve his/her interests as defined earlier. The interests the author dwells on relate to his/her use of land and the restrictions imposed on him/her by planning authorities. However, there is the legitimate question whether this homo and his/her interests remain stable for long periods or change as the economic and social system in which he/she is embedded is also undergoing change. Is the homo of the 1920s or 1960s the same person as that of the 1990s, years of economic optimism, or the 2010s, period of a sharply declining economy? Are his/her attitudes the same? This is, however, a dilemma facing the planning theorist in any national context or period. Can one develop a diachronic theory? The best answer one can give, is that the theorist has to rely on enduring characteristics which override the fluctuations of the economic environment. The behaviour of the individual will change but there will be a substratum of attitudes which remain relatively stable, precisely because they have a long past. It is valid, in the author's opinion, to claim that, in the case we are examining, radical changes have taken place, while the attitudes of our homo individualis have remained relatively stable and proved their resilience. Formulating a planning theory on the basis of this assumption remains the author's purpose.

ΕΦΤΑΣΑΕΔΩ

[11] Individualism/Wikipedia.

[12] The Myth of Individualism – Homo Stupidus.

10.7 Back to the Task of Building a Planning Theory: The Greek Planning Model

The author intends to remain faithful to his hypothesis that a theory exists which explains the planning model, he has so extensively examined so far. Economou (2002) considers that the factors included in the socio-economic framework of Greek spatial policy, which determined this model are: (a) the fragmentation and spatial dispersal of landed property of all social strata; (b) the clientelist structure of the political system; (c) the individual- and family-centred social behaviour, the mistrust of the state, and the social toleration of law-breaking; (d) the transition of the country from a less to a more advanced stage of development and from a rural to a service economy; (e) the pressing needs of a spatial character, which had to be addressed, e.g., infrastructures, and, (f) the international trends of planning practice and theory, as they evolved over time and influenced Greece. Always according to Economou, there were also other sub-systems with a critical influence. One of them was the government support of small land ownership, through planning policy, for instance advantageous planning regulations or building coefficients and other instruments which favoured the appropriation of land rent. There are of course others who contend that the issue of small-scale property has been over-stressed and used as an alibi to hide the failures of planning (Mantouvalou and Balla 2004). A second sub-system, in Economou's view, is the inferior status of spatial policy, in comparison with macro-economic, development, and fiscal policies, and, simultaneously, the absence of policies of social welfare, of which the protection of small property functioned as a substitute. The third sub-system was land trespassing and unlawful occupation and the widespread illegality in land use and building. Finally, a fourth sub-system was the weakness of urban and regional planning. There has been a pronounced gap between real practice and official policy, as expounded in policy documents and planning legislation. Typical symptom of this state of affairs was the excessive number of diverse planning instruments, the large number of plans which remain without final approval, and the limited implementation of those that reach the final approval stage. With rare exceptions, this model of spatial development and policy had negative consequences, often irreversible. Economou followed the steps of this model, from the interwar years to the 1990s, to conclude that all its structural characteristics remained unaffected, in spite of the addition of elements which produce a hybrid system of old and new components. All this constitutes an accurate analysis which, however was written in the early years of the twenty-first century. Almost 20 years later, and 10 years after the beginning of the economic crisis, its author found himself in a government position, as deputy minister of urban and regional planning, and drafted Law 4759/2020. The debate this law aroused brought to the surface all the conflicts which underlie spatial development. This debate was outlined in Chap. 8, but will be discussed again in this chapter.

The political and economic system of Greece, in which the planning model described by Economou was developed, is far from clear. We have here a paradox.

There are those who considered it as undiluted statism of the Soviet variety, while others picture it as a laissez faire system, especially in the land market. The truth is that it was never genuinely capitalist and liberal. Whether it was "deformed" or "shapeless" capitalism (Vergopoulos and Amin 1975), "crony capitalism" (Close 2002), a type of "development in a state of underdevelopment" (Filias 1980), or a crooked, sui generis system (Dertilis 2016), is an open question. The dominant role of the state and its immediate impact on all activities were a constant. This warped function, within a supposed capitalism, is given flesh and bones in the process of clientelism and in critical cases of bargaining between power and citizen, a person sketched earlier as homo individualis. These critical cases surely include land, its use, and the associated exploitation rights. The modernization of the state in general, and particularly in spatial planning, gradually diminishes the margins of bargaining, but traditional relations survive stubbornly. Modernization encourages a hesitant devolution of powers and a timid recognition of the importance of the "local". As Andrikopoulou (1993) wrote long ago the combination of central and local rule is a sensitive issue for the central state. Economic reorganization on a global scale leads to new state roles and a reduction of expenditure on welfare, which, in turn, places obstacles in the road of decentralization (see also Kafkalas and Komnenos 1993). It should be added that strong re-centralization tendencies appear, and continue to do so in conditions of crisis. There is no doubt that, in the Greek conditions, spatial planning is a kind of social policy, the operation of a sui generis welfare state, provided we accept that social policy supports the reproduction of society at large and acts as a stabilizing factor (Gravaris 1993). One can then legitimately classify spatial planning and regional development policy along with health, educational, and environmental policy (Getimis 1993). The argument can be expanded to similarities with low-income policies or those for the alleviation of poverty. The catch is that when spatial policies are interpreted in this vein there is always the danger that they will take the form of cynically clientelist concessions, like the legalization of unauthorized construction, urban plan extensions, or encouragement of out-of-plan building, which benefit all classes without exception, but certainly the propertied ones. Social and spatial policies are supposed to serve the public interest, but at the end of the day social policy, via spatial planning, takes the form of a buyoff process, an arrangement of exchange of privileges and votes. The end-recipient is homo individualis, as defined in this chapter, a complex individual with narrow interests, but also member of a social group.

10.8 The Public Interest as a Planning Compass

The earlier outline of the concentration of decision making, the primacy of the central state, the relegation to an inferior place of local government, and the theoretical model of Greek planning, takes us to the relation of state and civil society. The key here is the concept of public interest, as mentioned above and in Chap. 3, where the arguments were presented of Campbell and Fainstein (1996), Salet (2019),

Alexander (2002), Campbell and Marshall (2002), and very briefly Moroni (2018), all expressing doubts for the existence of a generalizable public interest. The legal reality of planning in the Greek institutional environment is diametrically different. It is worth mentioning here a report of the scientific, legal expert committee of the Greek Parliament, when a bill on urban and regional planning, later Law 4759/2020, was under discussion. The committee invokes a series of decisions of the Council of State, regarding the positive actions of the state administrations for the safeguarding of the natural, cultural, and urban environment. Such actions include legislative, administrative, precautionary, or suppressive measures, and all necessary interventions in individual or collective activity. In taking these actions, the expert committee concluded, the legislature and the executive must weigh all factors of *national and public interest*, but in doing so and in assessing legally protected goods the state must always care for environmental protection, consistent with sustainable development.

We should hesitate to enter into a legal argument, but with outside help (Prevedourou 2020), we can deduce that the public interest is, under Greek law, a legal concept strictly defined by the Constitution and rules of law, operational within the order of law. The person of public interest is the "people", hence its social character. The public interest is identified with the interests of all the members of civil society, and aims at the satisfaction of their basic needs, but there can be a distinction between "general public interest" and "special" public interest which applies to particular sections of society. The pursuit of the public interest is binding for the state legislative organs which are in charge of the regulation of a range of issues, like environmental protection. These organs have the discretion to specify the public interest, always within the limits of the Constitution, when it comes, e.g., to the goals and actions of public administration and to the public agencies in charge of these actions. This definition implies that the public interest is pliable, dynamic, and evolutionary, on the basis of the prevailing social, political, and economic conditions, always within the framework of the principle of legality.

The legal interpretation of the public interest leaves room for its dynamic evolution, in different historical periods and social conditions. Small wonder that it attracted the interest, not only of social scientists in general, but also of planning theorists in particular, for the simple reason that planning is an action of the public administration serving the general interest or that of particular sections of society, as, in modern Greek history, refugees, landless peasants, and internal migrants displaced in wartime periods. The reader is reminded (see Chap. 3) of the view of Ernest Alexander (2002) that the public interest is of undisputed value from the procedural perspective of planning, but not in terms of substantive content. According to him, the public interest acquired a triple role, of which the first is to legitimize planning as a state activity. The second role, is to provide a moral paradigm or professional model of conduct. The third is to be used as a criterion, alas frequently vague, for the evaluation of plans, policies, and projects. The question of whether or not the public interest is a reliable guide to urban and regional planning practice is at the centre of articles by Michael Brooks (2017) and Stefano Moroni (2018). Brooks wonders if the public interest is real or an illusion, but admits that, as a rule, the planners declare that it is their guiding principle, even though its operational definition remains a problem. Its relationship with the notions of utility, common goods, and welfare explains why great economists in the past have dwelled on the subject, in their effort to decide whether particular policies lead to a maximum of utility and an optimal result. The definition of personal preferences and collective benefit, let alone their mathematical formulation, have proved difficult, yet this concern has been transferred to spatial planning. It is common knowledge that the problem drew the attention of planning theoreticians since the days of attempts to employ evaluation techniques, like cost-benefit analysis, in plans of spatial development. The problem grew larger when plans did not simply concern a public project, like building a new motorway or installations of public benefit, but also extensive alteration of land use and regional development. How is the criterion of public interest used in the evaluation of a regional spatial plan or in drafting legislation for a new planning system? The difficulty is to prove that in such cases a general public benefit is achieved and to decide who is authorized to legitimize an assertion of this kind. A possible answer is to claim that legitimation exists as long as we follow the democratic processes, through which these decisions are made. The lingering doubt is whether this answer is convincing, when sectional and group interests cannot be easily added up to a total public interest. One can therefore object that what counts is not a general public interest, but that of specific communities with which urban and regional planning is concerned. Apart from the fact that this view is not particularly helpful when the creation of a national planning system or the formulation of a nationwide plan are at stake, what every planner knows is that within an urban or rural area there exist multiple communities, with different trajectories or connective bonds. Within an alleged community, the range of personal values is daunting. Summing up the values of communities and of their members is an insurmountable obstacle. But if this is the state of affairs, what still remains to justify the claim that the general public interest is the ultimate yardstick? Nothing at all, is the answer according to Brooks! An argument against the use of the notion of a general public interest is of course the existence of multiple publics. The state, writes Purcell (2016) "commonly arrogates to itself the duty of representing the public, of standing in for it, of speaking for it. The State puts itself forward as the guardian or guarantor of the public interest. It prosecutes criminals, and fines polluters, and enforces zoning codes, and builds roads, and educates children, and houses the poor *in the name of* the public and in order to defend the public interest". Purcell's aim "is to argue against our habit of equating public and State" and "that we should consciously resist that equation. We should seek instead those moments when the two terms are distinct, and we should actively ward off the State's attempt to reconnect them". In short, he urges us "to imagine and create *publics without the State*".

The criterion of the public interest risks remaining a dead letter, apart from offering a convenient excuse when a planner has run out of arguments available in his arsenal. As we have seen, for many thinkers this is exactly the deadlock of rational and comprehensive planning and the justification of a collaborative approach, which accepts the existence of several interest communities and, as a consequence, addresses not a unique "customer" but a multiplicity of interlocutors. Moroni

(2018), quoted in Chap. 3, said that not everyone agrees that the public interest exists as an *a priori* criterion: "This is precisely the argument advanced by planning theorists who stress the centrality of communicative practices". However, the incompatibility of sectional priorities and of communities which are not monolithic, leads to implementation difficulties. In addition, the whole approach runs the risk of falling back on procedural assumptions which it condemns and of opening up a front of conflict with dominant state and administrative structures. We are here face-to-face with the critical parameter of the state, its relations with the citizenry, and its historical evolution, as well as with the state functions which have developed over a long period of time.

Moroni (2018) discusses the public interest in its broad sense, which expresses public and common goods. In order to discuss its meaning, he advances four versions of the claim of the concept's inexistence. He claims that the four interpretations correspond to a different 'kind of argument' (empirical, ethical, meta-ethical etc.) and a different 'perspective' (divergentism, dialogical proceduralism, classical liberalism, value pluralism, respectively).

- The first version is that the public interest exists as a fact, with inherent substantive value. "The argument is usually advanced by those planning theorists and political scientists who want to evidence the marked *distributive* effects of political decisions in a society that *de facto* comprises multiple factions and groups with different interests".
- The second is that it is empty of any a priori meaning. "This argument is advanced by planning theorists who stress the centrality of communicative practices".
- The third is that it has no extra-individual value. "This argument is advanced in particular by liberal thought".
- The fourth, and more radical version, is that the public interest does not exist as an always overriding substantive value. This is the "meta-ethical, value pluralism" argument.

Moroni espouses an approach, arguably a variant of liberal ideology, which he calls nomocracy. Drawing on a variety of insights, he argues, "one could say that it is in the public interest *to improve in the long term the chances of unknown individuals of pursuing their equally unknown and continuously changing purposes in a complex social environment that is open-ended.* This view is still based on moral individualism, but in a quite particular sense. Here, the public interest is not the *real* interest of any one specific person but the *potential* interest of anyone at all in the *long run*". It is easy, Moroni insists, to reject the existence of the public interest, but we shall remain obliged to justify a huge number of public interventions in conditions of complexity and diversity.

It is clear that the notion of the public interest, often used with excessive ease and superficiality, is a nodal concept in planning. What matters to us here, is that the Greek planning system, redrawn again and again until today, is founded on a logic of public interest. As a goal it is translated into objectives of economic development, amelioration of living conditions, or protection of the natural and cultural heritage, although the sincerity of these objectives is often disputed, as happened recently when legislation was being debated in 2020. One could fall back on the convenient argument that the public interest is squarely embedded in public law and the case law of administrative courts. It is the author's view, however, that it is also entrenched in the reality of the Greek state, as it was formed historically and was established in the citizens conscience and in social practice. This reality is enveloped in a complex nexus of equilibria and social life processes, which provides a framework of compromise and mutual exchanges between political powers and civil society.

10.9 Spatial Planning

Spatial planning did not appear in a political and social vacuum. Elias Kourliouros and this author tried to place planning in a theoretical framework (Wassenhoven and Kourliouros 2007, 369–370). Spatial planning as we know it today is linked with the emergence and consolidation of a complex division of labour which generated a differentiation, multiplication and heterogeneity of human activity and their location in geographical space. This process was historically associated with industrial capitalism, urbanization, and the path of western societies toward modernity. Modern spatial planning, we concluded, is a system of public management of these multiple and heterogeneous phenomena in space. But, as repeatedly stressed here, the conditions that imposed the need of spatial planning in developed capitalist countries were a far cry from those prevailing in Greece when organized planning was first tentatively introduced in the 1960s and 1970s. Naturally, the same observation holds true, to an even greater extent, if we return to settlement planning in the 1920s. This is not to deny that long before the twentieth century, even in antiquity, there were forms of spatial intervention – planning would be a misnomer – which shared with their modern counterpart the key characteristic of a conscious transformation of geographical space (Wassenhoven 2019).

The analysis of spatial planning in Greek literature is to a considerable degree influenced by the conception and role of planning in a capitalist society. Although with reference mainly to urban space, the approach of Lagopoulos (2017) is quite relevant for the full spectrum of spatial planning. Lagopoulos keeps a safe distance from empirical and positivist theories and embraces, as in fact he had done before, the view of political economy. He maintains, in a Marxist perspective, that all processes of production, circulation and consumption of products and services are structurally fundamental for society and space under capitalism. To these processes must be added both the reproduction of individuals and social relations through the patterns of consumption and all public actions, like planning, for the regulation of these processes. The regulation of land is certainly among the latter. In his theoretical analysis of land as a produced coefficient of production and as an instrument of labour, in the Marxist sense, Sofoulis (1979) had remarked that as a social category, land can only be conceived as a product, from the moment that, in its specific form, is introduced in the realm of objects controlled by the human being. There cannot be land, or sheer soil, which was once appropriated by humanity, in whatever capacity, without becoming a product. It is this product which spatial planning produces, or flatters itself that it does. The unanswered question is whether, under special circumstances, the production of land is achieved in other ways, e.g., by the capitalist market or by an unofficial "market" of social or electoral compromises, between the state and its clients. It is at this point, of an informal market, that we find the difference between genuine legality and faulty legitimation, on which various authors, like Kotzias, commented as we saw in Chap. 7.

These positions betray a radical viewpoint which is not of course original. In the framework of a Marxist analysis, the system of planning was from very early criticized as part of the legal and political superstructure of the capitalist mode of production, from the needs of which it could not possibly escape. The interpretation of spatial development and land use planning, through a Marxist lens, is based on an understanding of the state as an arm of the ruling classes, as an instrument for the stability of class structure, through interventions which facilitate the reproduction of capitalism, and as a medium of the cohesion of social formation (Wassenhoven 2004a, b, 2005). Notwithstanding radical criticism, the fact remains that in the Greek planning system a version of rational and systemic planning is still dominant, even though Marxist and postmodern analyses are still present and still insist that spatial planning has been depoliticized. Hadjimichalis (2018), following Chantal Mouffe (see Chap. 4) repeats the distinction between "politics" and "the political", considers depoliticization as a product of neoliberalism, and blames the technocrats. To implement neoliberal policies, he argues, development issues have to be presented as purely technocratic, so as to enable their formulation in a non-political, even a-spatial, context. Thus, technocratic choices remain indifferent to existing socio-spatial inequalities and the political agenda of the ruling class is rationalized by ostensibly relegating decisions to technocrats, who are freed from the obligation to account for their actions which are considered above dispute. We can be certain that no practicing planner or "technocrat" would readily agree with this assessment, but it is true that their proposals are frequently and fiercely criticized. Planners are even considered as agents and pawns of the state and capitalist interests, even though they may not be conscious of their role! What must be acknowledged is that apart from their professional contribution, their work is essential for the construction of a theory, a sort of "theory-in-use" or "espoused theory" (see Chap. 3), that emerges from their direct or indirect assumptions.

The field of activity of the urban and regional planner, the "technocrat" in the opinion of Hadjimichalis, is not identical with that of the human geographer, i.e., the analysis and explanation of geographical space, as the present author stressed before. Geographical space, not just physical, undergoes transformation because of the effect of several forces: Natural phenomena and disasters, sectoral (national and cross-national) policies (health, defence, transport etc), individual, collective and business actions, international contingencies, conventions and initiatives, violent events (wars, civil conflicts, persecutions, ethnic cleansing, plunder, human ravage), and, finally, official urban and regional planning actions and policies. All the transformations of space are the object of human geography. The theory of human geography interprets the result of transformation and has a lot to offer to planning theory, but is *not* planning theory, in the same way that it is not a theory of business management or of international relations. On the contrary, spatial planning theory seeks to explain official urban and regional planning actions, policies and practices, which are steered by state agencies or by supra-state authorities, like the European Union, which are still routed via the national state. This author's conclusion is then that the critical variables of a planning theory, at least one adapted to the Greek situation, are the state, the public administration (central, regional, local), administrative justice, and the agencies (even the planners themselves), which the state mobilizes to promote its policies. Such a theory has to explain the "empirical phenomenon" of urban and regional planning praxis (see Chap. 9). The state, in this frame of analysis, is not just a body of legal rules, an executive, a legislature, an organizational administrative structure, a bureaucratic apparatus, and a set of procedures. It is also the embodiment of a historical heritage, of traditional practices, of a network of relations with citizens and voters, of connections with the dominant economic system, and of established practices of legitimation or exchange. The empirical phenomenon is all that mixed together. The state, even in conditions of globalization, dictates the priorities, the goals, the content and the methods of planning practice, even with the simple expedient of imposing guidelines for urban and regional plans. The instructions issued by the state delimit the scientific-professional practice, e.g., the range of social views and positions that the planner takes into account, as he balances between a rational-normative logic and a collaborative, postmodern approach.

10.10 Rationality and Exceptionalism

This author had remarked a long time ago (Wassenhoven 2002), that the public, rational, and comprehensive planning model was under siege, in Greece too. The challenge against it had reached the administration, in the form of a rising emphasis on public-private partnerships and on governance, and, even more so, the universities with a renewed emphasis on traditional urban, ad hoc design. The identification of the rational model with a top-down, authoritarian mentality, in contrast to a bottom-up participatory philosophy, had fed the suspicion even of those, who accepted its basic principles out of a progressive ideology and an inclination of friendliness for public planning. Gradually, the author continued, the basic assumption of rational planning, the clear logic of modernism, was judged by some as bearing the guilt of failure. What is often ignored is that a founding stone of rational planning is the distinction between collective and individual action. This implies a separation of "market rationalism" and "social rationalism". The individual, as a member of a society, ought to be engaged in a task of collective planning, with the intention to control his/her own self and contribute to the construction of a common future, by using – this is a critical point – the power of logical argumentation. In his/her action there ought to coexist scientific thinking and human creative power. This rational model, in its correct conception as a process of decision-making, was no doubt weakened by some of its assumptions, which were not essential in the first place.

One of them was the excessive emphasis on wholeness and universality, in other words the claim that it can conceive human life in its totality and design it as a single object. This was neither feasible, nor necessary. What was, however, necessary was the investigation of all implications and side-effects of planning, not so much with technical methods and techniques, but with systematic exchange of views and with meaningful, not narrowly bureaucratic, citizen participation. Here, in the author's interpretation, lies a second defect, i.e., the introversion which marked the rational model of planning, when the attempt to enhance its scientific integrity and to arm it with operational research and statistical tools led to an apotheosis of the simulation of just about everything in a claustrophobic environment. Sadly, those who identified rational planning with mathematical analysis, to equip it with higher scientific status, which would be beyond reproach, succeeded in alienating it from the political context of planning and in exposing it to the accusation of democratic deficiency. Nothing of all that was inevitable. If this slippery road was to a large extent avoided in the case of Greek planning, it is because it clashed with existing traditional, non-rational we might say, elements in the country's government practice which placed obstacles on the way of this rigid perception of planning. The "eastern" element of the Greek character was inimical to undiluted rationality and faithful to a *bazaar-like* solution of planning problems. To put it differently, it was prone to compromise. As often stressed earlier, it was subordinated to the logic of political clientelism, patronage, favouritism and *rousfeti* (incidentally a Turkish word), i.e., a favour for political gain. This did not prevent the official national planning system to evolve into an impressive, top-heavy, and *very* rational structure.

In 2020, the latest of a series of comprehensive urban and regional planning laws was aiming, once again, to bring the planning system to a stage of final integration. But it is true to say, this system under its previous mutations (see Fig. 8.1 on main planning laws in Chap. 8) had already achieved a remarkable level of acceptance. Notwithstanding the inevitable weaknesses and partial objections voiced by various commentators, its basic structure was not disputed. Since the first laws after the turning point of the 1975 Constitution, its key characteristics received the open or tacit approval of diverse political parties. The planning acts legislated by conservative, right-wing, socialist, liberal, and/or left-wing governments were similar in their broad outline. The most eloquent examples were the successive laws of 2014 and 2016, proposed and voted by a right-wing and a left-wing government. The second cancelled the first and put a new system in its place, but the similarity was astonishing. Even new planning instruments, like the Special Town Plans, which had attracted venomous attacks, remained in place. A possible explanation is the pressures of the lending international institutions in the context of the economic crisis and the urgent borrowing requirements of the country. The theoretical debates carried out in other countries, which doubted the supremacy of the comprehensive, rational model, never touched the model's acceptance in Greece, at least as far as its legal structure is concerned. Internationally, the rational model seemed to become vulnerable and obsolete, at least in theory, and a new conception took its place for a while, which placed emphasis on the multiplicity of social groupings and communities with diverging values and priorities. In Greece however, the practical impact of postmodern theoretical currents, like the communicative turn, did not expand beyond a limited circle of academics and intellectuals. Law 4447/2016 enacted under a left-wing government is a clear indication. The association of the communicative turn with the unrecognized values of minorities was evident in Greece too, but the dominant planning model was never threatened by the collaborative planning approach or by more radical currents. On the contrary, a return of the systems logic is already evident due to the incorporation of environmental priorities in the routines of planning. If nothing else, the Greek planning system tends to become even more inclusive and global. One explanation is the neoliberal emphasis on what we will call later *high-profile* planning (see Chap. 11), which favours a parallel and autonomous planning activity, totally subordinated to economic growth priorities. The point is that even this shift, exemplified by special legislation in the early years of the economic crisis around 2010–2012, was not reversed under left-wing rule in 2015–2019. On the contrary, glamorous projects, like the redevelopment of the former Athens airport at Hellinikon, received a final approval.

A more insidious explanation of the perseverance of an overall rational planning system, is the less voiced, at least openly, confidence that whatever the official and publicized features of the system, the reality of planning on the ground would remain unaffected: *Plus ça change, plus ça reste la même chose*! The official system projects a modernist façade, but the underground reality tells a different story, developed in a well-known network of political bargaining. This network affects intimately urban and regional planning, and remains active, regardless of the institutional system of planning. Modernist politicians, parliamentarians and administrators may sincerely wish to erect a westernized version of planning, but many of their colleagues make sure that the state ultimately tolerates, not to say encourages, the inclusion of spatial planning in a particular variant of social concessions. Such a carefully disguised approach is of course enveloped in arguments of protection of the disadvantaged classes. This allegedly "democratic", or crudely populist and vote-catching, attitude is embedded in the nexus of interpersonal and clientelist relations, which function as a party-political safety net, proclaimed to secure social stability and peace. Throughout the postwar years, the balance between a rational, modernizing, centralized, and regulatory tendency, on one hand, and a people-friendly, populist, clientelist, and concessionary one, on the other, has remained stable. Recognition of this balancing act is essential if one wishes to understand the evolution of the planning system, the conflicts that marked recent legislative initiatives in 2020, and the spate of subsequent law amendments which were aimed at tilting the balance in favour of populist inclinations. When, e.g., the 2020 law tried to put an end to out-of-plan construction, the unpopularity of the measure led to a policy reversal through urgent amendments which favoured derogations, dispensations, and deviations, and of course reopened loopholes.

Derogations, a real headache for those trying to interpret Greek planning legislation, are not an exclusive Greek invention. A quotation from Preteceille (1982, 145), to whom we referred earlier, is very eloquent: "It is the use which is made of plans which is important rather than their explicit content as such. The case of derogation is interesting because it dispels the distorted picture derived from a simplistic vision of urban planning. If undue attention is paid to the urban plan, one neglects contradictory tendencies which can be invoked in the form of derogation, or negative planning. Derogation must be empirically studied to sort out this problem. As yet it has hardly been touched upon, but one can already indicate that there are many very different types of derogation, certain of which can be relevant to the practice of urban planning (such as the application of urban regulations to certain types of plot, or their strict application which can impede all amelioration of the urban tissue). Others are essentially relevant to the distribution of rent and the effects of these vary from the planning point of view, though in most cases they aggravate the saturation of public facilities. In urban plans and derogations we can also see the determinations of politics and state intervention". The clear intention of derogations is to help the concerned property owner bypass the main stream of planning regulations. It is the paradise of creative illegality! It is also part of the routine work of many practicing planners, engineers, and architects struggling to obtain a building permit and to determine whether a particular plot is, in Greek planning terminology, "whole" (*artio*)[13] and "buildable".

Derogations may seem to be a fanciful example of the daily routine of planning at the critical level of property interests, but they are at the same time an epiphenomenon of the logic of exceptionalism, the logic that "my problem is a unique case" and cannot be treated as one more case among thousands. They are in addition an element of widespread resistance to the yoke of power and state prescriptions, which the average citizen refuses to obey; the neck of a good Greek cannot tolerate a yoke, is a popular expression, regardless of the field of application, be it the rigour of statutory planning or vaccination against a lethal virus or externally dictated policy. The citizen immediately smells a conspiracy of interests, which limit his/her freedom of choice. Whether this attitude betrays innate irrationalism or a form of inherited popular wisdom is hard to say. Historical experience has taught the citizen that, indeed, it was often a conspiracy of interests that had the upper hand in a range of affairs which affected his/her own interests and wellbeing. His/her instinctive reaction is to mobilize his/her own network of influence to achieve his/her goal and, in a manner of speaking, resort to a counter-conspiracy in which he/she can rely on accomplices, with whom he/she rushes to negotiate. In the process of negotiation, he/she might have to offer something in exchange – even to break the law – but expects to receive something in return: an exception, a derogation, a preferential treatment. Very briefly, he/she will have to compromise.

[13] The term *artio* is sometimes wrongly translated as "even". In fact, in implies a piece of land satisfying a set of criteria (e.g.. minimum size, accessibility by public road, minimum street frontage and plot depth, ability to accommodate a building of certain size etc). It is apparent that the designation "whole and buildable" is crucial in areas beyond the limits of statutory plans. For the hundreds of thousands of land owners in out-of-plan areas the designation is of make-or-break importance, as it greatly affects the value of their possession. In Greek legal jargon different terms are used to designate a piece of land, depending on whether it is in or out of a plan area.

10.11 The Theorist and the Theoretical Model

Important questions remain unanswered regarding the parameters of a planning theory, notwithstanding the author's effort at the end of Chap. 9. One well tried approach available to a theorist is to remain sequestered in the framework of the role of planning in the operation of the capitalist system. The answers are readily available and regularly reiterated in radical theory. The theorist can relate all policy on urbanization, settlement systems, and urban development to the needs of capital and labour reproduction, familiar to those studying advanced industrial societies. He/she can also adapt this approach to the particularities of a deformed or stymied capitalist system, as mentioned before. Can we extend our theoretical explanation of planning to such a vast field, well beyond the domain of planning practice? We can naturally resort to the well-rehearsed method of drawing our inspiration from a vast field of scientific literature, from physics to economics, or from mathematics to philosophy and psychoanalysis, in the hope that we shall stumble on some useful and untried concept. The danger is that this immersion in sophisticated sources will keep widening the gap between planning theory and planning practice. The objective of theory is not to propose prescriptions to practice, but it is undoubtedly enriching it, as much as it receives useful feedback. But writing planning theory treatises with endless references to Aristotle, Engels, Gramsci, Kuhn, Deleuze, or Foucault, is not for this author a practical way of securing the above enrichment.

There are already interesting attempts to explain the behaviour of the planning theoretical community itself, an introverted group with its own professional values and exclusive public, usually institutionalized in academic environments. These attempts are inspired by the peculiar needs of university careerism and of the "publish or perish" syndrome. The rapid transformation of urban and regional spaces in conditions of advanced capitalism provided ample opportunity for theorists to develop innovative explanations. Soon, this "academic professionalism" spread like a forest-fire to less developed countries. However, focusing on the personality of the theorist and on his/her experiential makeup is a necessity, because it is his/her personal journey that shapes his/her theoretical armour.

The theorist does not operate in a locked cellar, uninfluenced by his/her own personal evolution. He/she is, as argued in Fig. 10.1, a product of the processes he/she is anxious to explain.

The author's approach is, as expected, not that of a jurist. He looks at the planning system through its theoretical explanations, but also through its inner logic and the forces that produced it. In Greece, as in many countries of the East and South, this logic owes a lot to prototypes, regulations, methods and tools, originally conceived in advanced industrial countries. The adoption of foreign models is not limited to urban and regional planning in the narrow sense. It also extends to organizational and administrative formations of governance, from the early years of the modern Greek state. This is of the utmost importance because the "spatial cells" in which planning is later used (regions, prefectures, municipalities) were formed from the beginning, revised of course several times. What is also present ever since

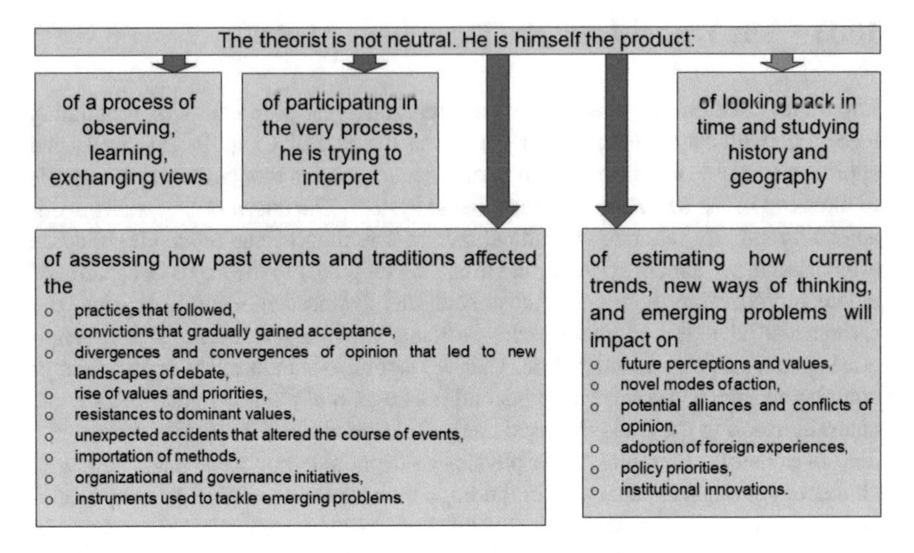

Fig. 10.1 The theorist as a "product". (Diagram drawn by the author)

the first days of independence is the close relationship of nation and state, which is a powerful symbolic signifier. If we recall that in planning theory, especially in the institutional approach (see Chap. 3), the nature of a planning agency is of equal, perhaps greater, importance than the process and content of planning, we realize why the institutions which are solidified from the beginning, particularly in a highly centralized state, are a strong determinant of the planning system. As explained by Amenta and Ramsey (2009), "iInstitutional arguments rely not on aggregations of individual action, or on patterned interaction games between individuals, but on 'institutions that structure action' … Institutions are emergent, 'higher-order' factors above the individual level, constraining or constituting the interests and political participation of actors 'without requiring repeated collective mobilization or authoritative intervention to achieve these regularities'".

The configuration of the Greek planning system is a clear dependent variable of the institutional structure that oversees and manages it. The planning system, as portrayed in the latest planning law of 2020, but also in previous acts, has a systemic character strongly reminiscent of systems theory and rational planning. Each planning level corresponds to a spatial and administrative sub-system, in a strict hierarchy. As in other countries, the national territorial system contains regional systems, subdivided in their turn into regional departments (former prefectures), municipalities, and municipal departments. The systemic character of the planning system does not find its expression only in the structure of the system. It is also evident in planning processes, typical of a rational-procedural model. The specifications or briefs of regional planning frameworks and local plans, very recently updated by the government, emphasize the process of goal setting, end-state targeting, scenario-building, selection of preferred plan, monitoring, feedback and revision, consultation, and participation. This methodology is a must for all planning studies and is accompanied by strategic environmental assessment studies, ad hoc environmental plans (e.g., for watercourses), and other technical studies, such as transportation plans. The methodology confirms the continued application of rational, process planning, a conclusion which is not denied by the requirement of consultation at various levels and the institutionalized – rather formalistic – participatory exercises or plan reassessments after a fixed period of years, which in fact are postponed ad nauseam. The introduction of environmental protection processes has strengthened the rationalist element of planning, at the expense, some would argue, of its political character. This trend seems destined to continue at an accelerated pace in novel forms of planning, like maritime spatial planning, disaster planning, risk prevention, resilience planning, and climate change mitigation. The socio-political content of all these forms of planning is by no means disputed, but it must not escape our attention that there are already protesting voices which object to a technocratic turn, which neoliberalism welcomes with open arms.

Regardless of the rational-systemic character of the planning system the reality of planning at a level lower than ambitious statutory plans, i.e., at the level of land use implementation plans and street layout plans or in areas which for the foreseeable future will remain out of the limits of such plans, belongs to a different universe. This is where noble intentions are compromised and derogations are voted in parliament in late night sessions to accommodate a variety of interests. As indicated in Fig. 10.2, this is where a Southern perspective at the contact point of East and West imposes a different theoretical explanation. The influence of the dominant currents of ideas is doubtful, hence the indication in the diagram with a question mark. It is only a country-specific, national context which can provide the material for a theory, which relies on the practicalities of pragmatism, socio-political complicity,

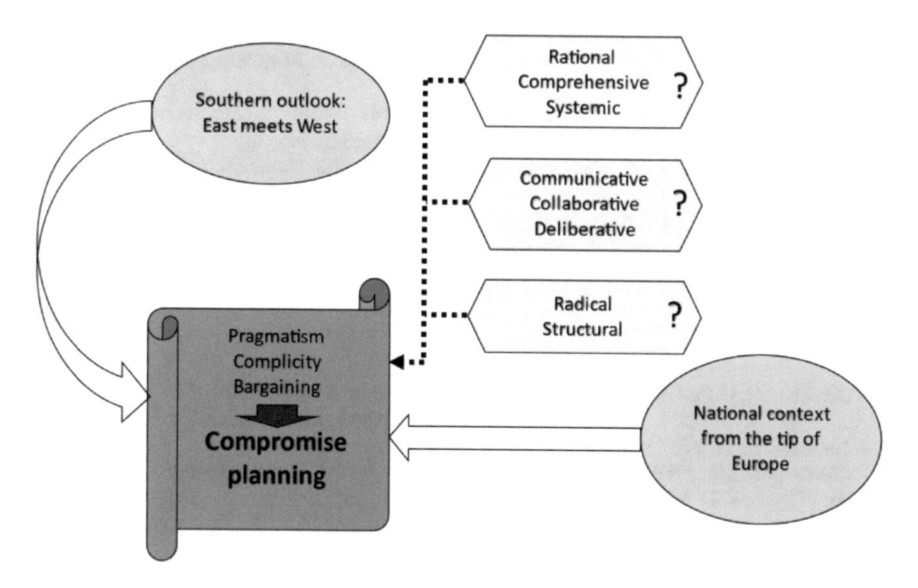

Fig. 10.2 Compromise planning – Factors of influence. (Diagram drawn by the author)

and state-citizen bargaining. It is what is described by the author as compromise planning.

Epilogue

The main purpose of this chapter was to place the proposed new theoretical model of compromise planning in the context of human geography concepts, of an analysis of the individual as the final recipient of planning action, and of the realities of planning. Compromise, adjustment, clientelism, bargaining, improvisation, coping and complicity were related to a pragmatist approach. Use was made of theoretical contributions from international and Greek sources. The chapter includes a presentation of the various conceptions of the individual person, from the *homo economicus* or *socialis* to the version of the *homo individualis*, as proposed and described by the author. This is the type of *homo* best suited to the approach of compromise planning and the empirical reality of the Greek spatial planning system, already extensively presented in previous chapters. In this chapter, the author returned to the issues of the public interest, rationality and exceptionalism because of their direct relevance for the understanding of the operation of the spatial planning system which serves as a case study. The position of the spatial planning theorist, who is himself considered as a product of social forces, is outlined. In the final section of the chapter the author discussed the factors which influence the theoretical model, as an introduction to the following and final chapter.

References[14]

Aegean, Seminars of the (1995) Geographies of integration, geographies of inequality in Europe after Maastricht. 1993 Syros seminar. Athens and Thessaloniki

Aegean, Seminars of the (1998) Space, inequality and difference: from 'radical' to 'cultural' formulations? 1996 Milos seminar. Athens

Afouxenidis A et al (eds) (2019) Geographies in an era of fluidity: critical essays on space, society and culture in honour of Lila Leontidou. Propobos, Athens (bilingual)

Alexander ER (2002) The public interest in planning: from legitimation to substantive plan evaluation. Plan Theory 1(3):226–249

Amenta E, Ramsey KM (2009) Institutional theory. In: Leicht KT, Jenkins JC (eds) Handbook of politics: state and society in global perspective. Springer

[14] **Note**: The titles of publications in Greek have been translated by the author. The indication "in Greek" appears at the end of the title. For acronyms used in the references, apart from journal titles, see at the end of the table of diagrams, figures and text-boxes at the beginning of the book.

Abbreviations of journal titles: AT Architektonika Themata (Gr); DEP Diethnis kai Evropaiki Politiki (Gr); EpKE Epitheorisi Koinonikon Erevnon (Gr); EPS European Planning Studies; PeD Perivallon kai Dikaio (Gr); PoPer Poli kai Perifereia (Gr); PPR Planning Practice and Research; PT Planning Theory; PTP Planning Theory and Practice; SyT Synchrona Themata (Gr); TeC Technika Chronika (Gr); UG Urban Geography; US Urban Studies; USc Urban Science. The letters Gr indicate a Greek journal. The titles of other journals are mentioned in full.

Andrikopoulou E (1993) New local development institutions and social policy problems. In: Sakis Karagiorgas Foundation (ed) Dimensions of present-day social policy. Conference proceedings, Athens, pp 527–542. (in Greek)
Beauregard RA (1996) Between modernity and postmodernity: the ambiguous position of U.S. planning. In: Campbell S, Fainstein S (eds) Readings in planning theory. Blackwell, Oxford, pp 213–233. (first published 1989)
Bolan RS (2017) Urban planning's philosophical entanglements: the rugged, dialectical path from knowledge to action. Routledge, London
Bowles S, Gintis H (2002) Homo reciprocans. Nature 415:125–127
Brennan NM (2010) Homo economicus', 'homo socialis', 'homo fabulans' and 'homo publicus': Conceptualising impression management. Conference paper (available on the web)
Brooks MP (2017) Planning theory for practitioners. Routledge, London. (first published 2002)
Campbell S, Fainstein S (1996) Introduction: the structure and debates of planning theory. In: Campbell S, Fainstein S (eds) Readings in planning theory. Blackwell, Oxford, pp 1–14
Campbell H, Marshall R (2002) Utilitarianism's bad breath? A re-evaluation of the public interest justification for planning. Plan Theory 1(2):163–187
Close D (2002) Greece since 1945: politics, economy and society. Longman/Pearson Education, London
Derruau M (1991) Human geography (Prevelakis G Trans). National Bank Cultural Foundation, Athens (French original: Géographie humaine, 1976) (in Greek)
Dertilis G (2016) Seven wars, four civil conflicts, seven bankruptcies: 1821–2016. Polis, Athens (in Greek)
Economou D (1994) Urban development and spatial structure of the settlement network. In: Getimis P, Kafkalas G, Maravegias (eds) Urban and regional development: theory, analysis and policy. Themelio, Athens, pp 43–93. (in Greek)
Economou D (2002) The structural features of the spatial policy and development model.[15] In: SADAS (ed) Architecture and the Greek city in the 21st century. Proceedings of the 10th Panhellenic architectural conference (1999). SADAS and TEE, Athens, pp 73–84 (in Greek)[16]
Falk A (2003) Homo Oeconomicus versus Homo Reciprocans: Ansätze für ein neues Wirtschaftspolitisches Leitbild? Perspektiven der Wirtschaftspolitik (available on the web)
Filias V (1980) Aspects of maintenance and change of the social system. Vol. B. Nea Synora – A. Livanis, Athens. (in Greek)
Getimis P (1993) Social policies and local state. In: Getimis P, Gravaris D (eds) Social state and social policy. Themelio, Athens, pp 91–121. (in Greek)
Getimis P, Kafkalas G (2001) Scientific thinking on Greek space 1974–2000: an evaluation attempt and thoughts for the future. Topos 2001(16) (in Greek)
Gintis H, Helbing D (2013) Homo socialis: an analytical core for sociological theory. Available at SSRN: https://ssrn.com/abstract=2362262
Giordano C (2005) Homo creator. The conception of man in social anthropology. Finance et Bien Commun 22(2):25–31
Gold JR (2019) Behavioural geography. In: Kobayashi A (ed) International encyclopedia of human geography, vol 1, 2nd edn. Elsevier, Oxford
Granovetter M (1985) Economic action and social structure: the problem of embeddedness. Am J Sociol 91:481–493
Gravaris D (1993) Elements for a critical theory of social policy. In: Getimis P, Gravaris D (eds) Social state and social policy. Themelio, Athens, pp 29–72. (in Greek)
Hadjimichalis C (ed) (2010) Greekscapes: Aerial photography atlas of modern Greek landscapes. Charokopeion University and I.S. Latsis Foundation (available on the web) (in Greek)
Hadjimichalis C (2018) Landscapes of crisis in South Europe. Alexandreia, Athens (in Greek)

[15] The title of the article in the table of contents was different.

[16] No pages are mentioned in the table of contents.

Hillier J (2008) Plan(e) speaking: a multiplanar theory of spatial planning. Plan Theory 7(1):24–50
Hirsch P, Michaels S, Friedman R (1990) Clean models vs. dirty hands: why economics is different from sociology. In: Zukin S, DiMaggio P (eds) Structures of capital: the social organization of the economy. Cambridge University Press, Cambridge, pp 39–56
Hoch C (2019) Pragmatic spatial planning: practical theory for professionals. Routledge, New York
Iosefidis T (2019) Epistemological and methodological dimensions of the critical realism approach to the social sciences and human geography. In: Afouxenidis A et al (eds) Geographies in an era of fluidity: critical essays on space, society and culture in honour of Lila Leontidou. Propobos, Athens, pp 45–58. (in Greek)
Kafkalas G, Komnenos N (1993) Local development strategies and social policy problems. In: Sakis Karagiorgas Foundation (ed) Dimensions of present-day social policy. Conference proceedings, Athens, pp 515–526. (in Greek)
Kourliouros E (2001) Trajectories of the theories of space: economic geographies of production and development. Ellinika Grammata, Athens (in Greek)
Lagopoulos AP (1981) Town planning manual – part B. planning interventions: Vol. II: space-functional planning and planning of built structures. Aristotle University of Thessaloniki (in Greek)
Lagopoulos APh (1982) An historical – materialist critique of the planning models for urban and regional scale. Working papers. University of Thessaloniki, Chair B of Urban Planning, Thessaloniki
Lagopoulos AP (1994) Analysis of urban practices and uses. In: Getimis P, Kafkalas G, Maravegias (eds) Urban and regional development: theory, analysis and policy. Themelio, Athens, pp 123–166. (in Greek)
Lagopoulos AP (2017) Urban planning theory and methodology: from political economy to the semiotics of space. Patakis, Athens (in Greek)
Lai LWC, Lorne FT (2015) The fourth Coase theorem: state planning rules and spontaneity in action. Plan Theory 14(1):44–69
Lancaster LW (1968) Masters of political thought (vol. 3): Hegel to Dewey. George G. Harrap, London. (first published 1959)
Lauria M (2010) Does planning theory affect practice, and if so, how? Plan Theory 9(2):156–159
Lazzarato M (2012) The making of the indebted man: an essay on the neoliberal condition, Semiotext(e) intervention series 13. MIT Press, Cambridge, MA
Leontidou L (2011) *Ageographitos Chora* [Geographically illiterate land]: Hellenic idols in the epistemological reflections of European Geography. Propobos, Athens (in Greek)
Mantouvalou M, Balla E (2004) Changes in the land and building system and planning stakes in today's Greece. In: NTUA (ed) City and space from the 20th to the 21st century. Volume in honour of Prof. A.I. Aravantinos. NTUA, University of Thessaly, and SEPOCH, Athens, pp 313–330. (in Greek)
Moroni S (2018) The public interest. In: Gunder M, Madanipour A, Watson V (eds) The Routledge handbook of planning theory. Routledge, London, pp 69–80
Needham B (2017) A renegade economist preaches good land-use planning. In: Haselsberger B (ed) Encounters in planning thought: 16 autobiographical essays from key thinkers in spatial planning. Routledge, New York/London, pp 165–183
Panayotatos E (1982) Introduction to regional spatial planning. NTUA, Athens (in Greek)
Parker G, Doak J (2012) Key concepts in planning. Sage, London
Preteceille E (1982) Urban planning: the contradictions of capitalist urbanisation. In: Paris C (ed) Critical readings in planning theory. Pergamon Press, Oxford, pp 129–146. (originally published in French 1974)
Prevedourou E (2020) The meaning of public interest. Internet University lectures on administrative law. Aristotle University of Thessaloniki
Purcell M (2016) For democracy: planning and publics without the state. Plan Theory 15(4):386–401
Relph E (2015) Placeness, place, placelessness. https://www.placeness.com/topophilia-and-topophils

Roy A (2017) Worldling the South: toward a post-colonial urban theory. In: Parnell S, Oldfield S (eds) The Routledge Handbook on cities of the global south. Routledge, London, pp 9–20. (first published 2014)
Salet W (2019) The making of the public. Plan Theory 18(2):260–264
Seamon D, Sowers J (2008) Place and placeness, Edward Relph. In: Hubbard P, Kitchen R, Vallentine G (eds) Key texts in human geography. Sage, London, pp 43–51
Skordili S (2007) Industrial geography. In: Terkenli T, Iosephidis T, Chorianopoulos G (eds) Human geography: man, society and space. Kritiki, Athens, pp 310–327. (in Greek)
Sofoulis C (1979) Land as produced factor of production. Papazisis, Athens (in Greek)
Subrt J (2015) Homo sociologicus revisited: how relevant could questions be that were formulated half-a-century ago? Scientific Cooperations International Journal of Arts, Humanities and Social Sciences 1(1)
Vergopoulos C, Amin S (1975) Shapeless capitalism. Papazisis, Athens (in Greek)
Wassenhoven L (2002) The democratic quality of spatial planning and the contestation of the rational model. Aeichoros 1(1):30–49. (in Greek)
Wassenhoven L (2004a) Spatial planning theory. Lecture notes. NTUA, Athens (in Greek)
Wassenhoven L (2004b) The theory of urban planning: what future is there in the 21st century? In: NTUA (ed) City and space from the 20th to the 21st century. Volume in honour of Prof. A.I. Aravantinos. NTUA, University of Thessaly, and SEPOCH, Athens, pp 75–102. (in Greek)
Wassenhoven L (2005) Spatial planning and the critique of political economy. Lecture notes. National Technical University of Athens (in Greek)
Wassenhoven L (2010) Territorial governance: theory, European experience and the case of Greece.[17] Kritiki, Athens (in Greek)
Wassenhoven L (2018a) The parallel universes of rhetoric and practice: lessons learnt from spatial and urban planning reforms in Greece. In: Farinós Dasí J, Peiró E (eds) Territorio y Estados/ Territory and states. Tirant Humanidades, Valencia, pp 347–372
Wassenhoven L (2018b) Hellinikon … passion: the development of the former Hellinikon airport, Athens. Sakkoulas, Athens (in Greek)
Wassenhoven L (2019) The ancestry of regional spatial planning: a planner's look at history. Springer, Cham
Wassenhoven L (2022) Putting our country in order: a history of spatial regional planning in Greece after the Second World War. Kritiki, Athens (in Greek)
Wassenhoven L, Kourliouros E (2007) Social and political dimensions of spatial planning. In: Terkenli T, Iosephidis T, Chorianopoulos G (eds) Human geography: man, society and space. Kritiki, Athens, pp 366–406. (in Greek)

[17] In collaboration with K. Sapountzaki, E. Asprogerakas, E. Gianniris and Th. Pagonis.

Chapter 11
Conclusions: The Parallel Worlds of Planning – Variants of Compromise

Abstract The main section of this relatively short chapter, which brings the book to a close, is the outline of a theoretical model based on the notion of compromise planning. This outline includes a presentation of the main issues and factors out of which the author's theoretical schema is constructed. However, before reaching this conclusion, he tries to explain the forms that compromise is taking, which have different origins, sometimes traditional and based on old clientelist bonds, sometimes modern and conceived in the neoliberal globalized conditions of the last few decades. The role of the state is once again briefly outlined and the ugly aspects of compromise planning are restated, whether they are low-profile or high-profile depending on the private interests benefiting from compromise arrangements.

Keywords Compromise theory · State · Planning system · High- and low-profile interests · Theoretical schema · Theory components

Prologue

There is no doubt that the *desideratum* of a scientifically well-founded theory, to which the author referred in previous chapters, has pushed planning theorists in the direction of basing their arguments on the work of great thinkers of the past, on whose insights and interpretations modern planning theories were constructed. The author did not intend from the start to plunge into this futile search but in some cases, he departed from this rule, e.g., in dealing with the theory of the state, to the extent of course permitted by his academic background. Being unable to return to the roots of philosophy and social science, he studied thoroughly the work of planning theoreticians who did and thus offered him a precious bridge to original knowledge. Of great assistance was naturally the experience and knowledge he accumulated during his career. It was from this base that he started gradually organizing his thoughts regarding a theory of a type of planning, which, while wearing a mantle of rational comprehensiveness, is supported by delicate balances of compromise between the state and the average citizen. In this final chapter we first look at the various faces of compromise, some of which have a traditional ancestry, while

L. C. Wassenhoven, *Compromise Planning : A Theoretical Approach from a Distant Corner of Europe*, https://doi.org/10.1007/978-3-030-94331-8_11

others are new and are typical of a new economic age. The author then summarizes the crucial role of the state and the form of the planning system, before outlining his, admittedly optimistic, hopes that new forms of reconciliation between state and citizens will emerge, which will replace the dark side aspects of planning. At the end of this final chapter, the author sketches the picture of the theoretical model of compromise planning and presents its critical components.

11.1 Janus-Faced Compromise

Throughout this book the author insisted on the need for a planning theory, which is native, appropriate, and tailor-made; a theory which has a southern complexion and odour and can blend together national characteristics and "westernized" orientation; a theory which reflects the seemingly irreconcilable reality layers of North-West and South-East. He explained to the best of his ability the juxtaposition of at least two worlds of planning, the official, modernist and rational planning system, and the shadowy world of planning in the trenches or planning "behind closed doors". He explored, with admiration and curiosity, the impressive theoretical constructions, which he tried, in Chap. 2, to classify into four currents of ideas on which were attached the tags of "rational, communicative, radical, and climate". He toyed from the beginning with the idea of a fifth current, as an umbrella under which would be accommodated pragmatism and his own view of compromise planning. It has been suggested by a well-meaning commentator that the label "compromise" is a pejorative term which betrays weakness, a lack of courage and determination, an abandonment of the ideal goal of consensus. It was an opinion over which the author pondered with concern, not only because he does not underestimate the importance of consensus – who does? – but also because it was not clear what the shortest route to consensus is. In the Greek case the labyrinth of negotiation between government and society follows paths which are a far cry from communicative and deliberative processes, praised by eminent planning theorists.

Greece is a country with a highly fragmented pattern of real estate property and a huge number of legal land titles, in relation to the country's size. This breeds a system of planning in practice, which is divorced from the image of official planning. The problem of out-of-plan areas and of the statutes that regulate them has been referred to repeatedly; statutes which are quite separate from statutory local urban plans. Illegal building is particularly associated with such areas, although partial illegality of construction also exists to an astonishing degree in densely populated urban zones. This is fertile ground for an informal or parallel activity, where compromise thrives. This is the "traditional" face of the Janus-faced process of compromise, a low-profile face. It is the "eastern" face, product of the development of the full 200 years of Greek nation building, not to mention the seeds of even older historical periods. It is the inheritance of a period of infant, or deformed, capitalism, and of a state formation constantly wavering between traditionalism and modernity.

Janus, however, has a second face, more modern, high-profile, openly capitalistic. Here too we have a process of compromise, but on a grander scale. For Greece, unlike advanced capitalist economies, this is a fairly new development the first traces of which can be found in the late 1990s, when Greece took the bold step of organizing Olympic Games in Athens. It is certain that others will disagree and point to large scale developments, infrastructures and industries which had appeared before. They were, however, state-initiated projects (as in the field of energy production or transport), public settlement plans (as in the case of refugee settlement), isolated examples of large hotel projects, or preferentially supported extractive and manufacturing plants. What emerges around the end of the twentieth century is a parallel and very prominent system of ad hoc planning legislation which exemplifies a different, high-profile mode of negotiation and compromise. The fact that this parallel system was soon incorporated in mainstream legislation does not negate the circumstances of its initial adoption. This process gained speed when the 2010 economic crisis erupted. Greece – once again – defaulted, and foreign loans as well as investment were urgently required. The stream of the legislation of exceptionalism is still under way, with new legislation on strategic investments in 2021. One can claim that this is a capitalist, neoliberal version of compromise, carried out at a higher level of politics.

To summarize, low-profile, traditional compromise revolves around a well-known set of planning situations: Out-of-plan areas, settlements of fewer than 2,000 inhabitants, settlements preexisting the 1923 first legislative planning decree, land use derogations and exceptions, illegal construction, "normalization" of unlicensed buildings, the outright "legalization" of which was declared unconstitutional. This is the standard vocabulary of exceptionalism and low-profile, humdrum compromise planning, which follows familiar, well-trodden paths, although always in the shadow of official shining proclamations. High-profile compromise planning is the field of innovations in ad hoc large projects, in special plans for public land properties or for projects of strategic importance, plans of areas of integrated tourism development, mixed tourist resorts, and special town plans, which can revise statutory town plans. The project of the redevelopment of the former Athens international airport at Hellinikon, based on an ad hoc planning act, is the most publicized case (see Text-Box 8.11).

11.2 The State and Its Pivotal Role

The state is at the epicentre of the compromise agenda. As explained in Chap. 10, the public interest is regularly invoked to justify the state's actions, but is amenable to all sorts of interpretations. As a result, the political actors within the state pursue their goals through a complex whole of choices which they take care to legitimize within the institutional framework. In all spatial matters, the planners are a vital instrument for this legitimation. It can be argued once more that it is impossible to explain the Greek planning system and its application in practice without focusing on the role of the state and of its multiple ties with society. Theoretical explanations founded on the conditions of other countries are of limited help. This simple truth is fortunately emphasized by those who approach this problem from a South-East – or Global South – perspective. Admittedly, this differentiated approach is frequently due to the distinctive population structure of countries of the South and the position of minorities; but it is also due to other parameters. Yiftachel (2006) rightly underlines that in the South, liberal democracy is not always a stable order, but rather an unstable sectoral economic agenda. Land property systems are fluid and prone to disputation. Alternative supposedly rational interpretations are conflictual and inconsistent and land policy is riddled with dark or grey assumptions. At the same time, the usual imported theories on the role of the state, in relation to land antagonisms, remain inadequate. In Greece too, land development is diverted into bypasses which constitute a grey planning zone and have proved remarkably resilient. In Chap. 10, the author discussed the notion of the public interest and the confidence of governments that hierarchical, rational planning serves it well.

More than that, the public interest is a central premise of the Constitution. It is the author's view that this faith in the existence of an overriding public interest can indeed be found in Greek society. This assertion may look unorthodox or out of date but the truth is that it is a residue of historical circumstances and is on the whole justified. Notwithstanding the existence of love-hate relations between citizens and state, from which everything is expected, these relations have been forged in a bond which withstood several waves of social turbulence. Apart from the state's identification with the nation, there exists in this bond a solid base of continuous interaction and negotiation with the institutions of the state, an interaction channelled through the political system and its administrative ramifications. Bargaining relieves the pressures and tensions in endless rounds of compromise.

Planning could be described as a typical specimen of compromise. On one hand, the state and its representatives, backed in their action by a solid legal framework, consider that they serve the public interest, the "good of the people", even though their position is determined by a nexus of interests and fragile equilibria. On the other hand, the average citizen weighs the pros and cons of this stance not as a rational homo economicus, or even as a homo socialis, who wishes to serve the general social interest. He/she is rather what the author called a homo individualis, a compound creature in whose mind is condensed his/her own, as well as the family or group, interests; it is with this compass in hand that he/she bargains with political power. This form of interaction is felt to be more efficient in his/her dealings with the state. Clientelist relations, on a personal or party-political plane, is the most effective negotiating tool, giving rise to a planning culture which favours this behaviour, rather than the openness or transparency of official planning (Knieling and Othengrafen 2015). The typical small land owner, the self-employed professional or petty-entrepreneur find in this culture a "way of doing", taught to them by past experience. If nothing else, this is a state of affairs, which secures a form of consensus, stability and balance, through which the innovations introduced by the state – or by international processes – can be absorbed and made acceptable. The wider social interest is, in this indirect way, inscribed in the personal balance of interests of the individual. The gradual social acceptance of a planning system, which evolves gradually over a period of decades, can be explained in this logical framework; it is internalized by the individual, assessed as tolerable, and to an extent considered positive. There are occasions of course when the spectre of the centralized authoritarian state rises in the public mind and causes negative reflexes, which can be fully justified when the state issues orders and ultimatums, without prior argument and consultation.

The centralized character of planning, once again evident in recent legislation, is undoubtedly a cause of democratic concern. The central government reserves for itself all the powers of approval of regional planning frameworks and local plans, local government is kept in the margin for "constitutional" reasons, but against all democratic logic, participation processes are usually seen as a bothersome and bureaucratic affair, and the central administration dictates asphyxiating plan specifications for every corner of the national territory. For advocates of a bottom-up approach this is a very unsatisfactory reality. It seems to be the outcome of a widespread conviction that the public interest is best served by the central state, provided that the latter is open to negotiation, based on conditions, which the citizen has tacitly accepted on the basis of accumulated compromise. When this understanding is challenged, even violently, the challenge is condescendingly and conveniently perceived as part of a political game consistent with the worldview which the citizens consider as a cultural heritage. Besides, the mechanisms of reaction are more effective when they follow the lessons of this heritage. There is naturally the alternative of appeals to the courts, often resorted to with facile excuses, but there is no denying that the traditional bargaining methods are more "productive". There was, in 2020, an excellent example of this circuitous method of reaction, when new urban and regional planning legislation had reached the stage of parliamentary approval. The government perceptions of the protection of the public interest and of a modernizing rationality, for some embarrassingly authoritarian, conflicted with interests and reactions, because they violated the unwritten laws of the sui generis Greek capitalist system. The clauses concerning out-of-plan building activity were those that brought the conflict to a peak, simply because they threatened sensitive age-long balances of compromise. The old grudges resurfaced and the government was soon in retreat, mobilizing the familiar expedient of amendment after amendment and derogation after derogation.

11.3 The Ugly Face of Planning and Hopes for the Future

The Greek-style reality of planning is not, theoretically, so paradoxical or unique. The official planning system, as found in the hierarchy of plans or as printed in the Government Gazette, may be impeccably rational and decently modern, but the system in practice is of another world, not dramatically different from incremental, piecemeal, step-by-step planning processes, which were the basis of some theories in the 1960s. This is a planning process which does not chase visions, but tackles problems, as and when they appear. Besides, at the stage of local plan implementation and of granting building permits, the ease of discreetly ignoring plan provisions and court decisions is well known. We are back to our familiar syndrome of compromise – let alone corruption – which begins in late-night parliamentary law amendments and reaches the stage of issuing building permits. This is an absurdity which, perversely, ensures a feeling of stability, until the next round of the modernity-tradition battle. One should not be impatient and complain. Not even half-a-century has passed since the 1975 Constitution which placed spatial planning on a firm base; or not even a century, since the historic legislative decree of 1923 on local planning of towns and settlements!

A pessimistic and cynical observer would certainly reach the conclusion that spatial planning is an elaborate smokescreen hiding an unpleasant reality. In certain conditions this unpleasant feeling is justified, especially when planning activity serves to conceal political anomalies, as in the case of the Greek military dictatorship of 1967–1974. The counter-argument is that ultimately planning is slowly consolidated as an expression of the public interest, especially when new parameters are introduced which an increasing number of citizens consider essential. Natural disaster prevention, risk mitigation and climate change adaptation are such parameters, which are rapidly finding a place in the planning agenda. Earthquake mitigation has of course a very long history in Greece, but the threat of natural disasters like devastating forest fires and floods is a relatively more recent concern. A few years ago, no one could predict the loss of dozens of lives during the 2007 fires or of more than 100 in the tragic fire of 2018 in Mati, in the vicinity of Athens. The devastating fires of August 2021, mainly in Euboea and Attica, fortunately with no loss of human beings, brought to everyone's lips the slogan of "prevention". A new ministry was created for Civil Protection and Climate Change, soon corrected to Climate Crisis. Climate change entered the everyday vocabulary and became a key ingredient of spatial planning activity.

Sustainable spatial development, according to the 2020 planning law, embraces all the spatial, territorial and environmental dimensions of sustainable development plus those which are related to the *rational* organization of geographical space. The use of the term "rational" in the law is not an accident. This widening of spatial planning goals consolidates the feeling that it is the public interest which is promoted, a feeling which, however, may require some time to cease being a mere ideological firework.

11.4 The Construction of a Theoretical Framework

The long series of parameters that enter a theory of compromise are summarized in Fig. 11.1 and in Text-Boxes 11.1 and 11.2. These parameters are grouped in eight groups, which hopefully represent the aspects of reality which the author tried to piece together in previous chapters. The groups in which he classified the components of compromise planning correspond to historical, economic and social developments which have been emphasized throughout the book because of their valuable explanatory potential. Rather than embarking on an analytical description, these

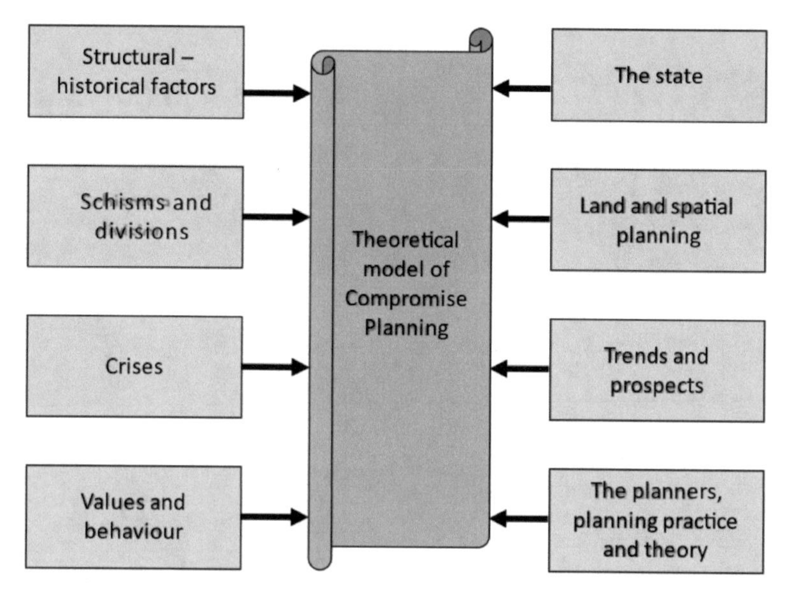

Fig. 11.1 Compromise planning – Theoretical model. (Diagram drawn by the author)

developments and contributory factors are listed in a condensed form. They should not be approached in isolation, because it is their combined effect which explains the form of planning which requires a new theoretical interpretation. They are the building materials out of which compromise planning is cobbled together. Planning in practice would have been different without the confluence of all these streams of events, conditions, institutions and values. It is the difference of these materials which causes our inability to theorize on the basis of imported theoretical work.

Text-Box 11.1: Compromise Planning: History, National Schisms, Crises, Values

Structural historical factors

- Geopolitical position: East meets West
- Gradual territorial annexations: North and South divisions
- The dominant "state of Athens"
- Historical evolution and nation-state building
- Nationalism and Hellenism
- Interests of the native population
- Westernized mentality of the Greeks of the diaspora
- Autochthonous and heterochthonous interests

(continued)

Text-Box 11.1 (continued)

National schisms and divisions

- Wars and domestic conflicts
- Political divisions and social cleavages
- Repeated national schisms
- Liberal or socialist v. conservative or royalist allegiances

Crises

- External shocks, settlement of Greek refugees and resulting sharp population increases
- Repeated economic crises, public debt defaults, bankruptcies, and foreign financial control
- Quasi-colonial dependency on foreign powers
- Economic and regional development and reconstruction priorities
- Foreign investment needs

Values and behaviour

- Traditional value systems
- Modernism v. tradition
- Contested European identity: Internationalism and localism
- Family and kinship networks
- Political clientelism and favouritism
- Disobedience, alienation, and irrational behaviour
- Politicians and patrons
- The solution of low-profile compromise

Text-Box 11.2: Compromise Planning: State, Planning, Trends, Planners
The state

- State-citizen love-and-hate relations
- The state as a father-figure, as an obstacle to be circumscribed and as a partner to be deceived
- Concentration of decision-making (and taking)
- Party-political control of the state
- Trends towards a technocratic state – Neoliberalism

(continued)

Text-Box 11.2 (continued)

- Business interests, pressure groups, and behind-the-scenes accommodation
- High-profile compromise
- Imported administrative structures and practices
- Interaction of political establishment and administrative practice
- Official modernism in state organization and contrast with policy in practice
- European Union membership

Land and spatial planning

- Land property fragmentation, land reforms, size of real estate property
- Spatial planning and management as a substitute of social welfare policy
- Statutory plans and land development in out-of-plan areas
- Division of territory into planned, quasi-planned, and unplanned land
- Land ownership as an insurance against economic upheavals
- Land use planning and political hypocrisy
- Planning systems, urban and regional plans, imported instruments, and tools and their foreign origins
- Erratic, inconsistent, and obscure planning legislation – Proliferation of statutes
- The parallel roles of the executive, legislature, and administrative justice in the interpretation and management of spatial planning

Trends and prospects

- Conflicting interests of development, private wealth, and care for the cultural and natural environment: An unstable balance
- Widening inequalities and injustices
- Economic instability and changing role of real estate property
- The inroads of modern technology and a changing social landscape
- The environmental and climate challenge

The planners, planning practice and theory

- The planning profession and its foreign profile
- The relative delay of academic education
- Under-staffed government, especially at local level
- Unpredictable cycles of planning activity and their impact on the planning profession
- Unemployment problems of planning professionals
- Lack of official recognition of the planning profession
- Dearth of theoretical thinking

Epilogue

After a long and perhaps tiresome journey, the effort to add a new planning theory to a long series of theories full of the wisdom of famous theorists, has come to an end. In this last chapter the author tried to shed additional light on the faces of compromise in planning, one traditional, dispersed and low-profile, the other high-profile and more akin to modern large-scale initiatives of neoliberal character. What was reiterated is the conspicuous role of the state, the main actor and arbiter in small and large decisions alike. The author tried to be fair in his judgments and not to hide the dark side of planning, while expressing an optimism that just, rational and democratic planning can instil in the public psyche an appreciation of what the pursuit of public interest can really offer. What is meant here is a public interest which can wipe out all the petty compromises and the shady arrangements for glamorous projects; a public interest arising out of a conviction that planning can be beneficial for all and replace the deals which are usually swept under the carpet. But while finding solace in this optimistic expectation, the author strived to picture the model of compromise planning as vividly as possible and present its essential components. The hope is naturally that compromise theory does justice to the reality it intends to explain. There is, however, a proviso that we should not ignore. As Mahatma Gandhi expressed it: "All compromise is based on give and take, but there can be no give and take on fundamentals. Any compromise on mere fundamentals is a surrender. For it is all give and no take".[1]

References[2]

Knieling J, Othengrafen F (2015) Planning culture – a concept to explain the evolution of planning policies and processes in Europe? EPS 23(11):2133–2147

Yiftachel O (2006) Re-engaging planning theory? Towards 'South-Eastern' perspectives. PT 5(3):211–222

[1] See brainyquote.com

[2] **Note**: The titles of publications in Greek have been translated by the author. The indication "in Greek" appears at the end of the title. For acronyms used in the references, apart from journal titles, see at the end of the table of diagrams, figures and text-boxes at the beginning of the book.

Abbreviations of journal titles: AT Architektonika Themata (Gr); DEP Diethnis kai Evropaiki Politiki (Gr); EpKE Epitheorisi Koinonikon Erevnon (Gr); EPS European Planning Studies; PeD Perivallon kai Dikaio (Gr); PoPer Poli kai Perifereia (Gr); PPR Planning Practice and Research; PT Planning Theory; PTP Planning Theory and Practice; SyT Synchrona Themata (Gr); TeC Technika Chronika (Gr); UG Urban Geography; US Urban Studies; USc Urban Science. The letters Gr indicate a Greek journal. The titles of other journals are mentioned in full.

Index

Note: The numbers to which the reader is referred correspond to chapter-sections.

L. C. Wassenhoven, *Compromise Planning : A Theoretical Approach from a Distant Corner of Europe*, https://doi.org/10.1007/978-3-030-94331-8

Printed in the United States
by Baker & Taylor Publisher Services